**Evolution**

# Evolution

## From Molecules to Ecosystems

EDITED BY

**Andrés Moya**

AND

**Enrique Font**

*Cavanilles Institute for Biodiversity and Evolutionary Biology, University of Valencia, Spain*



OXFORD
UNIVERSITY PRESS

*This book has been printed digitally and produced in a standard specification in order to ensure its continuing availability*

OXFORD
UNIVERSITY PRESS

Great Clarendon Street, Oxford OX2 6DP

Oxford University Press is a department of the University of Oxford.
It furthers the University's objective of excellence in research, scholarship,
and education by publishing worldwide in

Oxford New York

Auckland Cape Town Dar es Salaam Hong Kong Karachi
Kuala Lumpur Madrid Melbourne Mexico City Nairobi
New Delhi Shanghai Taipei Toronto

With offices in

Argentina Austria Brazil Chile Czech Republic France Greece
Guatemala Hungary Italy Japan South Korea Poland Portugal
Singapore Switzerland Thailand Turkey Ukraine Vietnam

Oxford is a registered trade mark of Oxford University Press
in the UK and in certain other countries

Published in the United States
by Oxford University Press Inc., New York



Reprinted 2007



ISBN 978-0-19-851543-2

# Contents

# Foreword

This volume commemorates a meeting that was held in Valencia, Spain, in October of 2000, on the subject of "Evolution: From Molecules to Ecosystems". The meeting celebrated two birthdays, as well as a coming of age. The birthdays that we celebrated were the 500th for the Universitat de València, and the birthing of the Institut Cavanilles de Biodiversitat i Biologia Evolutiva. The university founded this new institute to build on its strong faculty in ecology, genetics, and evolution. The namesake of the institute, Antonio Josef Cavanilles, started the university's botanical garden, which also happened to provide the conferees with a beautiful setting for coffee breaks and a delicious dinner of *paella*, the local specialty.

The meeting was also a celebration of the coming of age—the scientific maturation—of evolutionary biology. Maturation does not imply an end but, instead, reflects the increased vigor that comes with expanding horizons and new questions. Evolutionary research today spans such diverse approaches and perspectives as molecular biology and paleontology, genomics and ecological genetics, formal theory and critical experiment. The issues that are addressed range from the fate of mutations to the origin of new genes, from mechanisms of speciation to patterns of radiation after mass extinctions, from recent migrations to ancient relationships, from symbiosis to virulent disease, from the origin of play to perceptions of beauty. The reader will find all of these topics, and more, covered in this volume by leading experts from around the world.

The meeting's two plenary speakers, Francisco Ayala and Daniel Dennett, have each contributed to this volume. Their papers clearly show the tremendous range of evolutionary applications and ideas. Ayala and his coauthor Stephen Rich examine the population-genetic structure and evolutionary history of the agent of malaria, a scourge that infects millions of people. Dennett offers an explanation for human creativity in which our designs, as well as our capacity for design, arise from Darwinian selection working at multiple levels.

As someone who attended the meeting, it is difficult for me to say which was better: the scientific content of the conference, or the gracious hospitality of our hosts in that lovely Mediterranean city. This volume admirably captures the former but, alas, it cannot recreate the latter. For that pleasure, one must travel to Valencia.

*Richard E. Lenski*
*East Lansing, Michigan*

# Preface

There can be little doubt that, as we usher in the third millennium, Darwinism and evolution are on the rise. Less than 150 years have elapsed since the publication of Darwin's seminal work on evolution by natural selection, yet in this short period evolutionary theory has transformed our thinking in all aspects of human endeavor. One contributor to this volume, the philosopher Daniel Dennett, has described evolutionary thinking as a "universal acid" that eats through everything it touches. Much of the current interest in evolution has been sparked by recent conceptual, theoretical and empirical advances in several biological disciplines. As an example, the sequencing of complete genomes from multicellular organisms, including humans, is changing not only how we think about ourselves but also our connection with other life forms. Progress is taking place at such a fast pace that even practitioners feel the need for an occasional gathering to provide a broad perspective on the field as a whole.

During November 2–4, 2000, the University of Valencia hosted a workshop entitled "Evolution: From Molecules to Ecosystems." This workshop was an important part of the activities commemorating the fifth centennial of the University of Valencia and the launching of the Cavanilles Institute for Biodiversity and Evolutionary Biology.* The aim of the workshop was to create a stimulating atmosphere for open discussions of the current state of knowledge and the future of evolutionary biology. The ambition was not to be comprehensive and balanced in all dimensions, but rather to explore major areas of current interest where important advances have recently taken place. The venue for the workshop, the Botanical Gardens of the University of Valencia, provided a state-of-the-art facility for the workshop's formal conferences and a convivial atmosphere fostering informal discussions of all topics relating to the study of evolution. The workshop was attended by close to 200 participants from all over the world. Twenty speakers delivered 19 conference presentations and two public keynote addresses. Ten invited participants, selected on the basis of their expertise in evolutionary biology, also contributed to the discussions that took place after the presentations.

The high quality of the conferences delivered at the workshop prompted us to produce a multi-authored volume based in part on the contents of the presentations, but also incorporating ideas and controversies that arose during the ensuing discussions. Our aim was not to produce a typical, run-of-the-mill proceedings volume, but a book that would serve both as a useful reference for specific topics that are currently attracting great interest and as a companion text for introductory and upper level courses in evolutionary biology.

As with the workshop, the overarching theme of the book is evolution (from molecules to ecosystems). We instructed all authors to consider their chosen topics in a broad context and keep the depth and scope of their contributions at a level adequate for an audience of mainly graduate and advanced undergraduate students. Several authors that were not initially involved with the workshop were called in to touch on topics, such as development and speciation, that we felt were not adequately

* The Cavanilles Institute was founded in 1998 as part of the University of Valencia's strategy to foster research excellence by relocating research teams in biodiversity and evolutionary biology to an autonomous institute, and is one of the few university institutes in Europe devoted to the global study of evolution and biodiversity.

covered during the workshop, and they kindly agreed to do so. Each chapter was read critically by at least two outside reviewers, after which the authors and the editors revised them to address the reviewers' comments and suggestions.

The book is comprised of 23 chapters written by prominent biologists who have made significant contributions to the field of evolutionary biology. The topics included in the book do not encompass all aspects of evolutionary biology, but they do feature a sufficiently wide range of contemporary issues to illustrate many key advances in the field over the past 10 years. Indeed, we believe that this volume can stand alone as a sampler of the diversity of questions and research approaches that constitute the modern study of evolution.

The range of research expertise among the contributors to the book spans all major areas of evolutionary biology. However, it is evident that the contributors come to the discussion of evolutionary theory from many different points of view and do not speak with a single voice. This heterogeneity was not only unavoidable given the lack of constraints imposed on the participants, but also something we actively promoted.

Many chapters are readily accessible and require little prior knowledge of the topic. Others are more demanding, particularly in population and quantitative genetics. Throughout the text, we have identified important concepts and technical terms the first time they appear in a chapter in boldface. Many of these terms are briefly defined in the text and/or in the glossary included at the end of the book. Because of space limitations, the literature citations are more to offer a way into the current literature than to give full credit for discoveries. Several chapters include a section entitled "Further reading." The references listed in this section provide useful background information on the topic of the chapter.

# Acknowledgments

As editors of this volume we have received generous assistance from many people. We are, of course, very grateful to Ian Sherman and his colleagues at Oxford University Press for their encouragement at all stages of this project. Juli Peretó and the "Fundación General de la Universitat de València" (Patronato Cinc Segles) were instrumental in securing the funding necessary to organize the workshop upon which this book is based. The workshop was possible thanks to the generous support of the Caja Rural Valencia and the Conselleria de Cultura i Educació (Generalitat Valenciana).

Finally, we wish to express our sincere appreciation to the following colleagues for their careful and constructive reviews (other reviewers, not listed, asked to remain anonymous): Montserrat Aguadé (Universitat de Barcelona, Spain), Michalis Averof (IMBB-FORTH, Greece), Francisco Ayala (University of California, USA), Dave Begun (University of California, USA), Michael J. Benton (University of Bristol, UK), Neil W. Blackstone (Northern Illinois University, USA), Linda Boothroyd (University of St. Andrews, UK), Roger Butlin (University of Leeds, UK), Pierre Capy (CNRS, France), Andrew G. Clark (Penn State, USA), Jean Clobert (Université Pierre et Marie Curie, France), Teresa J. Crease (University of Guelph, Canada), Richard Dawkins (University of Oxford, UK), Luc De Meester (Katholieke Universiteit Leuven, Belgium), Claude Desplan (NYU, USA), Isabelle Dupanloup (Universita di Ferrara, Italy), Dan Dykhuizen (SUNY, USA), Dieter Ebert (Fribourg University, Switzerland), Brandon Gaut (University of California, USA), John J. Gilbert (Darmouth College, USA), Neil Greenberg (University of Tennessee, USA), Roy Gross (University of Würzburg, Germany), Stephan Harding (Schumacher College, UK), Aziz Heddi (INSA-INRA, France), Edward C. Holmes (University of Oxford, UK), Toby Johnson (University of British Columbia, Canada), Peter D. Keightley (University of Edinburgh, UK), Matthew Kramer (Beltsville Agricultural Res. Center, USA), Michel Laurin (Université Paris 7, France), Jessica L. Mark Welch (Marine Biology Lab., USA), Constance I. Millar (University of California, USA), Arne Mooers (Simon Fraser University, Canada), Guillermo Orti (University of Nebraska, USA), John Parsch (Ludwig Maximilian Universitat, Germany), Sergio M. Pellis (University of Lethbridge, Canada), David Perrett (University of St. Andrews, UK), Peter Price (Northern Arizona University, USA), Andy Purvis (Imperial College, UK), François Renaud (CNRS, France), Naruya Saitou (Natl. Institute of Genetics, Japan), Outi Savolainen (University of Oulu, Finland), Hanneke Schuitemaker (University of Amsterdam, The Netherlands), Manuel Serra (Universitat de València, Spain), Terry W. Snell (Georgia Inst. of Technology, USA), Richard Southwood (University of Oxford, UK), Mark Stoneking (Max Planck Inst., Germany), Diana Tomback (University of Colorado, USA), Jacques van Alphen (Leiden University, The Netherlands), Tom van Dooren (Leiden University, The Netherlands), David B. Wake (University of California, USA), Lawrence J. Weider (University of Oklahoma, USA), and Kenneth H. Wolfe (University of Dublin, Ireland).

# Contributors

**Montserrat Aguadé**, Departament de Genètica, Facultat de Biologia, Universitat de Barcelona, Diagonal 645, 08028 Barcelona, Spain.

**Francisco J. Ayala**, Department of Ecology and Evolutionary Biology, University of California, 321 Steinhaus Hall, Irvine, CA 92697-2525, USA.

**Casey M. Bergman**, Lawrence Berkeley National Laboratory, Genome Sciences Division, Berkeley Drosophila Genome Project, 1 Cyclotron Rd. Mailstop 64-121, Berkeley, CA 94720, USA.

**Jaume Bertranpetit**, Unitat de Biologia Evolutiva, Facultat de Ciències de la Salut i de la Vida, Universitat Pompeu Fabra, Doctor Aiguader 80, 08003 Barcelona, Spain.

**Gordon M. Burghardt**, Departments of Psychology and Ecology & Evolutionary Biology, 1404 Circle Drive, University of Tennessee, Knoxville, TN 37996-0900, USA.

**Armando Caballero**, Departamento de Bioquímica, Genética e Inmunología, Facultad de Ciencias, Universidad de Vigo, 36200 Vigo, Spain.

**Francesc Calafell**, Unitat de Biologia Evolutiva, Facultat de Ciències de la Salut i de la Vida, Universitat Pompeu Fabra, Doctor Aiguader 80, 08003 Barcelona, Spain.

**Alexandra M. Casa**, Institute for Genomic Diversity, Cornell University, Ithaca, New York, USA.

**Luc De Meester**, Laboratory of Aquatic Ecology, Katholieke Universiteit Leuven, Ch. De Bériotstraat 32, 3000 Leuven, Belgium.

**Daniel C. Dennett**, Center for Cognitive Studies, Tufts University, Medford, MA 02155-7059, USA.

**Robert F. Denno**, Department of Entomology, University of Maryland, College Park, Maryland 20742, USA.

**Douglas H. Erwin**, Smithsonian Institution, PO Box 37012, National Museum of Natural History, Department of Paleobiology, MRC-121, Washington, D.C. 20013-7012, USA.

**Antonio Fontdevila**, Departament de Genètica i Microbiologia, Universitat Autònoma de Barcelona, Campus, Edifici C, 08193 Bellaterra, Spain.

**Steven W. Gangestad**, Department of Psychology, The University of New Mexico, Albuquerque, New Mexico 87131, USA.

**Aurora García-Dorado**, Departamento de Genética, Facultad de Ciencias Biológicas, Universidad Complutense, 28040 Madrid, Spain.

**Africa Gómez**, Department of Biological Sciences, University of Hull, HU6 7RX, Hull, UK.

**Fernando González-Candelas**, Instituto Cavanilles de Biodiversidad y Biología Evolutiva and Departamento de Genética, Universitat de València, Apdo. 22085, 46071 Valencia, Spain.

**Godfrey M. Hewitt**, School of Biological Sciences, UEA, Norwich NR4 7TJ, UK.

**Julie C. Ho**, Institute for Genomic Diversity, Cornell University, Ithaca, New York, USA.

**Edward C. Holmes**, Department of Zoology, University of Oxford, South Parks Road, Oxford OXI 3PS, UK.

**Charles E. King**, Department of Zoology, Oregon State University, Corvallis, OR 97331, USA.

**Stephen Kresovich**, Institute for Genomic Diversity, Cornell University, Ithaca, New York, USA.

**Winfried Lampert**, Max Planck Institute of Limnology, Postfach 165, 24302 Plön, Germany.

**Amparo Latorre**, Instituto Cavanilles de Biodiversidad y Biología Evolutiva, Universitat de València, Apdo. 22085, 46071 Valencia, Spain.

**Carlos López-Fanjul**, Departamento de Genética, Facultad de Ciencias Biológicas, Universidad Complutense, 28040 Madrid, Spain.

**Michael Lynch**, Department of Biology, Indiana University, Bloomington, IN 47405, USA.

**Axel Meyer**, Department of Biology, University of Konstanz, D-78457 Konstanz, Germany.

**Richard E. Michod**, Department of Ecology and Evolutionary Biology, University of Arizona, Tucson, AZ 85721, USA.

**Andrés Moya**, Instituto Cavanilles de Biodiversidad y Biología Evolutiva, Universitat de València, Apdo. 22085, 46071 Valencia, Spain.

**Aurora M. Nedelcu**, Department of Ecology and Evolutionary Biology, University of Arizona, Tucson, AZ 85721, USA.

**Tomoko Ohta**, National Institute of Genetics, Mishima 411-8540, Japan.

**Nipam H. Patel**, University of Chicago, Department of Organismal Biology and Anatomy, Howard Hughes Medical Institute, MC 1028, N-101, 5841 S. Maryland Ave., Chicago, IL 60637, USA.

**Merrill A. Peterson**, Biology Department, Western Washington University, Bellingham, Washington 98225, USA.

**Stephen M. Rich**, Division of Infectious Disease, Tufts University, 200 Westboro Road, North Grafton, MA 01536, USA.

**Julio Rozas**, Departament de Genètica, Facultat de Biologia, Universitat de Barcelona, Diagonal 645, 08028 Barcelona, Spain.

**Menno Schilthuizen**, Institute for Tropical Biology and Conservation, Universiti Malaysia Sabah, Locked Bag 2073, 88999 Kota Kinabalu, Malaysia.

**Bronwen Scott**, School of Life Sciences and Technology, Victoria University of Technology, P.O. Box 14428, Melbourne City, MC 8001, Australia.

**Carmen Segarra**, Departament de Genètica, Facultat de Biologia, Universitat de Barcelona, Diagonal 645, 08028 Barcelona, Spain.

**Manuel Serra**, Institut Cavanilles de Biodiversitat i Biologia Evolutiva, Universitat de València, Apdo. 22085, 46071 Valencia, Spain.

**Jean-Christophe Simon**, UMR BiO3P, Biologie des Organismes et des Populations appliquée à la Protection des Plantes, INRA, Domaine de la Motte BP 35327, 35653 Le Rheu Cedex, France.

**Terry W. Snell**, School of Biology, Georgia Institute of Technology, Atlanta, GA 30332-0230, USA.

**Randy Thornhill**, Department of Biology, Castetter Hall, University of New Mexico, Albuquerque, New Mexico 87131-1091, USA.

**Roeland C. H. J. van Ham***, Centro de Astrobiología, INTA-CSIC, 28850 Torrejón de Ardoz, Madrid, Spain. *Current address: Plant Research International, Business unit Genomics, PO Box 16, 6700 AA, Wageningen, The Netherlands.

**Rafael Zardoya**, Museo Nacional de Ciencias Naturales, CSIC, José Gutierrez Abascal 2, 28006 Madrid, Spain.

PART I

# The genetic machinery of evolution

CHAPTER 1

# Near neutrality and its implications for evolution

**Tomoko Ohta**

By the middle of the twentieth century, neo-Darwinism was very popular among evolutionary scientists and random **genetic drift** was thought to be unimportant except in the context of Wright's shifting balance theory. Then, in the 1960s, Kimura applied population genetics theory to biochemical data. He compared amino acid sequences of cytochrome *c* and hemoglobin $\alpha$ among mammalian species available at that time, and found that the number of mutant substitutions in evolution was larger than expected according to Haldane's prediction based on **genetic load**. In 1968, Kimura proposed the **neutral theory** of molecular evolution (Kimura 1968). In the following year, King and Jukes published a similar idea based on biochemical considerations (King and Jukes 1969).

In the early 1970s, I thought that the borderline mutations between the selected and the neutral classes ought to be important, and published the slightly deleterious mutation theory (Ohta 1973), that was later called the **nearly neutral theory** (Ohta 1992). In the neutral theory, new mutations are classified into deleterious, neutral, and advantageous classes. In contrast, the nearly neutral theory gives importance to the borderline (i.e. nearly neutral) mutations. Figure 1.1 depicts the relationship between the selection, neutral, and nearly neutral theories.

Through DNA sequence comparisons, particularly through separate estimation of the numbers of **synonymous** and **nonsynonymous substitutions**, the nearly neutral theory is gaining support. The purpose of this chapter is to review progress in the nearly neutral theory and present recent results from simulation studies based on an interaction model.

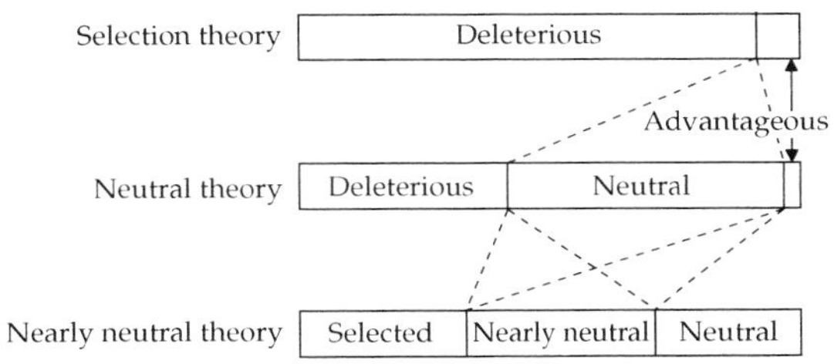


**Figure 1.1** A diagram showing the comparison between the selection, neutral, and nearly neutral theories based on the proportion of various classes of mutations.

## The molecular clock and near neutrality

It is well known that the rate of molecular evolution is equal to the neutral mutation rate under the neutral theory, which provides a basis for the **molecular clock** of protein and DNA evolution (Kimura 1983). Let the mutation rate be $\nu_g$ per gene per generation, and let $u$ be the probability of **fixation** of a mutant gene. In a population of size $N_e$ (diploid), the total number of mutations appearing in the population is $2N_e\nu_g$ per generation. A fraction $u$ of them becomes fixed in the population, and the rate of substitution per generation, $k_g$, is,

$$k_g = 2N_e\nu_g\, u. \tag{1.1}$$

The value of $u$ is equal to the initial frequency, $1/(2N_e)$, for a neutral mutant, and the rate of evolution becomes,

$$k_g = \nu_g. \tag{1.2}$$

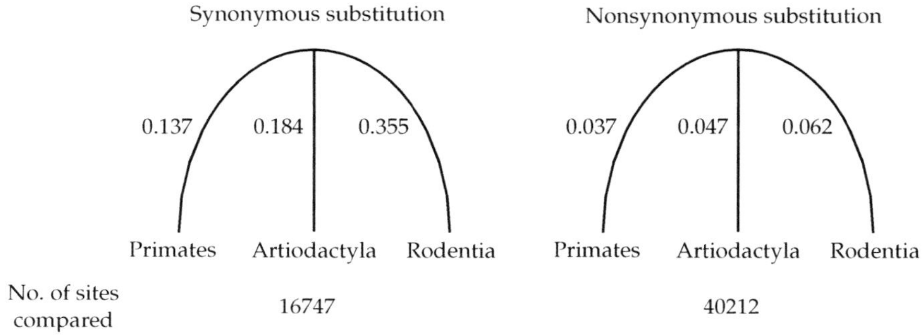


**Figure 1.2** Star phylogenies of 49 mammalian genes (from Ohta 1995). The numbers beside each branch are the estimated numbers of substitutions per site. Note that the number of synonymous substitutions may be underestimated, but the general pattern is unchanged (see Yang and Nielsen 1998).

If a mutant is not strictly neutral, but nearly neutral, $u$ becomes larger or smaller than $1/(2N_e)$ depending upon the selective value. The simplest case is that of a semidominant gene with selection coefficient, $s$, that is positive or negative. The crucial quantity is the product, $N_e s$, such that fixation probability is an increasing function of $N_e s$ (Kimura 1962). The important point is that $u$ takes a positive value even if $s$ is negative, provided that the absolute value of $N_e s$ is not large.

Under the nearly neutral theory that gives significance to the borderline mutations between the selected and neutral classes, very slightly deleterious mutations are expected to be common. For such mutations, there is a negative correlation between $u$ and population size, $N_e$. Therefore, the evolutionary rate of a gene by **drift** is higher in a small population than in a large population (Ohta 1973). This may be expressed as follows:

$$k_g \propto \nu_g / N_e. \qquad (1.3)$$

This relationship is extremely important for the molecular clock. The mutation rate is thought to depend on the number of cell generations, and hence on generation time. In general, large organisms tend to have short generation times and vice versa (Chao and Carr 1993). Therefore, for nearly neutral mutations, the generation time effect of mutation rate is expected to partially cancel with the population size effect of fixation probability, resulting in the molecular clock (Ohta 1973). In other words, the generation time effect should be stronger in neutral mutations than nearly neutral mutations.

This prediction can be examined by estimating the numbers of synonymous and nonsynonymous substitutions. Figure 1.2 shows **star phylogenies** for three mammalian orders (Ohta 1995). From the figure, it may be seen that the generation time effect is more conspicuous for synonymous substitutions than for nonsynonymous substitutions, that is, the result agrees with the nearly neutral theory. It is now commonplace to examine the divergence patterns of synonymous and nonsynonymous substitutions between closely related species and within populations. Such analyses are in the same line as the present study.

So far, we have discussed the average pattern of synonymous and nonsynonymous substitutions. Another interesting problem is the variance of evolutionary rate in relation to near neutrality. It has long been known that protein evolution displays large fluctuations in evolutionary rate (Ohta and Kimura 1971; Langley and Fitch 1974; Kimura 1983; Gillespie 1991). My analyses of 49 mammalian genes confirm this large variance. Following Gillespie (1991), attention was paid to the **dispersion index**, $R$, that is, the ratio of the variance to the mean number of substitutions among lineages. Results of the analysis with the same data as in Fig. 1.2 are given in Table 1.1. As can be seen from the table, the dispersion index is often 5–6 for

**Table 1.1** Average dispersion index

| | Synonymous | | Nonsynonymous | |
|---|---|---|---|---|
| | Unweighted | Weighted | Unweighted | Weighted |
| Primate–artiodactyl–rodent (39 genes[a]) | 25.01 | 5.89 | 8.46 | 5.60 |
| Primate–lagomorph–rodent (38 genes[b]) | 20.73 | 5.56 | 6.61 | 3.95 |
| *Drosophila melanogaster–subobscura–pseudoobscura* (24 genes[c]) | 4.52 | 4.37 | 1.65 | 1.64 |

[a] Ohta (1995).
[b] Ohta (1997).
[c] Zeng *et al.* (1998).

both synonymous and nonsynonymous substitutions. These results are supported by the maximum likelihood analyses of Yang and Nielsen (1998).

Zeng *et al.* (1998), also included in Table 1.1, found that the dispersion index is large for synonymous substitutions but not so for nonsynonymous changes in *Drosophila*. In plants, the chloroplast genes of rice, tobacco, pine, and liverwort have large dispersion indexes (Muse and Gaut 1994). However, Kusumi *et al.* (2002) found that the dispersion index is not large for the nuclear genes of conifer trees.

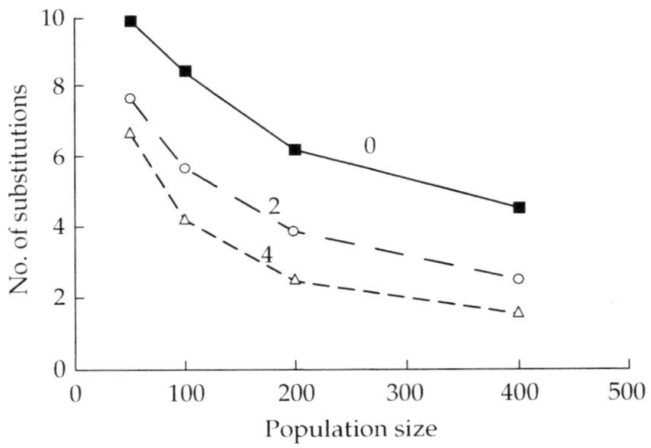


**Figure 1.3** The number of mutant substitutions per term ($12/\nu$ generations) as a function of population size (from Ohta 1997). Parameters are $N = 24$ and $\bar{s} = 0.25$. Numbers beside each line are $K$.

## Interaction systems: a simulation study of the NK model

The erratic pattern of evolutionary rates is thought to reflect interactions at various levels. The **mutational landscape model** was proposed to explain the episodic pattern of molecular evolution by Gillespie (1984, 1991). The model is based on strong selection, and environmental shifts are thought to be responsible for bursts of mutant substitutions. More recently, Kauffman (1993) proposed a generalized landscape model, the **NK model**, which may be extended to treat the nearly neutral mutations of a protein.

The NK model assumes that each amino acid makes a **fitness** contribution that depends on $K$ other amino acids among the $N$ that make the protein. In the original NK model, there are two states at each site. For each interacting combination, fitness is a uniform random number, and the fitness of an entire sequence is obtained by taking the average for all sites. In my simulations, I assumed nine states at each site. Also the fitness of each site is assumed to decrease with distance from the first site (for details, see Ohta 1997).

The results for the case of $N = 24$, and $K = 0, 2$, and 4 are given here. Let $\nu$ be the mutation rate per gene. The number of mutant substitutions that occurred in the period of $12/\nu$ generations, denoted one "term," was examined, with $\bar{s} = 0.25$. Figure 1.3 shows the

**Table 1.2** Results of simulations to show the effect of duration time of small size on the dispersion index ($R$). Parameters are $\nu = 0.0024$, $2N_e = 50$ or 200, and 100 replications

| Duration time of small size (terms) | Neutral[a] | | Selected[b] ($\bar{s} = 1/8$) | |
|---|---|---|---|---|
| | No. of substitutions | $R$ | No. of substitutions | $R$ |
| X1 | 3.04 | 1.90 ± 3.66 | 3.47 | 1.75 ± 2.29 |
| X0.8 | 2.78 | 1.29 ± 1.61 | 3.20 | 1.67 ± 2.80 |
| X(2/3) | 2.73 | 1.39 ± 1.60 | 2.99 | 1.33 ± 0.89 |
| X0.5 | 2.64 | 1.19 ± 1.24 | 2.84 | 1.34 ± 1.82 |
| X0.2 | 2.65 | 0.87 ± 0.39 | 2.53 | 0.95 ± 0.43 |
| X0.1 | 2.97 | 0.99 ± 0.38 | 2.72 | 0.99 ± 0.50 |

[a] Term = $3/\nu$.
[b] Term = $5/\nu$.

number of substitutions in a term as a function of population size. The figure shows that the number of substitutions is a decreasing function of population size, in agreement with the predictions of other models of near neutrality. It is also clear from the figure that, as $K$ gets larger, the number of substitutions decreases, that is, selective constraints becomes stronger as expected. The dispersion index was only slightly larger than unity except for the case of $K = 0$ (no interaction). Hence, it appears that the $NK$ interaction is not important in increasing the index.

Other factors that may inflate the variance of evolutionary rate are shifts of selective value, changing population size, and higher level (i.e. among genes) interactions. The results of simulations that incorporate shifting selective values indicate that the dispersion index becomes larger than in the case of no shifting. Its magnitude is often 2–3, and the shifting is not by itself sufficient to explain the observed values. My simulation study of the effects of changing population size also shows that the dispersion index becomes large such that the weaker the selection, the larger the effect of changing population size. This agrees with the observation that the index is larger for synonymous than nonsynonymous substitutions in *Drosophila* (Zeng *et al.* 1998).

The results of Araki and Tachida (1997) suggest that the effect of changing population size depends on the time spent at the same population size. Therefore, the effect of duration time was also examined by simulation. By keeping the duration time at a large population size at one term as before, the effect of six levels of duration time at small population sizes in the range of 0.1–1 terms were studied. Other procedures were the same as before. The results are given in Table 1.2. As can be seen from the table, the $R$ value depends on duration time. The longer this time is, the larger the dispersion index becomes. Roughly speaking, if duration time is less than the reciprocal of the mutation rate, the change of population size has no effect on inflating the value of $R$.

Why does the dispersion index become large when population size changes? When the population is large, more neutral mutations accumulate than when the population is small. When the population size changes from large to small, drift becomes more rapid and the rate of substitution tends to increase (Chakraborty and Nei 1977). Conversely, if population size changes from small to large, drift becomes slow and **substitution rate** decreases on average. Therefore, the index increases by changing population size.

So far, we have considered interactions within a gene. The NK model can be extended to deal with the case of a group of linked genes that interact with each other. Each site now represents a gene (locus), and $K + 1$ genes interact among $N$ genes. I conducted a simulation study to examine the number of accumulated mutations at each site. The period for mutant accumulation was longer than before to allow for the appropriate number of

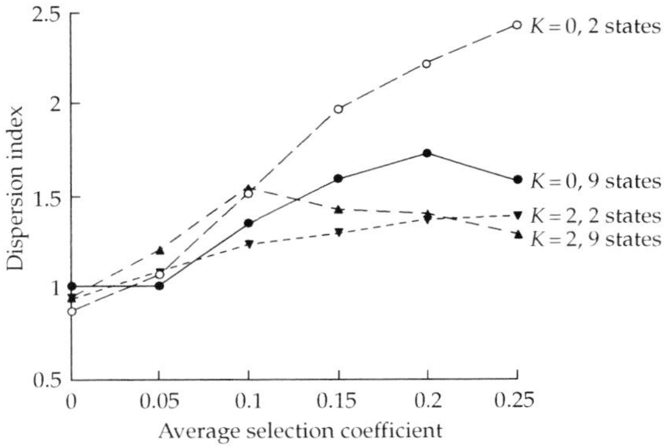


**Figure 1.4** Results of simulations of the relationship between selection intensity and dispersion index, with $N = 6$ (six loci). In the simulations, mutants were accumulated in a period of $18/\nu$ generations, so that the expected number of substitutions for the neutral case is three per locus. The dispersion index was obtained from the accumulated number of mutants at each locus with 100 replications. The figures were drawn using average values of six loci.

substitutions to take place, which enabled us to measure the dispersion index at each site. Also the selection model was modified so that the level of selective constraint was the same for all sites. The case of six sites ($N = 6$), or six interacting loci, was studied for $K = 0$ and 2. The number of allelic states of a gene was two or nine. Figure 1.4 gives the results for a dispersion index that is the mean of six loci. In the figure, the intensity of selection was taken as a variable from $s = 0$ to $s = 0.25$ (average value). For $s = 0$, all loci are neutral and the dispersion index is around unity. The condition of two allelic states may correspond to the ON–OFF states of genes, whereas that of nine states may be regarded as a more complicated interaction system. The states could correspond to a level or pattern of gene expression. It may be noted that the dispersion index monotonically increases for 2-state systems by increasing the selection intensity, but not for the 9-state system. Remember that the case of $K = 0$ and 2-state is the situation of an ordinary 2-allele and independent loci system with a smooth fitness landscape. Therefore, the variance becomes larger by increasing selection intensity. If there are nine states, the allelic difference tends to decrease as selection continues, that is, the population fitness approaches the adaptive peak. This effect is stronger for more intense selection, and the erratic pattern is reduced by increasing selection intensity.

Real examples of interacting systems are now available. Von Dassow *et al.* (2000) present a model which deals with the network of genes that makes body segments in the fruitfly *Drosophila*. This network is an interaction system involving the products of five segment polarity genes, including wingless, engrailed, and hedgehog. The effects of varying parameter values over small ranges are equivalent to those of minor mutations. The results indicate that the segment polarity network is robust to parameter variation and initial conditions. Therefore, the system appears to have properties that are reminiscent of the concept of "self-organization" postulated by the NK model (Kauffman 1993, pp. 198–221). However, Endy and Brent (2001) point out that, based on a model of bacterial chemotaxis, living systems appear to use different ways for the regulation of cellular processes. These authors define regulation as "the processing of information from the genome, from internal events, and from external events, by an amorphous architecture of diffusing molecular components."

Another notable result of the simulations is that the dispersion index is larger when $K = 0$ than when $K = 2$. This is thought to reflect the ruggedness of the fitness landscape, such that the larger the $K$ value, the more rugged the landscape is (Kauffman 1993). Hence, as the landscape becomes more rugged, the erratic pattern of rate variation by shifting becomes smaller.

The actual process of protein evolution is likely to be compounded, that is, a single step of the multiloci model described above is composed of several mutant substitutions. Huelsenbeck *et al.* (2000) have shown that the compound Poisson process is appropriate for analyzing mammalian mitochondrial sequences. Gillespie (1991) formulated the compound Poisson process and pointed out that in this case the dispersion index is given by

$$R = \mu_z + \frac{\sigma_z^2}{\mu_z}, \tag{1.4}$$

where $\mu_z$ is the mean and $\sigma_z^2$ is the variance of the number of substitutions for each step of the Poisson process. In our case, the original process has a larger variance than the Poisson process, and the above formula may be modified as follows:

$$R = (1 + a)\mu_z + \frac{\sigma_z^2}{\mu_z}, \qquad (1.5)$$

where $a > 0$. The formula implies that both $a$ and $\mu_z$ contribute to increasing $R$.

The preceding discussion raises a difficult question. If one observes that the dispersion index is larger than one at a certain locus, it is difficult to determine whether this large $R$ value is caused by shifting on the fitness landscape or by changes in population size. Additional data are urgently needed to clarify the mechanisms of evolutionary change in genes and proteins.

## Polymorphisms and near neutrality

An important prediction of the nearly neutral theory regarding polymorphisms is the excess of less frequent alleles in a population compared with the neutral prediction. In 1970s, I investigated this problem by analyzing electrophoretically detectable polymorphisms (Ohta 1975). I found that the observed and theoretical distributions of alleles agree well except for the excess of rare alleles in *Drosophila* and humans. Also, a theoretical analysis using the stepwise mutation model implied that the distribution depends on the values of $N_e\nu$ and $N_es$. When these values are much less than unity, the allele distribution is indistinguishable from the neutral case, whereas it approaches the deterministic equilibrium as they get larger. Actual data suggested that these values are larger in *Drosophila* than in humans (Ohta 1975). However, the issue of selection versus drift was not settled by these studies, because expanding population size may explain these deviations from neutrality (Nei 1987).

In the present DNA era, one can make more definitive statements about weak selection, particularly by separately measuring synonymous and nonsynonymous polymorphisms. A popular approach for detecting selection is to compare the pattern of within-population divergence with that of between-population divergence separately at synonymous and nonsynonymous sites (McDonald and Kreitman 1991). For mitochondrial genes, an excess of nonsynonymous within-population divergence is often found (e.g. Ballard and Kreitman 1994; Nachman *et al.* 1994; Rand and Kann 1996). Such an excess is thought to be due to slightly deleterious amino acid substitutions. For nuclear genes, both this pattern and the opposite, that is, an excess of nonsynonymous between-species divergence, are found (e.g. McDonald and Kreitman 1991; Long and Langley 1993). The latter pattern is probably caused by advantageous mutant substitutions. However selection may be weak and may include nearly neutral cases if there is shifting on the fitness landscape.

Rand and Kann (1996) defined a statistic, the **Neutrality Index** (N. I.) for measuring departures from neutrality according to the McDonald and Kreitman (1991) test. The index is

$$\text{N.I.} = \frac{\dfrac{\text{no. polym. replacement sites}}{\text{no. fixed replacement sites}}}{\dfrac{\text{no. polym. synonymous sites}}{\text{no. fixed synonymous sites}}}$$

Using this index, Weinreich and Rand (2000) found contrasting patterns between nuclear and mitochondrial genes of various animals.

It is now thought that even synonymous substitutions are not completely neutral. Akashi (1995) estimated that the selection intensity on codon bias is $N_es \approx -1$, for *Drosophila simulans*, and that the value may be smaller for *D. melanogaster*. In particular, he counted "preferred" and "unpreferred" **silent substitutions** separately for segregating genes of *D. simulans*, and found that the former segregates at higher frequencies and is more often fixed between species than the latter (Akashi 1999). He applied the statistics of Sawyer *et al.* (1987) and showed the presence of weak selection such that $2N_es \approx 1$ for the preferred codon substitutions and $2N_es \approx -1$ for the unpreferred ones. Eyre-Walker (1999) has also shown that very weak selection on codon usage bias has been at work for mammalian genes.

Li and Sadler (1991) analyzed the polymorphic pattern of 49 human genes and found low heterozygosity at nonsynonymous sites compared with the other sites. The result implies a high prevalence of slightly deleterious amino acid substitutions. Others have conducted similar analyses using data from the human genome diversity project. Recent genome diversity studies indicate that nonsynonymous **single nucleotide polymorphisms** (SNPs) are rarer than the random expectation as compared with synonymous SNPs, and therefore nonsynonymous SNPs are likely to be often slightly deleterious.

Note here that nonsynonymous SNPs were found in samples of 100 or so genomes. As such they can not be definitely deleterious, and are excellent examples of nearly neutral mutations or slightly deleterious mutations.

## Further thoughts on drift vs. selection

One decisive piece of evidence supporting the prevalence of nearly neutral mutations is the rapid sequence divergence found in endosymbiotic bacteria. Moran (1996) found accelerated sequence evolution in *Buchnera* compared to its free-living relative, and suggested that slightly deleterious mutant substitutions occurred because of the small population size. However, even in this case, another explanation is possible, that is, the relaxation of purifying selection. In order to discriminate between the two possibilities, Funk *et al.* (2001) studied intraspecific polymorphisms in *Buchnera*, and found a very low level of polymorphism. They, therefore, conclude that the rapid sequence divergence is the result of accumulation of slightly deleterious mutations, because one expects high polymorphisms under the hypothesis of relaxed constraints (see van Ham *et al.*, Chapter 9). In general, however, drift and selection cannot be separated at the molecular level. In particular, since interactions are so prevalent, the fitness of a mutation is dependent on its genetic as well as its environmental backgrounds, such that its overall effect would be small. On the other hand, even synonymous substitutions are weakly selected (Akashi 1995), so that very many mutations fall in the nearly neutral class.

In the mitochondrial genome of *Caenorhabditis elegans*, synonymous sites seem to be under strong selection. Denver *et al.* (2000) directly measured the mutation rate of the mitochondrial genome by accumulating deleterious mutations, and found that the rate is two orders of magnitude higher than the rate estimated from molecular evolution studies. They argue that purifying selection is responsible for the difference. Their results imply that only a minor fraction (*ca* 1 percent) of new mutations are accepted during evolution. If the mutants' effects are continuously distributed from definitely deleterious to selective neutrality, the distribution is such that only 1 percent of mutations can spread in the population. This selection is much stronger than the previous estimate based on polymorphism data (Akashi 1995). The difference probably reflects the effect of population structure such that local population size affects polymorphisms. Furthermore, the distribution should differ between various classes such as synonymous and nonsynonymous substitutions, and **deletions** and **duplications**. The strong selection on *C. elegans* mitochondrial genome seems to be exceptional, since the proportion of accepted mutations is much larger in genomes of other organisms (e.g. Kondrashov 1995).

An interesting case of the effect of genetic background on near neutrality is the activity of Hsp 90, a heat-shock protein from *D. melanogaster*. Rutherford and Lindquist (1998) reported that Hsp 90 has a large effect on genetic variation. Hsp 90 is a chaperone molecule that helps restore the native folding of proteins that have been disrupted by high temperatures. These authors argue that Hsp 90 helps in the accumulation of cryptic variations in ordinary conditions, but that under stress, these variations are expressed for selection to work. This may be related to the nearly neutral model. If high temperature is the stress, cryptic mutations may be expressed only when temperature is above some critical value. Therefore, selection coefficients would vary from time to time or from region to region. As I have argued elsewhere (Ohta 1972), the variance among the selection coefficients of mutants increases when the environment becomes more uniform, such that the probability of a mutant being advantageous or neutral becomes larger in more uniform environments. Also, the smaller the population size is, the more uniform the environment becomes. Therefore, the theory of near neutrality (Ohta and Tachida 1990; Tachida 1991) is applicable to the evolution of cryptic mutations via Hsp 90. Consequently, nearly neutral mutations and morphological evolution may be connected.

It has been argued that most mutant substitutions at the molecular level are neutral, and are irrelevant to adaptive evolution (e.g. Nei 1987, pp. 415–16). However, in view of the intricate interactions between drift and selection, some nearly neutral mutant substitutions could have been responsible for the critical shifting of adaptive peaks in evolution, and hence for morphological changes. This may be a modern version of Wright's shifting balance theory (Wright 1938).

The concept of near neutrality needs to be understood in relation to these processes. Of course, drift or random sampling of gametes at reproduction is the basis of near neutrality. As emphasized already, genetic backgrounds as well as ecological factors influence the effectiveness of natural selection.

## Summary

The nearly neutral theory contends that borderline mutations, whose effects are so small that both **random drift** and selection influence their behavior, are important. The theory predicts that there is a negative correlation between evolutionary rate and population size. This prediction was confirmed by separately estimating divergence patterns of synonymous and nonsynonymous substitutions of mammalian genes. The variance of the evolutionary rate was also examined by data analyses and by simulations based on an interaction model. Together with other available observations, the results suggest a prevalence of nearly neutral mutations at the molecular level. The pattern of changing evolutionary rates reflects interactions of gene function at various levels, and therefore the interplay of drift and selection is important for organismal evolution.

I thank three anonymous referees for their useful suggestions on the manuscript and Ms Yuriko Ishii for secretarial assistance.

CHAPTER 2

# Inferring the action of natural selection from DNA sequence comparisons: data from *Drosophila*

**Montserrat Aguadé, Julio Rozas, and Carmen Segarra**

Advances in molecular biology techniques over the last two decades have contributed to a dramatic increase in the number of DNA sequences available in databases. In addition, the development of more efficient and powerful hardware, software, and algorithms has facilitated the handling and analysis of these sequences. Population genetic studies have greatly benefited from these developments. Datasets based on **homologous** DNA sequences from a single species are now essential to understand the forces driving molecular evolution. Indeed, valuable information about the mechanisms involved in the evolution of particular gene regions can be obtained from studying intraspecific nucleotide variation. Our understanding of this information has also benefited from further development of theoretical models such as the infinite-sites model (Kimura 1983). According to this model, clear predictions can be made about the level and pattern of variation expected in a **panmictic population** at mutation–**drift** equilibrium. These predictions also apply to the properties of the gene genealogies of neutral alleles. In this sense, **coalescent theory** (see Box 2.1) is a useful approach to study the statistical properties of genealogies under neutrality (Hudson 1990; Fu and Li 1993). Neutral genealogies can, however, be distorted by natural selection or, alternatively, by historical factors that cause populations to deviate from equilibrium.

Detecting the action of natural selection has been one of the main objectives of population genetic studies. Selection at a particular site affects not only the target of selection but also nucleotide variation at closely linked sites. This effect, known as **linked selection**, causes a characteristic footprint on nucleotide variation at sites linked to the selected site. Therefore, the action of natural selection can be inferred through the detection of its footprint on linked nucleotide variation. However, several demographic factors, such as bottlenecks or geographic subdivision, may leave a similar footprint to that left by natural selection. Thus, it is not easy to discriminate between selective and demographic factors. The extent of the footprint along the entire genome constitutes the main criterion to discern them. Demographic factors affect the whole genome and, therefore, their genome-wide footprint should be detectable in a large number of genes. In contrast, selective factors affect particular genes; consequently, their footprint should be detectable only in these gene regions.

## The action of selection

The footprint left by natural selection on nucleotide variation differs with the kind of selection, that is, whether it is **balancing selection** or **directional selection**. In both cases, however, natural selection affects the evolutionary relationship of DNA sequences from a region encompassing the site under selection (Fig. 2.3). This effect is reflected in the gene genealogy of the DNA sequences sampled. The statistical properties of genealogies affected by selection differ from those of neutral genealogies.

## Box 2.1 The coalescent

The coalescent is a population genetic model that describes, in probabilistic terms, the **neutral evolution** of gene genealogies, or gene trees (for review, see Hudson 1990; Fu 1999; Donnelly and Tavaré 1995). This model is currently the most powerful approach to infer past evolutionary events from DNA sequence data, and provides a theoretical framework for developing new statistical tests. Among other useful features, the coalescent provides efficient computer simulation methods because only a sample of individuals, and not the entire population, is used.

A simple and efficient algorithm for generating gene genealogies that is appropriate for haploid and diploid organisms is that proposed by Hudson (1990). It is a fast algorithm and can be easily implemented on a computer. Genealogies are generated under the following assumptions: (i) the neutral infinite-sites model; (ii) a large and constant population size; (iii) no intragenic recombination; (iv) no migration; and (v) a homogeneous mutation rate along the DNA region studied. These genealogies allow the empirical distribution of several test statistics to be obtained.

The procedure to generate gene genealogies is illustrated by means of an example for a sample of five sequences ($n = 5$) from a population of $N$ individuals (Fig. 2.1). Table 2.1 gives a glossary of the symbols used. The steps followed to generate a simulated gene genealogy are,

**(1)** Determine the times of the $n - 1$ internal nodes of the genealogy, $T_j$ (Fig. 2.1(a)). As $T_j$ follow approximately an exponential distribution with parameter $\binom{j}{2}$, simulate $n - 1$ exponential random variables with that parameter.

**(2)** Generate the **topology** of the genealogy. As the coalescent process runs backwards in time (i.e. from the current sampled sequences to their most recent common ancestor, or MRCA), the process starts from the $n$ sampled sequences.

(2.1) As all lineages have equal probability to coalesce (to fuse in their common ancestor), choose at random (using a uniform random variable) the two lineages (or branches) that coalesce at time $T_5$. In the present example, this coalescent event involves lineages 4 and 5, where lineage $k$ represents the lineage starting at node $k$. Coalescence of these lineages generates internal node 6 (Fig. 2.1(b)); therefore, lineage 6 represents the common ancestor of lineages 4 and 5.

(2.2) Repeat step 2.1 (Figs 2.1(c–e)) until the number of lineages in the sample reaches one, i.e. until the last coalescent event (the MRCA; node 9). In the example, the second coalescent event occurs between lineages 1 and 2 generating internal node 7 (Fig. 2.1(c)), the next coalescent event involves lineages 3 and 7, generating internal node 8 (Fig. 2.1(d)), and the last, or the oldest coalescent event, joins lineages 6 and 8, creating internal node 9 (Fig. 2.1(e)).

**(3)** Incorporate mutations into the genealogy. As the number of mutations along particular lineages is Poisson distributed with mean $ut_b$ ($= \theta T_b / 2$), simulate $2n - 2$ (one for each branch) independent Poisson random variables with parameter $\theta T_b / 2$ (Fig. 2.1(f)). In the example, the true value of $\theta$ was considered to be 10, which results in 17 mutations in this particular realization. According to the infinite sites model, each mutation occurs at a different site of the sequence and, therefore, the number of segregating sites is equal to the total number of mutations.

The information obtained from this computer-generated gene genealogy (the topology, branch length, and number of mutations placed on each branch) allows the value of relevant statistics to be calculated. In the example, the value of Tajima's $D$ statistic (Tajima 1989) is 0.115. The empirical distribution of a particular test statistic can be obtained from multiple iterations (pseudoreplicates) of the above coalescent process (Fig. 2.2). The confidence intervals for the test statistic can be obtained from that distribution, which can be used to test for departures from the strictly **neutral model** (null hypothesis).

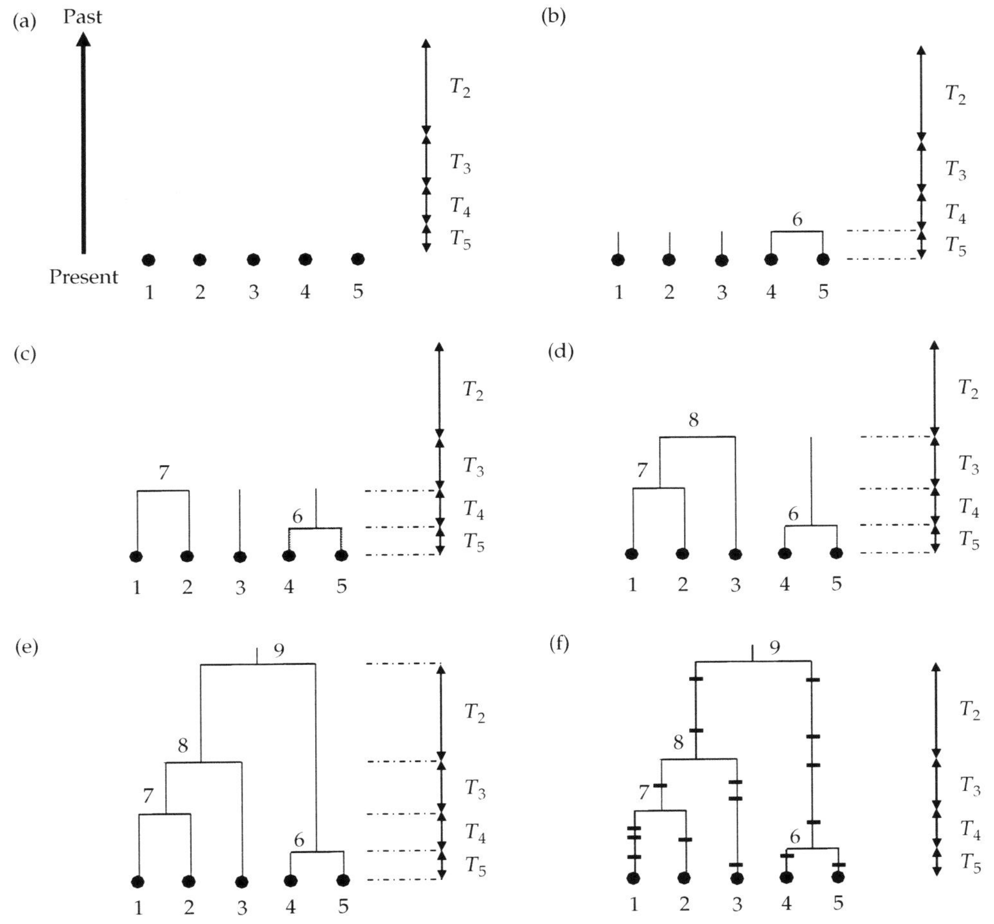


**Figure 2.1** The steps followed to generate a gene tree by computer simulations based on the coalescent (see text).

**Table 2.1** Glossary of symbols used

| | |
|---|---|
| $N$ | Population size[a] |
| $G$ | Number of gene (sequence) copies in the population[b] |
| $n$ | Sample size |
| $T$ | Time measured in $G$ generations |
| $T_b$ | Branch length measured in units of $G$ generations |
| $t_b$ | Branch length measured in generations ($t_b = GT_b$) |
| $T_j$ | Time, measured in units of $G$ generations, during which there are $j$ distinct ancestors in the sample |
| $u$ | Mutation rate per sequence per generation |
| $\theta$ | $2Gu$ |

[a] Total number of individuals (both males and females) in the population.

[b] Assuming equal numbers of males and females, $G = 2N$ for autosomal regions of diploid organisms, $G = 1.5N$ for $X$-linked regions of diploid organisms, and $G = 0.5N$ both for $Y$-linked regions of diploid organisms and for mtDNA regions.

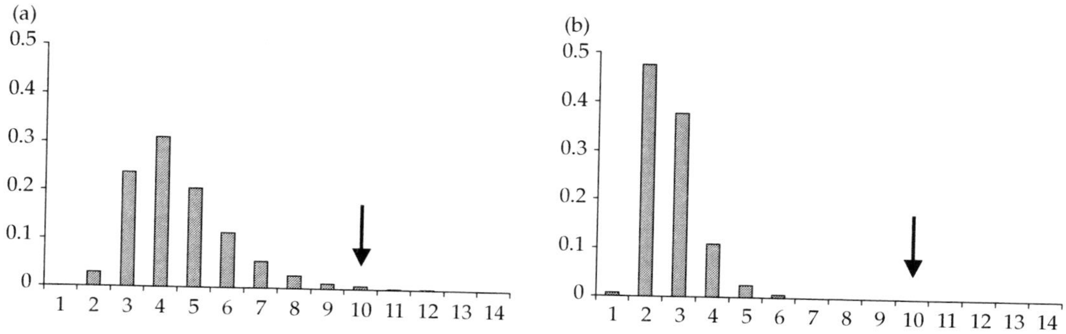


**Figure 2.2** Distribution of the maximum number of identical lines in a sample of 24 sequences with $\theta = 15.504$. (a) $R = 0$ (no recombination); (b) $R = 49.2$ (estimate of the population recombination parameter at the *rp*49 region based on the comparison of the physical and genetic maps). The value reported by Rozas *et al.* (2001) is indicated by an arrow.

The assumptions that underlie Hudson's (1990) algorithm are unrealistic for many purposes. However, this algorithm can easily be extended to include a range of population scenarios (population subdivision, population growth, etc.), recombination rates or nucleotide substitution models. Thus, it is possible to obtain the empirical distribution of several test statistics under more complex conditions (e.g. including recombination), for which there is no analytical solution. An adaptation of this algorithm is implemented in the DnaSP software (Rozas and Rozas 1999).

## Balancing selection

Natural selection that maintains two or more alleles in a population is called balancing selection. The action of balancing selection increases the sojourn time of the selected variants in the population. In the simplest model, this kind of selection maintains two alleles, or allelic classes, that at least differ at the particular nucleotide site affected by selection. The polymorphism present at the selected site is also known as **balanced polymorphism**. Balancing selection causes the two allelic classes to remain in the population for more time than expected under random **genetic drift**. During their sojourn in the population, the two allelic classes accumulate neutral variation. In regions with no recombination, there is no exchange of information between allelic classes and, therefore, variants that originate in each class contribute to their differentiation. The number of differences fixed between classes increases with time as a result of the eventual **fixation** of new variants. Thus, the degree of genetic differentiation between allelic classes is higher for older selected alleles. In regions with recombination, the higher the recombination rate, the shorter the DNA fragment affected by the independent accumulation of variation. Consequently, the degree of differentiation between classes is comparatively low.

Balancing selection affects gene genealogies. In fact, the time back to the most recent common ancestor (MRCA) of the two allelic classes depends on the sojourn time of the two selected alleles in the population. For old polymorphisms, this time is much longer than that expected for the coalescence of neutral alleles. Consequently, balancing selection generates a gene genealogy split into two clusters differing at many sites (Fig. 2.3(b)). Indeed, in this genealogy, many mutations are in the internal branch that separates these clusters, resulting in the segregation of a high number of polymorphisms at intermediate frequencies.

## Directional selection

Directional positive selection operates when a variant shows some advantage over other variants present in

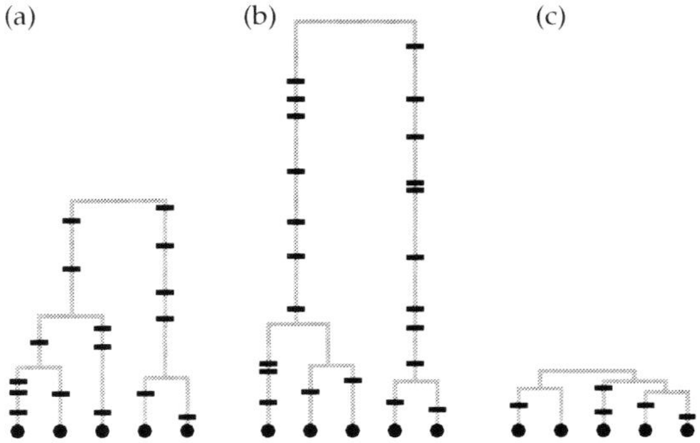


**Figure 2.3** Gene genealogies under different evolutionary scenarios: (a) neutrality, (b) balancing selection, and (c) directional selection. The estimated Tajima's $D$ values for the three genealogies are 0.115, 0.914, and −1.124, respectively.

the population. The advantageous variant can spread in the population and later become fixed. Fixation can be very rapid for strong selection. Variants linked to the selected site also become fixed. This process, known as **hitchhiking** or **selective sweep**, eliminates all previous nucleotide variation around the selected site (Kaplan *et al.* 1989). The length of the DNA fragment affected by the selective sweep depends on the recombination rate and on the strength of selection. Strong selection in regions of low recombination results in large DNA fragments with reduced variation. In addition, in regions with an extremely low recombination rate, the DNA fragments affected by different selective sweeps can overlap, which results in large chromosomal segments depleted of variation. Therefore, the footprint of directional selection—reduced variation—is more likely to be detected in regions with extremely low recombination rates.

Directional selection affects gene genealogies. For recent selective sweeps, the time back to the MRCA of a sample of sequences is limited by the time back to the sweep. This time is much shorter than the expected coalescence time for a neutral genealogy and results in a **star genealogy**, that is, a genealogy with very short internal branches compared with external branches (Fig. 2.3(c)). In this genealogy, most mutations are on external branches and, thus, are present only once in the sample of sequences (**singleton mutations**).

The hitchhiking model is not the only selective model that predicts a depletion of nucleotide variation around the selected sites. Purifying selection against deleterious mutations, or **background selection** (Charlesworth *et al.* 1993), has a similar effect. However, it is possible, although not easy, to discriminate between these models, as their effects on gene genealogies differ (e.g. background selection does not generate star genealogies).

## Detecting the action of selection

Theoretical population genetic models make clear predictions about the level and pattern of variation expected in a panmictic population in mutation–drift equilibrium. Thus, it is possible to infer the action of natural selection from its impact on nucleotide variation. Several tests have been developed to determine whether the pattern of variation observed within species conforms to that expected under neutrality (Tajima 1989; Fu and Li 1993; Hudson *et al.* 1994; Fu 1997). The application of most of these tests requires data on only intraspecific nucleotide variation.

The informative content of a sample of sequences from a single species greatly increases when the sequence of the homologous region from a closely related species is available. This allows the comparison of mutations that segregate within species (**polymorphism**) and those that have become fixed between species (**divergence**). Under neutrality, levels of polymorphism and divergence should correlate, as both are a function of the neutral mutation rate. Consequently, high neutral mutation rates result in high levels of neutral polymorphism and divergence. This expected proportionality between polymorphism and divergence in any genomic region, and even for different kinds of sites (**synonymous** versus **nonsynonymous**), is the basis of several neutrality tests. Among these tests are the Hudson, Kreitman, and Aguadé or HKA test (Hudson *et al.* 1987), and the McDonald and Kreitman or MK test (McDonald and Kreitman 1991).

Some neutrality tests allow the identification of the kind of selection involved in the evolution of a particular gene region. Indeed, the long sojourn of two allelic classes under balancing selection causes an excess of polymorphic sites with variants that segregate at intermediate frequencies. Consequently, positive values of Tajima's $D$ (Tajima 1989) and Fu and Li's $D^*$ and $F^*$ (Fu and Li 1993) statistics are expected. In addition, an excess of polymorphism relative to divergence is expected in the region encompassing an old balanced polymorphism, which should be detectable by the HKA test.

In contrast, directional selection has a different effect. An excess of singletons and low frequency variants are expected after a selective sweep. This excess results in negative values of Tajima's and Fu and Li's test statistics. Additionally, decoupling between polymorphism and divergence is expected soon after a selective sweep. Thus, reduced polymorphism relative to divergence is expected in the gene region affected by a recent selective sweep. This deviation from neutral expectations should be detectable by the HKA test.

Some demographic or historical factors can distort gene genealogies in a similar way to natural selection. Indeed, bottleneck events, population expansion or population subdivision may also affect the evolutionary relationship between sequences. For instance, population expansion causes a similar effect to that produced by directional selection. Population subdivision with no (or reduced) migration causes the independent (or quasi independent) evolution of two or more sets of sequences and, thus, has a similar effect to that of balancing selection. Additional information from other gene regions is required to ascertain whether the deviation from neutral expectations detected in a particular gene region is caused by selection or by demographic factors. The effect of demographic or historical factors should be genome-wide, while selection should affect only the particular gene region that is the target of selection.

## Empirical data

The number of population sequence datasets in *Drosophila* has steadily increased in the last decade, especially in *Drosophila melanogaster* and *D. simulans*, but also in other species of the melanogaster and obscura groups. *D. melanogaster* and *D. simulans* are cosmopolitan species that originated in Africa some 2.5 million years ago and recently expanded worldwide. Several studies have attempted to summarize the patterns of nucleotide variation detected in these species in order to discern between selective and demographic factors (Andolfatto and Przeworski 2000; Przeworski *et al.* 2001). However, it is premature to draw general conclusions from these analyses as they are based on population surveys that have used diverse sampling strategies. Indeed, many studies are based on species-wide samples that include only one or very few sequences from any given locality. On the other hand, studies based on larger samples from particular populations rarely include African and non-African lines, that is, lines from ancestral and derived populations of these species. Furthermore, most of the regions analyzed were chosen because they were considered candidates for adaptive evolution and, therefore, they were not random representatives of the genome. This is the case for genes that encode allozymes, male accessory gland peptides (*Acp* genes) or peptides involved in the immune response (e.g. the Cecropin or *Cec* genes). In other cases, DNA variation was surveyed in regions with drastically reduced recombination (or no recombination) because of the predicted relationship between the recombination rate and the effect of selection on linked neutral variation.

Natural selection leaves a characteristic region-specific footprint in DNA sequences that can be a priori differentiated from the similar but genome-wide effects of demographic factors. The footprint left by selection is dependent on the time back to the selective event. Only regions affected by old balanced polymorphisms are expected to exhibit the increased level of **silent variation** associated with the selected variants. Similarly, only regions affected by relatively recent selective sweeps exhibit the reduced level of variation associated with the fixation of an advantageous mutation by directional selection. Furthermore, the extent of the region affected by a selective event is a function of the recombination rate and of the strength of selection (i.e. the selection coefficient). Therefore, it is not easy to find strong and unambiguous evidence for molecular **adaptation**. Here, we present some examples in which either balancing or directional selection have been proposed to explain the patterns of nucleotide variation observed.

### Balancing selection

The presence in *D. melanogaster* of latitudinal clines for several allozyme loci in different continents pointed to balancing selection as the evolutionary factor responsible for maintaining these polymorphisms. In this species, the first survey of DNA sequence variation in such a locus, the alcohol-dehydrogenase, *Adh* gene, (Kreitman 1983; Hudson *et al.* 1987; Kreitman and Hudson 1991) revealed an excess of silent variation associated with the *Fast/Slow* allozyme polymorphism, which was compatible with the effect of balancing selection. Subsequent studies of nucleotide variation at other allozyme loci have failed to detect this pattern, and even in the case of *Adh* further data suggest that the excess of silent variation may predate the origin of the *Fast/Slow* polymorphism (Begun *et al.* 1999). Thus, strong empirical support for a role of balancing selection in the maintenance of allozyme polymorphism in *Drosophila* is lacking. This lack of evidence might be attributable to the fact that in *D. melanogaster* and *D. simulans* most of the allozyme genes surveyed are located in regions of high recombination, where the extent of the region affected by adaptive polymorphism may be very small. A more parsimonious explanation is that allozyme polymorphisms are due to relatively recent mutations, and thus the corresponding allelic classes have not had enough time to accumulate variation.

Analyses of DNA sequence variation in some genes of the mammalian immune system revealed a high level of nonsynonymous variation (Hughes and Nei 1988). This observation was considered to support the maintenance of diversity in the corresponding proteins by balancing selection. Although a similar situation was *a priori* expected in the genes involved in the *Drosophila* immune response, the first genes studied (*Cec* genes) exhibited a low level of nonsynonymous polymorphism. However, unlike the genes studied in mammals that are involved in their adaptive immune system, those studied in *Drosophila* are responsible for the innate immune system of the insect. Therefore, selection does not seem to favor diversity in the proteins that mediate the general antibacterial response of insects (Clark and Wang 1997; Ramos-Onsins and Aguadé 1998).

## Directional selection

In *Drosophila*, more empirical support is found for the action of directional rather than balancing selection. There is, however, strong debate about its pervasiveness. The examples below point to the action of directional selection and have been grouped according to the kind of deviations from neutral predictions.

### *Effect of recent selective events on nucleotide variation*

Recent selective events leave a footprint of reduced variation in the region surrounding the selected site. When there is no recombination, the selective sweep affects the whole molecule (e.g. mtDNA, the dot chromosome of many *Drosophila* species, etc.). When the recombination rate is not null but drastically reduced, the regions affected by different selective events can overlap and the footprint can extend over a large region. In regions of high recombination, the extent of the region affected is comparatively much smaller. Moreover, in these regions, the original association of the selected variant with a particular **haplotype** may be more easily lost by recombination, causing unusual haplotype structures in the surrounding regions.

*Reduced levels of variation in regions of null or highly restricted recombination.* Initial surveys of DNA sequence variation in regions with highly restricted rates of recombination, or with no recombination, revealed reduced levels of variation compared with those in regions with normal levels of recombination (as reviewed in Aguadé and Langley 1994; Aquadro *et al.* 1994; Stephan 1994). In the former regions, silent divergence was not reduced, indicating that the low levels of polymorphism were not caused by unexpectedly high functional constraint. The HKA test revealed significant decoupling between polymorphism and divergence, indicating that the level of polymorphism was significantly reduced. In addition, there is a general relationship between recombination and nucleotide variation as the level of polymorphism tends to increase gradually with recombination (Begun and Aquadro 1992). Although these results support directional selection, they prompted the development of an alternative theoretical model, the background selection model (Charlesworth *et al.* 1993). This model, like the hitchhiking model, predicts a direct relationship between levels of polymorphism and recombination rates. Given present estimates of both the deleterious mutation rate in *D. melanogaster* and rates of recombination across its major chromosomes, the background selection model cannot explain the very low level of nucleotide polymorphism observed in the telomeric region of the *X* chromosome, that is, in the *yellow–achaete–scute* region. Consequently, directional selection must be invoked to account for the reduced levels of variation detected in this region, and most probably in other telomeric and centromeric regions that exhibit drastically restricted recombination rates.

*Reduced levels of variation in some chromosomes or chromosomal regions.* Selection can cause the level of **nucleotide diversity** to differ between sex chromosomes and autosomes. The hitchhiking and the background selection models predict differential effects on the *X* chromosome and on autosomes. Under the assumption that deleterious mutations are partially recessive, the background selection model predicts a higher reduction of variation in autosomes than in sex chromosomes. On the other hand, positive selection acting on recessive advantageous mutations would cause the opposite. An extensive survey of nucleotide sequence variation in 40 genes distributed across chromosome arms *X* and *3R* of *D. simulans* revealed a significantly lower level of synonymous polymorphism on the *X* chromosome than on the right arm of the third chromosome (Begun and Whitley 2000). As the level of synonymous divergence did not differ between chromosomes, this observation cannot be explained by differences in the neutral mutation rate. The lower level of polymorphism on the *X* chromosome is also incompatible with the predictions of the background selection model, and is thus better explained by some form of positive directional selection, which exerts a greater effect on the *X* chromosome than on the autosomes.

The low level of nucleotide sequence variation at the *Sdic* gene, a newly evolved chimeric gene in *D. melanogaster*, suggested the action of positive selection. This reduction may also have been caused by background selection, as this gene is located at the base of the *X* chromosome where recombination is reduced. However, the study of nucleotide variation in 10 genes distributed at increasing distances on both sides of the *Sdic* gene revealed a trough of variation in the region encompassing the novel gene (Nurminsky *et al.* 2001). In *D. simulans*, where this gene is absent, analysis of nucleotide variation in the homologous surrounding regions did not reveal the equivalent trough of variation expected under background selection. These observations support the hypothesis that the newly evolved gene has undergone one or more selective sweeps in its short evolutionary history.

*Unusual haplotype structure in regions of high recombination.* An unusual pattern of variation was detected in the *Sod* (superoxide dismutase) gene region in European and North American samples of *D. melanogaster* (Hudson *et al.* 1994). Although the level of variation in this region was comparable to that present in other regions, a major haplotype harboring very little variation was present at high frequency in both samples. The presence of this haplotypic structure in a highly recombining region deviates significantly from neutral expectations in a panmictic population. This pattern is compatible with a rapid increase in the frequency of this haplotype caused by directional selection acting on an advantageous mutation (Hudson *et al.* 1994); however, it could also reflect recent admixture of differentiated populations. Although a similar haplotype structure was detected in two regions located 3.7 kb and 19.2 kb downstream of the *Sod* gene, this pattern was not observed in a region 12.7 kb upstream of this gene. This latter observation favors a selective explanation for the pattern of variation observed in the *Sod* region.

A similar pattern was detected in the *rp49* (ribosomal protein 49) gene region in European and African populations of *D. simulans* (Rozas *et al.* 2001). Indeed, 45 segregating sites and eight length polymorphisms were detected in 24 sequences; there were, however, 10 identical sequences despite this high level of variation. Again, the presence of such a high number of identical sequences in a highly variable and recombining region deviates significantly from neutral expectations under **panmixis** (Fig. 2.2). This unusual pattern of variation both in a derived and in a putatively ancestral population is very unlikely under a demographic scenario of recent admixture of differentiated populations. Thus, positive selection acting on an advantageous mutation located either within or very close to the region studied provides a better explanation of the pattern observed.

*Effect of old selective events on nucleotide variation*
The **substitution rate** of selectively neutral variants is equal to the neutral mutation rate, while the rate of substitution of positively selected variants is a function of the advantageous mutation rate, the effective population size and the strength of selection. It can be easily envisaged that the advantageous mutation rate for a particular genomic region might vary considerably over time as a consequence of environmental changes. Thus, the fixation rate of advantageous mutations will not only differ between genes but, more importantly, will vary over time. For example, colonization of new habitats may temporally increase the fixation rate for genes that contribute to the adaptation of the organism to the new habitat. Similarly, speciation may increase the fixation rate for genes involved in sexual reproduction. When adaptation is associated with changes in proteins, old bursts of adaptive changes cause decoupling between nonsynonymous polymorphism and divergence. This results in an excess of fixed nonsynonymous changes, which should be detectable by the MK test.

*Excess of fixed amino acid replacements in proteins involved in sexual reproduction.* In comparisons between *D. melanogaster* and *D. simulans*, Civetta and Singh (1998) were the first to report that the genes involved in sexual reproduction (and other sex-related genes) exhibited on average a higher $K_a/K_s$ ratio than other genes. In some of these genes, particularly in those encoding male accessory gland peptides (*Acps*), comparison of intra- and interspecific synonymous and nonsynonymous variation revealed an excess of fixed nonsynonymous changes. This excess was detected in the three largest *Acp* genes studied in *D. melanogaster* and *D. simulans*: *Acp26Aa*, *Acp29AB*, and *Acp32CD* (Aguadé 1998, 1999; Begun *et al.* 2000), a strong indication that positive selection has driven the evolution of the corresponding proteins.

## Conclusions

Natural selection leaves a characteristic footprint in DNA sequences as it affects their evolutionary relationship, and thus gene genealogies. This footprint can be detected from the analysis of intraspecific nucleotide variation. Balancing selection increases the sojourn in the population of the allelic classes selected, which results in a genealogy split into differentiated clusters. In contrast, directional positive selection causes a depletion of neutral variation around the selected site, which results in star genealogies. However, demographic factors and selection can have similar effects on gene genealogies. The extent of the footprint in the genome constitutes the main criterion to discern selective and demographic factors. Indeed, selection affects only particular regions of the genome whereas demographic factors leave a genome-whereas footprint.

In *D. melanogaster* and *D. simulans*, most surveys of nucleotide sequence variation have concentrated on genes that are considered candidates for adaptive evolution. The level and pattern of intra- and interspecific variation in some of these genes support the action of natural selection. Indeed, the adaptation of these species to the recently colonized temperate zones may have shaped variation in many genes. However, it has sometimes proved difficult to disentangle the effects of selective and demographic factors. Analysis of nucleotide variation at noncoding regions in derived populations might more easily reveal the signature of population history. Once this history is established, it may be easier to detect the locus-specific signature of natural selection. Moreover, high-throughput methods to survey nucleotide sequence variation in multiple regions—or even in complete genomes—for large samples will become available in the **genomics** era. In this framework, the quantification of the rate of adaptive evolution might be a realistic challenge.

CHAPTER 3

# Rates and effects of deleterious mutations and their evolutionary consequences

**Aurora García-Dorado, Carlos López-Fanjul, and Armando Caballero**

In a highly influential paper, Muller (1950) put forward the idea that deleterious mutations affecting the heterozygote appear at a high rate in populations, thus, imposing a reduction in **fitness (mutational load)** that could become unbearable. As shown by Haldane (1937), this load is mainly determined by the rate of deleterious mutations and is practically independent of their severity, although it is modulated by their degree of dominance. It took a long time until adequate experimental data for *Drosophila* viability were first obtained by Mukai (1964), who estimated a high genomic rate of mutations (**haploid rate** $\lambda > \approx 0.3$, the number of deleterious mutations appearing per gamete and generation) showing very little dominance (average **coefficient of dominance** $E(h) \approx 0.4$, where $h = 0$, 0.5, and 1 denotes recessive, additive and dominant gene action of mutations, respectively) and mild average effects (average **selection coefficient** $E(s) < \sim 0.03$, the relative reduction in fitness of the homozygous mutant). These values were corroborated by subsequent work (Mukai *et al.* 1972; Ohnishi 1977a) and became typical parameters extensively used in evolutionary genetic models, apparently supporting Muller's pessimistic views on the role of mutational load.

In the last decade, however, renewed interest on the matter prompted additional experimental work, the results of which raise questions about the validity of former estimates. The controversy is far from resolved, as shown by the contrasting opinions maintained in the latest reviews (García-Dorado *et al.* 1999; Keightley and Eyre-Walker 1999; Lynch *et al.* 1999). Here, we review the published estimates of $\lambda$, $E(s)$, and $E(h)$ obtained from spontaneous **mutation accumulation** (MA) experiments. We focus on deleterious mutations in eukaryotes (see Drake and Holland 1999, for a review on mutation rates among RNA viruses). We also infer the fraction of mutations that are undetected in MA experiments and examine some of the evolutionary consequences of the estimated rates of deleterious mutation.

## Review of estimates of $\lambda$ and $E(s)$ from MA experiments

Before proceeding, a brief description of the experimental and analytic methods seems appropriate. In MA experiments, deleterious mutations accumulate as a Poisson process under relaxed selection in lines derived from the same uniform genetic background. This allows the estimation of the per generation rate of decline for a **fitness component** trait, $\Delta M = \lambda E(s)$, and the rate of increase in the between-line variance, $\Delta V = \lambda E(s^2)$. Thus, lower and upper bound estimates of $\lambda$ and $E(s)$ can be computed. Such Bateman–Mukai (BM) estimates are calculated as follows: $\lambda \geq \Delta M^2/\Delta V$, $E(s) \leq \Delta V/\Delta M$. In parallel, the information contained in the observed distribution of line means can be more efficiently used by assuming a convenient family distribution of mutant effects (usually a gamma distribution) and finding point estimates, rather than estimates of bounds, of the mutational parameters that better account for the data. Two methodologies have been used: maximum likelihood (ML, see Keightley and Bataillon 2000) and minimum distance (MD, see García-Dorado and Gallego 2003).

Estimates have been obtained for several eukaryotic species: the fruitfly *Drosophila melanogaster*, the crustacean *Daphnia pulex*, the nematode *Caenorhabditis elegans*, the yeast *Sacharomices cerevisae*, and two plants, the crucifer *Arabidopsis thaliana* and the wheat *Triticum durum*. A summary of $\lambda$ and $E(s)$ estimates (generally rounded, for clarity, to the second decimal) for relative fitness and fitness component traits is given in Table 3.1. For plant and invertebrate animal species, recent BM estimates of $\lambda$ are about one order of magnitude smaller than classical Mukai/Ohnishi estimates, the corresponding $E(s)$ values being larger. This discrepancy vanishes when MD estimates are considered. On the whole, the data show remarkable concordance, indicating $\lambda$ and $E(s)$ values of about 0.01 and 0.1, respectively. We will now discuss the most salient aspects of the estimates.

**Table 3.1** Estimates of the haploid genomic rate $\lambda$ and the average homozygous effect $E(s)$ of deleterious mutations for relative fitness or fitness component traits

| Species | Character | BM bound estimates | | MD/ML direct estimates | |
|---|---|---|---|---|---|
| | | $\lambda$ | $E(s)$ | $\lambda$ | $E(s)$ |
| *D. melanogaster* (1) | Viability | 0.44[a] | 0.02[a] | 0.01 | 0.19 |
| *D. melanogaster* (2) | Viability | 0.14[a] | 0.03[a] | 0.01 | 0.23 |
| *D. melanogaster* (3) | Viability | 0.04[a] | 0.15[a] | 0.02 | 0.16 |
| *D. melanogaster* (4) | Viability | 0.02[a] | 0.10[a] | 0.02 | 0.10 |
| *D. melanogaster* (5) | Viability | 0.03[a] | 0.08[a] | — | — |
| *D. melanogaster* (6) | Viability | 0.003[a] | 0.25[a] | 0.005 | 0.16 |
| *D. melanogaster* (7) | Fitness | — | — | 0.03 | 0.26 |
| *D. melanogaster* (8) | Fitness | 0.03 | 0.14 | — | — |
| *A. thaliana* (9) | Fitness | 0.015 | 0.36 | — | — |
| *T. durum* (10) | Productivity | — | — | 0.05[b] | 0.20[b] |
| *C. elegans* (11) | Intrinsic growth rate | 0.01 | 0.05 | 0.003[b] | 0.10[b] |
| *C. elegans* (12) | Intrinsic growth rate | 0.007 | 0.22 | 0.02[b] | 0.06[b] |
| *S. cerevissiae* (13) | Fitness | — | — | 0.0001[b] | 0.22[b,c] |
| *S. cerevissiae* (14) | Fitness | — | — | 0.0005[d] | 0.09[d] |

(1) Mukai *et al.* (1972), (2) Ohnishi (1977*b*), (3) Fry *et al.* (1999), (4) Fernández and López-Fanjul (1996), (5) Chavarrías *et al.* (2001), (6) Caballero *et al.* (2002), (7) Houle *et al.* (1992), (8) Ávila and García-Dorado (2002), (9) Schultz *et al.* (1999), (10) Bataillon *et al.* (2000), (11) Keightley and Bataillon (2000), (12) Vassilieva *et al.* (2000), (13) Zeyl and DeVisser (2001), and (14) Wloch *et al.* (2001).

[a] Lethal and severely deleterious mutations excluded.
[b] ML estimates.
[c] Estimate of $E(sh)$.
[d] Direct estimate obtained by tetrad analysis.

## *Drosophila melanogaster*

The BM analysis of classical experiments by Mukai and Ohnishi indicated that mutations affecting egg-to-adult viability arise at a high rate ($\lambda > \sim 0.3$) and have small average homozygous effects ($E(s) < \sim 0.03$). However, it has been suggested that the viability decline observed in these experiments could be partially ascribed to causes different from spontaneous mutation (Keightley 1996; García-Dorado 1997). In fact, a more recent MA experiment consistently suggests $\lambda$ values about tenfold lower ($\lambda \sim 0.02$) and larger $E(s)$ estimates ($E(s) \sim 0.1$) through three repeated assays. The first and second assays correspond to non-competitive viability at generations 104–106 (Fernández and López-Fanjul 1996) and 210 (Caballero *et al.* 2002) while, in the third, viability scores were obtained at generations 250–255 in competitive conditions similar to those of Mukai (Chavarrías *et al.* 2001). The two latest assays used contemporary evaluations of the MA lines and the control, and the rate of decline in mean and the rate of increase in variance (and, hence, the BM estimates of $\lambda$ and E($s$)) were close to those observed earlier.

Natural selection could have acted during the course of the Fernández and López-Fanjul experiment, producing downward biases of $\lambda$ and $E(s)$. However, computer simulations accounting for selection within and between MA lines, as well as in the control population, suggested that elimination of mutations through selection is not the main cause of the discrepancy between the contrasting estimates of mutational parameters (Caballero *et al.* 2002). Two other results support the same conclusion. First, there was no indication of a temporal decline of the control average, suggesting that accumulation of mildly deleterious mutations was not important in this case. Second, the number of lines lost after 255 generations was only 9 percent larger than that expected from accidental losses (Chavarrías *et al.* 2001) and this was only concordant with a simulated model with few mutations of large effect (Caballero *et al.* 2002).

The discrepancy between Fernández and López-Fanjul's and Mukai's results can be ascribed to a much larger decline in mean in the latter, as the pertinent increase in variance was about the same in all *Drosophila* experiments. This may be due to the lack of a suitable control in Mukai's experiment that would allow unbiased estimates of $\Delta M$. This impediment has been obviated by computing MD estimates of mutational parameters unconstrained by the observed $\Delta M$, giving similar estimates of $\lambda$ for both experiments (García-Dorado 1997) (Table 3.1).

Viability estimates of $\Delta M$ obtained by an "order method," have been recently reported by Fry (2001) using both Mukai's original data and new data. With this method, $\Delta M$ is calculated as the relative difference between the means of "control" MA lines (those with the highest viability in the later assays, inferred to carry few or no mutations) and quasinormal MA lines (those with at least one-half the viability of the "controls", thus, excluding lethal and highly deleterious mutations). The average value of $\Delta M$ for the entire genome (1.1 percent) did not modify classical estimates substantially. However, the mutational nature of the viability difference between "control" and "quasinormal" MA lines is not clear, at least for Mukai's 1964–9 data, and the validity of the $\Delta M$ estimate relies on the arbitrary choice of an initial MA period, where viability is assumed to decline linearly. For example, using Mukai's (1969) data, the quadratic regression of viability on time suggests that the initial viability decay was five times smaller than previously calculated (García-Dorado and Caballero 2002).

Estimates of mutation rates for fitness have been obtained by Houle *et al.* (1992) and by Ávila and García-Dorado (2002). In Houle's experiment (Houle *et al.* 1994) the control was later shown to be contaminated, but MD estimates were nevertheless obtained (see García-Dorado *et al.* 1999). Both experiments gave very similar estimates (see Table 3.1).

### *Arabidopsis thaliana*

Schultz *et al.* (1999) and Shaw *et al.* (2000) reported results from 924 or 40 MA lines maintained by single-seed descent and derived from a single inbred founder, after 10 or 17 generations of MA, respectively. A control obtained from generation 0 stored seeds was assayed synchronously to the MA lines. In both experiments, plants were grown in benign conditions and a number of reproductive and quantitative traits were scored. For all traits and experiments, the mutational decay was very weak and generally non significant. Significant between-line variances were detected by Shaw *et al.* (2000), but none of those calculated by Schultz *et al.* (1999) significantly departed from zero.

At face value, the BM estimates obtained by Shaw *et al.* (2000) were $\lambda = 0.002$ and $E(s) = 0.47$ averaging over traits (mean number of seeds per fruit, number of fruits per plant, and dry mass of infrutescence). A later analysis using Markov Chain Monte Carlo Maximum Likelihood gave values of $\lambda = 0.06$ and 0.10 for fruit and seed number, respectively (R.H. Shaw *et al.* 2002). However, the distribution of mutational effects indicated an approximate equal number of positive and negative mutational effects. The behavior of these traits, therefore, is not typical of a fitness component. As only deleterious mutations are considered in this review, these estimates are not included in Table 3.1.

For a measure of fitness, Schultz *et al.* (1999) provided BM estimates $\lambda = 0.05$ and $E(s) = 0.23$. However, these values were calculated as the average of estimates obtained in a set of bootstrapped samples. While bootstrapping is very useful to compute errors of estimates, the estimates themselves must be calculated from the original sample. This can be done using Tables 3.1 and 3.2 in Schultz *et al.* (1999), resulting in estimates of $\lambda = 0.015$ and $E(s) = 0.36$.

### *Triticum durum*

Preliminary results have been reported by Bataillon *et al.* (2000). Starting from a homozygous base population, 135 MA lines were derived and subsequently maintained through enforced selfing. After seven generations, an experiment was performed where the total number of seeds per plant was scored both in the MA lines and in 60 control lines (from generation 0 seeds). ML estimates of mutational parameters (Table 3.1) were compatible with the *A. thaliana* and later *D. melanogaster* results mentioned above. However, the **polyploid** condition of *T. durum* casts doubts on the validity of a comparison with the mutation rate of diploid species, as the masking effect of mutations by different alleles in a polyploid species may lead to a larger mutation rate.

### *Daphnia pulex*

In the MA experiment by Lynch *et al.* (1998), BM estimates ranged from $\lambda > 0.8$, for mutations reducing viability, to $\lambda > 0.25$, for mutations increasing 3rd clutch size. However, the frozen control was only evaluated at generations 7 and 16 (it was later disregarded on the basis of its poor performance), and the lack of significant differences between control means at these generations does not imply environmental stability over the whole experiment (32 generations). Furthermore, the originally collected water (recycled in a diatomaceous earth filter and used throughout the experiment) could have undergone important changes, and the *Scenodesmus* culture used to feed the MA lines might have evolved in the laboratory. These circumstances could have affected the expression of fitness components in *Daphnia*, which is known to be a very sensitive organism, and could have reduced the final performance of the MA lines and the disregarded control. Therefore, these data should be taken with extreme caution and are not included in Table 3.1.

### *Caenorhabditis elegans*

Results from two MA experiments have been reported (Keightley and Caballero 1997; Vassilieva and Lynch 1999) and they have been reanalyzed by Keightley and Bataillon (2000), who also included estimates for intrinsic growth rate newly obtained from Keightley and Caballero's data. Across traits (lifetime productivity, longevity, and intrinsic growth rate), both experiments provided similar estimates of $\lambda$ (BM: mean = 0.013, range = 0.001–0.031; ML: mean = 0.005, range = 0.003–0.011). Average estimates of $E(s)$ were more divergent (BM: mean = 0.10, range = 0.05–0.23; ML: mean = 0.32, range = 0.07–0.68). However, the experiments agree in showing low mutation rates and large average mutational effects.

More recently, results covering 214 generations of MA for the Vassilieva and Lynch experiment were reported (Vassilieva *et al.* 2000). Estimates of $\lambda$ and $E(s)$ averaged over productivity and growth rate were comparable to the earlier ones (BM: $\lambda = 0.03$, $E(s) = 0.15$; ML: $\lambda = 0.02$, $E(s) = 0.16$). These results refer to an optimal temperature of 20 °C. In addition, a parallel assay at a stressful temperature (12 °C) was performed, resulting in BM $\lambda$ estimates sevenfold lower and E($s$) estimates fivefold higher.

### *Saccharomyces cerevisiae*

Zeyl and DeVisser (2001) have studied the accumulation of spontaneous nuclear mutations affecting competitive fitness. Data pertain to 50 MA lines, all derived from a homogeneous wild type base, propagated by the transfer of colonies established from single cells, and maintained in a rich medium. Competitive fitness relative to a marked strain, otherwise identical to the base strain, was obtained after 36 transfers (*ca* 600 generations). After excluding those MA lines carrying mitochondrial deleterious mutations, no overall fitness decline was detected, thus precluding BM estimates. However, ML estimates were computed, resulting in a very low $\lambda$ value (*ca* $10^{-4}$) and a large heterozygous mutational effect $E(sh)$ (*ca* 0.22). The experimental procedure used casts some doubts on the validity of those estimates, as transfers were performed every two days (*ca* 16 generations) and, thus, selection could have acted during these periods of colony growth, eliminating detrimental mutations.

Wloch *et al.* (2001) obtained direct estimates of mutational rates and effects by analyzing **tetrads** in a large number of yeast colonies. Basically, deleterious mutations were identified when the four meiotic products show a segregation pattern 2:2 between wild type haploids and those with a reduced growth rate. The estimates are in agreement with the previous ones, and the extrapolation to *Drosophila*, correcting by the number of genes and that of cell divisions, gives $\lambda = 0.04$ (Wloch *et al.* 2001), in agreement with the indirect fitness estimates for this species (Table 3.1).

### Indirect estimates from artificial mutagenesis

Artificial mutagenesis has been used as a shortcut to long spontaneous MA experiments. The method requires calibrating the number of MA generations equivalent to some mutagenic treatment, under the assumption that the distribution of effects of induced and spontaneous mutations are similar. For example, in *D. melanogaster* the number of MA generations "equivalent" to a given dosage of the mutagenic agent ethyl methasulfonate (EMS) can be calibrated as that giving the same frequency of accumulated recessive lethals, or the same increase in variance for some quantitative trait. However, the EMS treatment imposed by Ohnishi (1977a,b) produced as many recessive lethal mutations as 84 generations of spontaneous MA, and as much increase in variance for abdominal bristle number as 364 spontaneous MA generations. These contrasting figures show that EMS and spontaneous mutations are different regarding their relative effects on lethals and quantitative variation. This is not surprising, as EMS mostly induces C/G $\rightarrow$ A/T transitions, while an important amount of spontaneous mutation in *D. melanogaster* is due to **transposition** events (see Fontdevila, Chapter 16). Even if EMS and spontaneous mutations were similar regarding deleterious effects, it is unclear which of the above calibrations apply, if any. Thus, the EMS-induced viability decline in Ohnishi's data suggests a spontaneous $\Delta M = 0.21$ or 0.05 percent, depending on the calibration chosen. Similarly, the viability decline in the EMS *D. melanogaster* experiment by Yang *et al.* (2001a) gives $\Delta M = 1.3$ percent using the lethal frequency to calibrate the number of MA generations equivalent to the EMS dosage used, and $\Delta M = 0.13$ or 0.3 percent using the rate of increase in variance for bristle traits (0.13 percent using the average increase in bristle variance found by Yang *et al.* (2001a) in the EMS experiment and the MA estimates reported by Houle *et al.* (1996); 0.3 percent using the calibration obtained from Ohnishi's experiment). Interestingly, these EMS induced deleterious mutations did not show detectable **epistasis** or increased effect under harsh environmental conditions.

## Coefficient of dominance of spontaneous mutations

As each new mutation appears only in heterozygous condition for some time, and natural selection generally maintains deleterious mutations at low frequencies, the mutational impact on fitness depends on the fraction of their effects that is expressed in heterozygosis, that is, on their degree of dominance. If the estimation of the homozygous effects of mutations is very demanding, that of heterozygous effects is even harder, mainly due to statistical biases and the low resolution of experimental designs.

Estimates of $E(h)$ can be obtained from analyses of natural populations (see García-Dorado *et al.* 1999), but these estimates assume that allele frequencies are at mutation–selection balance and, therefore, are highly prone to several sources of bias. For this reason, this type of estimates will not be considered in the present review. Direct estimates from MA experiments are more reliable, but they are not free of shortcomings, as will be shown below. In the following, we will confine ourselves to estimates for mutations with small effects on fitness (i.e. excluding lethal or highly detrimental mutations) because they are the most interesting from an evolutionary viewpoint. Moreover, the class of mild to severe mutations corresponds to that detected in MA experiments (see below).

There are several procedures to estimate the average coefficient of dominance from the homozygous and heterozygous effects of MA lines. In the early analyses made in the 1960s and 1970s, the ratio of the observed reductions in mean viability in heterozygotes and homozygotes was used (Mukai and Yamazaki 1968). This provides an estimate of the average of $h$ values of mutations weighted by their corresponding selection coefficients. The main disadvantage of this method is that, if the reductions in viability are not only due to mutation but also to other genetic or environmental factors, the estimates obtained are biased upwards. Thus, the available evidence suggests that the estimates of $E(h)$ obtained by Ohnishi (1977c) by this method are biased (see García-Dorado and Caballero 2000). An alternative way is based on the regression of heterozygous on homozygous viabilities. This method is not affected by the bias mentioned, but gives estimates of the average $h$ values of mutations weighted by their squared selection coefficients. Although mutations with larger effects tend to have lower values of $h$, the method gives roughly unbiased estimates of $E(h)$ for subsets of mutations with small deleterious effects. Therefore, only if MA lines with slight reductions in fitness (quasinormal lines) are used, the bias incurred can be assumed to be small. In what follows, only regression estimates for quasinormal MA lines are considered, unless otherwise indicated.

Estimates of $E(h)$ obtained from MA experiments are summarized in Table 3.2. In the initial experiments carried out by Mukai and coworkers in the 1960s, the viability of homozygotes and heterozygotes for chromosomes that had accumulated mutations during some period of time were compared. Mukai made a distinction between two classes of heterozygotes: (1) "**repulsion heterozygotes**" (derived from crosses of two different MA lines, implying that different mutations accumulate on each homologous chromosome), and (2) "**coupling heterozygotes**" (derived from crosses between MA lines and either the "original" chromosome line or an independently sampled one, assumed to be free of mutations).

The results from these experiments were conflicting. Most estimates of $E(h)$ using coupling heterozygotes were negative or very low, implying overdominance at some loci or nearly complete recessive gene action (see Table 3.2). For the repulsion heterozygotes, however, the estimated $E(h)$ was 0.40. An additional complication was that a fraction of the repulsion heterozygotes (about one-fifth of the total) also indicated overdominance. For this fraction, one or both parental MA lines presented a high viability. This prompted Mukai and Yamazaki (1968) to interpret that these

**Table 3.2** Summary of regression estimates of the average coefficient of dominance $E(h)$ of deleterious mutations from MA experiments

| **References** | | **$E(h) \pm$ S.E.** |
|---|---|---|
| Viability in *D. melanogaster* | | |
| Mukai and coworkers[a] | Coupling heterozygotes (with "original" chromosome) | $-0.17$[b] |
| Mukai and coworkers[a] | Coupling heterozygotes (with non-isogenic chromosomes) | 0.11[c] |
| Mukai and Yamazaki (1968) | Repulsion heterozygotes | 0.40[d] |
| Ohnishi (1977c) | Coupling heterozygotes (with "original" chromosomes) | 0.10[e] $\pm$ 0.08 |
| | Repulsion heterozygotes | 0.03[e] $\pm$ 0.08 |
| Chavarrías *et al.* (2001) | Coupling heterozygotes (with "control" chromosomes) | 0.32 $\pm$ 0.36 |
| Other life-history traits in *D. melanogaster* | | |
| Houle *et al.* (1997) | Early fecundity | $-0.03 \pm 0.16$ |
| | Late fecundity | $0.12 \pm 0.32$ |
| | Male longevity | $0.37 \pm 0.60$ |
| | Female longevity | $0.26 \pm 0.18$ |
| | Male mating ability | $-0.07 \pm 0.68$ |
| *C. elegans* | | |
| Vassilieva *et al.* (2000) | Productivity | $0.64 \pm 0.18$ |
| | Survival to maturity | $0.05 \pm 0.14$ |
| | Longevity | $-0.10 \pm 0.28$ |
| | Intrinsic rate of increase | $0.55 \pm 0.18$ |
| | Rate of convergence | $0.48 \pm 0.19$ |
| | Mean age at reproduction | $0.69 \pm 0.29$ |

[a] See references in Simmons and Crow (1977).
[b] Average of five negative estimates.
[c] Average of two estimates, 0.09 and 0.13, obtained in different genetic backgrounds.
[d] Estimate obtained after excluding overdominant heterozygotes.
[e] Regression estimates obtained by García-Dorado and Caballero (2000).

heterozygotes were, in fact, coupling heterozygotes, and excluded them from the analysis (the estimate of 0.40 was obtained after that exclusion). The reason for the discrepancy between both types of heterozygotes is not evident, although García-Dorado and Caballero (2000) have discussed some plausible explanations. A possibility is that the same recurrent recessive mutations could have occurred in different chromosome lines. Thus, repulsion heterozygotes would be, in fact, homozygous for those mutations shared by both parental lines, while this would not occur in coupling heterozygotes. If this was the case, the large estimate obtained from the repulsion heterozygotes would be biased upwards.

In contrast to Mukai, Ohnishi (1977c) found similar results for coupling and repulsion heterozygotes. Both estimates were also close to 0.40, but there are reasons to believe that they are biased upwards. The estimation method used by Ohnishi was the ratio of viabilities, and a putative nonmutational decline in viability could have inflated the estimates to a large extent. The regression estimates obtained by García-Dorado and Caballero (2000) from the reanalysis of Ohnishi's data are free from such a bias, and indicate a low coefficient of dominance (Table 3.2). Out of 80 chromosomes assayed by Ohnishi (1977c) in heterozygous condition, 78 had homozygous viabilities larger than 0.85. Therefore, the bias incurred in these estimates due to weighting by the squared selection coefficient should not be very large.

In a more recent experiment, chromosomes extracted from full-sib lines after 250 generations of MA were analyzed by a method similar to that of Mukai and Ohnishi (Chavarrías *et al.* 2001). The regression estimate of $E(h)$ was 0.32. Because only 93 MA lines survived out of the initial 200, it is possible that some selection occurred during the experiment, eliminating mutations of moderately large effect. Thus, considering the negative correlation between $s$ and $h$, the above value is expected to overestimate $E(h)$ for new unselected mutations. Nevertheless, its large standard error precludes a clear-cut conclusion.

Houle *et al.* (1997) obtained regression estimates of $E(h)$ for different life-history traits in *D. melanogaster*, indicating varying degrees of partial recessivity. Weighting the estimates of the different traits by their corresponding squared mutational coefficients of variation, the average estimate across traits is 0.05 (García-Dorado and Caballero 2000). However, this analysis differs from all others reviewed in that all nonlethal lines—not just quasinormals—were considered. The possibility of chromosomes carrying mutations of very substantial effect being involved in the analysis implies that the estimates of $E(h)$ for mild mutations can be biased downwards.

To date, the only set of $E(h)$ estimates from MA experiments in a species different from *D. melanogaster* is that recently obtained by Vassilieva *et al.* (2000) in *C. elegans* (Table 3.2). Those estimates were generally compatible with additive gene action ($E(h) = 0.5$), but their large standard errors do not allow rejecting partial recessivity.

## Incorporating molecular information

The previous review of spontaneous mutational parameters for fitness (MD/ML/direct estimates in Table 3.1) indicates that the rate of mutation observed in MA experiments ranges from 0.0001 in *Saccharomyces* to 0.05 in *Triticum*. Likewise, the detected homozygous mutational effects range from 0.06 to 0.26. Despite the variation among species and traits, the classical view that there is a large rate of mutations ($\lambda > \sim 0.15$) with mild effect ($s < \sim 0.03$) is obviously questionable.

MA experiments have limited power. Regarding $\Delta M$, the more precise estimates should be those obtained from the longer MA experiments which, for eukaryotic organisms, are those of Vassilieva *et al.* (2000) in *C. elegans* and Chavarrías *et al.* (2001) in *D. melanogaster*, run for 214 and 255 MA generations, respectively. BM, ML, and MD analyses indicate that the pertinent mutational decays ($\Delta M = 0.0015 \pm 0.0003$ for intrinsic fitness and $\Delta M = 0.0023 \pm 0.0004$ for viability) should be mainly ascribed to moderate to severe deleterious mutations, although mutations with tiny deleterious effects making no significant contribution to the fitness decline could pass undetected. From the standard errors involved in the above experiments, an additional undetected rate of decline should be smaller than *ca* $5 \times 10^{-4}$. If we assume a high rate for those "tiny" mutations (say $\lambda_{tiny} > 0.5$), the corresponding deleterious effects should be very low ($E(s) < 10^{-3}$). The contribution of such mutations to the mutational load can be relevant in the very long term, and it has been argued that they could be responsible for the evolution of sex (see below).

The available experimental evidence concerning the rate and distribution of deleterious mutations is inconclusive, particularly when "tiny" effects are considered. However, given the evolutionary relevance of this issue, we will try to draw some inferences from the joint consideration of published molecular and MA evidences in *C. elegans*, and we will tentatively extrapolate the conclusions to other organisms.

### Deleterious mutations undetected in *Caenorhabditis* MA experiments

For *C. elegans*, 32 percent of all point mutations are **constrained mutations**, that is, they are subject to selection pressure (Shabalina and Kondrashov 1999). Furthermore, about 19 percent of all point mutations cause amino acid substitutions, and 90 percent of these changes are evolutionary constrained (Davies *et al.* 1999). Thus, $0.19 \times 0.90 =$ 17 percent of all point mutations cause amino acid constrained changes, so that about half of the constrained mutations occur in noncoding DNA. In addition, for this species, the product of the haploid genomic size ($8 \times 10^7$ bp) by the nucleotide mutation rate per generation ($2 \times 10^{-9}$) gives a total rate of point mutation per haploid genome and generation equal to 0.16 (Drake *et al.* 1998). Therefore, the rate of amino acid mutation is $\mu_a \approx 0.16 \times 0.19 \approx 0.0304$, that of constrained amino acid mutation $\lambda_a \approx 0.16 \times 0.17 = 0.027$, and that of constrained point mutation $\lambda_p \approx 0.16 \times 0.32 = 0.051$.

Since $\Delta M$ in MA experiments seems to be due to moderate or severe deleterious effects, we assume that in *C. elegans* it is mostly due to amino acidic mutations. Considering that the rate of deleterious mutation detected in *C. elegans* MA experiments is 0.0066, we can infer the fraction of amino acid mutations for the following three classes (Table 3.3): (1) "mild to severe," the only ones detected in MA experiments, representing the $0.0066/0.027 =$ 25 percent of the constrained amino acid mutations and, therefore, the $0.25 \times 0.9 = 23$ percent of the total rate of amino acid mutation; (2) "tiny," representing the remaining 75 percent of the constrained amino acid mutations and, therefore, a $0.9 \times 0.75 \approx 67$ percent of the total rate of amino acid mutation; (3) "neutral," embracing the remaining 10 percent of amino acid mutations.

**Table 3.3** Classification of amino acid mutations ($\mu_a$) in *Caenorhabditis elegans*

| Class | Evolutionary effect | MA experiments | % of $\mu_a$ | Range of effects |
|---|---|---|---|---|
| "Neutral" | Unconstrained | Undetected | 10 | $s < 5 \times 10^{-7}$ |
| "Tiny" | Constrained | Undetected | 67 | $5 \times 10^{-7} < s < 5 \times 10^{-4}$ |
| "Mild to severe" | Constrained | Detected | 23 | $5 \times 10^{-4} < s$ |

Using EMS, Davies *et al.* (1999) induced an average of 45 constrained amino acid mutations per line, equivalent to about 1700 generations of spontaneous MA, but only 3.6 (8 percent) of those mutations were detected in fitness laboratory assays (Keightley *et al.* 2000). This fraction is about one-third of that deduced above (23 percent; Table 3.3), and the difference cannot be ascribed to transposition, which was not active in the MA lines. A possible explanation for the discrepancy could be the loss of a fraction of moderate to severe deleterious mutations in the EMS lines, due to selection. The authors provide simulation results showing that at most three mutations could have been lost per line during the 10 selfing generations preceding the fitness assay, due to average plate mortality. However, using Vassilieva and coworkers' (2000) mutational estimates, the EMS treatment of Davies *et al.* (1999) (presumably equivalent to 1700 MA generations with multiplicative effects on fitness) is expected to reduce fitness average to only 5 percent of its original value. The proportion of plates surviving after the first generation of selfing was not given, but after the second generation it was about 60 percent that of the control, and raised afterwards. This suggests that **purifying selection** may have been important immediately after the EMS treatment, causing a reduction in the number of fitness mutations to be detected in the laboratory assay.

Table 3.3 also shows a tentative range of selection coefficients for each class of mutations. Davies *et al.* (1999) showed that the deleterious effect of mutations undetected in MA experiments should be very small (about $s < 5 \times 10^{-4}$), so we use this lower bound to indicate the minimum effect of the detected mutations ("mild to severe") in *C. elegans*. From **diffusion theory**, the probability of **fixation** of an additive deleterious mutation with an effect $s = 5/N_e$ (where $N_e$ is the long-term evolutionary size of the species) is 0.04 percent that of a neutral mutation. Thus, we take this arbitrary but conservative value as the lower bound for severely constrained mutations. Assuming that the $N_e$ of *C. elegans* may be as large as $10^7$, we consider that mutations with $s < 5 \times 10^{-7}$ are effectively "neutral." Therefore, the effects of constrained mutations undetected in fitness assays ("tiny" mutations) may be in the range $5 \times 10^{-7} < s < 5 \times 10^{-4}$.

## Inferences on the rate of mutations with "tiny" or "mild to severe" deleterious effects to other species

As explained above, there are grounds to consider a class of "mild to severe" deleterious mutations detected in *C. elegans* MA experiments, representing between 8 and 23 percent of amino acid mutations, and a class of undetected "tiny" deleterious mutations comprising at least 67 percent of amino acid mutations. We can tentatively extend this classification to other species, keeping in mind that the fraction of each class that would be selectively constrained in any given species will depend upon the corresponding effective population size.

As stated before, about half of the constrained point mutations in *C. elegans* are amino acidic ones and the proportion of amino acid mutations that are constrained is 90 percent. Thus, if $\mu_a$ is the total rate of amino acid mutations, $0.9 \times \mu_a$ are constrained amino acidic mutations and about the same amount are constrained non-amino acidic ones. Higher organisms often have larger proportions of noncoding DNA, but we will assume that this is mostly due to an excess of nonfunctional DNA, so that the ratio of noncoding to coding deleterious mutations remains roughly similar. We will also assume that non-amino acidic point mutations usually have "tiny" deleterious effects and, therefore, we will include all of them into this class. Therefore, the inferred rate of point mutations with "mild to severe" deleterious effects is $\lambda_{\text{p m-s}} \approx 0.23\,\mu_a$ (from Table 3.3), and that for "tiny" deleterious effects is about $\lambda_{\text{p tiny}} \approx (0.67 + 0.9) \times \mu_a \approx 1.57\,\mu_a$ (the first term within brackets refers to the amino acidic mutations and the second one to non-amino acidic mutations).

The total ($\mu_a$) and the constrained ($\lambda_a$) amino acidic mutation rates can be inferred from molecular data, and have been recently reviewed for different species by Keightley and Eyre-Walker (2000). The mutation rates for vertebrates should be corrected, however, as they were computed assuming 80 000 genes, instead of the currently accepted number (about 35 000). The corrected haploid rates are given in Table 3.4, averaged for different groups of organisms. The values for *C. elegans* are also included for completeness. As expected, the proportion of amino acid mutations that are constrained ($\lambda_a/\mu_a$) increases as we go downward in the table, since the effective population size presumably also increases. Table 3.4 also gives the rates of point deleterious mutation for the "tiny" and "mild to severe" classes, obtained using the above inferences.

For *D. melanogaster* $\lambda_{\text{p m-s}} = 0.009$, a value slightly lower than that obtained from recent MA experiments and MD estimates discussed in the previous sections (see Table 3.1). The difference can be ascribed to transposition, which is very active in this organism, up to 0.1 insertions occurring per gamete and generation (Maside *et al.* 2000). Insertions in coding sequences, which represent 10 percent of the *Drosophila* genome, could account for an important fraction of the rate of recessive lethals (about 0.015, Ohnishi 1977b) and other mutations with severe effects. Thus, most loss-of-function mutations detected for eye color loci have been found to be due to insertion or **deletion** events (Yang *et al.* 2001b). However, transposition events at noncoding sequences often have only "tiny"

**Table 3.4** Total ($\mu_a$) and constrained ($\lambda_a$) haploid rate of amino acid mutation estimated for different species, computed from Keightley and Eyre-Walker (2000) and Shabalina and Kondrashov (1999) (see text for explanation)

| Species | $\mu_a$ | $\lambda_a$ | $\lambda_a/\mu_a$ | $\lambda^a_{p\ tiny}$ ($5 \times 10^{-7} < s < 5 \times 10^{-4}$) | $\lambda^a_{p\ m-s}$ ($s > 5 \times 10^{-4}$) |
|---|---|---|---|---|---|
| Primates | 1.061 | 0.536 | 0.52 | 1.67 | 0.244 |
| Other mammals[b] | 0.394 | 0.273 | 0.68 | 0.62 | 0.091 |
| Mouse/Rat | 0.131 | 0.107 | 0.82 | 0.21 | 0.030 |
| Chicken/ Old world quail | 0.129 | 0.109 | 0.85 | 0.20 | 0.030 |
| *D. melanogaster* | 0.040 | 0.033 | 0.84 | 0.06 | 0.009 |
| *C. elegans* | 0.030 | 0.027 | 0.90 | 0.05 | 0.007 |

[a] Speculative inference of the haploid rate of point mutation ($\lambda_p$) for different ranges of deleterious effects using *Caenorhabditis* estimates.

[b] Sheep, cow, cat, and dog.

deleterious effects, and the frequency of those with "mild to severe" effects is unknown.

For humans, the effective evolutionary size is in the order of $10^4$ (Zhao *et al.* 2000), so that the "mild to severe" class ($s > 5 \times 10^{-4}$) will be included into the class of evolutionarily constrained mutations (conservatively, those with $s > 5/N_e = 5 \times 10^{-4}$). Accordingly, the fraction of amino acid mutations with mild to severe effects inferred for *C. elegans* (23 percent, Table 3.3) is somewhat smaller than the fraction of amino acidic mutations that are constrained (48 percent averaged for the human/chimpanzee genomes; Keightley and Eyre-Walker 2000). This suggests that the distribution of mutational effects could be roughly similar across taxa, and gives some support to the extrapolation of *C. elegans* results to other organisms. Thus, assuming $\mu_a = 1.38$ (obtained from Keightley and Eyre-Walker 2000 considering 35 000 genes) we obtain $\lambda_a = 1.38 \times 0.48 = 0.66$, and $\lambda_{p\ m\text{-}s} = 1.38 \times 0.23 = 0.32$.

## Evolutionary inferences

Previous sections have provided a description of the rate and effects of deleterious mutations, characterized by an increase of the haploid rate of the species with generation length (or the number of germ cell divisions). The values range from about 0.05 in *C. elegans* to about 1.8 in primates, with most mutations (*c*.85 percent) having only "tiny" deleterious effects ($s < 5 \times 10^{-4}$) (see Table 3.4). Although mildly deleterious mutations could have been frequent in some *Drosophila* experiments (possibly due to transposition events), they do not seem to be particularly common in most cases. The scenario for *Drosophila* and *Caenorhabditis* appears to be that of a low rate of deleterious mutations with "mild to severe" effects, whose average is about 0.2. The rate for higher organisms is unknown, but could be relatively high for primates (up to about 0.25) and other large mammals (up to about 0.1). The rate of deleterious mutation due to insertions and deletions should be added in order to compute mutational loads in sexual species. That rate is supposed to be small for most taxa, but may amount to 0.1 per gamete in *Drosophila*. Although the homozygous deleterious effects of "mild to severe" mutations is on the average high, the degree of dominance seems to be inversely related to the magnitude of the effect, so that their impact will generally be low in large populations.

In what follows we will discuss some of the more direct evolutionary consequences of this rough description. Unpublished diffusion predictions will be mentioned; these are based on a mean coefficient of dominance of $E(h) = 0.2$ and the distribution of deleterious effects (MD estimates) for *Drosophila* quoted above (García-Dorado 2003), adjusted for the $\lambda_{m-s}$ inferred for the species considered (Table 3.4).

## Evolution of sex

The role of deleterious mutations on the evolution of sex has prompted scores of printed pages (and continues to do so). Here, we discuss the more direct implications of our inferences for one popular hypothesis for the evolution of sex.

It has been shown that if, on average, more than one deleterious mutation occurs per genome and generation ($\lambda > 0.5$) and there is **synergistic epistasis**, sex could increase the rate at which the population gets rid of the mutational load, up to the point of compensating, in the long-term, for the twofold cost of sexual anisogamic reproduction (see Serra *et al.*, Chapter 12). This has been called the **mutational deterministic hypothesis** of sex evolution. However, Table 3.4 shows that the total rate of point mutation ($\lambda_{p\ tiny} + \lambda_{p\ m-s}$) is below 0.25 for groups of organisms with anisogamic sexual reproduction. Thus, the overall rate of deleterious mutation per gamete is below 0.5, even after allowing a comfortable margin for non-point deleterious mutations. Thus, the deterministic mutational hypothesis cannot be invoked as a general explanation for the evolutionary advantage of anisogamic sex (Keightley and Eyre-Walker 2000; Yang *et al.* 2001b).

Furthermore, even for large mammals, the rate of "mild to severe" deleterious mutation is below 0.5. Thus, the possible advantage of sex should be ascribed to "tiny" mutations. Theoretically, in sexual populations, these "tiny" mutations could put asexual **clones** to disadvantage, but it would take an extremely long time before such disadvantage is worse than the twofold cost of sexual reproduction, and it is likely that the asexual clones will invade the sexual population before this happens.

The conditions for the evolution of obligate versus facultative sex are even more restrictive, because sexual and asexual gene pools are mixed and there is individual within-population selection. If this is the case, the advantage of obligate sex requires large rates of genome degradation that, in turn, depend both on mutation rates and on deleterious effects. Therefore, the available data on deleterious mutation suggest that the deterministic mutation hypothesis cannot be held as a general cause of the evolution of anisogamy and obligate sexual reproduction.

## Mutational load in large populations

Another question of interest, both from the evolutionary and the conservationist point of view, is the impact of deleterious mutations on population fitness. The mutational load of an infinite equilibrium population due to segregating deleterious mutations is defined as $L_s = [W_{max} - E(W)]/W_{max}$ (where $W_{max}$ is the expected fitness of a genotype carrying none of the deleterious mutations segregating in the population, and $E(W)$ is the mean population fitness). It is well known that $L_s$ depends on the deleterious mutation rate ($\lambda$). Assuming additive gene action within loci, $L_s = 1 - \exp[-2\lambda]$ under a between loci multiplicative fitness model, and $L_s = 2\lambda/(1 + 2\lambda)$ under an additive one.

Since the mutational load does not depend on the magnitude of the deleterious effects, the high rate of constrained mutations found in some organisms (like humans, where $L_s$ could be as high as 0.97), raises concern about their future survival. However, this measure of load represents the population fitness loss relative to a hypothetical individual that, in populations with large deleterious mutation rates, does not exist. This assertion qualitatively holds for large finite populations. For example, with $N_e = 10^4$ and mutations with effect $5 \times 10^{-4}$ occurring at a rate $\lambda = 1$, the mutational load is $L_s \approx 0.7$, but the probability of the optimum genotype ($G_{opt}$ with $W_{max}$) is virtually zero ($P[G_{opt}] = \exp[-9.000]$). Thus, the relevance of this measurement of load is obscure. It has a clearer meaning for bottlenecked populations that have attained a new equilibrium after subsequent expansion, so that $G_{opt}$ is a genotype actually present after the bottleneck. However, if the population expanded to a large $N_e$, the new equilibrium would only be attained after a very long time, so that positive selection on new favorable mutations cannot be ignored. On the contrary, if the population expanded to a relatively low $N_e$, the threat from continuous random fixation of deleterious mutations, due to **drift**, would, in the long term, exceed that ascribable to the segregating ones.

Probably, the main source of potential danger for the survival of large populations is the load concealed in the heterozygous condition, due to the partially recessive effects of segregating deleterious genes. Paradoxically, this concealed load is greater in large populations, where many deleterious recessives can segregate at low frequency. When homozygosis increases, after a bottleneck or a sudden subdivision of the population, that threat will appear under the form of inbreeding depression. It has often been argued that the high rate of inbreeding depression ($d$) observed in *Drosophila* could be explained only if deleterious mutations were very common. However, this prediction is based on estimates of the degree of dominance obtained under the assumption of high rates of mildly deleterious mutation, which produces $E(h) \approx 0.4$. Thus, the argument is flawed by circularity. On the contrary, high rates of inbreeding depression are also predicted with low deleterious mutation rates and appropriate levels of recessivity for "moderate to severe" effects ($E(h) \approx 0.2$; see Table 3.2). For example, using *D. melanogaster* parameter values ($\lambda = 0.03$, with $E(s) = 0.22$ and $E(h) = 0.2$), the expected rate of inbreeding depression due to nonlethal mutations in an equilibrium population with $N_e = 10^5$, is about $d = 0.8$ percent decline in fitness per 1 percent increase in inbreeding.

Using the above rates of "mild to severe" mutation, inferred for different species by extrapolating the *C. elegans* estimates (Table 3.4) and the *D. melanogaster* distribution of deleterious effects, we can obtain diffusion predictions for the rate of inbreeding depression. For humans, assuming $N_e = 10^4$ and $\lambda_{\text{p m-s}} = 0.32$, we obtain $d = 5.9$. For other large mammals $d = 2.4$, and for rodents and birds $d = 0.8$ (assuming $N_e = 10^5$ in both cases). Thus, in the offspring of full sibs, multiplicative fitness would be reduced to about 23, 55, or 82 percent that of their parents, respectively. These estimates of inbreeding depression are in rough agreement with estimates of the number of **lethal equivalents** (a measure of the rate of inbreeding depression for viability or other fitness components), which is in the range of 1–4 per gamete (see table 10.4 in Lynch and Walsh 1997). Nevertheless, the actual inbreeding depression can be larger than the diffusion prediction if nonadditive variance for fitness is being maintained by mechanisms different from the mutation–selection balance.

## Mutational meltdown in small populations

In small populations, the most dangerous mutation load is that arising from fixation of deleterious mutations through drift. Essentially, this refers to mutations with $s < 1/N_e$ and, therefore, "moderate to severe" deleterious mutations rarely contribute to the fixation load, except in very small populations, where nonmutational causes of extinction can be more pressing. Mildly deleterious mutations are potentially dangerous for $N_e < 100$, inducing the so-called **mutational meltdown** (Lynch *et al.* 1995), that is, an accumulation of fitness-reducing mutations that lowers the population size, thereby further increasing their probability of fixation. This synergistic effect of mutation and drift may lead to population extinction. However, the process will only be effective in the very long term, unless the rate of mutations with mild effect is extremely large. For example, the fixation of mildly deleterious mutations occurring at a rate $\lambda = 0.5$, with $E(s) = 0.05$, would produce a rate of fitness decline of $5 \times 10^{-4}$ in populations of $N_e = 100$, but the decline would be two orders of magnitude lower using the rates given in Table 3.1.

It should be noted that many "tiny" deleterious mutations, constrained in very large populations, could eventually become fixed after an apparently irrelevant reduction of $N_e$. The habitat of many species has been recently restricted and fragmented, so that their $N_e$ could have been reduced by two orders of magnitude, say from $10^5$ to $10^3$. Thus, many mutations with $5 \times 10^{-5} < s < 5 \times 10^{-3}$ are no longer constrained and will accumulate, resulting in some fitness decline. At first, the frequency of segregating mutations increases but, after a very long time, a new equilibrium will be reached, characterized by a higher rate of deleterious fixation. The expected fitness decline cannot be predicted, since the rate for those mutations is unknown. For higher organisms, with long generation intervals, it could be of the order of $10^{-4}$, but it could also be much smaller. In the short term, the relevance of this decline is negligible. From an evolutionary viewpoint, however, those mutations make up a new "very slight deleterious" class (Kondrashov 1995), that could reduce the expected evolutionary life of species with moderate $N_e$. Thus, it is important to know the actual rate and effect distribution of this class of mutations, and whether selection on new favorable mutations (Poon and Otto 2000) or synergistic effects between deleterious ones (Kondrashov 1995) could compensate for the slow rate of decline it causes. A clue can be found in the human genome. Since the human long-term $N_e$ is only of the order of $10^4$ and the fraction of constrained mutations is considerably smaller than in other taxa, an important fraction of the load could have already accumulated. However, humans still show a high fitness level, supporting important demographic expansion in the harsh environments of third-world countries.

A different question is the future mutational impact on human populations (Crow 1999). Large rates of moderately deleterious mutation could have been efficiently purged in the past. However, mutation rates may have raised due to mutagenic agents and to increased parental age, and the selection coefficient against many deleterious mutations has recently been dramatically reduced due to beneficial environmental changes. Although even "tiny" deleterious effects can be constrained due to the large current population size, "tiny to mild" deleterious mutations could segregate for some time before being lost, causing fitness impairment.

## Summary

Current estimates of genomic rates of deleterious mutation, mean mutational effects and dominance coefficients are reviewed for a variety of fitness traits and species. The experimental evidence suggests that the rate of appearance of deleterious mutations detected in experiments is in the order of up to 0.1 per zygote per generation and, therefore, an order of magnitude smaller than classical *Drosophila* estimates obtained in the 1960s and 1970s. The average effect of detected deleterious mutations is in the order of 0.1 or greater, and it is concluded that the class of mutations with effects in the range 0.01–0.05 is not as common as previously indicated. Recent evidence also suggests that the coefficient of dominance of new mutations (*ca* 0.2 on average) is smaller than previously thought. Combining several pieces of information from *Caenorhabditis* experiments, we tentatively infer the rate of mutation for different deleterious classes and discuss the main evolutionary consequences of these inferences.

We are grateful to two anonymous referees for helpful suggestions on the manuscript. This work was supported by grants PB98-0814-C03-01 (Ministerio de Educación y Cultura, Spain), 64102C124 (Universidade de Vigo, Spain), PGIDT01PX130104PN (Secretaria Xeral de Investigación e Desenvolvemento, Xunta de Galicia, Spain), and BOS2000-0896 (Plan Nacional de I + D + I, Ministerio de Ciencia y Tecnología, Spain).

CHAPTER 4

# Gene duplication and evolution

**Michael Lynch**

Motivated in part by Ohno's (1970) influential book, substantial attention has been given to the idea that gene **duplication** is a major mechanism for the origin of new gene functions. A theoretical population-genetic framework for understanding the evolutionary mechanisms responsible for the success versus demise of gene duplicates has begun to emerge. Substantial evidence now exists that many of the key evolutionary lineages of multicellular eukaryotes have experienced one or more complete genome doublings (**polyploidization**) some time in the distant past (Wolfe 2001), and the newly unveiled genomic sequences of diverse species clearly indicate that gene duplication is an ongoing process in all organisms. The emerging picture is one in which the eukaryotic genome is a dynamic playing field in which new genes are continuously arising via duplication events, with most being eliminated by **drift** and/or natural selection, some simply replacing their ancestral copies, and a few being preserved along with their twins for long periods of time.

It is clear that the evolution of organismal complexity has been accompanied by a net growth in gene number. The genomes of the simplest single-celled prokaryotes appear to contain a minimum of 400 and a maximum of 7000 or so genes. The genomes of nonparasitic multicellular eukaryotes appear to contain no fewer than 10 000 genes, and the maximum probably exceeds 100 000. What remains unclear is whether the hallmarks of organismal complexity (e.g. the origin of new cell types, mechanisms of cell–cell communication, cell adhesion, etc.) *require* an amplification of genome size. There are, after all, many ways to wring multiple functions out of single-gene copies, including **alternative splicing**, modularization of tissue-specific expression patterns, post-translational modification, etc.

This is not to say that gene duplication is a minor player in the origin of new gene functions. Numerous compelling cases for **neofunctionalization** following duplication events have been lucidly outlined by Patthy (1999). But the broader view taken in this chapter is that gene duplication influences evolution via processes other than the evolution of new functions. Most notable among these processes are: (1) the elimination of pleiotropic constraints as duplicate descendants partition up the multiple functions of their ancestral single-copy gene, and (2) the passive origin of microchromosomal rearrangements when ancestral gene functions are reassigned to new locations. In this sense, the gene-duplication process provides fuel for both of the major engines of evolution—**adaptation** and speciation.

## The evolutionary demography of duplicate genes

The power of gene duplication as an evolutionary force depends on the rate at which duplicate genes arise. Although there is currently no simple way to estimate this rate, the complete genomic sequences for several species provide an indirect route (Lynch and Conery 2000, 2003). Through comparative sequence analysis, the total pool of duplicate genes within a genome can be identified, and the relative ages of the duplicate pairs can be estimated from the pairwise divergence of **silent sites** in coding regions under the assumption that such sites accumulate nucleotide changes at a relatively constant rate. The age distribution of duplicate pairs can then be used to estimate the average rates of origin and elimination of duplicate genes, in the same manner that demographers use age distributions to estimate birth and death rates. Under a steady-state birth–death process, advocated in a somewhat different context by Nei *et al.* (1997, 2000), the frequency of duplicate pairs will exhibit an exponential decline with age, with the abundance of identical to nearly identical duplicates providing information on the birth rate, and the rate of decline in abundance with age providing an estimate of the loss rate. Most eukaryotic genomes assayed to date exhibit the approximate pattern predicted by this model (Lynch and Conery 2000, 2001, 2003; Achaz *et al.* 2001).

Application of the preceding logic to the genomic sequences of a diverse set of eukaryotic species (*Arabidopsis thaliana, Caenorhabditis elegans, Drosophila melanogaster, Homo sapiens, Saccharomyces cerevisiae, Schizosaccharomyces pombe*, and *Encephalitozoon cuniculi*) yields estimates of the rate of birth of duplicate genes in the range of 0.001–0.03 per gene per million years, with an average of *ca* 0.01 per gene per million years (Lynch and Conery 2003). In other words, on a time scale of 100 million years or so, nearly every gene in a genome can be expected to have duplicated at least once. These indirect estimates are conservative in that they do not include contributions from large **multi-gene families**, of which there are many in eukaryotes. In addition, direct empirical estimates of gene duplication rates in *Drosophila* appear to be higher, on the order of $10^{-6}$ to $10^{-4}$ per gene per generation (Shapira and Finnerty 1986). Thus, the rate of duplication per gene is at least of the same order of magnitude as the rate of mutation per nucleotide site (Li 1997), and perhaps considerably higher. This implies that changes in gene content may often rival changes in gene sequence as a mechanism of phenotypic evolution. On the other hand, as is the case for most replacement nucleotide substitutions within genes, most duplicate genes appear to be evolutionarily short-lived. The half-lives for such genes in the previous list of species range from 1 to 17 million years, with an average of about 5 million years (Lynch and Conery 2003).

Although the individual estimates are very approximate, these demographic analyses highlight the dynamic nature of eukaryotic genomes with respect to gene content. As a consequence of a stochastic balance between gene birth and death rates, total genome size may remain approximately constant within specific lineages for long periods of time, but throughout such periods there is likely to be continual turnover with respect to the specific genes that are present in redundant copies. The precise mechanisms by which duplicate genes arise are not yet well understood, but they are probably diverse. Many newly arisen gene duplicates apparently arise via local events, as they are often tandemly associated with their parental copy. However, duplications to new chromosomal locations also occur. The most common mechanism for nontandem duplicates may be the capture of nascent DNA fragments during the repair of double-strand breaks, which arise several times daily per cell (Ricchetti *et al.* 1999; Lin and Waldman 2001), although the source of such DNA fragments is uncertain. Whether duplication spans will contain one or more fully functional genes is a matter of chance, and incomplete duplication events may result in products that are "dead on arrival."

## Mechanisms for the preservation of duplicate genes

All duplicate genes are expected to be initially carried by a single member of a population, and hence to be highly vulnerable to stochastic loss early in their history. To be successful in the long term, a duplicate gene must first drift towards **fixation**, and then having arisen to high frequency, the selective forces for its maintenance must be sufficiently large to prevent its subsequent loss by degenerative mutation. The precise mechanisms by which duplicate genes are preserved have a fundamental bearing on genome evolution. For example, the reciprocal preservation of both members of a pair of duplicates leads to an expansion in genome size, while the preservation of a new unlinked duplicate combined with the loss of the ancestral copy has no effect on genome size but does induce an alteration of the genetic map (the relevance of which is discussed below).

### Neofunctionalization

One of the more notable mechanisms for the joint preservation of a pair of gene duplicates is the process of neofunctionalization, whereby one copy acquires a beneficial mutation to a new function. Models of neofunctionalization via gene duplication generally assume that new beneficial functions are acquired at the expense of essential ancestral functions, the unspoken reasoning being that selectively advantageous mutations with no negative pleiotropic effects on essential wild-type function should have had no barriers to fixation prior to duplication. Gene duplication alters the selective landscape for the subset of beneficial mutant alleles with negative pleiotropic effects on the ancestral gene function by opening up the opportunity for one locus to experiment evolutionarily while the other retains the ancestral function. This model extends back at least to Haldane (1933), and some aspects of it were explored quantitatively by Walsh (1995) for the case in which the duplication is assumed to be initially fixed in the population.

Although most considerations of the neofunctionalization process have focused upon mutations arising subsequent to the duplication event, Spofford (1969) made the key observation that the arrival of new mutations may not be a prerequisite for neofunctionalization. The simple logic for this argument is that the spectrum of mutations arising subsequent to a duplication event are the same as those arising prior to duplication. Thus, a mutant allele endowed with a beneficial function at the expense of an essential ancestral function may be maintained at low frequency at the ancestral single-copy locus by **balancing selection**—even though homozygotes for such an allele will be inviable, the heterozygotes may have elevated **fitness**.

The presence of such ancestral polymorphisms can facilitate the route to neofunctionalization in two ways. First, if the duplicate locus is founded by a neofunctional allele, fixation at the new locus will be promoted by positive selection while the original locus retains the ancestral function. Alternatively, if the new locus is founded by a "wild-type" allele that rises to a high enough frequency, the selective regime at the ancestral locus will be altered to one of positive selection for fixation of the neofunctional allele. In either case, the final outcome is functionally equivalent to the fixation of an **overdominant gene** action, with one locus being essentially monomorphic for the "wild-type" allele and the other for the neofunctional allele.

The primary conditions necessary for the maintenance of neofunctional alleles at the ancestral locus have been worked out (Lynch *et al.* 2001). First, the mutation rate to null alleles must be less than the square of the selective advantage of the neofunctional allele in the heterozygous condition ($s^2$). Second, the effective population size ($N$) must be sufficiently large that the power of random **genetic drift** is less than the strength of balancing selection (approximately, $N > 4/s^2$). Provided these conditions are met, then the neofunctional allele will be present at the initial locus with approximate frequency $s$, for example, an allele that is lethal in the homozygous state but increases fitness by 5 percent in the heterozygous state would have an expected frequency of *ca* 5 percent. Thus, just a moderate heterozygous selective advantage combined with a moderately large effective population size provides a setting by which a population can be poised to proceed towards neofunctionalization following a duplication event.

The actual probability of preservation of a pair of duplicate genes by neofunctionalization depends on several additional factors (Lynch *et al.* 2001). First, in large populations, the probability of neofunctionalization increases with $s^2$ provided the duplicate loci are unlinked. This scaling can be understood most easily by noting that if the new locus is founded by a wild-type allele, that allele will have a selective advantage that depends on the frequency of neofunctional homozygotes at the ancestral locus ($\approx s^2$). This advantage occurs because "absentee" homozygotes at the new locus are lethal on this genetic background. Alternatively, if the new locus is founded by a neofunctional allele, the marginal selective advantage of the founder allele over the absentee allele is $\approx s^2$ (Lynch *et al.* 2001). In both cases, the probability of neofunctionalization scales with $s^2$ because the probability of fixation of a newly arisen mutation in large populations is equal to twice the selective advantage (Crow and Kimura 1970). A second factor that influences the probability of neofunctionalization is the degree to which the two loci are linked. In the case of complete **linkage** (as is approximately the case for a pair of tandem duplicates), neofunctional alleles segregating at the ancestral locus cannot be exploited for the simple reason that the complete absence of the essential ancestral function would prevent the fixation of a linked pair of neofunctional alleles.

The main lesson to be learned from the theory leading up to these results is that the preservation of duplicate genes by neofunctionalization is a large-population phenomenon. If $N$ is smaller than $4/s^2$, neofunctional alleles are expected to be absent from the ancestral locus, and even if a wild-type allele becomes fixed at the new locus, one of the two loci is almost certain to experience a silencing event prior to the fixation of a new neofunctionalizing mutation. If the loci are completely linked, the neofunctionalization process becomes largely limited by the rate of origin of neofunctional mutations and their selective advantage on the wild-type background, and the critical effective population size for neofunctionalization becomes even larger.

Finally, we note that under the theory discussed above, neofunctionalization evolves through the modification of an active allele. Ohno (1970) suggested an alternative scenario by which a silenced gene duplicate might provide the substrate for the origin of a novel function (see also Marshall *et al.* 1994). By accumulating a series of molecular changes in a neutral fashion, a completely inactivated locus might yield a beneficial product that would be impossible to acquire via natural selection alone. Imagine, for example, an inactivated allele separated from a more optimal state by intermediate mutational steps associated with low fitness (if expressed). If after a fortuitous concatenation of secondary mutations with beneficial **epistatic** effects, the unexpressed locus is then somehow reactivated, the new locus would be strongly favored by natural selection, possibly displacing the ancestral locus (in the absence of **negative pleiotropy**) or coexisting permanently with it (in the presence of a trade-off). In principle, this mechanism for vaulting an adaptive valley might be facilitated, without requiring any gene reactivation, by **gene conversion** between the silenced and expressed loci (Hansen *et al.* 2000). However, whether these sorts of Lazarus effects play an important role in evolution remains to be seen.

## The masking effect of duplicate genes

Because all loci harbor suboptimal alleles due to the recurrent introduction of deleterious mutations, one might imagine that duplicate genes have an intrinsic selective advantage associated with their ability to mask the effects of recessive deleterious mutations at the ancestral locus. However, the magnitude of such an *indirect* advantage may only rarely be strong enough to promote the permanent preservation of duplicate genes. Fisher (1935) realized that two genes with identical roles in an effectively infinite population will not be mutually maintained by selection unless their mutation rates to nulls are identical. If this is not the case, the gene with the higher mutation rate will eventually be silenced by the differential accumulation of **genetic load**. In fact, because random genetic drift will eventually lead to the stochastic loss of one locus in a *finite* population, not even identical mutation rates will be sufficient for the permanent retention of duplicate genes (Clark 1994; Lynch *et al.* 2001). Consider, for example, the masking of a deleterious recessive lethal. The equilibrium frequency of null homozygotes at a single-copy locus is equal to the null mutation rate $\mu_c$ in large populations (Crow and Kimura 1970), and this must also equal the selective advantage of a rare functional duplicate at a new locus. However, $\mu_c$ is also the rate of silencing of the new allele, so these two factors cancel exactly, rendering the new duplicate effectively neutral. Thus, in the absence of neofunctionalization, something additional is required for the maintenance of a duplicate gene.

One possibility is a *direct* selective advantage of a duplicate locus, that is, increased fitness in individuals with three versus two functional genes. However, this scenario must be tempered with the alternative possibility that the overexpression of duplicate loci may negatively influence the quantitative balance between the total expression of the pair and their interacting partners (except in the case of polyploidization events, which maintain the stoichiometric relationships among all pairs of genes). This different outcome of incremental duplications and whole-genome duplications may explain the apparently higher rates of retention of duplicate genes following polyploidization events (Lynch and Conery 2000). A second possible mechanism for the permanent retention of duplicate genes is a balance between the differential selective advantages and mutation rates of a pair of duplicate loci. This would require, for example, one gene to operate more efficiently and the other to have a lower mutation rate to nulls (Nowak *et al.* 1997). Whether the fine balance required under this scenario is likely to ever be met is unclear. Still another possibility is the maintenance of a duplicate gene by recurrent introduction of new copies by the duplication process itself. In this case, however, since independent introductions of new copies are expected to almost always appear in new genomic locations, this process would at best lead to an equilibrium level of overall genomic redundancy, but not to the permanent retention of any particular pair of loci (Wagner 2000a). In other words, although the total amount of redundancy in a gene family may remain relatively constant over time under this model, the specific members of the family and their genomic locations would be expected to turn over as a consequence of the steady-state birth–death process, presumably leading to L-shaped age distributions similar to those actually observed (Lynch and Conery 2000).

Although the masking effects described above are concerned with compensation for mutationally silenced alleles, it has also been argued that duplicate genes may provide a buffer against cellular mishaps that lead to localized absence of gene expression (from normally active genes) (Tautz 1992; Nowak *et al.* 1997). There are a number of potential sources for such developmental errors, including somatic mutations, errors in transcription and translation, and errors in the inheritance of **methylation** patterns. Following the logic outlined above, for this mechanism to maintain a duplicate gene by natural selection, the developmental error rate would have to exceed the rate of origin of null mutations at the duplicate locus.

In summary, although the various masking models for the preservation of duplicate genes cannot be formally rejected, they all require rather special sets of mutational conditions and enormous population sizes to enable the very weak selective advantages of redundancy to come to prominence. The general paucity of duplicate genes in haploid microbes raises the most serious challenge to the idea that masking plays a prominent role in duplicate-gene retention.

## Subfunctionalization

Given the limitations on the various masking hypotheses, neofunctionalization is often assumed to be the *only* mechanism by which duplicate genes can become permanently preserved. Under this assumption, because neofunctionalizing mutations are rare relative to degenerative mutations, the vast majority of new gene duplicates are expected to be lost within a relatively short period of time. In the absence of positive selection, a fraction $[1 - 1/(2N)]$ of newly arisen gene duplicates will be lost by random genetic drift in an average $2 \ln(2N)$ generations (Kimura and Ohta 1969), and for the small remaining fraction $[1/(2N)]$ that manage to drift to fixation, subsequent silencing of one copy by degenerative mutations will eventually ensue. Most of the theory on this matter has assumed a model for fitness in which all genotypes are equally viable except for the lethal double-null homozygotes. In this case, provided $N\mu_c << 1$, the average time to gene silencing is approximately equal to the mean waiting time until the appearance of a null mutation at one of the loci, $1/(2\mu_c)$ generations. On the other hand, for $N\mu_c >> 1$, the time to silencing is no longer limited by the mutation process, but is on the order of the time for an effectively neutral null mutation to drift to fixation, that is, $\approx 4N$ generations (Watterson 1983; Lynch and Force 2000a). These predictions of a relatively rapid demise of the vast majority of duplicate genes are inconsistent with the high levels of duplicate-gene retention observed in ancient **polyploid** lineages (Wagner 1998; Force *et al.* 1999). Since the evidence for the origin of new gene functions in these lineages is limited, alternative mechanisms for duplicate-gene preservation must be at work.

A potentially powerful mechanism becomes apparent when one considers a broader view of gene structure than assumed under the classical model. Because of the complex nature of regulatory

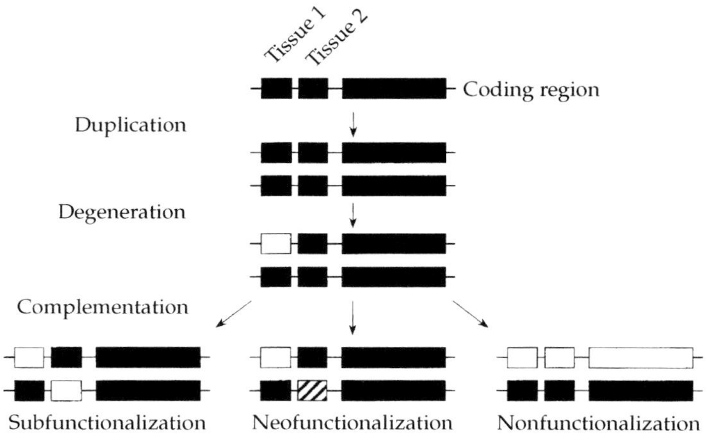


**Figure 4.1** The DDC (duplication–degeneration–complementation) model for the alternative fates of duplicate genes. The ancestral gene is depicted as having two independently mutable regulatory regions, each driving expression in a particular tissue. Solid boxes denote fully functional regulatory and coding regions, whereas open boxes denote loss of function, and the diagonal lines denote gain of a new beneficial function. Each group of genes reflects the fixed haploid state of the population. Following the duplication event, the first degenerative mutation eliminates the capacity of one of the copies to be expressed in tissue 1. The second mutational event then dictates the final fate of the pair: under subfunctionalization, the second copy acquires a complementary loss-of-subfunction mutation; under neofunctionalization, the second copy acquires a novel, beneficial expression pattern at the expense of the ability to be expressed in tissue 2; under nonfunctionalization, the first copy loses all functional ability.

regions, most genes in multicellular eukaryotes have independently mutable subfunctions. That is, a mutation that causes the loss of gene expression in one particular tissue or developmental period does not necessarily affect other tissue- or timing-specific aspects of expression. This more general view of gene expression leads to the prediction that duplicate-gene preservation must sometimes result from the partitioning of ancestral gene functions through complementary loss-of-function mutations in **paralogous** copies (Fig. 4.1). Under the duplication—degeneration—complementation (DDC) model of Force *et al.* (1999), this subfunctionalization process is driven entirely by degenerative mutations, which we know to be much more common than beneficial mutations.

Duplicate-gene preservation by subfunctionalization is a two-step process. First, one of the genes must become fixed for a mutation that eliminates a particular subfunction, an event that permanently preserves the second copy. Then, the second copy must lose an alternative subfunction, thereby reciprocally preserving the first copy. If the effective population size is sufficiently small that segregating null mutations are typically rare ($N\mu_c \ll 1$), then the fate of a newly arisen duplicate gene under the DDC model depends almost entirely on the relative rates of origin of subfunctionalizing and nonfunctionalizing mutations ($\mu_r$ versus $\mu_c$), and the probability of subfunctionalization can be approximated with combinatorial logic.

Consider, for example, a gene with two independently mutable subfunctions. The probability that a newly arisen duplicate will drift to fixation is $1/(2N)$, and having arrived at this point, the probability that the first fixed mutation eliminates a subfunction from one of the genes is simply the fraction of mutations that are of the subfunctionalizing type $2\mu_r/(2\mu_r + \mu_c)$. After such a fixation event, the intact locus is no longer free to lose its now unique subfunction whereas the partially debilitated locus is free to become completely silenced, so the total permissible mutation rate for the next step is $(2\mu_r + \mu_c)$, with $\mu_r/(2\mu_r + \mu_c)$ being the fraction of permissible mutations that eliminate the complementary subfunction at the intact locus. The probability of subfunctionalization ($P_{sub}$) is equal to the product of these three probabilities, $\alpha^2/(4N)$, where $\alpha = 2\mu_r/(2\mu_r + \mu_c)$ is the fraction of degenerative mutations that eliminate a single subfunction. Under this two-subfunction model, $P_{sub}$ approaches a maximum of $1/(4N)$ as $\alpha \to 1$.

A number of factors can increase the probability of subfunctionalization above this level, but none of them increases the upper bound to a level greater than $1/(2N)$. This is to be expected since $1/(2N)$ is the probability of initial fixation of an entirely neutral duplication, and the subsequent probability of permanent preservation cannot exceed one. For example, increasing the number of independently mutable subfunctions ($z$) increases $P_{sub}$ by increasing the number of paths by which complementary loss-of-function mutations can be acquired by the two copies (Force *et al.* 1999). Generalizing the definition of $\alpha$ to $z\mu_r/(z\mu_r + \mu_c)$, and keeping the ratio of subfunctionalizing to nonfunctionalizing mutation rates ($z\mu_r/\mu_c$) constant while allowing $z \to \infty$, it can be shown that the upper limit to $P_{sub}$ is $\alpha^2/(2N)$ (Lynch and Force 2000a). Thus, for the extreme case in which all mutations eliminate only single subfunctions and there are an effectively infinite number of such subfunctions, $P_{sub}$ attains a maximum value of $1/(2N)$.

Although the preceding theoretical results apply to the situation in which all mutations completely eliminate one or all subfunctions of a gene, mutations with partial effects on gene expression will further increase the probability of duplicate-gene preservation, even providing a retention mechanism for duplicate genes whose expression patterns cannot be subdivided (Lynch and Force 2000a). Such preservation occurs whenever the functional capacity of both loci is degraded to the extent that their joint presence is needed to fulfill the requirements of the single-copy ancestral gene. Consider, for example, a gene with a single function, with $s$ being the number of mutations with partial effects necessary to completely eliminate gene function. Letting $\mu_p$ be the rate of origin of such mutations and $\rho = \mu_p/(\mu_p + \mu_c)$ be the fraction of the total pool of mutations with partial effects, then the upper limit to $P_{sub}$ under this model, $\rho^2/(2N)$, is approached as the average effects of partially debilitating mutations decline to zero (i.e. $s \to \infty$). Again, this limit approaches a maximum value of $1/(2N)$ as mutations with partial effects become more predominant (i.e. as $\rho \to \infty$).

Averof *et al.* (1996) suggest still another mechanism by which the preservation of duplicate genes by subfunctionalization can be promoted. Newly arisen duplicates need not always be complete since, for example, critical regulatory elements in the flanking regions may be missing at the time of the duplication event. Recall that the probability of subfunctionalization of a newly arisen duplicate with two intact subfunctions is $\alpha^2/(4N)$ under a model in which single mutations completely eliminate aspects of gene expression. If a gene duplicate is missing one subfunction at birth, then the first step towards subfunctionalization has already been met, and the probability of subfunctionalization increases to $\alpha/(4N)$.

Finally, we note that the preceding theory applies to the case in which the effective size of a population is small (specifically, $N\mu_c << 1$). Provided these conditions are met, the simultaneous presence of polymorphisms at both loci is rare, and the probability of subfunctionalization is essentially independent of both the population size and the degree of linkage between duplicates. For larger populations, however, linkage plays a key role in determining the probability of subfunctionalization (Lynch *et al.* 2001). For unlinked duplicates, $P_{sub}$ asymptotically approaches zero at large $N$. Such behavior is a consequence of the long time (approximately $4N$ generations) required for a newly arisen duplicate to drift to initial fixation. If $N$ is sufficiently large, essentially all descendants of the initial duplicate will acquire silencing mutations by the time the lineage becomes fixed. On the other hand, for completely linked duplicates, $P_{sub}$ increases with increasing $N$, asymptotically approaching $1/(2N)$ as $N\mu_c \to \infty$. This behavior results from the fact that a linked pair of duplicates has a weak selective advantage over a single-copy gene, because complete inactivation of a "two-copy" allele requires the silencing of both members of the pair.

By postulating a preservational process driven entirely by degenerative mutations, the subfunctionalization model provides a null hypothesis for the interpretation of patterns of survival of duplicate genes. However, this preservational process may be the beginning, not the end, of new evolutionary pathways. Consider, for example, a single-copy locus that is a victim of a "jack-of-all-trades is a master-of-none" syndrome, such that an adaptive conflict exists between its multiple subfunctions. Under these conditions, complementary loss-of-subfunction mutations are expected to alter the selective landscape experienced by the two members of a duplicate pair, enabling each copy to become more refined to its specific subset of tasks (Piatigorsky and Wistow 1991; Hughes 1994). By this means, gene duplication combined with degenerative mutations may provide a unique mechanism for the creation of novel evolutionary opportunities through the elimination of pleiotropic constraints. Thus, although neofunctionalization and subfunctionalization may be viewed as independent preservational mechanisms, one involving positive selection for new beneficial functions and the other involving only the chance fixation of degenerative mutations, the processes of initial preservation and subsequent modification may become mutually blurred late in the evolutionary history of a pair of gene duplicates.

## The case for subfunctionalization

Prior to the formal development of the DDC model, circumstantial evidence for subfunctionalization as a mechanism for duplicate-gene preservation had already accumulated through a series of isozyme studies in polyploid fishes that repeatedly demonstrated the presence of tissue-specificity of expression of duplicated enzyme loci (Ferris and Whitt 1977, 1979). These observations have recently been supplemented by a substantial number of DNA-based investigations in zebrafish. Genomic analysis has shown the zebrafish to be a member of an ancient polyploid lineage (Amores *et al.* 1998), and on the order of 25 percent of the thousands of original pairs of duplicates arising from the polyploidization event are still functional (Postlethwait *et al.* 2000). A key to understanding the evolutionary modifications of these surviving zebrafish duplicates has been the availability of **orthologous** single-copy genes in tetrapods (usually chicken or mouse) as outgroups. Comparison of gene-expression patterns in the **homologous** tissues of these species provides a means of evaluating whether the differences among zebrafish **paralogues** is a result of the partitioning of ancestral functions or of the origin of new functions. In virtually every well-characterized case, subfunctionalization appears to be the most likely mechanism of preservation.

Consider, for example, the two zebrafish genes for microphthalmia-associated transcription factor, *mitfa* and *mitfb* (Lister *et al.* 2001). Only *mitfa* is expressed in neural crest, and only *mitfb* is expressed in the epiphysis and olfactory bulb. Although some of the molecular details remain to be worked out, the two zebrafish genes appear to be homologous to the two alternatively spliced forms of the single-copy locus found in tetrapods, with subfunctionalization (each copy adopting a single splicing variant) resulting from **deletions** in both regulatory and coding regions. As another example, consider the two zebrafish cytochrome P450 aromatase genes (Chiang *et al.* 2001b). One of these is expressed in the ovary and the other in the brain, whereas the single-copy gene in tetrapods is expressed in both tissues. Likewise, zebrafish has two *sox11* genes that are orthologous to the single-copy gene found in tetrapods. Although the two zebrafish paralogues overlap considerably in expression pattern, *sox11a* is expressed in the anterior and *sox11b* in the posterior somites, whereas the single *sox11* product in mouse is expressed in all somites (de Martino *et al.* 2000). Other zebrafish genes that appear to have experienced subfunctionalization following gene duplication include *dlx* (Quint *et al.* 2000), *en1* (Force *et al.* 1999), *hoxb5* (Bruce *et al.* 2001), *notch* (Westin and Lardelli 1997), *pax6* (Nornes *et al.* 1998), and *sox9* (Chiang *et al.* 2001a).

The process of subfunctionalization is by no means a peculiarity of polyploid fishes. For example, the nematode *Caenorhabditis elegans* has two β-catenin genes, one playing a role in cell signaling and the other in cell adhesion, whereas a single gene fulfills both functions in most other metazoa (Grimson *et al.* 2000; Korswagen *et al.*

2000). The functional differences between the paralogous *C. elegans* genes appear to be due to alterations in the coding region. In the barnacle *Sacculina carcini*, two *engrailed* duplicates are expressed late in development, one restricted to the nervous system and the other to the epidermis, whereas both expression patterns are fulfilled by a single gene in other arthropods (Gibert *et al.* 2000). In maize (*Zea mays*), two copies of the *p1 myb*-like transcriptional activator partition up expression patterns in male and female reproductive structures and leaves, whereas the single-copy orthologue in closely related teosinte fulfills all of these expression patterns (Zhang *et al.* 2000).

In addition to being facilitated by mutations in regulatory regions, subfunctionalization can occur via changes in coding regions, either through substantial modifications of alternative domains in proteins with multiple functions (qualitative subfunctionalization, as in the β-catenin genes of *C. elegans*) or through the accumulation of mildly deleterious mutations in both copies of a single-function gene (quantitative subfunctionalization). Insight into the types of selective pressures operating on the coding regions of duplicate genes may be acquired by comparing the number of nucleotide changes per amino acid replacement site ($R$) and per silent site ($S$). An $R/S$ ratio greater than one implies **directional selection** for change, as expected under a scenario in which one or both members of a pair have evolved a new function, whereas a ratio less than one implies that selection is predominantly purifying in nature, and a ratio not significantly different from one is consistent with an absence of selection. Large-scale surveys of duplicate genes arising from ancient polyploidization events in mammals (Li 1985), *Xenopus laevis* (Hughes and Hughes 1993), and zebrafish (Van de Peer *et al.* 2001) have consistently found $R$ to be substantially less than $S$, with the average ratio being similar to that observed in interspecific comparisons of single-copy genes.

One limitation of these types of comparative analyses is the cumulative nature of the nucleotide differences, which represent the culmination of tens of millions of years of evolution. In principle, newly arisen gene duplicates might experience an early phase of directional selection (with one member evolving a new function) or of relaxed selection (as a consequence of functional redundancy) followed by a longer phase of purifying selection once new gene functions have become established. A comparative analysis of distantly related sequences would reveal only the average of these patterns. To evaluate whether duplicate genes experience changes in intensities of selection as they age, Lynch and Conery (2000, 2003) estimated $R$ and $S$ for the complete sets of gene duplicates in fully to partially sequenced genomes of several plants, animals, and fungi. For all species, the $R/S$ ratio declines with $S$. Although a considerable fraction of newborn duplicates (those with $S \approx 0$) appear to accumulate replacement substitutions at rates not significantly different from the neutral expectation, and a few even exhibit the signature of directional selection ($R/S > 1$), the *average* $R/S$ for newborn duplicates in eukaryotes is approximately 0.7. This declines by about tenfold as $S$ increases, with the vast majority of pairs appearing to experience strong purifying selection by the time they have diverged by 10 percent at silent sites. Thus, the average intensity of purifying selection operating on a pair of duplicate genes does indeed increase with the age of the pair.

By comparing the coding regions of two duplicate genes to that of a single-copy outgroup, one may inquire further as to whether the $R/S$ ratio differs among the duplicate copies. Using this approach, Van de Peer *et al.* (2001) found that about half of the pairs of paralogous zebrafish genes exhibit significant heterogeneity in $R/S$ between the two members of the pair. Why the difference? An intriguing possibility concerns the observation that the rate of replacement substitution in vertebrate genes is inversely correlated with the number of tissues in which the gene is expressed (Hastings 1996; Duret and Mouchiroud 2000), that is, genes with more restricted tissue-specific patterns of expression evolve more rapidly at the amino acid level. Since a large fraction of zebrafish paralogues appear to have undergone subfunctionalization, it will be of considerable interest to learn whether the more rapidly evolving members of pairs are also the ones that have retained fewer functions. The one locus for which data are available, triosephosphate isomerase, follows the expected pattern (Merritt and Quattro 2001).

The analyses of genewide **substitution rates** noted above potentially hide a significant amount of information on protein divergence. For example, it is possible for the two members of a duplicate pair to evolve at the same average rate, with each member experiencing a different spatial pattern of evolutionary change, for example, one having a higher rate at the 3′ end and the other a higher rate at the 5′ end. To evaluate this possibility, Dermitzakis and Clark (2001) introduced a computational method for testing for spatial variation in the substitution pattern in the coding regions of different paralogues. Approximately 50 percent of their comparisons of mouse and human duplicate genes revealed significant regional variation among paralogous copies, some of which appear to be associated with functional domains.

## Speciation via the divergent resolution of duplicate genes

Most studies of duplicate genes have focused on their potential role in the origin of evolutionary novelties through the establishment of new gene functions. However, the high rate at which duplicate genes arise, move to unlinked positions, and become randomly silenced or subfunctionalized suggests that gene duplication may be an equally important contributor to the other major engine of evolution—the origin of new species (Lynch and Force 2000b). Consider an unlinked pair of duplicate **autosomal** genes in a diploid ancestral species. Divergent silencing or subfunctionalization of the duplicate copies in two descendent species results in a map change that can secondarily induce incompatibilities in hybrid progeny in several ways. Because the $F_1$ hybrids will be "presence–absence" heterozygotes at the two independently segregating loci, one-fourth of the $F_1$ gametes will contain null (or absentee) alleles at both loci. Thus, for a gene that is critical to gamete function, this single divergently resolved duplication would result in an expected 25 percent reduction in fertility. For a zygotically acting gene, 1/16 of the $F_2$ offspring from the interspecific cross would lack functional alleles at both loci, and another one-fourth would carry only a single functional allele. Thus, if the gene is haploinsufficient, 5/16 of the $F_2$ zygotes of such a cross would be inviable (or sterile).

Indirect evidence leaves little room for doubt that microchromosomal rearrangements resulting from gene duplication are significant contributors to the establishment of postzygotic isolating barriers among species (Lynch and Force 2000b), and direct support is beginning to emerge from genomic analyses (Lynch 2002). For example, comparative-mapping data from plants have identified pervasive microchromosomal rearrangements associated with duplication events (see Bancroft 2001 for a review). Consider the mustards *Arabidopsis thaliana* and *Brassica oleracea*, which are thought to have diverged from a common ancestor 10–20 million years ago (Yang *et al.* 1999). The two species exhibit many sets of paralogous chromosomal segments with substantial long-range colinearity in gene content, but small-scale lineage-specific rearrangements resulting from alternative losses of orthologous gene copies are common (O'Neill and Bancroft 2000; Quiros *et al.* 2001). Complicating matters is the fact that *B. oleraceae* is a triploid derivative of the older lineage containing *A. thaliana*, which itself experienced at least one ancient polyploidization event (prior to the divergence of *Arabidopsis* and *Brassica*). However, even this complexity is informative. For example, in an analysis of two triplicated paralogous chromosomal regions in *B. oleraceae*, O'Neill and Bancroft (2000) found that about two-thirds of the component genes had at least one member silenced on at least one paralogue (Fig. 4.2). These results as well as others (e.g. Ku *et al.* 2000) suggest that the dominant mechanism of chromosomal repatterning in plants is duplication of chromosomal regions followed by random gene loss. Although the data are less extensive for animals, recent chromosomal comparisons in mammals (Dehal *et al.* 2001) and nematodes (Coghlan and Wolfe 2002) indicate that plants are by no means unique in this regard.

To emphasize that the inheritance of ancestral duplications of autosomal genes is just one route by which reproductive incompatibilities can passively arise from map changes induced by the divergent resolution of duplicate genes, we now consider four additional mechanisms. First, consider the situation

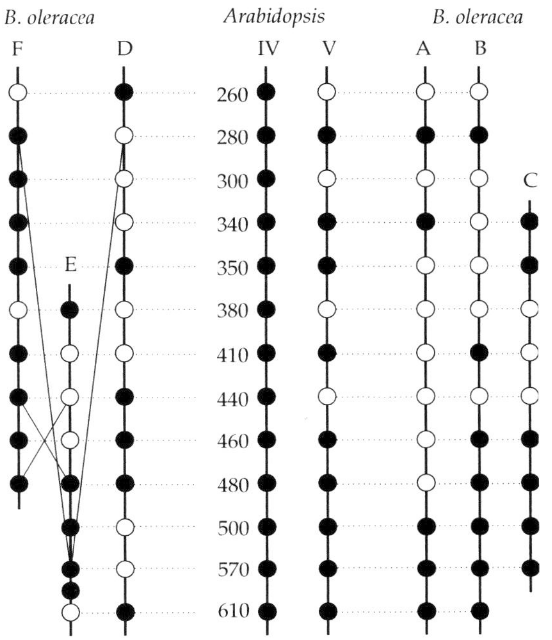


**Figure 4.2** Comparative physical maps of genes in chromosomal segments of *Arabidopsis thaliana* and *Brassica oleracea*. The segments from chromosomes IV and V in *Arabidopsis* represent a duplication event that preceded the divergence of *Arabidopsis* and *Brassica*, and these are each represented by paralogous segments in *B. oleracea* that arose from a subsequent triploidization event. The numbers along chromosome IV are abbreviated gene numbers. Closed circles represent functional genes; open circles represent silenced genes. Dashed lines connect the orthologous genes between the two species, with the diagonals indicating intrachromosomal rearrangements. Note that four of the active genes on chromosome IV were apparently lost from chromosome V prior to the divergence of the two species. Instances in which the paralogous copies within *Brassica* differ in state represent examples of nonfunctionalization of a duplicate gene within the *Brassica* lineage. The view is entirely from the perspective of chromosome IV, that is, genes that are present on V but absent from IV are not considered (modified from O'Neill and Bancroft 2000).

in which an ancestral duplicated gene for a male-specific function is initially present on both sex chromosomes, with the copy on the *X* becoming silenced in one descendent lineage and the copy on the *Y* being silenced in a sister lineage. A cross between females of the first population and males of the second would result in male progeny completely lacking in function, while the reciprocal cross would have active copies on both the *X* and the *Y*, potentially leading to dosage problems. Other complexities can arise with duplicated genes distributed on an autosome and a sex chromosome, particularly if **gametic imprinting** occurs (Lynch and Force 2000b). These types of imbalances are not expected in $F_1$ females (assuming the male is the heterogametic sex) because female progeny inherit identical genomic complements from both parents. Thus, duplication events involving genes on sex chromosomes are of potential relevance to understanding the mechanisms underlying Haldane's rule, which states that incompatibilities in interspecific crosses are most severe in progeny of the heterogametic sex (Orr 1997).

Second, a remarkable set of examples of map changes induced by gene duplication in plants involves the movement of genes between organelle and nuclear genomes. RNA-mediated transfers of several mitochondrial genes to the nuclear genome (accompanied by subsequent loss from the mitochondrion) have occurred on many independent occasions within recent lineages of flowering plants, with the overall rate in some cases rivaling the rate of nucleotide substitutions at silent sites (Adams *et al.* 2000, 2001). Details worked out for the mitochondrial respiratory protein gene *cox2* are particularly revealing. This gene was apparently duplicated to the nuclear genome of the ancestor of the Papilionoideae (a subfamily of legumes), transiently persisting as active copies in both genomes, with one or the other copy becoming randomly inactivated (with approximately equal frequencies, and by a variety of mechanisms) in almost all descendent lineages (Adams *et al.* 1999). These types of intergenomic gene transfers, initiated by gene duplication events, are not restricted to the mitochondrial genes, as a study of the chloroplast *infA* gene indicates large numbers of transfers to the nuclear genome (Millen *et al.* 2001).

What are the implications of such organelle-to-nucleus transfers? Consider the cross between a female of a species with an autosomal copy (A) of a gene (but no organelle copy, m) and a male of another species with the reciprocal arrangement. Letting small letters denote absentee alleles and assuming maternal inheritance of the organelle, the $F_1$ cross yields Aa/m progeny. If the gene is active in the gametic state, these individuals are expected to experience a 50 percent reduction in fertility due to the production of null (a/m) gametes. For a zygotically active gene, ignoring potential problems of gene dosage, there will be a 25 percent loss of progeny in the $F_2$ generation due to the segregation of aa/m genotypes. With $n$ such organelle–gene relocations, the fraction of viable $F_2$ progeny will be reduced to $(3/4)^n$. Thus, when one considers the very large number of organelle-to-nucleus transfers that apparently occurred soon after the colonization of early eukaryotes by the progenitors of mitochondria and chloroplasts, it is quite conceivable that divergent resolution of duplicated organelle genes played a significant role in the development of complete isolating barriers among the major eukaryotic lineages.

Third, although the previous arguments focus entirely on driving the divergent resolution of duplicate genes by degenerative mutation, map changes can also be induced by neofunctionalization, provided the copies acquiring new functions do so at the expense of the old function. As noted above, the probability of preservation of duplicate genes by neofunctionalization increases with population size because of the increased number of targets for rare beneficial mutations. Thus, unlike many genetic theories of speciation, the gene-duplication model works in very large populations. Moreover, contrary to the assumption of models that associate speciation with presumptive incompatibilities of independent adaptive changes acquired by sister taxa, under the gene-duplication model for speciation, neofunctionalization induces incompatibilities entirely through change in the map position of the ancestral gene function, that is, no complex mechanisms of **epistasis** associated with "coadaptive gene complexes" are involved.

Fourth, duplicate genes arising subsequent to prezygotic isolation of lineages are highly relevant to the process of divergent resolution. As noted above, duplication events are ongoing processes in all lineages, and because they are distributed over a large number of loci, those experienced in different lineages will almost certainly involve unique genes. To gain an appreciation for the rapidity with which map-change induced genomic incompatibilities can arise via ancestral and recurrent gene-duplication events, we now consider a simple model for genes with a single function. Letting $b$ denote the rate of origin of new gene duplicates and $d$ denote the rate of silencing, the dynamics for the number of excess copies of a functional locus are given by

$$dn/dt = b(n + 1) - dn, \qquad (4.1)$$

which yields the equilibrium number of excess copies of a gene, $b/(d - b)$. From the empirical results presented above, $b$ is on the order of $10^{-9}$ to $10^{-7}$ per generation, and for the situation in which all gene duplicates are subject to eventual silencing, $d$ is equivalent to the rate of origin of silencing mutations ($\sim 10^{-5}$ per generation). Thus, we expect on the order of 0.01–1 percent of the loci in a typical eukaryotic genome to be in a duplicated state at any point in time. These levels are consistent with the observed numbers of young duplicates in full-genomic sequences (Lynch and Conery 2000). From Watterson (1983) and Lynch and Force (2000a), the mean time to silencing one member of an established duplicate pair is approximately $[(2d)^{-1} + 10N]$ generations, where $N$ is the effective population size. We will approximate the rate of duplicate-gene loss, $L$, by the reciprocal of this time. Then, assuming that a random member of each pre-existing duplicate pair is silenced within each descendent sister species, there is a 50 percent probability that any pair will be divergently resolved. Assuming $b \ll d$, the expected number of duplicate loci in the base population to be divergently resolved after $t$ generations of isolation is

$$\delta_0 \approx Gb(1 - e^{-Lt})/(2d), \qquad (4.2a)$$

where $G$ is the fundamental number of genes per haploid genome (not including their duplicates).

Subsequent to the isolation of the two lineages, each population will continue to gain $2Nb$ new duplications per locus per generation, each with a probability of fixation equal to $1/(2N)$ under the assumption that the duplicates are neutral with respect to fitness. Noting again that each new fixed duplication has a 50 percent probability of being resolved in favor of the new locus, the rate of origin of new map changes ultimately becomes $Gb$ per generation. This asymptotic rate of divergence is approached on the time scale necessary for a newly arisen duplicate (initially in a single copy in a single individual) to first become established by fixation and then to become resolved by nonfunctionalization. To a first approximation, the expected number of map changes resulting from duplication events arising subsequent to geographic isolation is then,

$$\delta_{\mathrm{m}} = Gb\sum_{i=0}^{t}(1 - e^{-Lt}) \approx Gb\left(t - \frac{1 - e^{-Lt}}{L}\right), \quad (4.2b)$$

and the cumulative number of divergently resolved loci ($\delta$) is given by the sum of eqns (4.2a,b).

Under this model, tens to hundreds of map changes are expected after only a few million generations (Fig. 4.3). Letting $p$ denote the reduction in hybrid fitness per map change, then the expected hybrid fitness following the accumulation of $\delta$ map changes is $W = (1 - p)^{\delta}$. With $p = 1/16$, $W = 0.524$ when $\delta = 10$, and $W = 0.0016$ when $\delta = 100$. With $p = 5/16$, $W = 0.024$ when $\delta = 10$, and $W = 5 \times 10^{-17}$ when $\delta = 100$. Thus, although other factors undoubtedly contribute to species isolating barriers, gene duplication alone appears to be a sufficiently powerful mechanism for the origin of nearly complete genomic incompatibility within a few million years

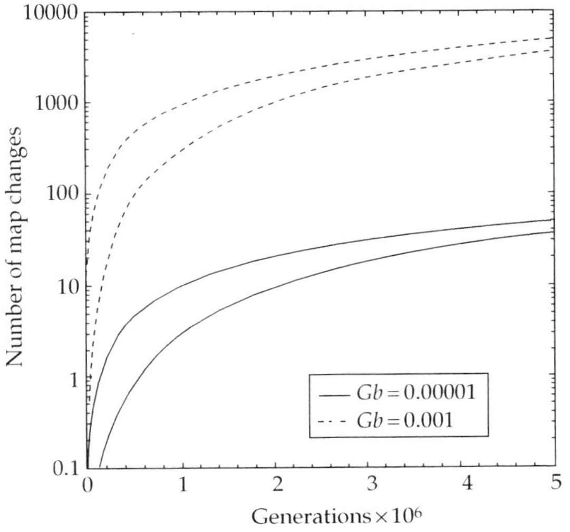


**Figure 4.3** Expected number of map changes induced by gene duplication for the situation in which the number of newly arising gene duplicates per genome per generation ($Gb$) is $10^{-3}$ or $10^{-5}$ and the null mutation rate per gene ($d$) is $10^{-5}$. For each pair of curves, the upper plot is for $N = 10^3$ and the lower is for $N = 10^5$.

of cessation of **gene flow**. This is the approximate time scale over which **postzygotic isolation** generally occurs in animals (Parker *et al.* 1985; Coyne and Orr 1997).

Genetic theories of speciation have traditionally focused around two competing hypotheses (for reviews, see Orr 1996; Rieseberg 2001). The Dobzhansky–Muller model postulates the accumulation of gene-sequence changes that are mutually incompatible when brought together in a hybrid genome, whereas the chromosomal model invokes the accumulation of rearrangements that result in mis-segregation in hybrid backgrounds. Both models are based on rather stringent assumptions, the general validity of which remains to be demonstrated. For example, the Dobzhansky–Muller model invokes the evolution of coadaptive complexes of epistatically interacting factors, none of which have been identified at the gene level, whereas the chromosomal model focuses on major chromosomal rearrangements, the fixation of which is greatly inhibited by the reduction in fitness in chromosomal heterozygotes.

A notable feature of the gene-duplication model for speciation is that it is consistent with *both* the chromosomal and the Dobzhansky–Muller models, while requiring fewer assumptions than either of them. The gene-duplication model is effectively a chromosomal model of speciation, but because the rearrangements are at the microchromosomal level, and hence unlikely to cause significant problems during meiosis, they accumulate passively without any alteration in fitness. The gene-duplication model is also effectively a Dobzhansky–Muller model, in that the map changes induced by divergent resolution result in pseudo-epistatic interactions without any changes at the gene level. Genomic incompatibilities that are simple consequences of reassignments of genes to new locations will appear superficially as epistatic interactions because the loss-of-function phenotype is a function of the number of active alleles at the two homologous loci in hybrid progeny. Thus, genetic analyses of species incompatibilities that fail to identify the specific underlying loci can lead to misinterpretations regarding the underlying genetic mechanism of postzygotic isolation.

Under the gene-duplication model, certain groups of organisms are expected to be more prone to speciation than others, the most notable of which are lineages that experience a doubling in genome size. One potential example of such a key event was noted above—the colonization of ancestral eukaryotic genomes by endosymbiotic organelles, with subsequent variation in transfers of organelle genes to the nucleus being experienced by many of the deeply diverging eukaryotic lineages (Martin *et al.* 1998). Polyploidization provides another enormous opportunity for the passive origin of isolation via divergent resolution of duplicate genes. Because of the stochastic nature of gene silencing, following the first map changes induced by divergent resolution, thousands of duplicate pairs are expected to remain in a functional state, and these then are free to become divergently resolved in subsequently isolated lineages. As a consequence, ancestral polyploidization has the potential to precipitate a large number of nested speciation events without requiring any underlying changes at the gene level. Adaptive divergence of the isolated lineages may then evolve secondarily. This mechanism for the origin of species may be particularly common in plants, which commonly experience polyploidization events. Indeed, more than a decade ago, Werth and Windham (1991) suggested that divergent resolution following polyploidization may be responsible for the thousands of species of modern-day ferns. Another spectacular example of an adaptive radiation following polyploidization is the relatively young Hawaiian silversword alliance, which is known to have arisen from a combination of genomes of two diploid North American species (Barrier *et al.* 1999).

Although polyploidization is much less common in animals, key genome doubling events may also have played a significant role in the diversification of the major animal lineages. First, given the apparent three- to fourfold increase in gene content in basal animals relative to fungi, genome amplification might very well have been involved in the origin of the major animal phyla. Second, although the mechanism of genome amplification remains controversial, it also appears that a substantial amount of gene duplication (perhaps equivalent to two polyploidization events) occurred prior to the radiation of the major vertebrate lineages (Postlethwait *et al.* 2000; McLysaght *et al.* 2002). Finally, the most speciose lineage of vertebrates, the ray-finned fishes, is a descendant of an ancient polyploidization event, and more recent secondary polyploidization events in fishes appear to be associated with enhanced rates of speciation (reviewed in Taylor *et al.* 2001).

## Summary

Surveys of the contents of completely sequenced genomes provide compelling evidence that gene duplication is an ongoing process in all organisms, with the approximate rate of duplication per gene being of the same order of magnitude as the mutation rate per nucleotide site, if not a little higher. Thus, there is little question that gene duplication plays an important role in genome evolution, and the association of several major phylogenetic radiations with ancient periods of genome amplification suggests an important role in phenotypic evolution as well. The specific mechanisms that are responsible for the preservation and proliferation of duplicate genes are less clear. Although it is tempting to conclude that the cellular complexity of large multicellular organisms required an amplification of gene content relative to that in simpler microbes, these groups of organisms differ in many ways other than phenotypic complexity. Most notable is the tendency for the effective size of a population to decline with an increase in body size. Because random genetic drift is an important determinant of the alternative evolutionary fates of duplicate genes, various aspects of genome-size evolution may be as much an indirect consequence as a cause of organismal complexity. That is, the growth of genome size in large, complex eukaryotic organisms may, in part, be a simple reflection of evolutionary forces that predominate in populations with relatively small effective sizes.

A number of hypotheses have been proposed to explain the long-term preservation of duplicate genes, and an array of comparative analyses point to something other than the evolution of new functions as the predominant mechanism. A substantial fraction of the data on gene-expression patterns supports the subfunctionalization model, wherein duplicate-gene preservation is driven entirely by the accumulation of degenerative mutations, and most effectively so in small populations. Results from coding-sequence analyses are less conclusive on this issue, but the primary observations are fully compatible with the subfunctionalization model: the fact that the average ratio of replacement to **silent substitutions** for duplicate pairs is always less than one; the apparent relaxation of selection against amino acid changes in newborn duplicates; and the spatial variation in evolutionary rates exhibited in paralogous copies. Although further functional analysis of duplicate genes and their **orthologues** in single-copy species will be required to fully evaluate the roles of degenerative and neofunctionalizing mutations in genome evolution, it should be kept in mind that the mutations responsible for the initial preservation of duplicate genes may often be obscured by subsequent mutational refinements to the performance of well-established duplicates. The formal separation of these changes raises considerable challenges for future molecular evolutionary studies.

Despite the remaining uncertainties in our understanding of duplicate-gene evolution, an important central role of degenerative mutations is now firmly established. Although such mutations are normally viewed to be contrary to adaptive evolution, this is not the case with gene duplication. By preserving duplicate genes, subfunctionalizing mutations ensure the continued exposure of both members of a pair to natural selection, thereby increasing the likelihood of occurrence of rare neofunctionalizing mutations. In addition, subfunctionalization provides a potentially powerful mechanism for eliminating the pleiotropic constraints that are unique to single-copy genes, thereby opening up new evolutionary degrees of freedom. Finally, divergent silencing and subfunctionalization of duplicate genes provides a simple passive mechanism for the origin of post-reproductive barriers between isolated populations. Thus, gene duplication may be just as relevant to the origin of new branches in the tree of life as it is to adaptive phenotypic evolution within lineages.

CHAPTER 5

# The evolution of gene regulation: approaches and implications

**Casey M. Bergman and Nipam H. Patel**

Since the earliest days of molecular biology, there has been increasing interest in understanding the role of transcriptional regulation in the morphological evolution of higher organisms. This trend owes its origin to the pioneering insight from studies of the *lac* operon which demonstrated that modern genomes contain DNA sequences devoted solely to the regulation of other genes (Jacob and Monod 1961). These sequences were shown to be closely linked in "*cis*" to the gene(s) under transcriptional regulation and function as the targets of "*trans*" acting factors. These ***cis*-regulatory sequences** were immediately regarded as being potential targets of functional change exerted by evolutionary forces such as mutation, selection and **genetic drift**. Further speculation concerning the role of *cis*-regulatory evolution has emerged more recently with better understanding of eukaryotic genome structure and organization. For instance, with the essentially complete sequences of eukaryotic genomes, it has become clear that complex organisms harbor vast amounts of unique DNA sequences that do not encode proteins. Some amount of this DNA appears to contain *cis*-regulatory information, and may provide an abundant source of material for the evolution of developmental processes and genomic complexity (Carroll 1994; Shabalina *et al.* 2001).

Over the last 30 years, various molecular and genetic methods have been used to amass evidence for various forms of regulatory evolution (including *cis*-regulatory evolution) in higher organisms. Each of these methods has virtues and shortcomings, and different lessons have been learned from each approach, which we discuss generally in the first section of this chapter. More recently, attention in the field of regulatory evolution has shifted to directly analyzing the sequences involved in transcriptional regulation—transcription factors and the *cis*-regulatory sequences to which they bind. Thus, in the second section we describe an emerging, sequence-based approach for the molecular evolutionary analysis of *cis*-regulatory evolution that uses tools borrowed from the field of bioinformatics. Using this approach, we describe a model of transcription factor **binding site turnover** to characterize the evolution of *cis*-regulatory sequences. In the third section, we discuss implications of this model of binding site turnover for evolutionary analyses of *cis*-regulatory sequences. Finally, we conclude with prospects for the analysis of developmental systems and their evolution in light of the lessons learned from the analysis of *cis*-regulatory evolution.

The topic of regulatory evolution has received considerable attention in the literature starting with a section in the classic paper of Wilson *et al.* (1977). MacIntyre (1982) and Dickinson (1991) treat the issue of regulatory evolution explicitly and fully in their chapters devoted to the topic. There have also been a number of more recent reviews of regulatory evolution, for which we point the reader to the references in the section entitled Further Reading. These reviews capture the essential classical and recent results, as well as the diversity of viewpoints and approaches in the field of regulatory evolution. Here, rather than attempting a comprehensive review of regulatory evolution, we wish to contribute to this discussion by describing some of the tools available for the analysis of *cis*-regulatory evolution. Our goal is to address conceptual issues in the study of *cis*-regulatory evolution by discussing the methods commonly used in the field. By cataloguing the "tools of the trade," we hope to frame what questions can and cannot be addressed by various approaches in the field of *cis*-regulatory evolution.

## Tools for the analysis of *cis*-regulatory evolution

As with many fields within molecular evolution, tools to study regulatory evolution in general, and *cis*-regulatory evolution in particular, have closely paralleled methodological advances in molecular biology. The first evidence used to infer regulatory evolution was based on *indirect* studies using **microcomplement fixation**, one the earliest methods used to measure evolutionary distances at the molecular level. These studies sought to make arguments about regulatory evolution based on rates of morphological evolution or speciation as a function of molecular distance. For instance, discordance in the molecular distance between hybridizing species pairs from different lineages (mammals versus frogs) was used to argue that factors besides protein evolution—namely, regulatory evolution—restrict the ability of species to hybridize (Wilson *et al*. 1974). This influential conclusion did not provide direct evidence for regulatory evolution, since these result can be explained by other factors, such as lineage-specific rates of protein evolution.

The earliest *direct* studies of regulatory evolution used allozyme electrophoresis to monitor gene expression in a population genetic or phylogenetic context. Such studies address whether evolved changes in gene regulation have occurred within or between species at the level of protein expression. For example, Dickinson (1980) demonstrated that *c*.30 percent of loci surveyed for allozyme expression exhibited changes in gene regulation between species of Hawaiian *Drosophila*. In these studies, whole animals or dissected tissues were analyzed for the expression levels of well-characterized enzymes. These methods can reveal quantitative shifts in protein expression levels, and qualitative changes in protein expression patterns when accompanied by tissue-specific dissection. Thus the anatomical resolution of dissection limits the analysis of changes in protein expression pattern. These methods also do not discriminate among various molecular mechanisms of regulatory evolution, since they conflate transcriptional and post-transcriptional mechanisms of gene regulation. Furthermore, they rely on the conserved functional properties of the enzyme and high levels of expression for detection. Thus, in studies of this type, amino acid substitution leading to divergence of protein activity can be (mis)interpreted as quantitative changes in the level of protein expression.

More recently, electrophoresis-based methods have been replaced by methods that examine gene expression more directly in whole or dissected preparations. These approaches commonly use antibody staining to detect protein expression and/or *in situ* RNA hybridization to detect levels or patterns of mRNA expression. These techniques allow improved analytical precision and a greater diversity of genes to be analyzed for changes in gene regulation. Importantly, these methods can also reveal qualitative changes in expression pattern in the absence of tissue-specific dissection, and in some instances have been used to reveal quantitative changes in expression levels. For example, Averof and Patel (1997) demonstrated altered patterns of Ultrabithorax/abdominal-A protein expression using antibody staining between different crustacean species, and correlated these changes with subsequent developmental differences in appendage morphology. Like allozyme methods, antibody-staining approaches conflate transcriptional and post-transcriptional changes in regulation; *in situ* RNA hybridization can more accurately detect changes in transcription, with the caveat that some observed alterations might be caused instead by changes in mRNA stability. Application of antibody staining to study gene regulation in an evolutionary context requires a cross-reactive antibody which recognizes conserved features of the **orthologous** protein without spurious cross-reactivity to other proteins, the development of which is often a major limitation of this method. And like allozyme-based methods, it is assumed that quantitative shifts in antibody staining reflect real changes in gene regulation, and not simply changes in detection of the protein resulting from amino acid divergence. *In situ* RNA hybridization across species requires cloning of **orthologues** from each species analyzed for probe preparation. Fortunately, highly conserved regions of the genes being studied often provide convenient sequences for their isolation by the **polymerase chain reaction**.

Electrophoretic and histochemical techniques have led to substantial progress in our understanding of regulatory evolution, but neither of these approaches can provide specific information about the genomic location of regulatory changes. When analyzing a gene by either of these approaches, changes linked in *cis* to the gene and/or from *trans*-acting factors located elsewhere in the genome may affect the level or pattern of expression observed. To overcome this limitation of electrophoretic and histochemical techniques, manipulative approaches employing interspecific genetic or transgenic analysis have been developed. These genetic approaches can uncover the location of evolved regulatory changes, and therefore offer deeper insight into the molecular mechanisms by which regulatory evolution occurs.

The genomic location of factors responsible for evolved regulatory changes can be genetically mapped using interspecific crosses. If the factors responsible for a regulatory difference map to the location of the gene whose regulation has evolved, changes in *cis*-regulatory sequences are inferred. Conversely, if explanatory factors map elsewhere in the genome, changes in *trans* acting factors are typically implicated. The work of Stern and others in *Drosophila* (Stern 1998; Sucena and Stern 2000) illustrates the use of interspecific genetics by mapping regulatory changes in *cis* to the *Ultrabithorax* and *ovo-shavenbaby* genes by complementation and **deletion** analyses. However, since species are often defined by their inability to hybridize, interspecific genetic approaches are naturally limited. Ironically then, as the chance of discovering regulatory changes grows with increasing evolutionary distance, the difficulty of interspecific genetic analysis also increases. Furthermore, these genetic approaches suffer from problems general to interspecific hybridizations, such as hybrid vigor or heterosis in the F1 generation, which may complicate analysis of the specific regulatory change under analysis (Dickinson *et al.* 1984). Nevertheless, interspecific genetic analysis represents one of the most powerful strategies to dissect regulatory evolution among closely related species.

The problems associated with interspecific crosses among divergent species can be overcome by interspecific transgenic analysis (Mitsialis and Kafatos 1985). In this approach, evolved regulatory changes are assayed by reporter gene expression driven by the regulatory sequences of a donor gene in a different species. The logic of these experiments is that if a reporter gene shows the expression pattern of the donor species when introduced into the host, then the transgene contains the DNA sequences responsible for evolutionary change and the evolved regulatory changes have occurred in *cis*. If the reporter gene shows the expression pattern of the host species when introduced into the host, then sequence evolution outside the transgenic construct is responsible for the evolved regulatory change. The usual interpretation of a host-like pattern is that the regulatory change occurred in a *trans*-acting regulatory factor. However formal exclusion that the regulatory change occurred in *cis*, but outside the region analyzed, is often limited by the size of recombinant constructs that can be transformed. In addition, these experiments can also be complicated by potential altered specificities between the *cis*-regulatory sequences of the donor and *trans*-regulatory factors of the host (P.J. Shaw *et al.* 2002). Finally, it must be noted that changes mapped in *trans* may result from changes in the factor itself or in its expression.

Interspecific transgenic analyses primarily have been used to demonstrate that the function of a *cis*-regulatory sequence is conserved across taxa, a step necessary to substantiate detailed functional analysis of *cis*-regulatory elements by comparative sequence analysis (see below). However, and more importantly for our discussion, transgenic analysis has also been used to prove that novel expression domains arise by evolution of *cis*-regulatory sequences. For example, Brady and Richmond (1990) showed that lineage-restricted eye and hemolymph expression of the *D. pseudoobscura Esterase-5B* gene can be recapitulated in *D. melanogaster* using a transgenic construct that contains only *cis*-regulatory information. In addition, claims of conserved expression patterns often overshadow subtle changes in the level or timing of transgene expression or divergent aspects of secondary expression patterns (e.g. Tautz and Nigro 1998), thus the incidence of *cis*-regulatory divergence may be more common than is usually emphasized in the literature. Functional divergence of *cis*-regulatory sequences is also implicated by more sensitive analysis of interspecific transgenes in hosts with null genetic backgrounds for the gene of interest. Such experiments often reveal that interspecific transgenes fail to fully rescue mutant phenotypes, potentially implicating some level of *cis*-regulatory divergence (e.g. Maier *et al.* 1990).

## Identifying changes in *cis*-regulatory sequences

Despite the elaboration of methods and functional evidence for *cis*-regulatory evolution, the development of quantitative methods to analyze *cis*-regulatory sequence evolution has received surprisingly little attention. This can be explained in part by the historical precedence of isolating and sequencing peptides and protein-coding DNA long before noncoding *cis*-regulatory DNA sequences were abundantly available. As a consequence, many statistical methods in molecular evolution were specifically designed for protein-coding sequences and explicitly use the constraints of the genetic code to construct and test various metrics of sequence evolution. For instance, molecular evolutionary methods that use $K_a/K_s$ (and related) ratios or measures of codon bias are not applicable to the analysis of *cis*-regulatory evolution (see Aguadé *et al.*, Chapter 2). Fortunately, some population genetic tests of neutrality make no explicit reference to the type of sequence under consideration (e.g. Hudson *et al.* 1987), or can be modified to fit a simple model of *cis*-regulatory structure (e.g. Jenkins *et al.* 1995; Ludwig and Kreitman 1995), and thus can be adopted for analyzing *cis*-regulatory sequence evolution. Nevertheless, there is currently a paucity of quantitative methods specifically designed for the analysis of *cis*-regulatory molecular evolution.

Recently, several independent efforts have borrowed an approach from the field of bioinformatics—**binding site prediction**—to study *cis*-regulatory molecular evolution (Chuzhanova *et al.* 2000; Liu *et al.* 2000; Ludwig *et al.* 2000; Kim 2001; Dermitzakis *et al.* 2003). By binding site prediction, we refer to a suite of computational approaches that make quantitative predictions about the location of transcription factor binding sites in **unannotated** genomic DNA using probabilistic representations to model transcription factor–DNA interactions (reviewed in Stormo 2000). The application of binding site prediction in an evolutionary context is a natural extension of computational methods designed for the analysis of single target sequences to multiple **homologous** sequences. The logic behind using binding site prediction to study *cis*-regulatory evolution is that, since binding sites are the fundamental unit of *cis*-regulatory structure (Arnone and Davidson 1997), they are likely to be the fundamental unit of *cis*-regulatory evolution as well. Therefore, quantitative methods that can accurately predict the occurrence of individual transcription factor binding sites should also be useful for analyzing binding site evolution in *cis*-regulatory sequences. By way of illustration, we now summarize one application of binding site prediction in an evolutionary context using a model enhancer sequence in *Drosophila* to show how binding site turnover can occur in a *cis*-regulatory sequence (Ludwig *et al.* 2000). More detailed consideration of the construction of binding site models and thresholds for prediction can be found in Bergman (2001) and Dermitzakis *et al.* (2003).

In this approach to binding site prediction, the typical "consensus" sequence representation of transcription factor binding site specificity is replaced with a more flexible probabilistic representation called a **position weight matrix** (PWM) (Fig. 5.1). PWMs capture more information about the detailed base usage in binding sites than do consensus sequences, yet are simple statistical objects that sufficiently describe the sequence preferences of eukaryotic transcription factors (Frech *et al.* 1997; Schug and Overton 1997). PWMs can be used to transform windows of target DNA sequence into random variables which measure

(a)

| | 1 | 2 | 3 | 4 | 5 | 6 | 7 | 8 | 9 | 10 |
|---|---|---|---|---|---|---|---|---|---|---|
| A | 1.6 | 2.8 | 2.5 | 1.8 | 0.1 | 0.6 | 0.6 | 0.1 | 0.2 | 1.8 |
| C | 0.7 | 0.4 | 0.6 | 1.1 | 0.1 | 0.1 | 0.4 | 0.1 | 0.5 | 0.3 |
| G | 1.3 | 0.3 | 0.3 | 0.5 | 4.1 | 3.6 | 3.4 | 0.4 | 0.7 | 0.8 |
| T | 0.4 | 0.1 | 0.3 | 0.5 | 0.4 | 0.2 | 0.2 | 2.9 | 2.3 | 0.8 |

(b)

**Figure 5.1** Position weight matrix (PWM) for the *Drosophila* transcription factor *Kruppel*, a zinc-finger containing repressor of *even-skipped* stripe two expression. (a) Each cell in the matrix represents the likelihood ratio for a given nucleotide to be recognized by *Kruppel* relative to background nucelotide frequencies. The likelihood that a ten base pair sequence is recognized by Kruppel as a binding site is obtained by multiplying the individual ratios across each position of the PWM. (b) Sequence logo representation of Kruppel binding site usage showing the highest likelihood nucleotide at each position of the PWM in (a). The sequence logo was generated at http://www.bio.cam.ac.uk/seqlogo/logo.cgi using the methods of Schneider and Stephens (1990).

(a)

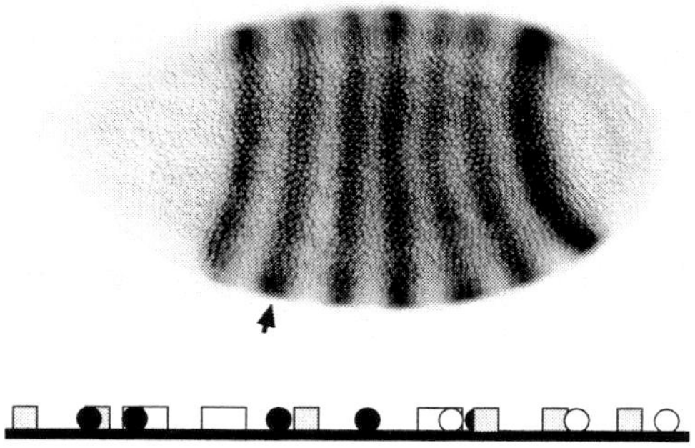

(b)

**Figure 5.2** (a) *even-skipped* (*eve*) mRNA expression in a blastoderm stage *D. melanogaster* embryo. *eve* stripe 2 is indicated by the arrow. (b) Schematic representation of the enhancer sequence responsible for generating the second stripe of *eve* expression. Grey boxes, white rectangles, black circles, and white circles represent *Kruppel, giant, bicoid*, and *hunchback* binding sites, respectively.

the likelihood that a sequence is recognized by a given transcription factor, a process which we refer to as a *binding site likelihood scan*. Binding site likelihood scans provide a convenient way to visualize the potential binding site composition of *cis*-regulatory sequences and capture many features of *cis*-regulatory evolution including conservation of binding sites across species, as well as putative gains/losses of binding sites in any lineage sampled. Importantly, these methods are *alignment-independent* and, thus, circumvent many of the difficulties associated with alignment of multiple divergent *cis*-regulatory homologous sequences (see below). On the other hand, methods for binding site prediction suffer from biases in the training data sets used to model transcription factor usage, as well as problems in estimating the statistical significance of predicted sites (Claverie and Audic 1996; Frech *et al.* 1997; Dermitzakis *et al.* 2003). Nevertheless PWM based methods have yielded accurate functional predictions (Berman *et al.* 2002) and are among the most useful tools available for *cis*-regulatory prediction and **annotation** (Fickett and Wasserman 2000).

The utility of quantitative tools that can predict binding sites in an evolutionary context can be shown by the analysis of model eukaryotic *cis*-regulatory elements. For instance, blastoderm *D. melanogaster* embryos express the pair-rule gene *even-skipped* (*eve*) in seven transverse stripes, each of which has been shown to be governed by unique or shared enhancers (Gilbert 1994) (Fig. 5.2). The *cis*-regulatory sequences responsible for the second-most anterior stripe of *eve* expression have been the subject of extensive genetic, biochemical, and transgenic studies and represent a model eukaryotic enhancer. *eve* stripe two expression is achieved by the combined action of the activators *bicoid* (*bcd*) and *hunchback* (*hb*) and the **repressors** *giant* (*gt*) and *Kruppel* (*Kr*). Each of these transcription factors has been shown by DNAse 1 **footprinting** to interact directly with a 670 base pair fragment located *c*.1 kb from the transcription start site (Stanojevic *et al.* 1989; Small *et al.* 1991). These footprints have been demonstrated to be bona fide protein binding sites by mutagenesis and transgenic analysis (Stanojevic *et al.* 1991). Interestingly, however, the precise location and configuration of functionally characterized binding sites within the *eve* stripe two enhancer is amazingly flexible. This has been shown by experiments in which mutated binding sites can be complemented in *cis* by the addition of novel binding sites elsewhere in the enhancer (Arnosti *et al.* 1996).

These sort of *cis*-complementation experiments suggest that turnover in the composition or configuration of binding sites within an element may be tolerated, and even expected, during *cis*-regulatory evolution.

The possibility of binding site turnover in functionally constrained *cis*-regulatory sequences has recently been substantiated empirically in both flies and mammals. First, it is clear that polymorphism exists in functionally characterized transcription factor binding sites, demonstrating that the requisite intraspecific molecular variation necessary for binding site turnover exists in natural populations (Jenkins *et al.* 1995; Ludwig and Kreitman 1995). Polymorphism in binding sites clearly can lead to the *loss* of binding sites during *cis*-regulatory divergence, since functionally characterized binding sites are not always conserved (Dickinson 1991), even when the function of a *cis*-regulatory sequence is conserved (Ludwig *et al.* 1998). Reciprocally, experimental dissection of enhancers in multiple Dipteran lineages proves that the *gain* of binding sites can occur as well (Moses *et al.* 1990; Bonneton *et al.* 1997). Moreover, *de novo* gain of binding sites in *cis*-regulatory sequences by point mutation or small tandem **duplication** is highly probable considering that transcription factors typically recognize short, degenerate sequence **motifs** (Loomis and Kuspa 1992; Stone and Wray 2001). This dynamic view of the ongoing gain and loss of binding sites during *cis*-regulatory divergence predicts the structural reorganization of *cis*-regulatory sequences during evolution, even in the absence of functional change (Fig. 5.3a). This model can be contrasted with a model of constant structural constraint where the composition and configuration of binding sites is conserved during *cis*-regulatory sequence divergence (Fig. 5.3b). Experimental support for such a model of binding site turnover during *cis*-regulatory divergence under stabilizing

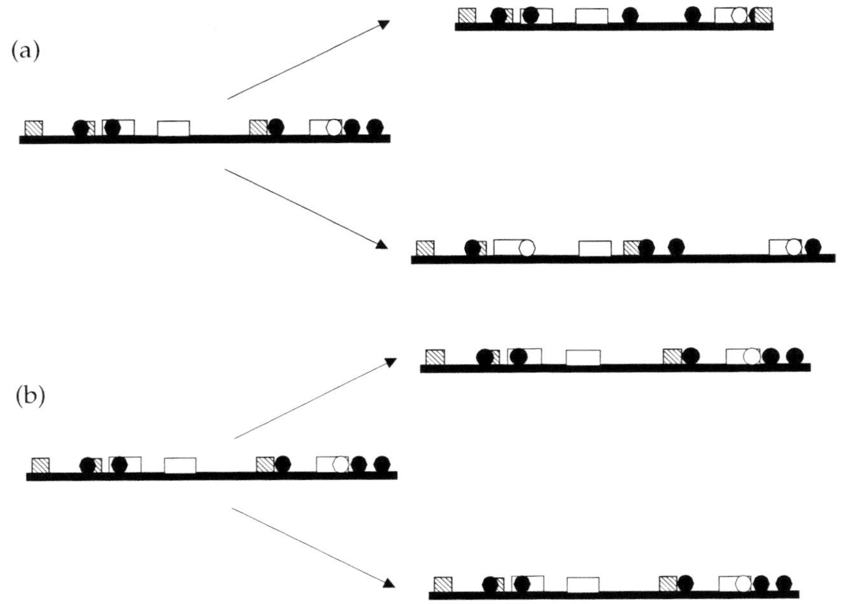


**Figure 5.3** Two alternative models to explain divergence in *cis*-regulatory sequence with conserved functional constraints. (a) Model of *cis*-regulatory sequence divergence under changing structural constraint resulting from binding site turnover. The binding site composition of an ancestral *cis*-regulatory sequence is reorganized independently in two descendent sequences. Note that the number, configuration, and spacing of binding sites can change as a result of binding site turnover. (b) Model of *cis*-regulatory sequence divergence under constant structural constraint; under this model the number and configuration of binding sites is conserved.

selection has been reported using transgenic analysis of interspecific chimeras and *in vivo* footprinting of homologous *cis*-regulatory sequences (Ludwig *et al.* 2000; Cuadrado *et al.* 2001). These studies suggest that multiple interacting substitutions resulting from the binding site turnover occur during *cis*-regulatory evolution (Ludwig *et al.* 2000). Thus, just as experimental analysis of model *cis*-regulatory sequences would indicate, flexibility in the composition and configuration of binding sites is likely to be a common theme in evolution of *cis*-regulatory sequences.

This model of binding site turnover can be studied at the sequence level using binding site prediction to reveal a number of interesting features of *cis*-regulatory evolution (Fig. 5.4). To begin, it is possible to identify all functionally characterized binding sites in the *D. melanogaster eve* stripe two enhancer using only a PWM-based binding site prediction (Ludwig *et al.* 2000; Bergman 2001). This result demonstrates that application of PWM-based binding site prediction can successfully recapitulate *in silico* functional results derived from *in vitro* analysis. This positive control indicates that binding site prediction may be a reliable way to identify the functional binding site composition for a set of homologous *cis*-regulatory sequences, thereby bypassing the need for comprehensive functional characterization. Putative binding sites are predicted in addition to the functionally characterized binding sites for the three transcription factors examined in the *eve* stripe two analysis (Fig. 5.4). These putative sites may represent false positive predictions or a too lenient threshold of the current method (for a discussion of cutoffs for binding site prediction, see Dermitzakis *et al.* 2003).

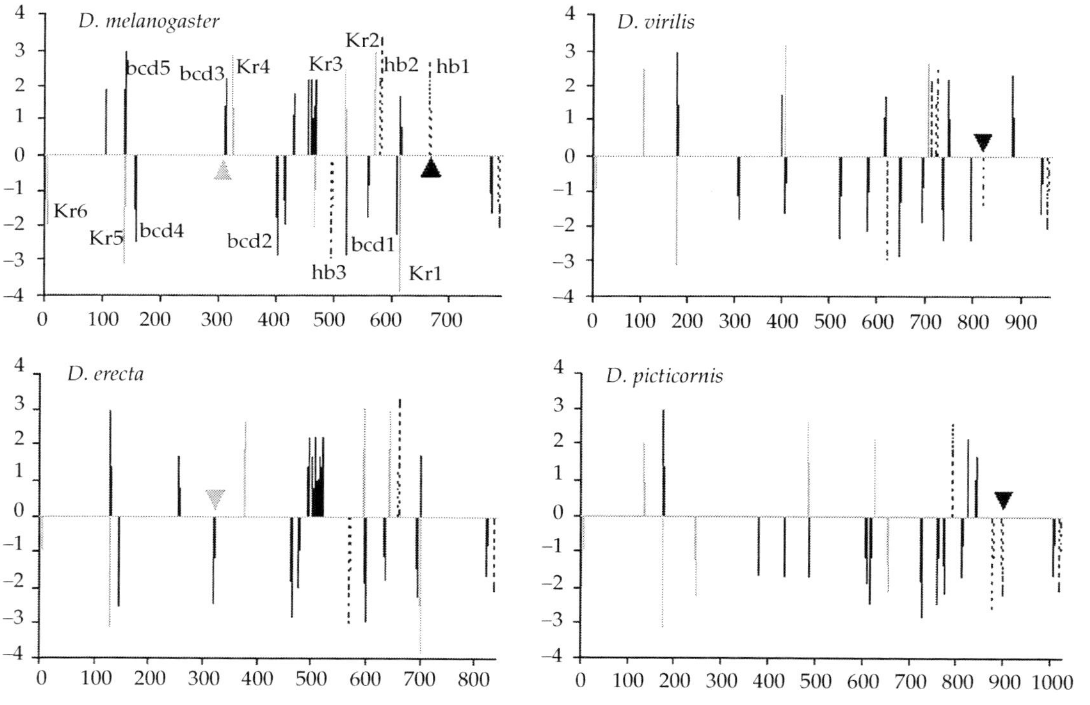


**Figure 5.4** Binding site likelihood scans of homologous *eve* stripe two enhancer sequences in *Drosophila*. PWMs for *bcd*, *hb*, and *Kr* were constructed by aligning samples of footprinted sequences (for details, see Bergman 2001). Positive and negative values in the plots correspond to putative binding sites on the plus strand and minus strand, respectively. The ordinate represents the log-likelihood value of a given sequence under the PWM model of binding site usage. The abscissa represents position in the sequence and shows differences in the total sequence lengths. Grey solid: *Kruppel*; black solid: *bicoid*; black dotted: *hunchback*. Experimentally verified binding sites are labeled for *D. melanogaster*.

Alternatively, however, these sites could be weak binding sites or sites not observed in DNAse cleavage studies (for evidence of weak *Kr* binding sites in the *eve* stripe two enhancer see Stanojevic *et al.* 1991).

Nevertheless, important aspects of *cis*-regulatory divergence can be studied using binding site likelihood prediction in an evolutionary context. For instance, it appears that overlapping binding sites (such as the overlapping *bcd5/Kr5* sites) are highly conserved across distantly related taxa. This observation can be explained by the possibility that point mutations in sequences bound by multiple proteins would have more drastic pleiotropic consequences than mutations in sequences bound by a single transcription factor. As expected from the discussion above, binding site prediction reveals potential gains and losses of binding sites during enhancer divergence (Fig. 5.3, arrowheads). Interestingly, it appears that both strong and weak sites can be either highly conserved or lost during *cis*-regulatory evolution. This observation highlights the possibility that there may be *strong selection* maintaining *weak affinity* binding sites, and vice versa. More generally, it is not clear that the strength of binding affinity is in any way related to the strength of selection acting on a given binding site. Binding site likelihood scans also reveal that binding sites for both activators and repressors exhibit turnover during evolution. Moreover, as expected from studies of enhancer function that suggest that transcription factors act at a distance from the transcription initiation site, there appears to be no strand specificity in the orientation of putative binding sites. Finally, there are several apparent cases of convergent evolution of binding sites in similar positions in the *eve* stripe two enhancer but on the opposite strand (see legend to Fig. 5.4). Such observations suggest either local **inversions** of binding sites or more intriguingly, potential constraints on the position of novel binding sites during evolution.

In summary, experimental dissection of a model eukaryotic enhancer sequence predicts the possibility of binding site turnover during *cis*-regulatory evolution. Direct analysis of the molecular evolution of such sequences reveals the necessary variation and functional evidence required for a model of binding site turnover. Finally, the process of binding site turnover can be studied in an efficient manner using binding site prediction in an evolutionary context.

## Implications of binding site turnover for the analysis of *cis*-regulatory evolution

The ongoing process of binding site turnover described in the previous section has important implications for evolutionary analyses of *cis*-regulatory sequences. Among these we will discuss the implications of such a model for (1) attempts to dissect *cis*-regulatory function by analysis of conserved sequences, and for (2) phylogenetic analysis using *cis*-regulatory sequences. Binding site turnover itself clearly deserves much further attention to understand the evolutionary forces shaping this pattern of molecular evolution. However, current evidence is sufficient to demonstrate that some model of binding site turnover must be considered when performing quantitative evolutionary analyses of *cis*-regulatory sequences.

One of the most common evolutionary analyses performed on *cis*-regulatory sequences is the identification of highly conserved sequences to assist functional dissection—an approach that has come to be known as **phylogenetic footprinting** (Tagle *et al.* 1988). Phylogenetic footprinting has been a long-appreciated method in the dissection of *cis*-regulatory sequences, and is based on a basic tenet of the **neutral theory** that rates of molecular evolution are inversely correlated with functional constraint (see Ohta, Chapter 1). Typically, phylogenetic footprinting is performed by pairwise analysis of homologous *cis*-regulatory elements from distantly related species (such as mice and humans) for which the function of the *cis*-regulatory element under scrutiny is assumed to be conserved. Species are chosen such that ample neutral divergence has occurred, so that sequence conservation is inferred to be a result of functional constraint and not insufficient divergence time. The recent availability of large amounts of noncoding DNA sequence data has driven the development of alignment and visualization tools specifically designed for phylogenetic footprinting (Miller 2001).

What effect would changing structural constraints resulting from binding site turnover have for phylogenetic footprinting of *cis*-regulatory sequences? Most importantly, binding site turnover will cause phylogenetic footprinting methods to underestimate the proportion of nucleotide sites that are under functional constraint in a *cis*-regulatory element. This is true because phylogenetic footprinting methods based on comparative sequence alignment can only reveal information that is *common* to the set of sequences under analysis. Sequence alignment can be viewed as being equivalent to finding the intersection among a set of sequences, and thus structural or functional information that is restricted to only one lineage cannot be revealed using alignment based comparative methods. A model of binding site turnover predicts that lineage-restricted structural information often contributes to *cis*-regulatory function, and thus the phylogenetically footprinted intersection of a set of *cis*-regulatory sequences will underestimate the number of functionally important elements. Thus, it will be impossible to define all functional *cis*-regulatory sequences solely on the basis of evolutionary conservation, and it will be necessary to complement phylogenetic footprinting with data obtained from *in vitro* and *in vivo* functional characterization. Put another way, the possibility of binding site turnover precludes the use of an evolutionary analogue of the minimalization criterion typically used to define *cis*-regulatory sequences.

It is often claimed that phylogenetic footprinting benefits from the simultaneous comparison of more than a single pair of species (Dubchak *et al.* 2000; Miller 2001). Especially for more closely related species, the addition of sequences offers the possibility of rejecting spurious false positive sequence conservation since sites may not be conserved in all pairs of sequences, thereby enhancing the signal to noise ratio of functional predictions (see Leung *et al.* 2000, for example). This argument makes sense under a model of constant structural constraint (see Fig. 5.3b), where the sequences to be identified are the same in all taxa. However, under a model of changing constraints due to binding site turnover (Fig. 5.3a), the inclusion of additional taxa may exacerbate the underestimation of functional sequences by phylogenetic footprinting. Since each lineage is likely to have its own configuration of binding sites, the multiple alignment of sequences from several species will only reveal those that are common to all taxa. Thus, for multiple sequence analysis to contribute to the enhanced detection of functional sequences, a *multiple pairwise* alignment framework, rather than a true multiple alignment, might be the appropriate approach for phylogenetic footprinting.

The ongoing process of binding site turnover, if taken to its logical extreme, would lead to complete primary sequence divergence, even for orthologous *cis*-regulatory sequences with conserved function (Ludwig *et al.* 2000). Thus attempts to phylogenetically footprint very distantly related *cis*-regulatory sequences (e.g. comparing flies and mammals) are unlikely to provide meaningful functional results. The same can also be said for **paralogous** divergence among distantly related gene duplicates. Moreover, when functional **homologies** are claimed for *cis*-regulatory elements at extreme evolutionary distances it will be difficult, if not impossible, to substantiate such claims based on primary sequence conservation [see for instance González-Crespo and Levine (1994), Xu *et al.* (1999), or Rincón-Limas *et al.* (1999)].

Whereas the process of binding site turnover may complicate attempts to dissect *cis*-regulatory structure and function *via* phylogenetic footprinting, it may benefit attempts to derive phylogenetic relationships among divergent species using *cis*-regulatory sequences. Phylogenetic analysis is typically performed using coding (or at least transcribed) sequences, however *cis*-regulatory sequences contain phylogenetic information that can be used to reconstruct species relationships. For instance using both maximum likelihood and maximum parsimony analysis, Kim (2001) demonstrated that conserved sequences in an enhancer element controlling stripe expression of the *hairy* gene in *Drosophila* generates accurate species topologies, even though the most highly conserved segments of the enhancer may not evolve in a clock-like manner. The accurate reconstruction of species relationships from *cis*-regulatory sequence may be expected, considering that substitution within highly conserved regions of enhancers is thought to be rare and typically occur only once per site on a given phylogeny (Ludwig *et al.* 1998). Given these results, it is conceivable that future attempts to reconstruct phylogenetic relationships will take advantage of *cis*-regulatory sequences, considering the large amounts of noncoding DNA available for such analyses in eukaryotic genomes.

Current phylogenetic analyses using *cis*-regulatory sequences take advantage of different aspects of the same information decoded in phylogenetic footprinting studies. Both types of evolutionary analysis are alignment-dependent and first require that the sequences studied be present in all taxa before the data can be converted into footprints or distance matrices/tables of informative sites. Viewed from this perspective, it is perhaps not surprising that phylogenetic footprints contain phylogenetic information. However, as we have seen above, the process of binding site turnover leads to lineage-restricted information that cannot be captured in a standard multiple alignment framework. The possibility of lineage-restricted structural information suggests that there may be an additional form of lineage-restricted phylogenetic information contained in *cis*-regulatory sequences that is not captured in multiple alignment based phylogenetic inference. Using the structural rearrangement of binding sites in *cis*-regulatory sequences for phylogenetic inference can be thought of as analogous to phylogenetic inference using larger-scale genomic rearrangements such as chromosome inversions of **intron** gain/loss.

To illustrate this point, we return to the *eve* stripe two enhancer in the four species previously analyzed from the standpoint of binding site prediction. Our claim is that the ongoing process of binding site turnover should lead to a unique form of phylogenetic information contained in the simple presence or absence of sequences under constraint. If lineage-restricted gain of binding sites occur on internal branches and become functionally incorporated into the *cis*-regulatory sequence, they will be shared by descendants and therefore provide shared derived characters. The possibility of unique, shared–derived functional constraints can be visualized by multiple pairwise comparison of

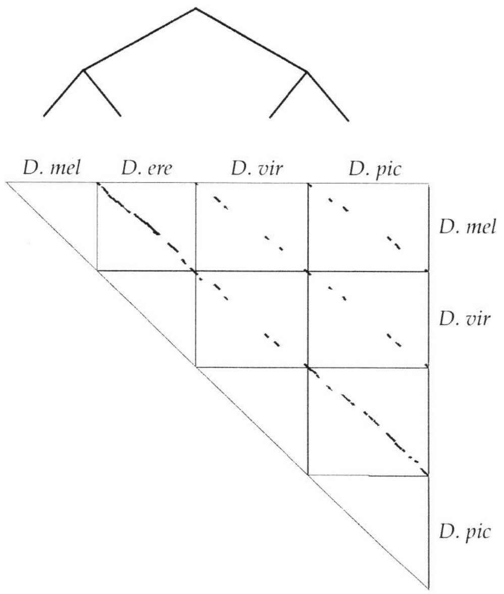


**Figure 5.5** Multiple pairwise analysis of homologous *eve* stripe two enhancer sequences in *Drosophila*. Each cell representes a different pairwise dotmatrix comparison between the species indicated by column and row. The four species compared have the following phylogenetic relationship: ( (*D. melanogaster, D. erecta*), (*D. virilis, D. picticornis*) ). Comparisons that transit the internal node of the phylogeny reveal the set of sequences that are common to all four species. Comparisons between *D. melanogaster* and *D. erecta* or between *D. virilis* and *D. picticornis* reveal sequences which a potentially under lineage-restricted constraint. The presence of lineage-restricted sequences may be used as shared derived characters in phylogenetic analysis.

the *eve* stripe two enhancer (Fig. 5.5). This diagram shows all pairwise dot-matrix comparisons of the *eve* stripe two enhancer between four *Drosophila* species with the following phylogenetic relationship: ((*D. melanogaster, D. erecta*), (*D. virilis, D. picticornis*)). As can be seen in Fig. 5.5, comparison of the two pairs of species that are most closely related (*D. melanogaster* versus *D. erecta*, and *D. virilis* versus *D. picticornis*) yields more shared sequences than any of the comparisons around the internal node of the phylogeny. These sequences are likely to be at least weakly constrained, given the fact that there are sequences in both comparisons which have completely diverged and are thus evolving at a faster rate. Thus, in addition to the set of sequences that are shared by all four species (shown by comparisons which transit the internal node of the phylogeny in Fig. 5.5), there is evidence of sequences under constraint that are unique to each **clade** of the phylogeny. Moreover, closer inspection reveals that these lineage-restricted sequences appear to be unique, suggesting that this phylogenetic information encoded by binding site turnover may not be seriously undermined by **homoplasy**. If it can be demonstrated that *cis*-regulatory sequences generally exhibit this pattern of divergence, it may be possible to develop a new class of phylogenetic methods to specifically analyze this information. These methods would seek to identify phylogenetic information in the *shared presence* of sequences in multiple pairwise alignments rather than in substitutions found in sequences present in a multiple alignment.

## Prospects for the analysis of developmental systems

The analysis of the *eve* stripe two enhancer in closely related *Drosophila* species, and other similar studies, have revealed the complexity of *cis*-regulatory evolution in the context of expression patterns that are more or less unchanged. More significantly, these studies presage the difficulties that are yet to be encountered in studying *cis*-regulatory evolution that has resulted in changes in patterns of gene expression. It is clear that alignment-based approaches, which are only able to capture a subset of the sequences responsible for conserved patterns of expression, will have difficulty identifying sequences responsible for altered patterns of expression. Moreover, a model of binding site turnover despite conservation of expression pattern suggests that the functionally relevant changes responsible for the divergence of expression patterns might be buried within or masked by other sequence changes that have no net effect on the pattern of expression. Clearly, genetic and transgenic analysis will remain necessary tools to identify the relevant change(s) responsible for the divergence of expression patterns. The ongoing process of binding site turnover suggests that, as the divergence between species compared increases, *cis*-regulatory changes will become increasingly difficult to analyze. Nevertheless, such analyses will be necessary if we are to understand more precisely how changes in *cis*-regulatory sequences contribute to evolutionary changes in development and morphology.

C.M.B. would like to thank Steve Arnold for inspiring the structure of this chapter with his approach to teaching evolutionary biology by presenting the "Tools of the Trade." The authors would also like to thank Roger Hoskins, Bruce Lahn, Jorge Vieira, and two anonymous reviewers for helpful comments on the manuscript.

## Further reading

Dover, G (2000). How genomic and developmental dynamics affect evolutionary processes. *Bioessays*, **22**, 1153–1159.

Gibson, G (2000). Evolution: hox genes and the cellared wine principle. *Current Biology*, **10**, R452–R455.

Johnson, NA and Porter, AH (2001). Toward a new synthesis: population genetics and evolutionary developmental biology. *Genetica*, **112–113**, 45–58.

Marshall, CR, Orr, HA and Patel, NH (1999). Morphological innovation and developmental genetics. *Proceedings of the National Academy of Sciences USA*, **96**, 9995–9996.

Palopoli, MF and Patel, NH (1996). Neo-Darwinian developmental evolution: can we bridge the gap between pattern and process? *Current Opinion in Genetics and Development*, **6**, 502–508.

Purugganan, MD (2000). The molecular population genetics of regulatory genes. *Molecular Ecology*, **9**, 1451–1461.

Stern, DL (2000). Evolutionary developmental biology and the problem of variation. *Evolution*, **54**, 1079–1091.

Tautz, D (2000). Evolution of transcriptional regulation. *Current Opinion in Genetics and Development*, **10**, 575–579.

True, JR and Haag, ES (2001). Developmental system drift and flexibility in evolutionary trajectories. *Evolution & Development*, **3**, 109–119.

Zuckerkandl, E (1997). Neutral and non-neutral mutations: the creative mix—evolution of complexity in gene interaction systems. *Journal of Molecular Evolution*, **44**(Suppl. 1), S2–S8.

CHAPTER 6

# Genomics and evolution: the path ahead

**Fernando González-Candelas, Julie C. Ho, Alexandra M. Casa, and Stephen Kresovich**

The draft of the human genome sequence published by two separate groups in February 2001 (International Human Genome Sequencing Consortium 2001; Venter *et al.* 2001) marks the beginning of a new era for biology and promises applications that should benefit humankind for centuries to come. Thousands of researchers from different countries have mobilized towards a common goal and now share multimillion-dollar budgets. Although initially criticized by organismal and environmental specialists for its concentration of resources and research efforts, this initiative is now widely acknowledged for its far-reaching, positive impact on all areas of biology, including evolutionary biology.

**Genomics**, the systematic study of all or a large portion of an organism's genome, can provide analytical and methodological tools to identify, document, and measure the extent of genetic variation inherent to living systems at an unprecedented pace. In contrast to single gene analyses, genome-based approaches provide a more comprehensive understanding of the evolutionary history of organisms. Because it is unfeasible to study all gene sequences from any given organism, information, on the evolutionary relationships among organisms is essential in order to interpret DNA sequence data. Therefore, post-genomics biology depends on critical inferences drawn from evolutionary history. Integration of information from both genomics and evolutionary biology is necessary to catalogue, document, and preserve biodiversity, the result of evolutionary processes that have occurred for over 3500 million years. We believe that genomics research will reach beyond basic science to include applications in the conservation of the evolutionary potential of species and populations for the future. In the rest of this chapter, we review the synergism between genomics and evolutionary biology, and how the use of genomic approaches to study the breadth of life may influence evolutionary biology in particular.

## Contributions from evolutionary biology to genomics

Genomics is one of the latest biological disciplines. The beginnings of this field date from 1980 when Botstein and colleagues proposed the study of whole genomes as opposed to the analysis of individual genes (Botstein *et al.* 1980). The discipline gained momentum in 1986, when the Human Genome Initiative, later designated as the Human Genome Project, was launched. In 1995, the first complete genome sequence from a free-living organism, the bacterium *Haemophilus influenzae*, was published (Fleischmann *et al.* 1995). Since then, organisms from the three domains of life, Archaea, Bacteria, and Eukarya, have had their genomes sequenced.

Genomics was initially perceived as an extreme form of reductionism, that is, the theory that complex phenomena can be explained by dissection and analysis of their simplest, most basic underlying mechanisms. The reductionistic approach has provided a deep understanding of many biological processes and systems. However, the central role of genomics in biological research—the integration of studies in gene function, inheritance, and evolution—will be accomplished only through a synthetic approach for the understanding of complex systems. Because evolutionary theory offers a comprehensive framework to biology, genomics will certainly benefit from this discipline, while simultaneously offering new information that will help us refine evolutionary theory. Evolutionary biology, however, has much more to offer to genomics than just a theoretical framework. Here, we outline several directions in which we believe there is a real chance of significant advance in the foreseeable future. These contributions can be broadly grouped in terms of (1) comparative mapping and analysis, (2) introduction of population-level approaches, and (3) choice of model systems.

## Comparative mapping analysis

One of the basic principles of evolutionary theory is that all life forms are derived from a single organism. Many of the applications and strengths of genomics as a research tool are based on this tenet and it would be impossible to fulfill many of the declared goals of genomic research without applying this principle. In order to be useful, a genome sequence must incorporate information on its structure (coding or noncoding) and function (protein or RNA-encoded). The process of appending structural, functional and other pertinent information (e.g. comments, literature references) to a genome sequence is referred to as "**annotation**." A sequence that contains all of these elements is said to be "**annotated**." Increasingly often, the process of annotation starts by submitting a raw sequence of DNA bases to a database from which a list of closest matches is generated with the support of programs such as Basic Local Alignment Search Tool (BLAST), FastA, or any of their variants. Assuming the annotations already in the database are correct, inspection of previously characterized sequences offers a first clue as to the likely function of the newly obtained sequence. It should be noted, however, that this description is clearly an oversimplification of a more complex task. In eukaryotic organisms, for example, identification of coding regions and **exon/intron** boundaries can be complicated by **alternative splicing** as well as post-transcriptional and translational modifications. Regardless of its weaknesses, however, comparative genomic analysis still provides the most powerful, simplest, and safest tool both to identify genes and other components of genomes and to predict their functions and interactions (Koonin 2001). Correct application of the **comparative method** requires recognition of **homologous**, in this case **orthologous**, sequences and an appropriate phylogenetic analysis from which inferences on evolutionary history can be made (Thornton and DeSalle 2000).

Functional–phylogenetic analysis allows inferences to be made regarding the preservation or divergence of function among members of gene families over evolutionary time scales. It also helps to identify **motifs** (i.e. short, conserved regions in protein sequences possessing important functional or structural roles), and to understand the basis for functional specialization of different gene families, gene subfamilies and specific phenotypes based on presence or absence of specific genes (Kawashima *et al.* 2000). This approach can also be useful for the study of the biology of difficult-to-culture microorganisms for which only the genomic sequence is available (Wagner 2000b).

**Comparative genomics** is a key foundation for the interpretation of data gathered by genome sequencing projects (Clark 1999). The following are some of the research areas that will benefit from comparative mapping:

***Linkage*** *analysis.* Construction of a genetic or physical map is a necessary step in the use of an organism as a model system, as it facilitates comparison with other species. Maps contain information regarding the relative positions of genes and the distance between them, inferred from their recombination rate (**genetic maps**) or physical location (**physical maps**). Once a map is available, it is then possible to compare the results of mapping experiments with different species and to test whether the environment of a certain gene has been altered or has remained constant since divergence from the last common ancestor of both species (Nadeau and Sankoff 1998). These comparisons may provide information on gene control regions, which are not free from selection despite belonging to the noncoding fraction of the genome (Wang *et al.* 1999).

***Polygenic** inheritance.* The study of multigenic, quantitative traits greatly benefits from comparative genomics (Walsh 2001). Given the large number of crosses and/or progeny necessary to identify **quantitative trait loci** reliably and the genetic basis underlying phenotypic variation, the study of polygenic traits is particularly difficult in species with limited numbers of offspring. The joint application of genomics and quantitative genetic analysis will enable the dissection of the contribution of each individual quantitative trait locus to variation in the corresponding trait.

*Identification of genes and regulatory sequences.* In every newly sequenced genome there is a substantial fraction of potential coding regions whose products have an unknown function. Furthermore, for many of those genes with identified function there is no experimental verification of their functional assignment. Application of prediction programs that use phylogenetic information is still rare but can allow more accurate predictions in automated searches. This kind of information could also help to establish the limits of coding regions and the location of regulatory sequences (Karlin *et al.* 1998). Annotation of sequencing projects will greatly benefit from more thorough phylogenetic analyses, using tools such as PYPHY (Sicheritz-Pontén and Andersson 2001), than simply searching for similarities in databases using BLAST or FastA.

*Gene function.* One of the newest computational methods for establishing functional links between proteins is the **phylogenetic profile** (Eisenberg *et al.* 2000). This is a description of the pattern of presence/absence of a particular protein in the genomes of different organisms. When two proteins have the same phylogenetic profile, the inference is made that there is a functional link between them. This apparently simple method is complemented by others, such as analysis of fusion patterns or the gene neighbor method that bring more power to the former. Although it is unlikely that all genes from a target species will have **orthologues** in a phylogenetically close model species, we can compare gene and Expressed Sequence Tags (EST) databases and assign function based on conservation of functional domains or motifs (Sapir *et al.* 2000).

## Population-level approaches in genomics

A second important contribution from evolutionary biology to genomics derives from the introduction of a population-based approach to understanding the relationship between genetic and phenotypic variation (Templeton 1999). For more accurate predictions of individual risks this approach emphasizes the extent and nature of variation in different populations before a given variant is associated with a certain disease or risk of disease. This population-level perspective will also help to elucidate the effect of the environment on the phenotype (i.e. genotype × environment interactions), hence allowing a wider range of therapies and treatments to prevent progress of disease in individuals with "risky" genotypes.

A population-based approach is also valuable for analyzing the genomes of prokaryotic organisms that, despite reproducing asexually, still retain remarkable levels of genetic variation. For example, based on population studies, Lan and Reeves (2000) have proposed that a bacterial species could be defined in terms of a shared set of genes, called "core" genes, with a given degree of similarity. Once a set of core genes for the species is defined, it could be used to establish whether this core set has similar or different composition in different groups of prokaryotes. Technologies such as subtractive hybridization or DNA microarrays will expedite the identification of this set of genes in any genome.

## Choice of model organisms

Finally, evolutionary biology can contribute to genomics in the choice of model organisms to be sequenced, for instance in developing models of gene function. In this sense, there is a continuous, positive feedback, since the better, most complete genomic information from a variety of organisms is available, the better phylogenetic reconstructions and inferences on the evolutionary processes that have operated in them will be possible. In consequence, better informed decisions may be taken and efforts allocated more efficiently. A neat example is provided by the sequencing of the puffer-fish (*Fugu rubripes*) genome, suggested by S. Brenner and almost completed, as this organism has the most compact genome (merely 350 Mb) among vertebrates.

## Contributions from genomics to evolutionary biology

Genetics has played a central role in evolutionary thinking since the gap between Mendelians and Darwinians was closed by the work of Fisher, Haldane, Wright, and others. Will genomics make a similar contribution to evolutionary theory in the future? While genetics stimulated research on a few model organisms (man, mouse, fruit fly, the nematode *Caenorhabditis*, yeast, and the plant *Arabidopsis*), the breadth of genomics research encompasses hundreds of species. Hence, genomics and evolutionary biology will strengthen links between studies of biological systems at very different time and organizational scales (Murray 2000). Although this fusion will contribute to a more integrated study of biology, data gathered using the tools of comparative genomics can be combined with traditional morphological and ecophysiological data to study evolutionary processes.

Despite recent technological advances in genomic analyses, the impact of genomic technology on evolutionary studies remains minimal (see Box 6.1 for some genomic techniques relevant for evolutionary studies). The comparative method, however, represents one of the most powerful approaches for the study of biological systems and, in the near future, it will certainly be used at the genomic level at an increasing pace. What kinds of hypotheses will we be able to propose and test based on this new information? Will they affect evolutionary theory and to what extent? We believe that the answer to these questions is "yes" and "to a very large extent." Among the main accomplishments of large-scale evolutionary genomics are: complete sequencing of large regions from many taxa, more precise phylogenetic reconstructions and inferences of evolutionary processes within and among specific genomic

**Box 6.1 Genomic tools for the evolutionary biologist**

Advances in the sequencing of the human genome have only been possible through the development of new techniques and analytical tools, both for data production and processing. Many of these have a wide range of applications, and some are especially relevant to evolutionary biology. Weaver (2000) provides an updated review of many of these techniques, of which we will here mention the most relevant.

Genomic data relevant for evolutionary biology are those based on variation within and between different taxonomic levels. From gene order and chromosomal location to composition and frequency of **haplotypes**, there are hundreds of techniques that reveal variation, but the ones most directly derived from the genomic revolution are those showing variation at the nucleotide level. Desirable features in the new techniques are high-speed, high-throughput, affordability, relatively low technical demand, and ease of transfer to different organisms. As with DNA sequencing, many of the new techniques were first developed for the analysis of variation in the human genome and to map or characterize specific genes of medical interest. The first significant development was the shift from gel-based to capillary-based automated sequencers, widely used not only for sequencing but also for genotyping and screening of different single and multiple loci markers. The combination of different fluorophores and multiplexing enables high-speed genotyping of several loci from a single individual simultaneously.

There are two main alternatives to capillary sequencers for high-throughput genotyping.

The first one is based on micro-arrays with the development of techniques such as variant detector arrays (VDAs) (Haluskha *et al.* 1999) and micro array hybridization data (MAHD) (Wu *et al.* 2001). Also, capillary sequencing and micro-array technology have been combined in capillary array electrophoresis (CAE) (Shi *et al.* 1999), a new high-speed, high-throughput method for DNA sequencing, and in capillary conformation sensitive gel electrophoresis (capillary CSGE) (Rozycka *et al.* 2000), used to detect mutations with the potential of screening many mega bases of DNA per day.

Variation detected by these techniques usually corresponds to **single nucleotide polymorphisms** and/or to short **indels**. For both, it is necessary to detect and design appropriate primers or probes, which are usually derived from genome sequencing projects. However, for those species with little or no genome sequence information, there are some alternatives to reveal potential polymorphic sites. Denaturing HPLC (dHPLC) (Underhill *et al.* 1997), MALDI-TOF mass spectrometry based on chips (Buetow *et al.* 2001), and diversity array technology (DarT) (Jaccoud *et al.* 2001) are among the most popular methods for the rapid identification, development, and characterization of single nucleotide or short indel polymorphism.

The need for new methods to reveal variation, however, has to be matched by more powerful and faster methods and equipment for data handling and analysis. This is becoming the most pressing bottleneck in the production line, because most developments are being oriented to specific applications with a high potential return, and because evolutionary biology interests are usually different in scope. Nevertheless, inexpensive personal computers integrated into a single cluster (Beowulf or similar), popularization of parallelization techniques, and new algorithms for faster searching and evaluation of alternative topologies have enabled the application of computationally demanding phylogenetic reconstruction methods such as parsimony and maximum likelihood to increasingly large data sets, thus providing more accurate phylogenetic trees based on more and larger DNA sequences.

sites, comparative analyses of evolutionary processes in different genomic sites and regions, linkage of sequence evolution with structure and function, increased accuracy of functional genomic predictions due to larger data sets, identification of variation for population genetic studies, and evaluation of alignment and phylogenetic reconstruction techniques (Pollock *et al.* 2000).

As a consequence, not only more detailed and accurate phylogenetic reconstructions will be obtained, but it will also be possible to ascertain the role of different processes in evolution. For instance, it will be possible to investigate what genomic events triggered or accompanied differentiation at different levels and in different of groups of organisms. In addition, we will be able to address questions about the genomic regions that have driven evolutionary processes and the relative importance of changes in structural and regulatory regions in early stages of speciation. Comparison of eukaryotic genomes will also allow us to study the evolution of gene families, the molecular basis of **adaptation**, and rates of deleterious mutations (Charlesworth *et al.* 2001). We will also be able to determine genome differences that are due to selection (adaptation at the molecular level) and/or random **fixation** of neutral variants (Charlesworth *et al.* 2001). Kondrashov (1999) has suggested other fundamental questions in evolutionary biology that may be clarified by comparison of genomic sequences and the application of genomic technology to the study of intraspecific variation, including

- Causes of suboptimality, that is, the presence of a large number of slightly deleterious mutations in organisms as a consequence of **drift** and mutation pressure
- inference of **homology** or nonorthologous gene displacements;
- origin of nonhierarchical distributions in character states, mainly due to horizontal gene transfers (i.e. movement of genetic material between separate species) and **homoplasies**, that may confound

phylogenetic reconstructions or be indicative of adaptive processes;
• inference of ancestral character states to root phylogenetic trees;
• relative importance of **duplications**, **transpositions**, and gene and genome shuffling in generating novel allelic and phenotypic variation;
• identification and estimation of the relative importance of neutral mutations;
• interaction between directional and stabilizing selection based on evolutionary rates around the targets of selection.

We should also be able to establish the extent of structural and functional redundancy (see Lynch, Chapter 4), and whether there is any correlation between redundancy and organismal complexity. The understanding of both external and internal evolutionary forces modeling the development of organisms will help us to obtain a more comprehensive view of evolution. As a consequence, there is interest in learning what genetic systems are redundant and why, and whether redundancy is restricted to certain genes (e.g. Dover 2000).

The patchy composition of orthologues can be attributed to lineage-specific gene losses and horizontal gene transfer, the importance of which was appreciated in full only after comparison of complete prokaryotic genomes (Koonin 2001). At a genomic scale, these complexities make phylogenetic reconstruction a particularly difficult task. For example, widespread horizontal gene transfer among prokaryotes, or even among representatives of the different domains of life, blurs the delineation of microbial taxa (Doolittle 1999). Comparative genomics can be applied to better understand the distribution of horizontal gene transfer events and to test the **continuing horizontal transfer** (Jain *et al.* 1999) and **massive early transfer** hypotheses (Woese 1998). It is widely recognized that the most appropriate methods for inferring horizontal gene transfer are those based in phylogenetic analysis (Ochmann *et al.* 2000).

The integration of information on gene expression and protein and genetic interactions among organisms will certainly shed light upon many fundamental questions in evolutionary biology, such as how have certain pathways attained a similar structure while presenting different function, whereas others have evolved very different structures for a similar function? It will also be possible to design field and laboratory experiments to determine how selection for a certain phenotype has influenced gene expression. Moreover, we should be able to use genomic approaches, such as microarray analysis, for detecting DNA sequence polymorphisms and to establish associations between mutations and phenotypes (Murray 2000). Such studies will address the fundamental question of how organisms balance **robustness** with **evolvability** (Murray 2000). Understanding how evolutionary pressures put restrictions on the **modules** (a module is an autonomous folding and functional unit in a protein) can help us to understand why we find a limited number of pathways for a given function. Does this represent a historical contingency or a requirement to balance robustness with evolvability? Availability of large numbers of complete genome sequences will also shed light into the sequence of events that have led to the establishment of intimate relationships and coevolution among different species (e.g. pathogenicity or symbiosis; see van Ham *et al.*, Chapter 9), and in the evaluation of the footprints of these phenomena in the genomes of extant organisms.

The availability of genome sequences from different species will ultimately provide a catalogue of the genes present in each of them, and possibly allow the establishment of the minimum number of genes necessary to support different forms of life, from free living to endosymbiotic organisms. The comparison of minimal gene catalogues from the three domains of life will also allow the establishment of a minimum set of genes for sustaining cellular life. Comparisons between pairs of completely sequenced genomes place this number at about 250 genes (Koonin 2000). However, when multiple genomes are compared, the common set is reduced to about only 80 genes with a clear **orthology**. How many of these were present in the last common ancestor and which subsets of genes are eventually lost (or gained) with changes in lifestyle are two of the many questions to be ascertained from these comparisons.

Comparative analysis of the genomes of pathogenic organisms and their nonpathogenic relatives has already revealed the importance of the so-called "pathogenicity islands" in the determination of features associated with pathogenicity (Groisman and Ochman 1996). Occasionally, these pathogenicity islands have been acquired by horizontal gene transfer from phylogenetically distant organisms with a common habitat (Hentschel and Hacker 2001).

Ultimately, the joint comparison of a large number of genomes will provide a more complete and refined view of the tree of life, on which it will be possible to appreciate how much "pruning and grafting" has occurred throughout history. This tree will represent a very solid foundation for future biology, thus fulfilling one of the long-term desires of its founder as a scientific discipline, Charles Darwin.

## Joining forces: the preservation of biodiversity

One of the most pressing problems faced by humankind is the disappearance of **biodiversity**, i.e., the series of variants at all levels of biological organization. **Conservation genetics**, the use of genetic markers and DNA sequence information together with population genetic and ecological theory for designing and implementing conservation strategies, has achieved recognition during the last decade as an important tool for conservation efforts at the population and species levels. Aware of the importance of genetics in their programs, breeders and germplasm curators were among the first to adopt and apply genomics to their research. Evolutionary biology provides a theoretical framework to understand how species become extinct and how populations lose or maintain specific genes (Amos and Balmford 2001; Hedrick 2001). Unfortunately, many factors other than genetic variation are needed to prevent a population or a species from becoming extinct, thus leaving conservationists with an even more difficult task, since the rate of loss of biodiversity cannot be matched by preservation funding. Consequently, optimization of resources and efforts in conservation is a necessity.

As previously noted, evolutionary biology will benefit tremendously from new developments in genomics. As a consequence, the evolutionary perspective on the importance and extent of genetic variation in endangered species will gain in depth and detail. This will allow a better evaluation of the significance of evolutionary events and genetic causes of population decline, with better genetic and statistical resolution through the analysis of more loci with a better compliance of **neutral theory** postulates (see Ohta, Chapter 1). These new tools derived from genomics (i.e. markers, data for comparative analyses from DNA sequences and microarrays, etc.) have already provided examples of improved power in the analysis of past and current population events (bottlenecks, **gene flow**, nature and extent of adaptation, selection). They also allow more accurate estimates of parameters of fundamental importance for conservation, such as effective population sizes, individual relatedness, levels of inbreeding and heterozygosity, population structuring, and past and current migration rates. Furthermore, as these studies become technically easier and less expensive, they will be applied to a growing number of organisms, including many difficult-to-culture species for which there is currently little or no information at all on their genetic variability.

Similarly, a gain in quality and amount of information should also benefit germplasm collections. These collections have a dual goal: preserving genetic variation and using available genetic variants for breeding purposes in closely related or even unrelated species by either conventional or molecular breeding methods. In both cases, marker-assisted breeding relies on the availability of large numbers of highly polymorphic genetic markers that can only be obtained by using genomic methods. In addition, the availability of large numbers of molecular markers will allow precise functional analyses, thus providing a better understanding of the genetic basis for differences among strains or races for traits of agronomic interest. Also, the combination of genomic information with pedigree and morphology will permit the identification of relevant groups within a species and the easy and reliable assignment of individuals to those groups.

The joint application of complete genome sequences and evolutionary analysis will enable the launching of environmental genomic projects in which surveys of biological diversity in different ecosystems will permit not only the identification of organisms but also the discovery of relevant biological features from their complete or partial genome sequences (Cole *et al.* 1998; Wagner 2000b). This is the goal of the project proposed by Pollock *et al.* (2000), which will use genomic technology to produce large amounts of sequence data of direct interest to evolutionary or conservation biology. These authors postulate to complement comparative analysis with the collection of genome-scale data sets of relevant species with no need to previously isolate, **clone**, or sequence. Powerful computational tools and equipment, already available and tested in the reconstruction of large genome sequences, will be used to reconstruct the original sequences and to analyze them (Pollock *et al.* 2000).

Biology is entering a new age, one in which technical and conceptual developments will provide a truly unified understanding of life, its processes and history, and in which function, development and evolution will be as integrated in a single discipline as they are in living organisms. Genomics and evolutionary biology will certainly be at the core of this new biology.

F.G.C. was supported by project PB1998-1436 from Ministerio de Ciencia y Tecnología (Spain).

## Further reading

Brown, TA (2002). *Genomes*. 2nd edition. Wiley-Liss, New York.

Campbell, AM and Heyer, LJ (2003). *Discovering genomics, proteomics and bioinformatics*. Cold Spring Harbor Laboratory Press, Cold Spring Harbor, New York.

Gibson, G and Muse, SV (2001). *A primer of genome science*. Sinauer, Sunderland, MA.

Koonin, EV and Galperin, MY (2002). *Sequence–evolution–function: computational approaches in comparative genomics*. Kluwer Academic Publishers, New York.

Sankoff, D and Nadeau, JH (eds) (2000). *Comparative genomics*. Kluwer Academic Publishers, New York.

Sensen, CW (ed.) (2002). *Essentials of genomics and bioinformatics*. John Wiley & Sons, New York.

PART II

# Molecular variation and evolution

CHAPTER 7

# The evolution of virulence in AIDS viruses

Edward C. Holmes

**Acquired Immune Deficiency Syndrome** (AIDS) was first described in 1981 when clusters of gay men living in a number of American cities were found to be suffering from a variety of rare opportunistic infections caused by a deficiency in **CD4+ T-helper cells**, a crucial part of the human immune system. Within two years the cause of AIDS, a retrovirus later christened the **human immunodeficiency virus type-1** (HIV-1) was isolated, with a related virus, HIV-2, discovered a few years later. Since this time HIV has spread worldwide causing a global pandemic. Current estimates are that approximately 20 million people have already died of HIV with a further 36 million people infected worldwide, and most transmissions occurring through heterosexual contact (Piot *et al.* 2001). Sub-Saharan Africa is worst affected, where the **prevalence** in adults averages almost 9 percent, although in Lesotho, Swaziland, and Zimbabwe the figure rises to 25 percent, and to 36 percent in Botswana.

Despite the great strides made in understanding the basic biology, epidemiology, and origin of HIV, we still lack a full knowledge of some of its most fundamental features and an effective vaccine has yet to be developed. One question which forms a backdrop to a great deal of AIDS research is why some HIV sufferers can develop AIDS within a couple of years of acquiring the virus, while others appear to be healthy after almost 20 years of infection? The question of what explains differences in clinical outcome can also be asked on an evolutionary time scale; why does human HIV cause immune collapse, yet the viruses closely related to HIV and found in various nonhuman primates—the simian immunodeficiency viruses (SIVs), also known as the **primate lentiviruses** (PLVs)—cause no disease in their hosts? Hence, revealing the mechanisms that underpin HIV **virulence** is essential for a full understanding of the etiology of AIDS disease and for how we might eventually control it. On a more speculative level, knowing what controls HIV virulence may make it possible to predict how the virus might change in the future, for the better or for the worse.

Although definitions of virulence abound, which differ in their precision, virulence in the primate lentiviruses can simply be taken to mean the rate of progression from initial infection to immunodeficiency disease. The aim of this chapter is to judge the various theories that have been put forward to explain the evolution of HIV virulence, particularly through the use of evolutionary genetics. This will raise more questions than answers, but does highlight areas for future research. Finally, although it is important to remember that virulence is the outcome of an evolutionary interaction between both virus and host, this chapter will focus most on the virus where the relevant genetic data are more forthcoming. More general discussions of the evolution of virulence are available elsewhere (e.g. Ebert 1999).

## HIV dynamics and diversity

Notwithstanding the uncertainty over why progression times to AIDS vary so extensively, it is clear that the twin engines of replication and mutation play a fundamental role in HIV life history. On the replication side, HIV preferentially infects cells bearing CD4+ and chemokine receptors, most notably macrophages and T-helper cells, both of which are key components of the human immune system. It is the long-term loss of these critical cells—often by apoptosis—that eventually results in AIDS. The rapidity of HIV replication is staggering. Each day some $10^{10}$ viral particles are produced within an infected individual and the average generation time of HIV (from infection of cell, replication, assembly of mature virus, to infection of new cell), appears to be only about 2.5 days, so that an infected individual experiences roughly 150 new viral generations each year (Perelson *et al.* 1996). After a sustained period of viral replication the number of T-helper cells eventually declines to such low levels that the body is unable to see off the opportunistic infections that ultimately result in full-blown AIDS.

HIV is equally impressive in the speed with which it mutates. Mutation rates for retroviruses like HIV have been estimated to be at approximately 0.2 errors per genome replication and are due to the low fidelity of replication with reverse transcriptase (Drake and Holland 1999). This means that many viral progeny will differ by mutation from their parental genotype and many of these mutations will be deleterious. One intriguing proposal is that the mutation rate of HIV is so high that antiviral drugs which increase this rate even further may lead to severe **fitness** losses and even eradication because the virus is taken over a critical "error-threshold" (Loeb *et al.* 1999). Such a high mutation rate, coupled with the short generation time, results in very high rates of nucleotide substitution, usually in the region of $10^{-3}$ substitutions per nucleotide site per year, which is approximately one million times faster than the rate seen in human nuclear DNA. This "fast-forward" evolutionary process has clearly given HIV enormous adaptive potential and the genetic diversity of the virus within individual hosts can be substantial, encompassing variants that differ in their capacity to respond to antiviral therapy, in their cell tropism and also their virulence.

The genetic diversity of HIV within hosts is predictably magnified at the population level. On a global scale, this is reflected in the presence of phylogenetic clusters of viral isolates known as "groups" or "subtypes" depending on their level of divergence (Fig. 7.1). Three groups of HIV-1 have been identified, the most common of which is denoted "M" (main), with two lower prevalence groups from West–Central Africa denoted "O" (outlier) and "N" (non-M, non-O). These phylogenetic groups most likely represent independent transfers from nonhuman primate reservoirs (Hahn *et al.* 2000). While groups O and N have restricted distributions, the M group has spread worldwide and has been further subdivided into a series of smaller clusters known as subtypes (denoted A–D, F–H, J, and K). The phylogenetic structure of subtypes seems to reflect dispersal patterns of the virus over the last 30 years. For example, subtype B predominates in North America and Europe and contains the first viral strains isolated in the 1980s, while subtype C is extremely common in Southern Africa and India. A wider range of subtypes are found at high frequencies in sub-Saharan Africa, most notably A, C, and D (as well as various recombinant forms—see later), which reflects the longer history of HIV-1 in this region. Although subtypes have provided a useful framework to study genetic diversity in HIV-1 and track dispersal patterns, more recent phylogenetic studies have indicated that their structure to a large extent reflects **founder effects** and sampling artifacts (Rambaut *et al.* 2001). Hence, there is in reality a continuum of genetic diversity in HIV-1 in West-Central Africa where the virus first emerged, but the emigration of viral strains to other localities led to isolated regional evolution that in turn produced the discrete phylogenetic groups that we recognize today as subtypes. Subtypes have also been described for HIV-2, although this virus is generally restricted to West Africa. To date, seven subtypes have been identified and, like the groups of HIV-1, some clearly represent independent transfers from species of nonhuman primate to humans.

## Long-term trends in the evolution of HIV virulence

### Has HIV virulence decreased or increased?

The evolution of virulence has received considerable attention from population biologists. One view of virulence evolution, now perceived as overly simplistic, is that virulence gradually declines as the pathogen becomes better adapted to its host.

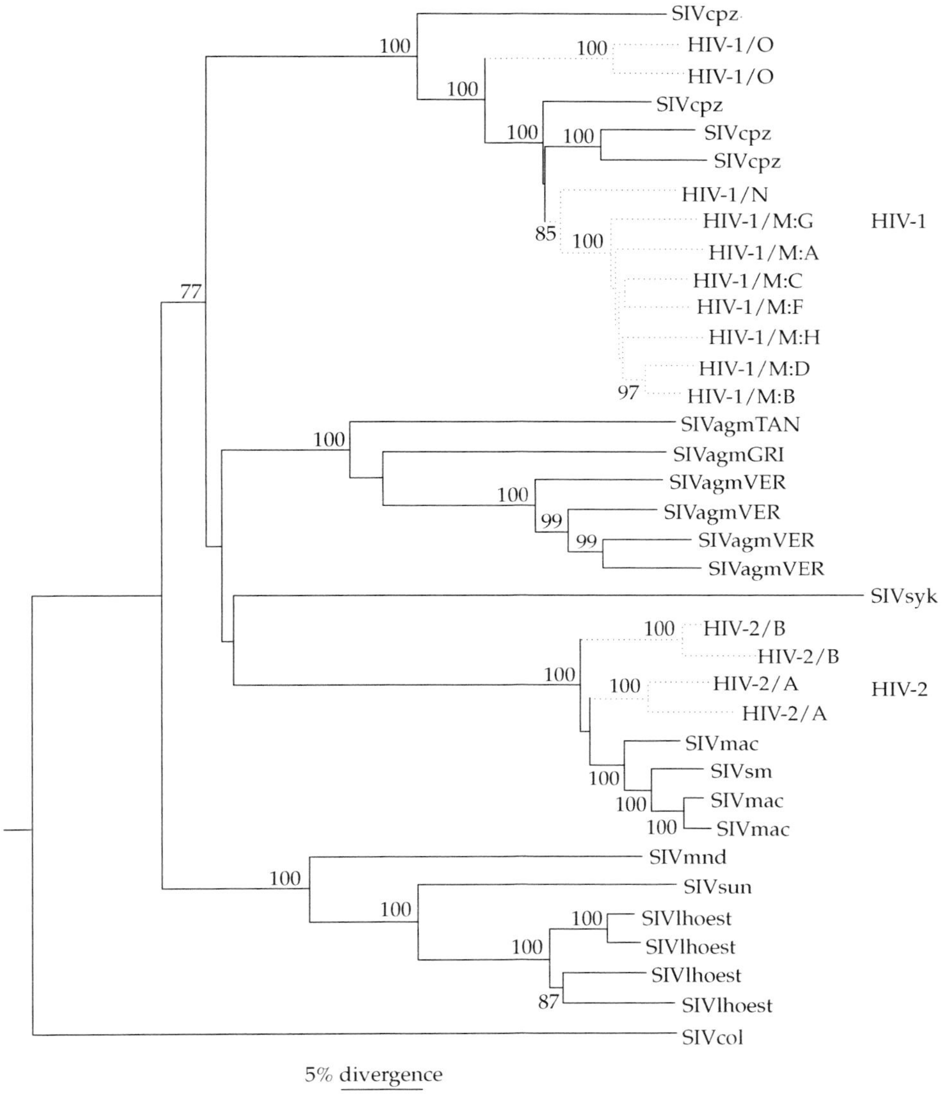


**Figure 7.1** Phylogenetic tree of a representative sample of primate lentiviruses. The tree was inferred from *pol* (polymerase) gene amino acid sequences using the neighbor-joining method under the JTT distance correction statistic. Bootstrap values for key nodes (percentages from 1000 replications) are shown and the tree is mid-point rooted. Lightly shaded branches depict the human forms of the virus. Abbreviations for viruses are as follows: HIV-1/M, HIV-1 group M (various subtypes shown); HIV-1/N, HIV-1 group N; HIV-1/O, HIV-1 group O; HIV-2/A, HIV-2 subtype A; HIV-2/B, HIV-1 subtype B; SIVagmGRI, grivet monkey; SIVagmTAN, tantalus monkey; SIVagmVER, vervet monkey; SIVcpz, chimpanzee; SIVcol, guereza colobus monkey; SIVlhoesti, L'Hoest monkey; SIVmac, macaque; SIVmnd, mandrill; SIVsm, sooty mangabey; SIVsun, sun-tailed monkey; SIVsyk, Sykes' monkey.

Evolution follows this course, it was assumed, because host death is maladaptive for the virus, resulting in a simple negative correlation between the length of time a virus has infected a host species and the severity of disease. A counter hypothesis is that virulence is simply a coincidental or chance by-product of viral infection, and hence not a trait that has been set by natural selection. Although it is impossible to fully rule out this hypothesis for HIV, that lentiviruses are so common in African primates and differ in their ability to cause disease hints that virulence is likely to be an evolved property.

Most studies of the evolution of virulence in recent years have focused on the relationship between virulence and transmission rate. Under these models, the currency of evolutionary success for the pathogen is the **basic reproductive rate of the pathogen** ($R_0$), defined as the average number of infections caused by a single index case in an entirely susceptible population, rather than simply host survival. In some cases, virulence may be directly beneficial to the virus because it increases transmission rate, and hence reproductive success. Alternatively, there may be a trade-off between virulence and transmission rate, as both are key components of viral fitness and linked by functional constraints, so that the level of virulence attained is determined by a more complex evolutionary interaction (Ebert 1999).

Given this theoretical backdrop, it is perhaps surprising that the most basic observation regarding long-term patterns of virulence in the primate lentiviruses is that there does appear to be a negative correlation between the length of time a species has been infected and the level of virulence, although there are few independent data points. In particular, the most serious manifestation of primate lentivirus infection—AIDS—is only known in humans and macaque monkeys. In both these cases the virus is clearly a recent acquisition from another primate host; HIV-1 from common chimpanzees (*Pan troglodytes*) and both HIV-2 and the macaque form of the virus (SIVmac) from sooty mangabey monkeys (*Cercocebus torquatus atys*). All other primate lentiviruses seem to have no adverse affects on the health of their hosts, although studies of disease in wild populations are sorely lacking. As such, these data are compatible with the simple view that virulence in primate lentiviruses gradually declines through time as the virus becomes better adapted to its host. Consequently, AIDS may signify a period of short-term **maladaptation** between the virus and its human host and HIV virulence may decline in the future as the evolutionary process plays out.

While this hypothesis seems to fit the data, there are a variety of reasons why it is likely to be overly simplistic. The first cautionary tale concerns HIV-2. Because individuals infected with HIV-2 progress to AIDS more slowly than those infected with HIV-1 (Marlink *et al.* 1994) it can reasonably be claimed that HIV-2 has a lower virulence than HIV-1, although the virological basis of this difference is unknown. This being the case, it could also be argued that HIV-2 has been associated with humans for longer, and hence is further along the evolutionary trajectory that leads to avirulence. However, the depth of HIV-2 lineages on phylogenetic trees is roughly equivalent, if not less, than those of HIV-1 (Fig. 7.1), indicating that HIV-2 is unlikely to have been associated with human populations for significantly longer than HIV-1. More fundamentally, there is no evidence from epidemiological records that the numbers of AIDS deaths associated with HIV-2 has declined through time, which would be expected if virulence has also decreased.

The second warning shot is that with the data available at present it is entirely possible that the virulence of HIV has actually *increased* since its initial appearance in human populations. This can be argued from the disparity between the inferred time of origin of HIV-1, which may predate the Second World War (Korber *et al.* 2000), and the well documented appearance of AIDS in Africa in early 1980s. Given the usual rates of progression from primary infection to AIDS, which are rather shorter in Africa than in the Western world, the first AIDS cases should have started to appear 5–10 years after the initial cross-species transmission event(s). Although there are a small number of case reports from the late 1950s to the 1970s that have been claimed to represent AIDS, the general absence of this disease in Africa until the 1980s could mean that HIV was unable to cause AIDS for the early years of its spread, so that virulence has increased toward the present. Such a notion gains some support from epidemiological theory which states that higher levels of virulence can be attained in larger and denser host populations, so that HIV virulence could have increased when the virus moved from rural to urban populations in Africa. While this "reverse" model of virulence evolution is a theoretical possibility, it is equally plausible that HIV has always had the potential to cause AIDS and that early cases of serious disease were simply misdiagnosed as one of a myriad of other infections that afflict African populations, or only occurred in isolated localities. Even in Africa today it is estimated that less than 5 percent of all AIDS cases are reported (De Cock 2001). Consequently, for the first decades of the HIV epidemic in Africa the virus may have been spreading in isolated rural communities with such a low prevalence that it went unnoticed. It was only when it reached urban Africa, and there was a specific disease to look for, that AIDS was recognized.

## HIV virulence and transmission rate

As discussed above, a potentially powerful way to understand the long-term evolution of HIV virulence is to establish the nature of the relationship, if any, between virulence and transmission rate. For HIV, this to a large extent means an examination of the link between virulence, transmission rate and **viral load** (intra-host viral population size). There are two peaks of viral load during an HIV infection at which point the host is most infectious; one at the initial **primary infection** which lasts for up to 3 months and corresponds to the time when the host has yet to mount an effective immune response, and the second during AIDS itself, which may last for years, when viral loads are high because the host immune system has collapsed. During the long asymptomatic phase between these two peaks the host immune response is relatively powerful, which means that viral loads are orders of magnitude lower than their peak values, so that the probability of transmission is likewise reduced. As a high viral load will evidently increase the probability of transmission it could be argued that AIDS directly increases viral fitness because it corresponds to a sustained period of high infectiousness. AIDS would then be an adaptive mechanism to increase the reproductive rate of HIV. However, while some HIV transmissions will have undoubtedly involved people in the AIDS stage of disease, the adverse clinical affects of symptomatic HIV infection will in general inhibit behaviors which might lead to viral transmission. As such, it is difficult to envisage that AIDS is a selectively determined trait to increase viral reproductive rate.

A very different relationship between virulence and transmission rate in HIV has been proposed following a consideration of human sexual behavior. Ewald (1991) suggested that HIV-2 has evolved lower virulence because the virus originated in West Africa where there is reportedly less sexual partner exchange than in other parts of this continent. Under these circumstances, natural selection will favor a longer progression time to AIDS, as this will increase the chance that new sexual partners are encountered during the lifetime of an infected host (Ewald 1999). Although this hypothesis has been questioned by many, perhaps the most compelling evidence against it is that HIV-1 types N and O are also of West-Central African origin, and presumably infecting populations with similar rates of sexual partner exchange as HIV-2, but as yet there is no evidence that they differ in virulence from viruses of the highly pathogenic M group of HIV-1.

Overall, the data available at present suggests that there is no direct relationship between virulence and transmission rate in HIV. This can be argued from an analysis of when during the lifetime of an infected host most transmissions occur. In particular, there are now a number of pieces of evidence which suggest that the initial weeks of primary infection may be the key period for HIV transmission. During this time the infected host experiences a very high viral load but with no symptoms that may thwart transmission, apart from the short influenza-like "seroconversion illness" that often accompanies the first immune response. If most transmissions take place during this time then there is little selective cost in later host death. Consequently, the development of AIDS does not reduce the reproductive success of the virus and the evolution of virulence in HIV can then be thought of as "short-sighted" (Levin and Bull 1994). Although it is unclear what proportion of HIV transmissions actually take place during primary infection, this model has strong theoretical support (Levin *et al.* 2001) and epidemiological studies in some Western populations reveal that rates of viral transmission are particularly high during primary infection and may be responsible for the very rapid spread of HIV through networks of gay men and injecting drug users. Such a model could also apply to sexual populations if the rate of partner exchange is sufficiently rapid.

There are also ways in which the relationship between virulence and transmission rate can be investigated directly. If there is indeed an evolutionary trade-off between virulence and transmission rate, then we might expect the lineages of highest virulence to be the least transmittable. Such a prediction could be tested using newly developed methods based on **coalescent theory** (see Aguadé *et al.*, Chapter 2) which can determine whether the subtypes of HIV-1, or indeed any cluster of viral lineages, differ in their rates of population growth by examining the branching structure of phylogenetic trees. Application of these methods to the different subtypes of HIV-1 revealed varying population dynamics (Grassly *et al.* 1999; Pybus *et al.* 2000). This is most clear-cut for comparisons involving subtypes A and B, with the former experiencing the pattern of exponential population growth typical of many RNA viruses, and the latter exhibiting logistic growth, in which growth rates have slowed toward the present. Although this could signify subtype-specific variation in transmission rate, differences in epidemiological circumstances seem a more reasonable interpretation. Specifically, the initially rapid growth of subtype B, followed by a later slowing down in the rate of transmission, could be because the virus first spread in Western populations of gay men and injecting drug users in which networks of transmission were already established. Under these conditions, the virus could spread rapidly along all the connections of the network until there was a saturation in the number of susceptible hosts, at which point the growth rate declined (Levin *et al.* 2001).

## HIV virulence and host population size

One complicating factor in establishing long-term trends in the evolution of HIV virulence is its relationship with host population size and density. Epidemiological theory states that there are two important parameters determining whether a new infection will emerge in a population. The first, $R_0$, we have encountered already. A virus will only be able to emerge in a population if $R_0 > 1$. Moreover, the greater the $R_0$, the higher the eventual prevalence of the virus and the larger the proportion of the host population that must be vaccinated for successful control. The second parameter of interest is the **threshold density of hosts** that must be reached if a viral infection is to be sustained (i.e. for $R_0 > 1$). This can be denoted $N_T$. What value this threshold takes for any virus depends on various aspects of its biology, such as the duration of infectiousness, transmittability and virulence. Hence, viruses of short duration, low transmittability and high virulence require the largest and most dense host populations to sustain their spread. For HIV, $N_T$ will critically depend, once again, on when during the infection cycle most transmissions occur. If most transmissions do indeed occur during primary infection, then HIV will behave in the same manner as any number of acute viral infections of short duration so that sustained transmission will not occur in a population without a relatively high $N_T$. Conversely, if HIV patients transmit virus for the entire time they are infected, as is the case for most other sexually transmitted diseases, then a lower $N_T$ is required to sustain the epidemic spread of the virus.

What is often forgotten when considering the evolution of HIV virulence is that the virus was probably circulating in small rural human populations for many years before finally entering denser urban populations. This extended period of evolution in small populations will have two effects. First, if there is a link between virulence and transmission rate then, smaller, only less dense host populations will only be able to sustain infections of relatively low virulence. This again points to HIV virulence being lower in the past than it is today; otherwise susceptible hosts would be lost too rapidly and the infection would burn out. Indeed, it is possible, if not likely, that the pandemic of HIV-1 we see today was caused by a small sample of a much larger number of viral lineages that have crossed from primates to humans during evolutionary history. Most of these would have suffered extinction, perhaps because host populations were too small to sustain their transmission, and only those lineages represented by groups M, N, and O have continued their spread as they were fortunate enough to enter populations with sufficiently large transmission networks. Alternatively, it could be that the surviving lineages were those that had sufficiently low virulence that host death did not occur before the virus had been transmitted to others in the population.

A second important, although rather more complex, outcome of initial evolution in small host populations is that under these conditions **genetic drift**, which in this context may be defined as the random sampling of strains at viral transmission, will have a major impact on evolutionary dynamics. Extensive genetic drift will mean that there is a stochastic element in the types of viruses that emerge and the phenotypic properties they possess. First, in the case of sexually transmitted diseases like HIV, genetic drift will be particularly strong if there is great variability in rates of partner exchange, perhaps with a core group of individuals responsible for most transmissions. This will mean that the **effective population size** ($N_e$) of HIV is far less than its overall prevalence, as appears to be the case in nature (Grassly *et al.* 1999), thereby giving **drift** a key role in shaping dynamics. Second, the large population bottleneck which accompanies the transmission of viruses to new hosts, where viral population size may decline to just a few virions, will mean that high fitness strains have no guarantee of evolutionary success. Indeed, it is even possible that chance sampling at HIV transmission is so strong that highly infectious strains are lost by drift when they first emerge, and that the level of virulence attained is not that which maximizes reproductive success. In sum, although natural selection appears to drive the intra-host evolution of HIV, it may only ever be a weak force in evolution at the population level where genetic drift may dominate (Fig. 7.2).

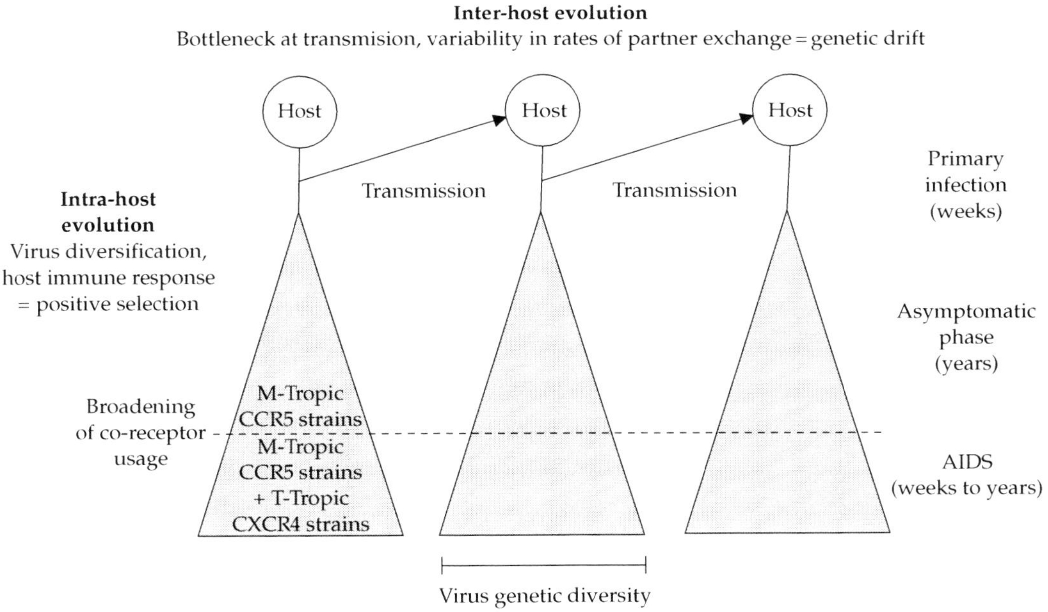


**Figure 7.2** Schematic representation of the relationship between virulence and HIV life history assuming that evolution is "short-sighted" with respect to virulence. Under this model, most HIV transmissions occur when the donor host is suffering primary infection. The strains that are transmitted tend to be of the M-tropic, CCR5 phenotype. Because of the bottleneck at viral transmission and variability in rates of partner exchange, the movement of strains among hosts (inter-host evolution) is dominated by genetic drift. In contrast, intra-host evolution is dominated by a selectively driven arms race between the virus and the host immune system. After years of replication the host immune system becomes so depleted that it can no longer suppress the more virulent T-tropic CXCR4 phenotype viruses from dominating the population. However, the late appearance of these strains means that they are rarely transmitted.

Evolutionary genetic analysis may offer a productive way to explore the impact of genetic drift on HIV emergence. This stems from the observation that the evidence for adaptive evolution is often strongest in host–pathogen systems, in which those pathogen proteins which interact directly with host cells, such as viral envelope proteins, have very high numbers of **nonsynonymous** ($d_N$) to **synonymous** ($d_S$) **substitutions** per site (as reflected in the ratio $d_N/d_S$, with $d_N/d_S > 1$ indicative of positive selection; see also Aguadé *et al.*, Chapter 2). If sequences of HIV and SIV are compared an even more revealing pattern emerges; the higher the virulence, the higher the $d_N/d_S$ ratio, presumably because a more virulent virus (HIV) stimulates a stronger immune response against it, so that host and virus are locked in an intense arms race. Consequently, one possible way to measure the change in HIV virulence through time is to examine the extent to which the $d_N/d_S$ ratio has also changed. The analyses of this sort undertaken to date (Sharp *et al.* 2001) point to the same conclusion; that proportionally more nonsynonymous changes have accumulated towards the present on the HIV-1 phylogeny, so that the $d_N/d_S$ ratio is increasing. At face value, this could again be taken to mean that HIV virulence has increased toward the present. There are, however, two important caveats. First, as stressed above, it is possible that the size of the pool of susceptible hosts has also increased towards the present. Under these circumstances, the lowest $d_N/d_S$ values would be associated with the smallest host population sizes, so that many advantageous mutations that arose early in HIV evolution have simply been lost through genetic drift, which would also reduce $d_N/d_S$ but without a change in virulence. More pertinently, the same increase in $d_N/d_S$ has occurred on SIV lineages where virulence is unlikely to have increased (Sharp *et al.* 2001). This points to the true explanation for the elevation in $d_N/d_S$ being that many nonsynonymous changes are mildly deleterious (particularly on transmission to new hosts) and have yet to be removed by purifying selection in lineages that have only appeared recently. This would inflate the apparent rate of nonsynonymous substitution towards the present, especially if the pressure of purifying selection is relaxed during periods of rapid population growth. Hence, although the impact of genetic drift on the evolution of HIV virulence cannot be fully determined at present, it is clearly an area that merits further study.

## Genetic determinants of HIV virulence

### Do current HIV strains differ in virulence?

While there is little direct evidence detailing how HIV virulence might have changed through time, rather more can be said about the differences in pathogenic potential exhibited by present day viruses. As discussed previously, there is compelling evidence that HIV-1 and HIV-2 differ in virulence (Marlink *et al.* 1994). Furthermore, these viruses may also differ in transmission rate, with HIV-2 spreading more slowly than HIV-1 (Kanki *et al.* 1994). Consequently, at this broad phylogenetic level, there seems to be a direct, and in this case positive, correlation between virulence and transmission rate, although the underlying mechanisms are unknown. This is particularly interesting given the evidence from West Africa that HIV-2 is being replaced by HIV-1 in regions where the two viruses co-circulate (cited in Ewald 1999). This example also indicates that in some circumstances there may be heritable genetic variation in reproductive output, a fundamental prerequisite for natural selection.

It has also been suggested that the subtypes of HIV-1 may differ intrinsically in virulence. Although there is some epidemiological evidence that individuals infected with viruses of subtypes C and D progress to AIDS more rapidly than those with subtype A strains (Kanki *et al.* 1999), the numbers of patients studied was small and counter evidence has been provided. Other evidence for subtype-specific differences in virulence involves subtype C, which is very common in sub-Saharan Africa and globally is now the most prevalent of all subtypes. Because subtype C strains seem to have a preference for CCR5 (macrophage) over CXCR4 (T-cell) chemokine co-receptors (Peeters *et al.* 1999), it could also be that subtype C has a lower virulence than other subtypes but is rather more transmittable (see below). However, it is equally clear that this subtype is responsible for many millions of AIDS deaths in Africa and Asia. Furthermore, it is not known whether all subtype C strains exhibit the same behavior, which is an important consideration given the arbitrary nature of subtype classifications.

Another factor that needs to be taken into account when considering the genetic determinants of virulence in HIV is recombination. It is now clear that HIV is able to recombine freely in nature. This is most notable when comparing HIV-1 strains assigned to different subtypes; it has been estimated that as many as 50 percent of the viruses currently spreading in sub-Saharan Africa may be inter-subtype recombinants, denoted **circulating recombinant forms** (CRFs). However, it is currently unclear whether these recombinant viruses differ in virulence (or in any other phenotypic property) from their parental forms, or whether certain **recombination break-points** are favored over others. However, the fact that many of these recombinants are mixtures of multiple subtypes indicates that mixed HIV infections must occur at high frequency, especially as most recombinants are likely to be deleterious and so will never be sampled. It is also unclear whether these mixed infections take place through (near) simultaneous **co-infection** or sequential **super-infection**. The latter has important repercussions because it means that there is little protective immunity among HIV subtypes, which will make the development of an effective vaccine all the more difficult.

More controversial is how frequently recombination occurs between viruses currently classified in the same subtype. To date, this process has only been reported occasionally. There are two possible explanations for the apparently low frequency of intra-subtype recombination. First, the methods of sequence analysis commonly used to detect recombination, which generally search for conflicting phylogenetic signals, may be weak when viruses are closely related. Conversely, while there is clearly little protective immunity among viruses assigned to different subtypes, strains from the same subtype may be able to confer some cross-protection because they exhibit fewer differences in key viral antigens. This would block super-infection, and hence recombination. Because there is, in reality, a continuum of genetic variation within and among subtypes, it is likely that closely related viral strains will confer immune protection against each other, as appears to be the case in experimentally infected primates, while more divergent viruses but still assigned to the same subtype will fail to stimulate a blocking immune response. Consequently, it is likely that intra-subtype recombination has occurred far more frequently than it has been detected to date.

How will recombination affect the evolution of virulence in HIV? On theoretical grounds, the most important conclusion is that the super-infection of viral strains will tend to increase virulence above that which maximizes $R_0$ because there will be increased intra-host competition among strains for cells to infect (Nowak and May 1994). Hence, it might be expected that the more diverse the super-infecting strains are, the more intense the competition and the quicker the progression to AIDS. On a more pragmatic level, recombination between diverse viral strains might infrequently produce novel genotypes with a much higher or lower virulence than either parental virus, although whether they will be successfully transmitted is another matter. Conversely, if a host is infected with only a single strain, intra-host competition will be less intense and the genetic variants produced by mutation alone will be more genetically homogeneous so that any subsequent recombination is less likely to produce genotypes with very different phenotypic properties. Although these predicted effects of HIV super-infection are of great importance, to date there have been no empirical studies that have provided the key discriminatory data. Given the potential importance of recombination in HIV, this is clearly a key area for future study.

### Co-receptor usage and virulence in HIV

By far the best evidence that HIV strains differ in virulence involves a polymorphism in co-receptor usage. As mentioned already, as well as CD4+, HIV makes use of a variety of chemokine co-receptors to infect host cells, most notably CCR5, the CC chemokine receptor and CXCR4, the CXC chemokine receptor. Some "**T-tropic**" strains of HIV-1 have a preference for T-cells and are unable to replicate in macrophages. These strains utilize the CXCR4 receptor. Conversely, other "**M-tropic**" strains of HIV-1 preferentially replicate in macrophages and primary T-cells and use the CCR5 receptor. M-tropic viruses dominate during primary infection and the early years of the asymptomatic phase, implying that they are important in viral transmission, whereas T-tropic strains are commonly found in AIDS patients (Bjorndal *et al.* 1997). Although both M- and T-tropic strains are equally cytopathic for their cell of preference, T-tropic strains can be considered as possessing higher virulence because they are associated with an accelerated rate of CD4+ cell decline, as both "naïve" T-cells and T-helper cells are infected, and hence a faster progression to AIDS (reviewed in Rowland-Jones 1998). Although the transition from low virulence M-tropic strains to high virulence T-tropic strains cannot itself be the cause of AIDS, because not all AIDS patients possess CXCR4-utilizing strains, this change in co-receptor usage is clearly of fundamental importance, particularly when seen in context of the avirulent primate lentiviruses (see later). Furthermore, it may also reflect a trade-off between virulence and transmission rate; while the M-tropic strains are less often associated with AIDS, their high frequency at primary infection suggests that they are intrinsically more infectious, such that natural selection has favored their appearance at this stage in virus life-history when the conditions for transmission are most favorable (i.e. high viral load, no immune response). On the other hand, the high virulence T-tropic strains appear to be transmitted less frequently and may simply arise because of the short-sighted evolutionary process that occurs after the bulk of viral transmissions have taken place (Fig. 7.2).

While the distinction between CCR5 and CXCR4 usage is clearly a key aspect of HIV virulence, one puzzle remains. Only a small number of amino acid changes in the V3 region of the envelope glycoprotein are required to broaden co-receptor usage from CCR5 to encompass CXCR4, in effect the transition from low to high virulence, yet this usually only takes place late on in infection despite the rapidity of viral replication and mutation. The simplest explanation for the delayed appearance of the T-tropic strains is that they generate a stronger immune response against them than the M-tropic strains and so are more efficiently removed by the immune response. T-tropic strains are therefore only able to dominate the viral population after there has been a substantial reduction in CD4+ levels which may take many years (Callaway *et al.* 1999). Evidence for this differential immune response is that the $d_N/d_S$ ratio is higher in T-tropic than M-tropic strains, indicating that the former are indeed subject to stronger immune selection (Shiino *et al.* 2000).

The distinction between CCR5 and CXCR4 is also important for host genetics. In particular, individuals who carry a 35 bp **deletion** (Δ35) in the CCR5 gene have a lower rate of progression to AIDS (Huang *et al.* 1996), while those homozygous for Δ35, and hence who lack a functional CCR5 receptor, are at greatly reduced risk of HIV infection (Dean *et al.* 1996). Similarly, a polymorphism in the CCR2 chemokine receptor has been associated with a delay in disease progression (Kostrikis *et al.* 1998). Human leukocyte antigen (HLA) **haplotypes** have also been implicated in determining disease progression. As a case in point, some female prostitutes in Gambia possessing the HLA-B35 allele have a powerful HIV-specific **CD8+ cytotoxic T-lymphocyte** (CTL) response such that they are effectively immunized against infection (Rowland-Jones *et al.* 1995). The ability of both CTL and CD4+ T-helper cell responses to control HIV viral load has likewise been demonstrated in a number of longitudinal studies, with long-term "non-progressors" to AIDS fortunate to have particularly strong CTL responses. Although these studies hint that differences in host genetics go a long way to explaining differences in progression times, direct competition experiments indicate that the viruses associated with long-term non-progressors often have intrinsically lower fitness than those sampled from patients who progress rapidly to AIDS (Quiñones-Mateu *et al.* 2000).

## Co-receptor usage and virulence in primate lentiviruses

As discussed at the outset of this chapter, the high virulence of HIV is especially striking when compared to its apparently benign relatives that infect a large variety of other primate species. As such, a detailed analysis of the similarities and differences between HIV and SIV is likely to be highly informative. Once again, differences in cell tropism and the strength of immune selection appear to be of fundamental importance.

To date, more than 20 nonhuman primate lentiviruses have been identified, although more doubtless exist in nature (reviewed in Hahn *et al.* 2000). With one exception—SIVmac from Asian macaque monkeys—these viruses all circulate in simian primates of African origin. However, as SIVmac is only found in captive animals, and the source of this virus was clearly a species of African monkey—the sooty mangabey—kept with them in captivity, it is safe to conclude that primate lentiviruses are exclusive to African primates.

One of the earliest discoveries from the comparative analysis of primate lentiviruses was that HIV-1 and HIV-2 have their origins in different primate species. HIV-1 is most closely related to those viruses found in populations of the common chimpanzee (SIVcpz). Given the intermixing of HIV-1 lineages with those of SIVcpz on molecular phylogenies, it must also be that there have been multiple transfers of virus from chimpanzees to humans (Fig. 7.1), although the exact manner by which this occurred is still the source of some debate. Conversely, the ancestry of HIV-2, like SIVmac, lies with the viruses found in sooty mangabey monkeys (SIVsm), a commonly infected West African primate species. As with HIV-1/SIVcpz, there is mixing of human and monkey lineages on molecular phylogenies that again indicates the transfer of multiple viruses from monkeys to humans.

If the HIVs and SIVs are so phylogenetically similar, so that their divergence is relatively recent, why are they so different in virulence? Unfortunately, the absence of data concerning the natural history of SIV in wild-living African primates makes it difficult to answer this question in the broadest sense. More progress has been made at the virological level. Perhaps surprisingly, all the evidence available to date indicates that avirulent SIVs replicate and mutate at about the same rate as the virulent human strains, although SIV infected African green monkeys may experience somewhat lower viral loads (reviewed in Holmes 2001). Where the virulent HIVs and avirulent SIVs do seem to differ is in the types of nucleotide substitutions that accumulate, reflecting a very different virus-host interaction. Most notably, SIVs accumulate proportionally fewer nonsynonymous to synonymous substitutions per site during intra-host viral evolution compared to HIV. The most compelling evidence comes from the V3 region of the envelope (*env*) glycoprotein which is strongly conserved in SIVs, but a highly variable immunodominant region in HIV-1, often experiencing strong positive selection (Hahn *et al.* 2000). The most obvious interpretation of these results is that the SIVs are subject to weaker immune selection pressure than the HIVs, so that there is less selective requirement to fix amino acid changes in regions like V3 that mediate immune evasion. It is even possible that virulent T-tropic strains do occasionally appear in SIVs, but that we have yet to sample wild-living animals with immunodeficiency disease.

Differences in co-receptor usage also appear to be of fundamental importance. As described previously, HIV-1 shows tropism for both the CCR5 and CXCR4 chemokine receptors. The same also appears to be true of HIV-2, although a broader set of tropisms are observed in this case. In contrast, the non-pathogenic SIVs rely exclusively on CCR5 and consequently only harbor the low virulence M-tropic strains (Chen *et al.* 1998). This again points to the infection of T-cells through the CXCR4 chemokine co-receptor as a key mechanism controlling virulence. The outstanding question is why SIVs do not harbor strains capable of infecting T-cells? Perhaps the most reasonable explanation, stemming from the observation that the immunodominant V3 region is conserved in the SIVs but variable in the HIVs, is that strong structural or functional constraints on protein structure prevent the **fixation** of mutations capable of using CXCR4.

In sum, the comparative analysis of HIV and SIV strains again reveals that the broadening of cell tropism to encompass both naïve T-cells and T-helper cells is perhaps the mechanistic key to viral virulence, although this is evidently an area that requires more intensive research.

## And the future?

In all discussions of HIV virulence it is tempting, although dangerous, to speculate on how the evolutionary process might pan out in the future. Perhaps the only safe answer is that unless the underlying causes of HIV virulence can be determined, particularly any relationship between virulence and transmission rate, then such predictions have little power. As a case in point, if virulence is simply the short-sighted by-product of intra-host viral evolution then there will be no heritable variation for this trait and virulence will be invisible to the action of natural selection.

Likewise, uncertainty over the relationship between virulence and transmission rate, and more fundamentally over when in virus life-history most HIV transmissions take place, makes it difficult to assess how the increasingly widespread use of anti-retroviral therapy will affect the nature of the HIV pandemic. As explained more fully by others (Levin *et al.* 2001), the effect of therapy critically depends on how it affects transmission rate; if there is no net effect, as expected if most infections occur during primary infection before patients are routinely given treatment, then anti-retroviral therapy will similarly have no major influence on HIV prevalence. Conversely, if the overall transmission rate is reduced with therapy, which in turn presumes that hosts transmit during the long asymptomatic phase, then the number of new infections will decline with the use of antiviral therapy. Moreover, it is possible that AIDS patients with suppressed immune systems may act as good reservoirs for a variety of opportunistic microbial infections which could then spread to the general population (Weiss 2001).

One area where predictions can be more safely made concerns how the human population will evolve under the selection pressure imposed by HIV, particularly at those loci which determine susceptibility to infection and progression to AIDS. Despite the rapidity and magnitude of the HIV/AIDS pandemic, it will take a long time for the virus to have a significant affect on the genetic structure of human populations. In one recent study using parameters from the South African HIV epidemic, it was predicted that given current selection pressures alleles that delay the onset of AIDS would increase from 40 to 53 percent over a 100-year period, with the average length of progression to AIDS extended by just one year (Schliekelman *et al.* 2001), and that this will involve many AIDS deaths along the way. Evolution in Western populations under weaker selection pressures will evidently be slower still. As elegantly summed up by one research team; "we cannot count on evolution in our population to save us from the AIDS epidemic" (Levin *et al.* 2001).

## Summary

Although our knowledge of the biology of HIV continues to improve, relatively little is known about the evolution of virulence in AIDS viruses. Most fundamentally, it is unclear whether HIV has always been associated with an AIDS-like illness, or whether this is a more recently evolved trait. An often ignored element in this argument is genetic drift, which may have played an important role in shaping the early evolution of virulence because HIV most likely first emerged in a small host population, in which there was no guarantee of continued viral transmission. The relationship between demographic history, evolutionary process, and virulence should be an area of study for pathogens in general.

Given that AIDS appears after an extended period of intra-host evolution, perhaps the best supported model for HIV virulence is that AIDS is the outcome of a "short-sighted" evolutionary process. Under this model, most new transmissions occur when a donor is still suffering the initial primary infection, when viral loads are high, so that later host death has no affect on the overall reproductive rate. The intra-host arms race between the virus and the immune system that follows will eventually result in immune collapse and the appearance of AIDS. The broadening to cell tropism from the infection of macrophages through the CCR5 chemokine co-receptor to encompass the infection of T-cells through the CXCR4 chemokine co-receptor also appears to be central to virulence, particularly as it reflects large-scale phylogenetic patterns; the avirulent SIVs only use CCR5 and are subject to weaker immune selection pressure, while the transition from CCR5 to CXCR4 usage in HIV is associated with increased virulence and intensified immune selection. Understanding the evolution of virulence in HIV is clearly an area where evolutionary genetics will play a fundamental role in the future.

I thank David Robertson, Andrew Rambaut, and two anonymous reviewers for valuable comments.

CHAPTER 8

# Evolution and population structure of parasitic protozoa: the *Plasmodium* model

**Stephen M. Rich and Francisco J. Ayala**

Human malaria is the cause of tremendous morbidity and mortality in the world. The complexity of its causative agents, *Plasmodium* species, has thwarted efforts to eradicate the disease (for review, see Spielman *et al.* 1993; Bradley 1999). Many methods have been proposed for intervening in parasite transmission, and hence reducing human malaria, including development of vaccines to protect individuals against infection and new drug therapies, as well as efforts that target the mosquito vector of the parasite. The extensive antigenic diversity of the parasite and its widespread resistance to existing anti-malarial drugs call for a better understanding of the parasite's biology as a prerequisite to enhance the likely success of control efforts. One important step is to determine how genetic variation arises and is maintained in the parasite's populations. We herein review work done to date on the molecular evolution of malarial parasites, with particular attention to the population structure of *P. falciparum*, the most virulent among the four human *Plasmodium* parasites.

## Phylogenetic origin of human malaria

The genus *Plasmodium* consists of nearly 200 named species that parasitize reptiles, birds, and mammals. *Plasmodium* belongs to the Apicomplexa, a large and complex phylum with about 5000 known species and, likely, some 60 000 yet to be described (Levine 1988; Vivier and Desportes 1988; Corliss 1994). The Apicomplexa are all parasites, characterized by the eponym structure, the apical complex. The taxonomy and phylogeny of the phylum have been the subject of controversy and frequent revision. At issue are whether *Plasmodium* evolved directly from **monogenetic** (i.e. with a unique host species) parasites of the ancient marine invertebrates from which the chordates evolved, or whether they originated by lateral transfer from other, already **digenetic** (i.e. one intermediate and one definitive vertebrate host species) parasites (Huff 1938; Manwell 1955; Garnham 1966; Barta 1989). There is no fossil record of apicomplexans (Margulis *et al.* 1993), but molecular investigations indicate that the phylum is very ancient, perhaps as old as the multicellular kingdoms of plants, fungi, and animals, and thus somewhat older than 1000 million years (Ayala *et al.* 1998).

The genus *Plasmodium* itself is also quite ancient, perhaps as old as the Cambrian, or about 600 million years old. Four species of *Plasmodium* are traditionally considered parasitic to humans: *P. falciparum*, *P. malariae*, *P. ovale*, and *P. vivax*. The phylogeny of the genus has been elucidated through molecular studies (e.g. Barta *et al.* 1991; Morrison and Ellis 1997), including our own investigations of three genes, the small subunit ribosomal RNA gene (*SSUrRNA*) (Escalante and Ayala 1994, 1995), the nuclear gene coding for the circumsporozoite protein (*Csp*) (Escalante *et al.* 1995), and the mitochondrial gene coding for cytochrome *b* (Escalante *et al.* 1998a). Tree topologies and genetic distances based on these several genetic loci are quite similar (for additional details, see Ayala *et al.* 1998), and so for present discussion we focus on one example, that is, the gene

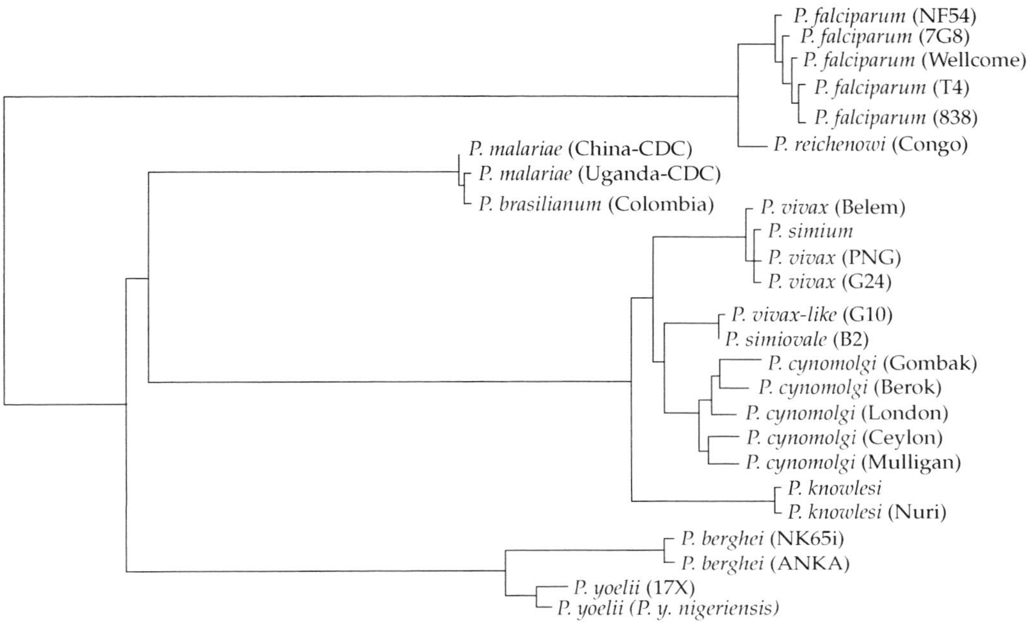


**Figure 8.1** Phylogeny of 12 *Plasmodium* species inferred from *Csp* gene sequences (from Ayala *et al.* 1998).

tree based on *Csp* genes (Fig. 8.1). Phylogenetic studies the *Csp* genes (and others) of the genus warrant three general conclusions:

1. The four human parasites, *P. falciparum*, *P. malariae*, *P. ovale*, and *P. vivax* are remotely related to one another, so that the evolutionary divergence of these human parasites greatly predates the origin of the primates. It follows that their parasitic associations with humans are phylogenetically independent; that is, at least three of these species have been laterally transmitted to the human ancestral lineage from other, nonprimate hosts. This conclusion is consistent with the diversity of physiological and epidemiological characteristics of the four *Plasmodium* species (Coatney 1976; López-Antuñano and Schumunis 1993).
2. *P. falciparum* is distinct from, although closely related to the chimpanzee parasite, *P. reichenowi*. Indeed, *P. falciparum* is more closely related to *P. reichenowi* than it is to any other *Plasmodium* species. The time of divergence between these two *Plasmodium* species is estimated at 8–11 million years ago, which is roughly consistent with the time of divergence between the two host species, human and chimpanzee (Escalante and Ayala 1995). A parsimonious interpretation of this state of affairs is that *P. falciparum* is an ancient human parasite associated with our ancestors at least since the divergence of the hominids from the great apes, and that the divergence of *P. falciparum* and *P. reichenowi* was concurrent with the divergence of their host species, humans and chimps.
3. The highly divergent human parasites *P. malariae* and *P. vivax* are genetically indistinguishable from two primate counterparts, the New World monkey parasites, *P. brasilianum* and *P. simium*, respectively. We infer that a lateral transfer between hosts has occurred in recent times, either from monkeys to humans or vice versa. The *Csp* genetic distance between *P. malariae* and *P. brasilianum* is $0.002 \pm 0.002$, not greater than the distance among the various isolates of *P. malariae* ($n = 2$), *P. vivax* ($n = 4$), or *P. falciparum* ($n = 8$) (Ayala *et al.* 1998). This suggests that *P. malariae* (isolated from humans) and *P. brasilianum* (isolated from

New World monkey) have not long diverged from one another and might be considered a single species exhibiting **host polymorphism**, that is, able to parasitize more than one host species (Escalante and Ayala 1994). A similar hypothesis might be put forth with respect to *P. vivax* and *P. simium*, since these two are also genetically indistinguishable.

## Host-switch in the primate malaria parasites

Whether or not the two species in each human–primate parasite pair (*P. vivax–P. simium* and *P. malariae–P. brasilianum*) should be considered the same or distinct species is merely a matter of taxonomy and nomenclatural convenience, and hence is not biologically substantive. What is more important is the conclusion that two of the four known human malaria parasites have nearly identical platyrrhine (New World monkey) parasite relatives. This is a strong indication that a **host-switch** has occurred in recent times (or even continues to occur). A host-switch is defined as a horizontal shift of a parasite from one host species to another distantly related host species. This is in stark contrast to the observed relationship of *P. falciparum* and *P. reichenowi* that have evolved vertically (i.e. from a common ancestor), in parallel with their respective human and chimpanzee host lineages.

Determining the direction of the host-switch between human and platyrrhine—either from monkey to human, or human to monkey—holds great biological relevance to understanding the evolution of the genus and understanding the origin of disease. Human and platyrrhine monkeys are distantly related and have only been geographically associated after the first human colonization of the Americas, which occurred within the last 15 000 years. Indeed, the host-switch may have occurred following the second influx of humans when Europeans began to colonize the America's in the sixteenth century. Whether 500 or 15 000 years have passed since the host-switch, it would be a mere moment in evolutionary time and so it is not surprising that the human and platyrrhine parasites are so little diverged. Epidemiological serosurveys of humans and monkeys in French Guiana indicate that platyrrhines may actually serve as zoonotic reservoirs for human disease (Fandeur *et al.* 2000), thus suggesting that host-switching may be a recurrent phenomenon.

Host-switches are a common occurrence among avian and reptile malaria parasites (Ricklefs and Fallon 2002). Unlike the Old World primate parasite—*P. reichenowi*—that thrives exclusively in chimpanzees, the platyrrhine malaria parasites are quite capacious in their host preference, and so that these New World parasites appear quite susceptible to host-switches. *P. simium* infects at least three, and *P. brasilianum* has been identified in as many as 26 species of New World monkeys (Gysin 1998). We have argued in the past (Ayala *et al.* 1998), on the grounds of evolutionary parsimony, that the host-switches observed in *vivax/simium* and *malariae/brasilianum* were most likely from primates to humans. Based on the observed host distribution, the alternative explanation of a switch from humans to primates seems less plausible since it would require that multiple independent switches have occurred. For example, in the case of the *malariae–brasilianum* host switch, it is improbable that no less than 26 human to platyrrhine (or platyrrhine to platyrrhine) host-switches would have occurred in the 15 000 year history of human habitation of South America.

*P. vivax* and *P. malariae* have widespread global distributions, while the complementary *P. simium* and *P. brasilianum* are restricted to South America. This is not inconsistent with a host-switch of either of the platyrrhine parasites to human hosts since humans have been remarkably vagile in the past several centuries and could have carried their parasites wherever they may have traveled. For example, a survey of ribosomal DNA sequences revealed the occurrence of a *P. brasilianum* isolate in Southeast Asia (Kawamoto *et al.* 2002). The most plausible explanation for a New World monkey malaria in Southeast Asia is that it was carried there by an infected human. This pattern of host transmission may become more evident as additional molecular genotypes of malaria isolates collected from around the globe become available.

Geographical distribution records are somewhat ambiguous with respect to determining the direction of host-switch in the case of *P. vivax* and *P. simium*. *Plasmodium vivax* is the most cosmopolitan of the human malarias, but it is notably absent from sub-Saharan Africa. Absence of *vivax*-malaria in the region has been attributed to the widespread occurrence of a genetic mutation of the Duffy blood group proteins in indigenous sub-Saharan peoples. Duffy proteins are expressed on the surface of erythrocytes, and are necessary for receptor-ligand mediated invasion of *P. vivax* into red blood cells (Miller *et al.* 1976). The sub-Saharan individuals with a $Fy^{a-b-}$ mutation do not express the Duffy receptor, suggesting an adaptive response for resistance to *P. vivax*. This **adaptation** is found primarily in these particular parts of Africa, and its occurrence is inconsistent with a recent introduction of *vivax*-malaria to humans. Nonetheless, the possibility that the Duffy mutation may have arisen in response to some other selection pressure cannot be discarded. Indeed, Duffy negative individuals are resistant to *P. knowlesi* (Mason *et al.* 1977), which is an Old World monkey parasite, and hence one with which human ancestors have shared a common geographical range for millions of years.

Historical documentation of nonmalignant (i.e. non-*P. falciparum*) malaria in humans is similarly equivocal. There is no record of nonmalignant malaria in South America prior to European colonization. This would be consistent with the interpretation that *P. vivax* (as well as *P. malariae* and *P. falciparum*) was introduced to the New World by the European colonizers and their African slaves. The weakness of this argument is that it relies on negative evidence, particularly unreliable when there are few records or studies that would have likely manifested the presence of malaria in the New World before the year 1500, even if it had indeed been present.

In the Old World, historical records are more complete. Chinese medical writings (dated 2700 BC), cuneiform clay tablets from Mesopotamia (c.2000 BC), the Ebere Egyptian Papyrus (c.1570 BC), and Vedic-period Indian writings (1500–800 BC), mention severe periodic fevers, spleen enlargement, and other symptoms suggestive of malaria (Sherman 1998). Spleen enlargement and the malaria antigen have been detected in Egyptian mummies more than 3000 years old (Miller *et al.* 1994; Sherman 1998). Hippocrates' (460–370 BC) discussion of tertian and quartan fevers, "leaves little doubt that by the fifth century BC *Plasmodium malariae* and *P. vivax* were present in Greece" (Sherman 1998, p. 3). If this interpretation is correct, the association of *malariae* and *vivax* with humans could not be attributed to a host-switch from monkeys to humans that would have occurred after the European colonizations of the Americas. This would be definitive evidence, so long as one accepts the interpretation that the fevers described by Hippocrates were indeed caused by the two particular species *P. vivax* and *P. malariae* (rather than, say, *P. ovale*).

The matter can, in any case, be resolved by comparing the genetic diversity of the human and primate parasites. Genetic diversity will be greater in the donor host than in the recipient host of the switch. If the transfer has been from human to monkeys, the amount of genetic diversity, particularly at **silent nucleotide sites** and other neutral polymorphisms, will be much greater in *P. vivax* than in *P. simium*, and in *P. malariae* than in *P. brasilianum* (including in each comparison the polymorphisms present in the various monkey host species). A transfer from monkey to humans should yield much lower polymorphism in the human than in the monkey parasites.

## Extant distribution of *Plasmodium falciparum*

By any estimation, *P. falciparum* is the most consequential of the human malarias, infecting 300–500 million people and accounting for some 2 million deaths per year (Trigg and Kondrachine 1998). While we have established that *P. falciparum* (or, more appropriately, its immediate ancestors) has been a parasite of the human lineage since before the split between humans and chimpanzees, we need now consider the extent of variation in the global *P. falciparum* populations. Innumerable epidemiological studies have indicated that these populations are remarkably variable; however, most studies of genetic diversity have focused on the genes encoding either antigenic determinants or drug resistance factors. Moreover, much of the observed diversity is attributable to changes that have occurred in nucleotide positions that lead to changes in the encoded amino acid, that is, **nonsynonymous** replacement changes. Replacement polymorphisms in antigenic genes are likely to be under strong diversifying selection imposed by the host's immune system. Indeed, Escalante *et al.* (1998b) have shown that polymorphisms in commonly studied antigenic genes are maintained by selection, and thus do not readily lead to estimation of their age. **Silent nucleotide polymorphisms** are more appropriate for estimating the age of genes because they are often adaptively **neutral** (or very nearly so) and, thus, they are determined by the mutation rate and the time elapsed since their divergence from a common ancestor (see Ohta, Chapter 1).

We previously examined single-copy, coding regions of 10 genes, nine nonantigenic plus the gene encoding the circumsporozoite protein (*Csp*), and found a complete absence of polymorphisms at silent nucleotide sites, that is, within those codons for which no amino acid replacement has occurred. The sample encompassed 10 912 fourfold and 20 061 twofold silent sites. An independent study of 10 loci, most encoding antigenic determinants, showed a similar scarcity of silent polymorphisms (Escalante *et al.* 1998b).

As we have expounded elsewhere (Rich and Ayala 1998, 2000; Ayala *et al.* 1999), five possible hypotheses may account for the absence of silent polymorphisms in *P. falciparum*: (i) persistent low effective population size, (ii) low rates of spontaneous mutation, (iii) strong selective constraints on **silent variation**, (iv) one or more recent **selective sweeps** affecting the genome as a whole, and (v) a **demographic sweep**, that is, a recent population bottleneck, so that extant world populations of *P. falciparum* would have recently derived from a single ancestral strain.

Hypothesis (i) can be excluded for the recent history of *P. falciparum*, given that the parasite occurs in many millions of infected individuals (each infected human may harbor as many as $10^9$ parasites, but these may derive from few, perhaps very few, independent infections, each transmitting only a few parasites). If the effective worldwide population of *P. falciparum* would have been very small (tens or at most hundreds of individuals) for very many generations until not long ago, this would effectively amount to a population bottleneck, as in hypothesis (v), with respect to neutral polymorphisms. A persistent small population size leads ultimately to a population that derives from a single ancestral gene. With respect to **balanced polymorphisms**, the outcomes are different. An extreme bottleneck of one single diploid individual entails the elimination of all but two alleles at most, whereas a small number of individuals (say, in the tens or hundreds) may carry tens of different alleles, even if the small number persists for many generations.

There seems to be no reason to suspect that spontaneous mutation rates are exceptional in *P. falciparum* (hypothesis ii), and there are two arguments against it. One is the high incidence of polymorphisms at antigenic and drug-sensitivity sites, both in worldwide samples (Escalante *et al.* 1998b) and in laboratory selection experiments with mice (Cowman and Lew 1989). The other argument is that there is divergence, in synonymous as well as nonsynonymous sites, between *P. falciparum* and other *Plasmodium* species (Hughes 1993; Escalante and Ayala 1994).

It has been suggested that the observed paucity of synonymous substitutions is attributable to the extreme AT-content of the *P. falciparum* genome that may reflect extraordinary codon-usage bias (Saul 1999). While AT bias may affect **substitution rates**, we have noted that it cannot account for the complete absence of polymorphism (Rich and Ayala 1999). Three lines of evidence support this claim: (1) intra- and interspecific comparisons of other *Plasmodium* species show that synonymous substitutions have occurred, even in the lineages leading to *P. falciparum* and *P. reichenowi* (Hughes and Verra 2001); (2) among fourfold redundant codons, AT bias may restrict A/T ↔ G/C changes, but a survey of 312 coding regions shows that A ↔ T changes are not restricted (Rich and Ayala 1999); and (3) while determining the age of the *P. falciparum* bottleneck, we estimated synonymous and nonsynonymous substitution rates empirically among

*Plasmodium* species, and these estimates are corrected for differential rates among two- and fourfold codons (Rich and Ayala 1998, 1999). Therefore, hypotheses (ii) and (iii) cannot account for a complete lack of synonymous substitution in the >10 000 fourfold redundant codons examined.

The selective sweep hypothesis (iv) is, in a way, a special case of the demographic sweep hypothesis (v); that is, a particular strain may have spread throughout the world and replaced all other strains impelled by natural selection. Selection of this kind could account for the absence of synonymous variation at any one of the 10 loci (see table 2 in Rich *et al.* 1998), if the particular gene itself (or a gene with which it is linked) has been subject to a recent worldwide selective sweep without sufficient time for the accumulation of new synonymous mutations. However, the 10 genes surveyed are located on at least six different chromosomes, and thus six independent selective sweeps would need to have occurred more or less concurrently, which seems prima facie unlikely. Nonetheless, a selective sweep simultaneously affecting all chromosomes could happen if one accepts the hypothesis that the population structure of *P. falciparum* is predominantly **clonal**, rather than sexual (see Ayala *et al.* 1998). This hypothesis is controversial, although we have argued that it may indeed be the case, the capacity for sexual reproduction of the parasite notwithstanding (Rich *et al.* 1997; Ayala *et al.* 1999).

We have earlier concluded that hypothesis (v) is most consistent with the observations and used **coalescent theory** to estimate the age of the ancestral strain or **cenancestor**. Based on estimates of silent substitution rates between species, we concluded with 95 percent confidence, that the set of *P. falciparum* isolates examined must have derived from a "Malarial Eve" within the last 57 000 years, but that the actual time of this coalescence might be an order of magnitude more recent (Table 8.1).

Recently, Conway *et al.* (2000) have presented further evidence in support of a rapid expansion of *P. falciparum*, starting from a small ancestral population. They have examined the entire mitochondrial DNA (mtDNA) sequence of *P. falciparum* isolates originating from Africa, Brazil, and Thailand, as well as the chimpanzee parasite, *P. reichenowi*.

**Table 8.1** Estimated upper-boundary times to the cenancestor of the world populations of *P. falciparum*

| **Divergence estimate**[a] | **Mutation rate ($\times 10^{-9}$)** | | **Estimated upper-bound (years)**[b] | |
|---|---|---|---|---|
| | **Fourfold ($\mu_a$)** | **Twofold ($\mu_b$)** | $t_{95}$ | $t_{50}$ |
| *Plasmodium* radiation | | | | |
| 55 | 7.12 | 2.22 | 24 511 | 5670 |
| 129 | 3.03 | 0.95 | 57 481 | 13 296 |
| *falciparum-reichenowi* | | | | |
| 5 | 3.78 | 1.86 | 38 136 | 8821 |
| 7 | 2.70 | 1.33 | 53 363 | 12 343 |

[a] Lower and upper estimates of divergence in millions of years based on *SSUrRNA* differences (Escalante *et al.* 1995).

[b] $t_{95}$ and $t_{50}$ are the upper boundaries of the 95 and 50 percent confidence intervals, respectively. Thus, in the first row, the cenancestor lived less than 24 511 years ago with a 95 percent probability, and less than 5670 years ago with a 50 percent probability.

The alignment of the four complete mtDNA sequences embraces 5965 bp, among which 139 sites contain fixed differences between *falciparum* and *reichenowi*, whereas only four sites are polymorphic within *falciparum*. The corresponding estimates of interspecific divergence ($K$) and intraspecific diversity ($\pi$, among *P. falciparum* strains), are 0.1201 and 0.0004, respectively. In short, the divergence in mtDNA sequence between the two species is 300-fold greater than the diversity within the global *P. falciparum* population. If we use the rRNA-derived estimate of 8 million years as divergence time between *P. falciparum* and *P. reichenowi*, then the estimated origin of the *P. falciparum* mtDNA lineages is 26 667 years (i.e. 8 million/300), which corresponds quite well with our estimate based on the 10 nuclear genes (Table 8.1). In a subsequent survey of a total of 104 isolates from Africa ($n = 73$), Southeast Asia ($n = 11$), and South America ($n = 20$), Conway *et al.* (2000) determined that the extant global population of *P. falciparum* is derived from three mitochondrial lineages which started in Africa and migrated subsequently (and independently) to South America and Southeast Asia. Each mitochondrial lineage is identified by a unique array of the four polymorphic mtDNA nucleotide sites.

Volkman *et al.* (2001) have conducted a comprehensive study of 3417 bp distributed among 25 **intron** sequences of *P. falciparum*. They chose intronic sequences to overcome the problem of reduced polymorphism that may be attributable to codon bias. Among the eight *P. falciparum* isolates they examined (a total of 27 336 bp), they found an extreme dearth of **single nucleotide polymorphism**. When they combined their data with our previous report (Rich *et al.* 1998), they came to the conclusion that the global *P. falciparum* population shares a recent common ancestor from 3200 to 7700 years ago.

The weight of evidence strongly supports the virtual absence of synonymous-site nucleotide polymorphism in nuclear (Escalante *et al.* 1998b; Rich *et al.* 1998; Volkman *et al.* 2001), mitochondrial (Conway *et al.* 2000), and plastid (Rich *et al.*, unpublished) genomes of *P. falciparum*. Nonetheless, Hughes and Verra (2001) have reported a number of synonymous polymorphisms among GenBank sequences of 23 protein-encoding loci in *P. falciparum*. At first this appears paradoxical, but closer evaluation of the data suggests that their conclusion that *P. falciparum* has had a "very large, long-term effective population size" is based on some flawed assumptions. The authors chose to restrict their analyses to loci that do not show significant levels of natural selection. However, the extremely low levels of nonsynonymous and synonymous nucleotide polymorphism (less than 0.2 and 0.3 percent, respectively), indicates that the power of neutrality tests will be greatly diminished, so that natural selection may have acted on some of their loci. Indeed since many of the loci examined by Hughes and Verra (2001) encode antigenic determinants, we might expect that even a single nonsynonymous change may be sufficient to confer some selective advantage. One of the loci, the dihydrofolate reductase (*Dhfr*) gene is known to have several amino acid polymorphisms (and no synonymous nucleotide polymorphism) which act through known biochemical mechanisms to confer resistance to the anti-malarial drug pyrimethamine. These polymorphic mutations are known to increase in frequency where pyrimethamine is applied as anti-malarial therapy, and hence it seems extraordinarily unlikely that this gene is not under selection.

Moreover, nine of the 23 genes examined by Hughes and Verra (2001) have no synonymous site polymorphisms. Among those loci for which they report synonymous polymorphisms, most are contained within repetitive regions. For example, the GLURP alleles, which comprise 36 percent of the 49 268 synonymously variant codons examined, have several short repeats and all of the synonymous variants are located within these repetitive regions. As expounded upon below, these repetitive DNA regions evolve at rates far greater than point mutations (Rich and Ayala 2000; Volkman *et al.* 2001). For these several reasons, we submit that Hughes and Verra's (2001) estimate of at least 300 000–400 000 year age of the most recent common ancestor is overinflated, although it is certainly much closer to our estimate of several thousand years than Hughes' original claims of a 35 million-year ancestry of some *P. falciparum* merozoite surface protein alleles (Hughes and Hughes 1995).

Historical records provide a plausible explanation for the recent worldwide expansion of *P. falciparum* (Sherman 1998). Hippocrates describes quartan and tertian (i.e., non-*P. falciparum*) fevers, but there is no mention of severe malignant tertian fevers, which suggests that *P. falciparum* infections did not yet occur in classical Greece, as recently as 2400 years ago. How can we account for a recent demographic sweep of *P. falciparum* across the globe, given its long-term association with the hominid lineage? One likely hypothesis is that human parasitism by *P. falciparum* has long been highly restricted geographically, and has dispersed throughout the Old World continents only within the last several thousand years, perhaps within the last 10 000 years, after the Neolithic revolution (Coluzzi 1997, 1999). Three possible scenarios may explain this historically recent dispersion: (1) changes in human societies, (2) genetic changes in the host–parasite–vector association that have altered their compatibility, and (3) climatic changes that entailed demographic changes (migration, density, etc.) in the human host, the mosquito vectors, and/or the parasite.

One factor that may have impacted the widespread distribution of *P. falciparum* in human populations from a limited original focus, probably in tropical Africa, may have been changes in human living patterns, particularly the development of agricultural societies and urban centers that increased human population density (Coluzzi 1997, 1999; Sherman 1998). Genetic changes that have increased the affinity within the parasite–vector–host system also are a possible explanation for a recent expansion, not mutually exclusive with the previous one. Coluzzi (1997, 1999) has cogently argued that the worldwide distribution of *P. falciparum* is recent and has come about, in part, as a consequence of a recent dramatic rise in vectorial capacity due to repeated speciation events in Africa of the most anthropophilic members of the species complexes of the *Anopheles gambiae* and *A. funestus* mosquito vectors. The biological processes implied by this account may have, in turn, been associated with, and even dependent on the onset of agricultural societies in Africa (scenario 1) and climatic changes (scenario 3), specifically the gradual increase in ambient temperatures after the Würm glaciation, so that about 6000 years ago climatic conditions in the Mediterranean region and the Middle East made possible the spread of *P. falciparum* and its vectors beyond tropical Africa (De Zulueta *et al.* 1973; De Zulueta 1994; Coluzzi 1997).

The three scenarios are likely to be interrelated. Once the demographic and climate conditions became suitable for the propagation of *P. falciparum*, natural selection would have facilitated the evolution of *Anopheles* species that were highly anthropophilic and effective *falciparum* vectors (De Zulueta *et al.* 1973; Coluzzi 1997, 1999). Studies of the evolutionary history of innate immunity factors conferring resistance to *P. falciparum* seem to support these hypotheses. One such resistance factor is a glucose-6-phosphate dehydrogenase (*G-6pd*) genetic deficiency that occurs in high frequency in areas of malaria endemicity, particularly in sub-Saharan Africa. Tishkoff *et al.* (2001) have argued cogently that these *G-6pd* mutants are only 3330 years old (1600–6640, 95 percent confidence interval), which further supports the notion that *P. falciparum* has only presented itself as a disease burden to humans in our very recent history.

## Evolution of antigenic loci

There is an apparent contradiction between the paucity of synonymous polymorphisms and the abundance of replacement changes observed in antigenic loci. Natural selection may offer an explanation since strong positive selection, particularly where immune evasion is at stake, can very likely fix even rare mutations. However, Hughes and colleagues proposed a model that requires that variants of genes encoding *P. falciparum* surface proteins may be as old or much older than the species itself (Hughes 1993; Hughes and Hughes 1995). They estimated that the ages of the most divergent alleles of *Msp-1* and *Csp* alleles are 35 million and 2.1 million years, respectively. It must first be noted that the polymorphisms in these antigenic genes, whether or not they are of ancient origin, do not contradict the recent origin of *P. falciparum* current world populations. As with the misinterpretation of the "mitochondrial Eve" model of human origins, it should be noted that the hypothetical "malarial Eve" does not represent a single ancestral individual, but rather some extreme constriction in effective population size. Ancient polymorphisms at certain loci under strong **balancing** (diversifying) **selection**, can be maintained through a severe constriction in population numbers, or even through a number of generations with small populations that would lead to the virtual complete elimination of neutral allelic polymorphisms, as noted earlier to account for the scarcity of silent polymorphisms. For example, although the mitochondrial lineage of modern humans is only 100 000–200 000 years old, natural selection has maintained extensive polymorphisms among human major histocompatibility complex molecules (involved in the immune response against invading foreign substances), some of which predate the split between humans and chimpanzees (Ayala 1995b). The *P. falciparum* antigenic genes are under strong diversifying selection for evasion of human immune response (McCutchan and Waters 1990; Miller *et al.* 1993; Escalante *et al.* 1998b), and so they too could be maintained despite a demographic bottleneck. The lack of silent site differentiation among dimorphic forms of several of these antigenic determinants further supports this hypothesis.

In order to better understand the evolutionary history of the antigenic alleles in *P. falciparum*, it is imperative to utilize a model that incorporates all biologically relevant information. A high level of amino acid polymorphism is evident in several antigenic genes that have been examined. Most of these amino acid changes have been be mapped directly to B and T cell **epitopes** (Anders *et al.* 1993). At the nucleotide level, the disproportionate number of nonsynonymous substitutions relative to synonymous substitutions, indicates that these regions are under positive diversifying selection (Escalante *et al.* 1998b). The requisite assumption of this model is that point mutations are equally likely to occur at any site, but only those that favorably alter phenotype (amino acid) will be selected, and hence become maintained, while the neutral sites (nonselected) will be lost or fixed at random. When deleterious mutations occur, negative selection will remove them, and the amino acid sequence will be conserved. This is the basic model of molecular evolution and, with various corrections for multiple nucleotide substitutions at individual sites, its validity has been confirmed in innumerable protein-coding genes in a great diversity of living species (Li 1997).

However, not all DNA sequences adhere to this model. Consider for example, the DNA repeat regions that make up **micro**- and **minisatellite** loci in various plant and animal species, including humans. Most of the variation within these repeats originates by a slipped-strand process that yields **duplication** and/or **deletion** of the repeated units. This process leads to rapid differentiation of alleles, wherein an individual mutational event can change several nucleotides at once, with greater impact on sequence divergence than the typical single nucleotide mutation process. Moreover, these mutations occur at rates that are orders of magnitude greater than that of single nucleotide substitutions (for review, see Hancock 1999). For this reason, even closely related individuals in a population, which may be identical in their coding DNA at most loci, may show marked differentiation at microsatellite loci. That these loci are so highly polymorphic reflects that repetitive elements are susceptible to frequent slipped-strand mutations. The diversification of DNA satellite sequences typically takes place in the absence of selection, since the loci themselves do not encode a protein product. For example, among the 27 336 bp of intronic sequence examined by Volkman *et al.* (2001), there was only one single nucleotide polymorphism, but several mutations were detected in microsatellite repeats within the introns. The authors concluded that microsatellite mutations occur at rates that are much higher than that of single nucleotide substitutions.

A notable feature of *P. falciparum* surface proteins is the presence of certain repeating nucleotide sequences, which encode short iterative amino acid sequences (Dame *et al.* 1984; Anders *et al.* 1988). These antigenic repeat regions are highly polymorphic, yet the repeat regions are known to be in many instances under immune selection. This presents a novel situation to the molecular evolutionist, in that these loci behave as one would expect satellite DNA to behave with respect to the rapid mutation process and the generation of variable-length sequences, while the repeat portions encode part of the functional protein, and so are subject to selection pressure.

As an example, let us consider the alleles of a known *P. falciparum* antigen encoding gene: the merozoite surface protein-1 (*Msp-1*) locus. Tanabe and colleagues subdivided the protein into 17 *blocks*, which were labeled as "conserved," "semiconserved," and "variable," based on the degree of polymorphism among various *Plasmodium* strains (Tanabe *et al.* 1987). By examining the polymorphic blocks, that is, the semi-conserved and variable blocks, Tanabe identified clear dimorphism that distinguished two groups, which we will refer to herein as Group I and Group II. We examined numerous *Msp-1* sequences of Group I and Group II strains from the GenBank database to determine the distribution of polymorphisms within each block. Our findings are summarized in Table 8.2. We looked at the amount of synonymous and nonsynonymous polymorphisms. It is clear from this analysis that the amount of nucleotide polymorphism is not uniform

**Table 8.2** Nucleotide diversity within and between group I and II alleles of the *P. falciparum Msp-1* genes

| Block | Length (codons)[a] | $N^b$ | $d_s$ | | | $d_n$ | | |
|---|---|---|---|---|---|---|---|---|
| | | | Group I | Group II | Group I + Group II | Group I | Group II | Group I + Group II |
| 1 | 55 | 33 | 0.019 | 0.021 | 0.017 | 0.017 | 0.010 | 0.013 |
| 2 | 55 | 33 | 0.106 | 0.185 | 0.150 | 0.449 | 0.497 | 0.553 |
| 3 | 202 | 33 | 0.038 | 0.006 | 0.042 | 0.018 | 0.000 | 0.023 |
| 4 | 31 | 29 | 0.031 | 0.000 | 0.020 | 0.307 | 0.000 | 0.215 |
| 5 | 35 | 29 | 0.000 | 0.000 | 0.070 | 0.000 | 0.000 | 0.026 |
| 6 | 227 | 8 | 0.000 | 0.000 | 0.282 | 0.004 | 0.001 | 0.300 |
| 7 | 73 | 8 | 0.000 | 0.000 | 0.361 | 0.003 | 0.000 | 0.072 |
| 8 | 95 | 8 | 0.000 | 0.000 | 0.338 | 0.000 | 0.003 | 0.711 |
| 9 | 107 | 8 | 0.000 | 0.023 | 0.409 | 0.005 | 0.043 | 0.126 |
| 10 | 126 | 8 | 0.008 | 0.000 | 0.448 | 0.011 | 0.000 | 0.394 |
| 11 | 35 | 11 | 0.000 | 0.000 | 0.128 | 0.000 | 0.000 | 0.068 |
| 12 | 79 | 11 | 0.000 | 0.000 | 0.000 | 0.000 | 0.000 | 0.000 |
| 13 | 84 | 11 | 0.000 | 0.042 | 0.040 | 0.005 | 0.007 | 0.052 |
| 14 | 60 | 11 | 0.000 | 0.018 | 0.212 | 0.002 | 0.005 | 0.371 |
| 15 | 89 | 19 | 0.000 | 0.000 | 0.216 | 0.001 | 0.003 | 0.089 |
| 16 | 217 | 19 | 0.002 | 0.032 | 0.277 | 0.005 | 0.027 | 0.185 |
| 17 | 99 | 19 | 0.002 | 0.019 | 0.007 | 0.010 | 0.027 | 0.016 |

The values are the mean numbers of synonymous ($d_s$) or nonsynonymous ($d_n$) substitutions per site.

[a] Block length may vary between group I and II alleles; the given value is the average length of group I and II alleles.

[b] Only partial *Msp-1* sequences are available for some strains in GenBank; however, our analysis includes eight complete coding sequences: three of Group I and five of Group II.

across the length of the molecule, and so we conclude that the different blocks may have quite distinct evolutionary histories.

For nearly every block, the degree of intra-group polymorphism is less than 0.05, and most are less than 0.01. The exceptional case is block 2, which shows markedly higher intra- and intergroup differences than any other block. This can be attributed to the tripeptide repeats in block 2 which create particular difficulty in determining appropriate alignment, and therefore render these deceptively high values of $d_s$ and $d_n$. Nucleotide repeats, and hence their corresponding peptide **motifs**, are susceptible to mutational mechanisms that occur at much greater frequency than singular point mutations. To discern the evolutionary history of regions containing these repeats, for example, *Msp-1* block 2, we must consider the most likely model by which mutations accumulate. In Fig. 8.2(a), we have presented such a model. Note that in the hypothetical ancestor, no repeats are present. Repeats arise first by duplication of a short sequence following mutation during replication of the DNA strand. In subsequent generations, additional copies of the repeat accumulate in some alleles by **slipped-strand mutation** (SSM), so that in a few generations various length polymorphisms may arise among the alleles (we would also point out that the same SSM process can lead to loss of a repeat copy, see Levinson and Guttman (1987) for details of the SSM process). Novel repeats may also arise by either substitutions within the ancestral repeat (as with the open circles) or alternatively new repeats may be born by precisely the means of the original repeat birth (shown here as gray triangles). This model is not particularly novel, since an analogous process is

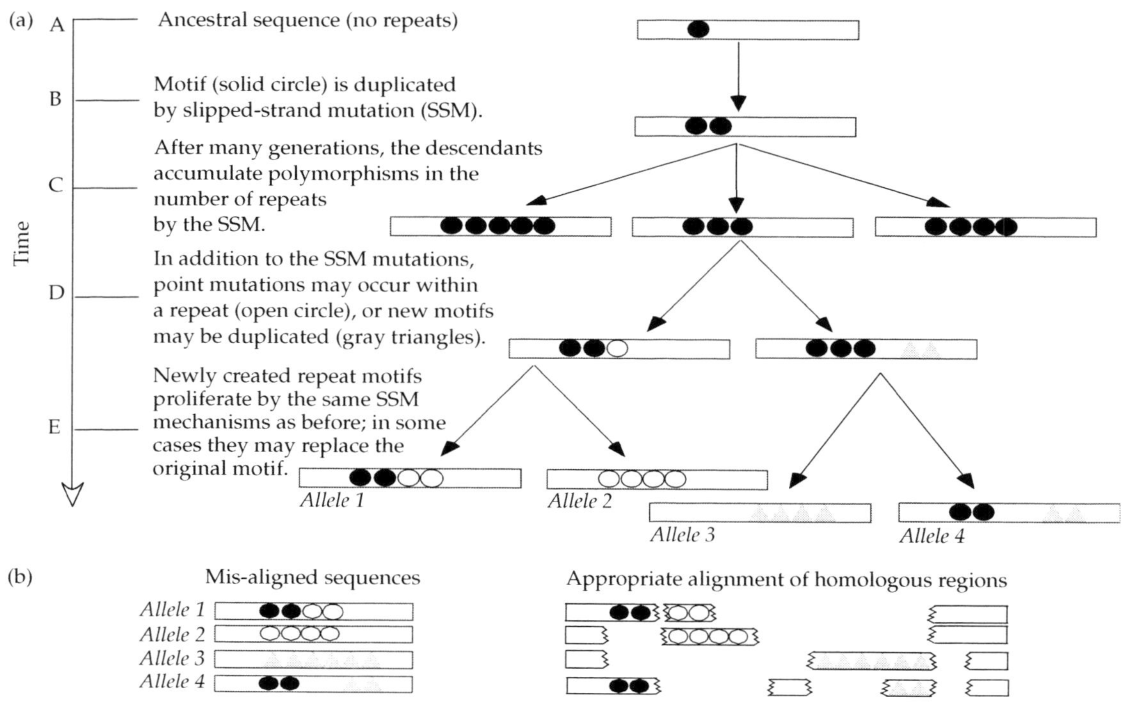


**Figure 8.2** (a) Model of nucleotide-repeat evolution. Rectangles represents the entire gene, which has a finite length such that proliferation of a repeating unit leads to loss of other repeats or loss of single-copy, nonrepeat regions. Three distinct repeat sequences, which differ from one another by at least one nucleotide, are shown as solid circles, open circles, and triangles. An arbitrary time scale is shown on the left. (b) An alignment of the four alleles from time point-*E*.

invoked to explain evolution of microsatellite loci. However, spurious conclusions arise when aligning the set of alleles that result from this process. This is demonstrated in Fig. 8.2(b), where we show the alignment of the four sequences from time point-*E*. Note that in the alignment on the left, the various repeat motifs are aligned according to their position, but as we have shown in Fig. 8.2(a), these repeats are not **homologous**. In fact, the most likely **homologue** of a given repeat is an adjacent repeat from the same allele. The alignment on the right of Fig. 8.2(b) shows how certain regions of individuals alleles may lack areas of **homology** in other alleles due to the duplication and replacement of repeat units. It is exactly this kind of alignment artifact that explains the extraordinary intra-group polymorphism in block 2. We have shown similar patterns in nucleotide repeat regions of other antigenic genes as well, including *Csp* and *Msp-2* (Rich *et al.* 1997, 2000; Rich and Ayala 2000).

With very few exceptions, the differences between group I and group II far exceed the amount of polymorphism within either group for both synonymous and nonsynonymous changes. As we stated above, the remaining 16 blocks (i.e. other than block 2) of the *Msp-1* show very little within group polymorphism. However, between groups, there is considerably more nucleotide polymorphism, both synonymous and nonsynonymous. Hughes (1993) and Hughes and Hughes (1995) have argued that polymorphisms observed between the two groups in blocks have been maintained within the species for millions of years by **balancing selection**.

For Hughes' model to be correct, one must invoke extraordinarily high rates of recombination and extreme selection coefficients. Given the biology of the parasite, and propensity of past and recurrent bottlenecks, we conjectured that this seemed rather implausible. This led us to propose that it is rate of evolution, and not the age of these blocks that is so vastly different. We hypothesized that the SSM processes that caused the extreme polymorphism and inflated estimates of nonsynonymous **nucleotide diversity** ($d_n$) in block 2, may also have occurred elsewhere in the molecule. Our suspicions were confirmed when we identified repeats within several of the most polymorphic *Msa-1* blocks, in particular, blocks 4, 8, and 14, which were previously characterized as nonrepeat blocks.

Consider block 8, which is the block identified by Tanabe *et al.* (1987) as showing the lowest amino acid similarity between groups (10 percent), and which in our analysis is the most polymorphic in terms of nonsynonymous nucleotide diversity ($d_n = 0.711$). We have identified three group-specific repeats within this block (see also Rich *et al.* 2000). One 9-bp repeat (R2a) is present in all group II alleles; and two repeats, of 6 bp (R1a) and 7 bp (R1b), are present in all group I alleles. We have hypothesized that the occurrence of these repeats within this very short stretch of DNA is a highly significant departure from chance. To test this hypothesis, we have searched the recently completed genomic sequences of *P. falciparum* chromosomes 2 and 3. The nucleotide sequences of repeats R1a, R1b, and R2a appear 25, 116, and 11 times, respectively, within the 947 kb of chromosome 2. Within the 1060 kb of chromosome 3; the R1a, R1b, and R2a are present 39, 52, and 7 times, respectively. None of the three-nucleotide repeats ever appears in tandem on either chromosome 2 or 3. Moreover, the average distance between each occurrence on these chromosomes is >20 kb, demonstrating that their repeated occurrence in the short 147-bp segment of *Msp-1* block 8 strongly departs from random expectation. It would, therefore, appear that, like block 2, this region has undergone rapid differentiation, driven by the SSM process outlined in Fig. 8.2(a). This rapid differentiation is the mechanism by which the parasite is able to generate such great antigenic diversity even in the face of recurrent demographic sweeps. The widespread occurrence of similar genomic regions of high mutability among potential vaccine targets will surely have some bearing on predicting the probability of success of protective vaccines.

## Summary

Parasites of the genus *Plasmodium* comprise a diverse array of species. Several of these species are important human pathogens with great impact on human health. *P. falciparum* is the most pathogenic of these species. *P. falciparum* shares a long evolutionary history with our own species, but some drastic events in its recent population history have genetically homogenized the species. There is strong evidence that a major demographic sweep or bottleneck occurred within the past 57 000 years, but probably more recently (Rich *et al.* 1997). The evidence is based on a lack of synonymous polymorphisms among nuclear genes, including both antigenic and nonantigenic loci, among introns, and among mitochondrial loci. Coalescence calculations for silent nucleotide variation converge into a single allele at the bottleneck. This, however, does not imply that the bottleneck consisted of one single individual. It is possible that certain polymorphisms under strong balancing selection could have survived the bottleneck by being present in several populations. However, a more plausible explanation is that certain highly polymorphic antigens do not reflect extremely old polymorphisms, but rather they are the result of extremely rapid rates of evolution associated with the repetitive DNA elements that they contain. Understanding these and other evolutionary aspects of the biology of *Plasmodium* provides important insights into the population structure and likely evolution of the parasite, with potential large implications for the control of human malaria.

## Further reading

Splendid coverage of current knowledge concerning malarial evolution is found in two multiauthored volumes:

Coluzzi, M and Bradley, D (eds) (1999). *The malaria challenge after one hundred years of malariology*. Lombardo Editore, Rome.

Sherman, IW (ed.) (1998). *Malaria, parasite biology, pathogenesis, and protection*. American Society of Microbiology Press, Washington, DC.

CHAPTER 9

# The evolution of symbiosis in insects

**Roeland C. H. J. van Ham, Andrés Moya, and Amparo Latorre**

Prokaryotic microbes are nearly everywhere: from the atmosphere at altitudes as high as 77 km to within subsurface sediments as deep as 4 km, from benign and **eutrophic** places like freshwater lakes, tropical rainforest soils, and mammalian intestines, to the most harsh environments of deep ocean hydrothermal vents, acidic rivers, desert soils, and the Antarctic ice shell. But not only are they "everywhere": in many habitats prokaryotes occur in astronomic numbers. Given their ecological ubiquity and abundance, it is not surprising to find many prokaryotic species in close physical association with members of the most speciose group of eukaryotes on earth: the insects. Microbes and insects not only show an amazing biodiversity by themselves, but they often come together and take evolutionary paths to persistent physical association (Buchner 1965).

In the following we will first address **symbiosis** or "living together" as a general phenomenon in organismal evolution and complexity, and we give a brief account of common molecular themes in the establishment and maintenance of symbiotic associations. We then focus on the biodiversity and evolutionary impact of microbial symbiosis in insects, and conclude with an overview of the most recent genomic and molecular population genetic studies of microbial symbioses and the insights these reveal into the evolutionary processes governing symbiosis and its long-term evolutionary fate.

## Symbiosis in the biotic world

As there has been considerable semantic debate over the issue, the word symbiosis requires a definition (Douglas 1994; Paracer and Ahmadjian 2000). Here, we use it in a broad sense to refer to all cases of microorganisms (symbionts) living together with a host, irrespective of the **fitness** effects on either partner. As such, it covers the canonic spectrum of physical associations between organisms, ranging from **mutualism**, in which partners are in association throughout most or all of their life cycle and are, therefore, assumed to be mutually beneficial, through **commensalism**, in which associations are labile and clear benefits or detrimental effects to partners have not or can not be established, to **parasitism**, in which there is a contingent physical association from the perspective of the host and a positive effect only to the parasite (Douglas 1994; Moran and Wernegreen 2000). Parasites that have clear negative effects on the host are called pathogens.

All symbioses considered in this chapter are **endosymbioses**, meaning that one partner, generally a prokaryote, is located inside the body of the other. Further distinctions, particularly relevant to symbioses of microorganisms carried by multicellular hosts, include the occupation of either extracellular or intracellular habitats, and whether symbioses are obligate or facultative from either partner's perspective. Finally, a key factor in the evolution of symbiosis is the mode of transmission of symbionts. In **vertical transmission**, symbionts are transferred from parent to offspring through the infection of eggs or of embryos within the mother's body. In **horizontal transmission**, host progeny is infected *de novo*, from outside the parent's body.

**Symbiogenesis**, the evolutionary process of establishing a symbiotic association, has been one of the dominant forces in the early evolution of life on earth. By some, it is believed to have laid down the very foundation of the nuclear lineage among the three domains of life, through a symbiosis between a eubacterium and an archaebacterium (Margulis 1992). The mitochondria and chloroplasts (the plastids) of modern eukaryotes are beyond doubt the product of symbiogenic events between prokaryotes and primitive eukaryotes that took place around 2000 and 1000 million years ago, respectively (Margulis 1992). Three developments were fundamental to these primordial symbioses. First, besides massive gene loss, a one-way transfer of genetic information contained in the symbiont to the host nuclear genome took place. Second, as a compensatory process, a protein import machinery evolved in the symbiont to obtain back the protein-products of those transferred genes that fulfill an essential metabolic role. And finally, under the selective pressure of increasing interdependence, a secure mechanism of vertical transmission of symbionts to the host's offspring evolved. These developments are believed to have proceeded rapidly in evolutionary time. For the symbionts, or the organelles they are at present, they implied an irreversible loss of control over their own cellular processes. Such a surrender of autonomy of formerly independently replicating entities is a common feature to all major transitions in the evolution of life (Maynard Smith and Szathmáry 1995; see Michod and Nedelcu, Chapter 17).

Beyond shaping the primal eukaryotic cell, symbiogenesis has continued to play a major role in the evolution of life. On one side of the spectrum, pathogenesis has been proposed as a driving force behind the evolution of immune systems in higher eukaryotes. Pathogenesis in this context is understood as the evolutionary process leading towards specialized exploitation of invaded hosts. On the other side, mutualistic interactions are thought to underlie the ecological and evolutionary diversification of many taxonomic groups. Prokaryotic symbionts in particular bring from their previous free-living state a broad range of metabolic capabilities with them. Some of these are specifically utilized by the host, as in the exemplary case of plant-sap sucking insects described below, and enable it to occupy an ecological niche otherwise inaccessible to it.

Many symbioses have reached a stage of complete interdependence (**obligate mutualism**), or complete dependence of one partner on the other (**obligate parasitism**). However, genomic integration of newly acquired symbionts into hosts, such as seen in the primordial eukaryotic symbioses, is not known from any group of multicellular eukaryotes. This marks a fundamental difference in the potential of symbiogenesis in uni- and multicellular eukaryotes. As we will discuss hereafter in the context of symbiont genome evolution, lack of genomic integration, together with certain population dynamic characteristics of obligate symbioses, evoke dramatic genome changes, that in a long-term evolutionary perspective, may even drive this particular kind of symbionts to extinction (Ochman and Moran 2001).

Symbioses involving prokaryotes are common among protozoa. Many of these unicellular eukaryotes feed on bacteria, which they swallow whole through phagocytosis. Once inside the cell, some bacteria may be of greater benefit to the host when still metabolically active than when killed and digested. This is also thought to be the way in which the eukaryotic symbioses with present-day mitochondria and chloroplasts were initiated. The former were once free-living purple bacteria that conferred respiration and efficient energy metabolism onto their hosts, while chloroplasts, once free-living cyanobacteria, rendered their hosts capable of photosynthesis. A prime example of extant protozoan symbioses is the association of methanogenic prokaryotes with anaerobic protozoa. Up in the ecological hierarchy, such protozoa in turn play a symbiotic role in the intestinal tracts of many herbivorous animals, from termites to mammals (Paracer and Ahmadjian 2000).

Mutualistic symbioses with microorganisms are also widespread in the plant kingdom and most importantly involve associations localized in root tissues with either nitrogen-fixing bacteria or mycorrhizal fungi. Among the best-studied symbioses is the association between leguminous plants and bacteria of the *Rhizobiae* group (Paracer and Ahmadjian 2000). Infection of the roots of these plants with *Rhizobium* leads to the formation of root nodules. The bacteria find shelter in these tissues, in return for which they are specifically induced to the ecologically important process of nitrogen fixation. The host plant benefits from this natural process of fertilization and has a selective advantage over other plant species in nitrogen-deficient soil habitats.

A most renowned form of symbiosis in the fungal kingdom is represented by the lichens. These are fungi that contain photosynthetically active algae or cyanobacteria. Lichens are a prime example of **facultative mutualism**: the majority of the fungal, algal, and bacterial species involved in lichen formation are also capable of free growth. Environmental conditions like temperature, moisture, and nutrient-availability either induce their cooperation or dissolution, with harsh conditions favoring lichen formation.

Apart from the arthropods and especially the insects, in which microbial symbioses are most common, many other animal phyla comprise groups that maintain mutualistic associations with prokaryotes. These include the Porifera (sponges), Cnidaria (jellyfish, anemones, corals, etc.), Nematoda, Annelida (worms), Mollusca (snails, clams, squids, etc.), and vertebrates. Of great ecological importance are the symbioses between dinoflagellate algae (*Symbiodinium*) and many benthic marine invertebrates, including sea anemones and corals. Similar to the role played by nitrogen-fixing bacteria in plants, the photosynthetic *Symbiodinium* provide nutrients to their hosts which, for example, makes reef-forming by corals possible in nutrient-poor tropical waters (Paracer and Ahmadjian 2000).

## Establishment of symbiosis

Whereas theoretical approaches to the study of symbiosis have always considered mutualism and parasitism as two sides of the same coin, experimental research on either pathogenic or mutualistic bacteria have long remained separate fields. Recent work, however, has stressed common features in the underlying genetic and molecular mechanisms of these two types of symbioses (Goebel and Gross 2001). That there should be parallels, at least at the onset of physical interaction with eukaryotic cells, seems obvious, as parasitism and mutualism are different outcomes of an evolutionary process that starts from presumably similar free-living conditions or **preadaptations** to intracellular life (Moran and Wernegreen 2000). In any case, before either road can be taken, hosts have to become infected. Infection for any would-be symbiont means that it successively has to enter the host body, gain access to host tissues and cells, subvert their defense mechanisms, and finally, secure maintenance and multiplication within the host.

Without regulation or cooperation from a mutualistic host, successful completion of an infectious cycle requires a certain genetic potential from the invader. Comparative studies of pathogenic and nonpathogenic bacterial strains have identified and characterized such genetic elements, which are known as **virulence factors**. These include a diverse array of genes involved in the production of cell-surface components, secretion systems, siderophores, toxins, or in the regulation of their expression. It has also been shown that virulence factors are particularly prone to horizontal gene transfer. They are often found in association with genetic elements that mediate this process, such as plasmids, **transposons**, or phages (Ochman *et al.* 2000).

One of the most exciting examples of infectious mechanisms used by different kinds of symbionts, parasitic or mutualistic, and most relevant to insect symbiosis, is described in a recent study of *Sodalis glossinidius*, the secondary symbiont of tsetse flies (Dale *et al.* 2001). Dale and co-workers demonstrated that attachment to and invasion of host cells by *Sodalis* requires functional components of a type III secretion system. A number of genes involved were shown to be most closely related to the **homologous** systems from the pathogenic bacteria *Salmonella* and *Shigella*. Type III protein secretion systems are one category in a family of bacterial transport systems that are used to export proteins or to deliver toxins or modulators directly to the host cells. They are used by a wide variety of bacteria pathogenic to animals as well as to plants. Moreover, a homologous system is present in *Rhizobium* where it plays a role comparable to the one now established in a mutualistic insect symbiosis.

Other examples of similar cellular systems employed for host–symbiont interaction by both pathogens and mutualists include quorum sensing mechanisms, which are involved in regulation of **virulence** expression in various animal and plant pathogens as well as in the regulation of bioluminescence in *Vibrio fischeri* symbioses, two-component regulatory systems associated with various virulence factors but also involved in the regulation of gene expression of symbiotic nitrogen fixation in *Rhizobium*, the presence of type IV secretion systems in both the insect symbiont *Wolbachia* and human and plant pathogens such as *Bordetella, Helicobacter*, and *Agrobacterium*, the presence of a flagellar export apparatus in both the aphid symbiont *Buchnera* and the pathogen *Yersinia enterocolitica*, and the presence of similar "genomic islands" associated with tRNA loci that are involved in pathogenicity in *E. coli* or in symbiosis in *Mesorhizobium*.

Upon successful multiplication within an invaded host, completion of the infectious cycle ultimately requires exiting the host, spreading to new ones and recurrent infection. At this stage, a potential symbiotic life-style becomes subject to natural selection. Parasitism and mutualism are commonly considered as the only stable outcomes of this evolutionary process. What are then the forces that determine symbiotic strategy? Two key factors, operating jointly, have been proposed by theoretical studies (Ewald 1995). These concern differences in fitness under infection among each of the partners involved and the mode of transmission of symbionts. Host susceptibility and ecological circumstances are determinants of relative fitnesses under infection. Considered from the perspective of host response to infection, modeling indicates that parasitism by the symbiont is (1) favored strongly in systems where host defense is weak and transmission occurs horizontally with genetically diverse symbionts that compete among each other, and (2) favored weakly in systems where host defense is strong and transmission occurs horizontally irrespective of the symbionts' genetic variability. Cooperation by the symbiont, on the other hand, is favored strongly (3) in systems of vertical transmission, irrespective of the strength of host defense, and (4) in systems with weak host defense and horizontal transmission of symbionts without genetic variability (Maynard Smith and Szathmáry 1995). In this scheme, host defense may either be inherently weak, due to the absence or faltering of specific cellular defense mechanisms, or because the invading symbiont contingently invokes increased fitness, for example, by increasing resistance to other pathogens or by the release of a nutrient essential to the host.

The model is in agreement with the longstanding notion that mutualistic symbionts are in fact attenuated parasites, but not with the assertion that pathogenic bacteria will ultimately evolve towards benign cooperation. One crucial question is, what invokes the evolution of mechanisms of vertical transmission? The benefit of vertical over horizontal transmission in symbioses that are in the process of evolving towards mutualism and increased interdependence seems obvious, as it is more secure to inherit what one is dependent on than having to rely on *de novo* acquisition of such a factor from the environment. Horizontal transmission is a safer strategy from the perspective of a virulent parasite: If the host is exploited to a point where resources get exhausted it pays to spread and infect new, healthy hosts. However, the model described above predicts that vertical transmission will decrease the virulence of a parasite. It is, therefore, likely that infections of hosts with small numbers of genetically homogenous parasites will evolve into mutualistic associations. Particularly in insects that already contain populations of vertically transmitted bacteriocyte symbionts, hitchhiking of benign parasites on the established mechanisms of symbiont transmission may promote the appearance over evolutionary time of secondary symbionts (see below).

Mutualistic symbioses that rely on horizontal transmission are to our knowledge rare among insects. In many insects, however, as in mammals, a vital extracellular gut microflora is acquired horizontally through exchange and consumption of infected food. Honey bees are known to inoculate food sources with *Bacillus* spp., which play a role in maturation and preservation of food stored in their hive. A prime example of horizontal acquisition of symbionts in other invertebrates are luminescent bacteria of the genus *Vibrio* that inhabit the light organ of the marine squid *Euprymna*. These bacteria are taken up from the environment on a daily basis. Their bioluminesence plays a role during feeding behavior of the host at night. Failure to acquire the bacteria in the early developmental stage results in defective light organ formation, which is believed to reduce life expectancy of the squids (McFall-Ngai 1999).

## Symbiosis in insects

Symbioses, in general, appear to have evolved around nutritional interactions. In mutualistic associations, nutrients may flow in both directions, from symbiont to host and vice versa, while this flow is notoriously unidirectional in parasitic associations. A conspicuous feature of the diversification of insects is that many groups have specialized in feeding on nutritionally inadequate diets (Douglas 1998). Homoptera, for example, feed on plant-sap, which is of unbalanced nitrogen/carbon content and deficient in a number of amino acids that insects, like other animals, cannot synthesize; termites and various beetles are renowned for feeding on wood, a diet low in proteins and amino acids and mainly composed of cellulose and lignin, which most insects are unable to digest; various Heteroptera and Diptera feed solely on animal blood, which is devoid of certain B vitamins. Concurrently, each one of these groups is known to harbor obligate, symbiotic microorganisms. A nutritional basis to these associations, therefore, seems obvious, indeed has long been suspected and in some cases has been confirmed through physiological (Douglas 1998), genetic (Baumann *et al.* 2000), and genomic studies (see below).

Insects arose from the arthropod lineage in the middle Devonian, about 385 million years ago, and rapidly diversified in the ensuing 100 million years. Estimates of the age of some prokaryotic symbioses with insects go back at least 300 million years (Moran and Baumann 1994). Given the group's proclivity to specialized feeding behaviors, which are enabled through nutritional interactions with prokaryotes, the presence of such associations throughout most of the insects' evolutionary history suggests that symbiosis has been a driving force in the diversification of the group.

Extant insect prokaryotic symbioses vary widely in characteristics like persistence, tissue distribution of the symbionts, their mode of transmission, and the level of integration between both partners. As many insects have been found to persistently harbor more than one prokaryotic species, each with a different host tissue distribution, a categorization is often being made into **primary** and **secondary symbioses** (Table 9.1). Associations with primary symbionts are obligate for both partners, permanent, and mutualistic. These symbionts are confined to a single, specialized cell type and are transmitted vertically from mother to offspring. A nutritional role has been experimentally established for only a few of these primary symbionts. A more appropriate designation of primary symbiosis is **bacteriocyte symbiosis**, which makes reference to the cells of the insect specialized for harboring bacteria (bacteriocytes). Bacteriocyte symbioses show a broad phylogenetic distribution among insects and have been estimated to occur in 10 percent of the known species. Much less is generally known about secondary symbionts. They are not confined to a specialized cell type and may occur in gut tissues, glands, body fluids, or in cells surrounding the bacteriocytes (Fukatsu and Ishikawa 1993). Although they have been shown, through both phylogenetic and experimental studies, to be subject to vertical transmission in various groups, their sporadic distribution among populations implies that they also undergo horizontal transmission. Their effects on host fitness are obscure, but despite the fact that they are in less stable association with their hosts, secondary symbionts are not generally considered to be commensals. Commensalism is inherently contingent, and for this the associations, often intracellular, show too much persistence among populations.

Commensalism *per se* has not been studied in insects, although one report on aphids (Harada *et al.* 1996) gives a glance at what kind of bacteria might be encountered in an insect. Some of these may be commensals, others obligate or facultative pathogens. Although harmless to the host, commensalism in insects and other arthropods plays an important, and often devastating, role in disease transmission in animals and plants. Plant-sap

**Table 9.1** Features of symbiotic prokaryotes in insects

| Host | Food source | Symbiont | Symbiosis | Proposed role | Related bacteria |
|---|---|---|---|---|---|
| Many insects | | *Wolbachia* | p | Sexual alterations | *Rickettsia* |
| | | *Rickettsia*-like | p | ? | *Rickettsia* |
| | | *Serratia* | p/c | | Enterobacteriaceae |
| Aphids | Plant sap | *Buchnera* | m | Essential amino acids, vitamins | Enterobacteriaceae |
| | | S-symbiont | c? | ? | Enterobacteriaceae |
| White flies | Plant sap | P-symbiont | m | Essential amino acid provisioning | γ-Proteobacteria |
| | | | | ? | |
| | | S-symbiont | | | Enterobacteriaceae |
| Psyllids | Plant sap | *Carsonella* | m | Essential amino acids, vitamins | γ-Proteobacteria |
| | | S-symbiont | | ? | Enterobacteriaceae |
| Mealybugs | Plant sap | P-symbiont | m | ? | β-Proteobacteria |
| | | S-symbiont | see text | ? | γ-Proteobacteria |
| Tsetse fly | Animal blood | *Wigglesworthia* | m | Vitamins | Enterobacteriaceae |
| | | *Sodalis* | m? | ? | Enterobacteriaceae |
| Carpenter ants | Plant nectar, honey dew | *Blochmannia* | m | ? | Enterobacteriaceae |
| Ticks | Animal blood | *Francisella*-like | p | | |
| | | *Coxiella*-like | p | | |
| Cockroaches | Omnivores | *Blattabacterium* | m | ? | *Flavobacterium* |
| Termites | Dead wood | *Blattabacterium* | m | ? | *Flavobacterium* |
| | | Many species inhabiting the hindgut | m, c | Cellulose degradation, nitrogen fixation | |
| Weevils | Grain | P-symbiont | m | Vitamins | Enterobacteriaceae |
| Bedbugs | Animal blood? | S-symbiont | m | Host reproduction | Enterobacteriaceae |
| Triatomine bug | Animal blood? | *Arsenophonus* | m | Vitamins | γ-Proteobacteria |

P: primary endosymbiont; S: secondary endosymbiont; c: commensalism; m: mutualism; p: parasitism.

sucking insects, with their highly evolved, needle-like mouth-parts, are efficient transmitters of plant pathogenic viruses and bacteria that cause major damage to agriculture. Over 40 percent of viruses and many bacteria and protozoa pathogenic to mammals have life cycles that include a commensal or benign parasitic passage through an arthropod, mostly blood-feeders. Those notorious for their pathogenicity in humans include, among others, West Nile virus, Dengue fever, and Yellow fever, viruses transmitted by mosquitoes, *Rickettsia prowazekii*, the agent of epidemic typhus and transmitted by lice, *Yersinia pestis*, primarily a rodent pathogen but occasionally transferred to humans through infected fleas and causing plague, *Borrelia burgdorferi*, the causative agent of Lyme disease and transmitted by ticks, various *Plasmodium* species, the agents of malaria and transmitted by mosquitoes, and African trypanosomes, transmitted by tsetse flies and causative agents of sleeping sickness (Crampton *et al.* 1997).

A most special kind of symbiont, occurring in many insects as well as in other arthropods and even other animal phyla, are bacteria of the genus *Wolbachia* (O'Neill *et al.* 1997). These have specialized on the infection of host reproductive tissues. They are primarily cytoplasmically (i.e. maternally) inherited. Nevertheless, horizontal transmission has played an essential role in their evolutionary history, as is evidenced by their widespread distribution. Together with a few other groups of prokaryotes and eukaryotic microbes, *Wolbachia* are known as **sexual parasites** after the effects they have on their hosts' reproductive systems. The reproductive alterations inflicted result in female-biased sex ratios and, thus, promote their own maternal transmission rate. The alterations include induction of **parthenogenesis** (female asexual reproduction), cytoplasmic incompatibility between strains and species, in which infected males cause sterility in uninfected females, and feminization of male offspring.

Molecular phylogenetic studies based on 16S rDNA sequences have contributed to establishing the phylogenetic position of symbiotic prokaryotes among the lineages of their free-living relatives (Fig. 9.1). It is apparent that stable symbioses have evolved many times independently in diverse groups of eukaryotes. The Proteobacteria (purple bacteria), the largest and physiologically most diverse group of the Eubacteria, are particularly rich in both symbiotic and (facultative) pathogenic lineages and are associated with a wide variety of animal and plant groups. This points to a common genetic potential of these bacteria to associate with eukaryotic cells. Nevertheless, one important conclusion that can be drawn from the bacterial phylogenies is that obligate symbionts mostly appear to have evolved from distinct and often ancient lineages that are either strictly parasitic or strictly mutualistic (Moran and Wernegreen 2000). In contrast, symbionts of less stable association, such as many insect secondary symbionts or facultative

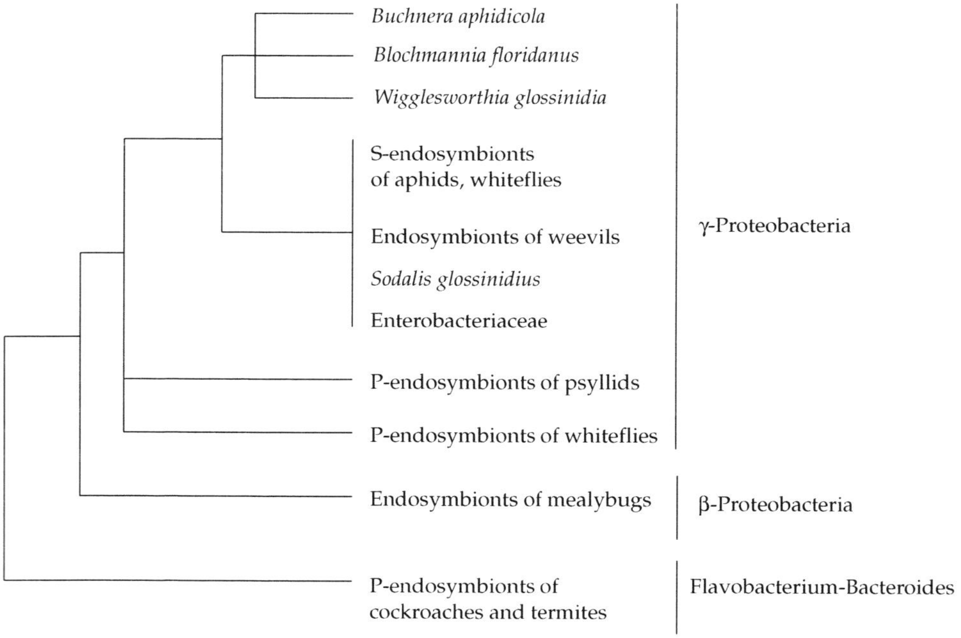


**Figure 9.1** Evolutionary relationships obtained by parsimony analysis of 16S ribosomal DNA sequences from insect symbionts. *Buchnera aphidicola*, *Blochmannia floridanus*, and *Wigglesworthia glossinidia* are the primary endosymbionts of aphids, carpenter ants, and tsetse flies, respectively. *Sodalis glossinidius* is the secondary endosymbiont of tsetse flies. Greek letters represent subdivisions of the Proteobacteria. P: primary endosymbiont; S: secondary endosymbiont.

pathogens of vertebrates, are found scattered over the evolutionary tree among more recent branchings of free-living bacteria.

Many insect symbionts belong to the $\gamma$ subdivision of the Proteobacteria, which includes the well-known Enterobacteriaceae (Fig. 9.1). Concordant with the conclusion drawn above, bacteriocyte symbioses are generally ancient and constitute distinct phylogenetic lineages, and they show a distinct pattern of co-speciation between host and symbiont as a consequence of their strictly maternal mode of inheritance. A number of those within the $\gamma$ subdivision, including the aphid-*Buchnera*, the carpenter ant-*Blochmannia* (Schröeder *et al.* 1996), and the tsetse fly-*Wigglesworthia* symbioses (Aksoy 1995), cluster outside though near the base of the Enterobacteriaceae. Most secondary symbionts, on the other hand, cluster within the Enterobacteriaceae, for example, those of aphids, white flies, tsetse flies, and psyllids (Baumann *et al.* 2000).

One remarkable finding of the molecular phylogenetic studies is the very close phylogenetic relationship that exists, within the Enterobacteriaceae, between the secondary symbiont of tsetse flies (*Sodalis*) and the bacteriocyte symbiont of weevils (Heddi *et al.* 1998). The hosts belong to different insect orders (Diptera and Coleoptera) and feed on animal blood and stored grain, respectively. The close affiliation of their symbionts implies a relatively recent divergence and illustrates two important aspects of symbiont acquisition in insects: first, that, in contrast to most symbioses described to date, there are also bacteriocyte symbioses of apparent recent evolutionary origin, and second, that the common ancestor of both symbionts must have been able to spread among widely divergent hosts.

A notable exception to Proteobacteria as the common reservoir of insect symbionts are the obligate intracellular symbionts of cockroaches and "lower" termites. These have been assigned to the genus *Blattobacterium* and belong to the *Bacteroides-Flavobacterium* group, a deeply branching lineage among the Gram-negative bacteria. The *Blattobacterium* symbiosis also holds the record of the oldest age for any mutualistic symbiosis in insects (300 million years ago; Bandi *et al.* 1995).

## Genomics and population biology of obligate symbionts

In the past 8 years, complete genome sequences have been determined for several Eukaryotes and many Prokaryotes, among the latter a considerable number of obligate parasites. In the wake of these projects, researchers working on insect prokaryotic symbiosis soon recognized that complete genomic information of symbiotic microorganisms had great potential to rapidly advance our understanding of such associations. Insect symbionts are generally uncultivable and not amenable to experimental research. A number of genome sequencing projects were initiated in the late 1990s and the first complete genome sequence of a mutualistic symbiont, that of *Buchnera aphidicola* BAp, the primary endosymbiont of the aphid *Acyrtosiphon pisum*, was published in the year 2000 (Shigenobu *et al.* 2000; see Table 9.2). More recently, two more *B. aphidicola* genomes have been reported, one from the aphid *Schizaphis graminum* (Tamas *et al.* 2002) and another from the aphid *Baizongia pistacea* (van Ham *et al.* 2003). Given that *B. pistacea* belongs to a subfamily phylogenetically unrelated to the other two aphids species, and that the comparison of the three *Buchnera* genomes reveals a nearly perfect gene-order conservation, it is likely that the onset of genomic stasis in *Buchnera* coincided closely with the establishment of the symbiosis with aphids, *c*.200 million years ago (van Ham *et al.* 2003). Akman *et al.* (2002) have also described the genome sequence of *Wigglesworthia glossinidia*, the primary endosymbiont of tsetse flies, the vectors of African trypanosomes (Table 9.2). The genome analyses give support to the previously proposed nutritional role of these bacterial endosymbionts. *B. aphidicola* have genes coding for essential nutrients that are lacking in the aphid diet. *W. glossinidia*, on the other hand, retains genes involved in the biosynthesis of vitamin metabolites.

Comparative genomic analysis of obligate symbionts has revealed a set of features shared by parasites and mutualists. These include genome size reduction, mutational bias towards AT, and accelerated sequence evolution. The most distinctive of these is genome size reduction (Table 9.2).

**Table 9.2** Genomic features of obligate prokaryotic symbionts compared with their free-living relatives

| Symbiont | Free-living relatives | Host | Type | Genome size (kb) | Genes | GC-content % |
|---|---|---|---|---|---|---|
| *Arabidopsis thaliana* (chloroplasts) | Cyanobacteria | Plants | m | 154 | 132 | 36.3 |
| *Arabidopsis thaliana* (mitochondria) | α-Proteobacteria | Plants | m | 367 | 57 | 44.8 |
| *Homo sapiens* (mitochondria) | | Animals | m | 16.6 | 37 | 44.2 |
| *Rickettsia prowazekii* | | Animals | p | 1100 | 834 | 28.9 |
| *Buchnera aphidicola* BAp | γ-Proteobacteria | Aphids | m | 641 | 563 | 26.2 |
| *Wigglesworthia glossinidia* | | Tsetse flies | m | 703 | 661 | 22.0 |
| *Mycoplasma genitalium* | Gram-positive bacteria | Animals | p | 580 | 483 | 31.6 |
| *Borrelia burgdorferi* | | Animals | p | 911 | 855 | 28.5 |
| *Chlamydia pneumoniae* | Spirochaetes | Animals | p | 1230 | 1052 | 40.5 |

m: mutualist; p: parasite.

Like the primal eukaryotic symbionts, mitochondria and chloroplasts, obligate symbionts have undergone massive gene loss (Ochman and Moran 2001). Their phylogenetic relationships reveal, for each of them, close affiliates with much larger genomes, indicating that the process occurred independently multiple times. A comparison of cellular functions lost in these reductive processes reveals an overall convergence across functional classes of genes, but also differences that reflect specific **adaptations** characteristic for each symbiosis (Fig. 9.2). A striking example is the retention of genes involved in essential amino acid biosynthesis in *Buchnera*, while most of these have been lost in intracellular parasites, for example, *Rickettsia* (Tamas *et al.* 2001). The very basis of the symbiosis between *Buchnera* and aphids is thought to be the production of these essential amino acids, which aphids cannot synthesize and which are in short supply in their diet of plant sap. *Rickettsia*, on the other hand, parasitizes on the metabolic products of its hosts and has retained a higher fraction of genes involved in their transport.

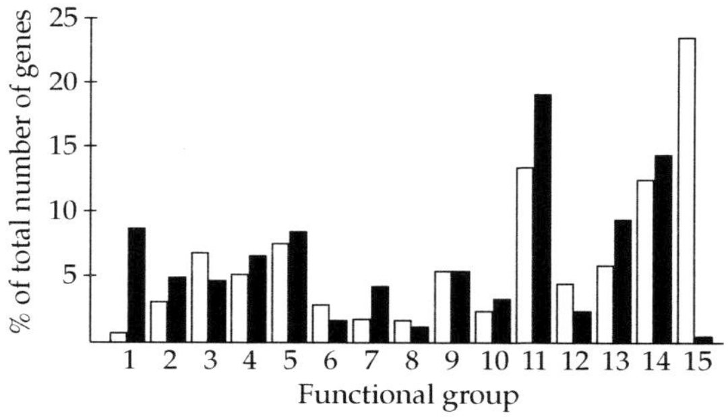


**Figure 9.2** Comparison of gene number distribution (%) from *Rickettsia prowazekii* (□) and *Buchnera aphidicola* BAp (■) according to the following functional categories of genes. 1: amino acid; 2: cofactor; 3: cell envelope; 4: cell processes; 5: energy; 6: fatty acids; 7: purines, pyrimidines; 8: regulatory; 9: replication; 10: transcription; 11: translation; 12: transport, binding; 13: other; 14: hypothetical; 15: species-specific.

The main selective force that slims symbiont genomes seems all too apparent at first sight: It has often been stated that the metabolic load of sustaining a large genome in prokaryotes promotes the removal of genes that have become redundant in a nutrient-rich, intracellular environment. Also, in intrapopulational competition, smaller genomes may confer an advantage because they replicate faster. Genome reduction described in these terms implies an adaptive process positively directed by selection (Silva *et al.* 2001). However, it is likely that other genetic mechanisms and selective forces affect the evolution of genome size (Ochman and Moran 2001). Furthermore, the merely qualitative reference made to the metabolic load of a large genome seems wholly at odds with the recent finding of **polyploidy** of the small genome of *Buchnera* (Komaki and Ishikawa 1999). One symbiotic *Buchnera* cell may contain up to 200 copies of its 640-kb genome, implying it contains as much as 25 times the amount of DNA in a free-living *E. coli* cell. Polyploidy also occurs in mitochondria and chloroplasts. A different hypothesis on the underlying selective process is given by Ochman and Moran (2001), who attribute symbiont genome reduction to a lack of effective selection for maintaining genes in highly specialized niches. In these terms it is, as the authors state, "a neutral or even deleterious consequence of long-term evolution under the conditions imposed by these life-styles."

Indeed, the obligate, intracellular life-style does not merely provide the benefits of a stable and nutrient-rich environment. The flip side is spatial confinement throughout most or all of their life cycle. This has two important consequences for the symbiont. First, its recombinational potential is strongly reduced, that is, they reproduce asexually. Within their secluded microhabitat, obligate symbionts have limited opportunities to come into contact with other genotypes, strains, or prokaryotic species, with which, in principle, they could exchange or recover lost genetic material. Even if they come close, cellular mechanisms must be in place for such exchange, and if genes from distant relatives are eventually incorporated, the symbiont should still be able to express them efficiently, matching for instance its codon usage pattern or regulatory signals. To date, there is only one report of horizontal gene transfer in an obligate mutualistic symbiont (van Ham *et al.* 2000), perhaps illustrating foremost the unlikelihood of the process to occur in such associations. In contrast, horizontal transfer is thought to play a prominent role in the evolution of free-living prokaryotes, a view supported by numerous instances of the phenomenon recently discovered in complete bacterial genome sequences, and quantified in an *E. coli* strain, in which 17 percent of the genes are estimated to be of exogenous origin (Ochman *et al.* 2000).

Second, spatial confinement profoundly affects the population dynamics of symbionts, resulting in often small population sizes and severe bottlenecking during transmission. Theory predicts for such populations with decreasing effective size that the efficacy of selection on mutations is reduced and that their **fixation** rate is increasingly determined by random **genetic drift**. Together with the general prevalence of mutations with adverse fitness effects and a lack of recombination, which can counterbalance these effects, **drift** renders such populations subject to the accumulation of deleterious mutations and results in a progressive loss of fitness (Moran 1996). This population genetic process is known as **Müller's ratchet**. A click of the ratchet can be seen as the fixation event of a less fit genotype in a population and the click rate is determined by the effective population size, the mutation rate and the distribution of their effects, the strength of selection, and the recombination rate.

The operation of Müller's ratchet has experimentally been demonstrated only in RNA viruses and some bacterial species, but in comparative DNA sequence analysis it is mostly inferred from aberrant mutational patterns, such as increased rates of **nonsynonymous** versus **synonymous substitutions** in protein coding genes, or from rates of destabilizing substitutions in structural RNAs and proteins. Through this kind of inference, the process has indeed been shown to act upon organellar genomes and on those of more recent prokaryotic symbionts, pathogens and mutualists alike (Moran 1996). For insect symbionts, elaborate demonstrations of this process come from the work of Moran and colleagues. They observed faster rates of sequence evolution in the 16s rDNA genes of symbionts than in their free-living relatives, and showed that the substitutions have a destabilizing effect on the predicted structure of the ribosomal RNA. Substitutional patterns in protein coding genes in *Buchnera*, the obligate symbiont of aphids, revealed a significantly increased rate of replacement versus silent substitutions.

An exacerbating effect of the accumulation of deleterious mutations occurs when the very genes responsible for replication, recombination, and repair themselves become affected by **mutational**

**load** and functional degeneracy. An inefficient, faltering replication machinery will produce more mutations during replication, which in turn can not or only inefficiently be repaired through deficient recombination and repair machineries. In this respect, comparison of obligate symbiont genomes has revealed a most striking pattern of convergent evolution. Many genes involved in the replication, recombination, and repair pathways have been lost independently in *Mycoplasma, Borrelia, Rickettsia,* and *Buchnera* (Fig. 9.2).

In the absence of recombination, **deletions**, as one category of deleterious mutations, are particularly unlikely to be reversed by back mutations. Without effective selection counteracting the loss of DNA, symbiont genomes are expected to progressively shrink. The massive and irreversible loss of genes and associated metabolic functions confines symbionts to their specialized niche. Like eukaryotic organelles, they may eventually cross the point-of-no-return in their lineage-specific process of genome degradation. However, according to the **nearly neutral theory** of molecular evolution (see Ohta, Chapter 1), although most fixed mutations will be deleterious, a small proportion of positive ones should also be present, which could compensate for the detrimental effects of previously fixed mutations. This positive selection has been demonstrated for GroEL, a protein that participates in the correct folding of many damaged proteins and that is overexpressed in endosymbiotic bacteria (Fares *et al.* 2002 a,b).

## Perspectives

The question arises what will happen once a seemingly irreversible path to genome reduction is taken? Theory predicts that the ultimate consequences of long-term degradation, with ever-increasing rates of mutation accumulation, are **mutational meltdown** and population extinction (Lynch *et al.* 1993). There are a number of examples in which organelles, or rather their genomes, seem to have gone extinct, or have been reduced to extremely small size. The most speculative ones include eukaryotic organelles like peroxisomes, most hydrogenosomes, and mitochondria of certain anaerobic protozoa. Except for those in flowering plants, mitochondrial genomes in general are of extremely reduced size, and have got as small as 6 kb in *Plasmodium falciparum*, the human malaria parasite, in which it encodes a mere five genes. *Mycoplasma genitalium*, a parasite of humans, and *Buchnera aphidicola* are, to date, the two prokaryotic symbionts with the smallest known genomes (Table 9.2). *Mycoplasma* has been estimated to be in association with eukaryotic cells for at least 150 million years and the aphid–*Buchnera* symbiosis has been dated at approximately 200 million years. Do these estimates imply that their symbioses are stable, in contrast to what Müller's ratchet predicts? The most likely answer is that genomic degradation, in any case, is a long-term process, the exact rate of which may vary among lineages and is determined by parameters specific to the type of interaction and its population dynamics.

For obligate, mutualistic symbionts of insects, a replacement theory was first postulated by Moran and Baumann (1994), to explain the frequent occurrence of secondary symbionts alongside the highly specialized bacteriocyte symbionts. The growing body of evidence that genome reduction in the latter is an ongoing process (Gil *et al.* 2002) is providing strong support for this hypothesis. Indeed, *Buchnera* is known to have been replaced by a different symbiont in at least one lineage of aphids (Baumann *et al.* 1995). In many others, in which secondary symbionts appear to be in a process of establishment, selection on both host and symbiont could favor them to take over the essential metabolic functions carried out by the bacteriocyte symbiont. This will, in turn, contribute to a relaxation of selection on those genes ultimately maintained in the bacteriocyte symbiont and will further accelerate its descent to mutational meltdown and extinction. From the perspective of host lineage evolution, a view emerges of metabolically potent prokaryotes being lured into mutualistic associations, at the expense of their cellular autonomy. Once they become genomically exhausted, and the host persist in its ecological niche, replacement occurs with a new species. With such a strategy the host could be viewed as the ultimate parasite.

## Summary

Many bacteria have succeeded in conquering the intracellular habitat of eukaryotic cells. In the long term, two evolutionary strategies prevail among invaders: parasitism and mutualism. The former is harmful to host fitness and consequently results in a prolonged evolutionary arms race. Mutualistic associations are overall beneficial to both partners. They may drive diversification and the exploitation of new ecological niches by the dominant partner. The biology of interactions between insects and microorganisms has gained much interest in the past decade. Here, we review the biodiversity and kinds of interactions found in insect–bacterial associations and seek to integrate new insights gained from diverse fields of study for our understanding of the evolutionary process of symbiosis.

This work has been funded by grant BMC2003-00305 from the Ministerio de Ciencia y Tecnología (Spain) to AM.

PART III

# The ecological and biogeographic context of evolutionary change

CHAPTER 10

# Evolutionary ecology: natural selection in freshwater systems

**Winfried Lampert**

Evolutionary biology studies *how* organisms evolve on the molecular, genomic, and organismic level. It looks at the origin and maintenance of genetic variability, the mechanisms leading to changing gene frequencies by chance or selection, and the constraints inhibiting this process. **Evolutionary ecology** studies *why* organisms evolve. It identifies the biotic and abiotic environmental factors that drive natural selection. Textbooks view evolutionary ecology from two sides, either as an approach to general ecology (i.e. explaining ecology with evolutionary ideas) or as a subdiscipline of evolutionary biology (i.e. explaining evolution by the ecological framework). Modern ecology texts mostly use the concept of natural selection. Colinvaux (1993) states: "Ecology is the study of how life was fashioned by natural selection." Pianka's (2000) rather general ecology text is even named "*Evolutionary ecology*." The alternative view is represented by Cockburn's (1991) book, which is not a general ecology text. This view is probably best represented by a quote from Bradshaw (1984): "Ecology is crucial for understanding evolutionary mechanisms."

I will adopt the more stringent definition of evolutionary ecology trying to point out how ecology can contribute to the understanding of evolution. The contribution of ecology can only be through natural selection, which can be summarized in the slogan "natural selection is ecology in action." Following this, I will look at **adaptations** of organisms, the ecological factors that shaped them and the consequences of these adaptations for their interactions with other components of the ecological system they are part of. The only way to a causal explanation of the adaptive value of a trait is the measurement of **fitness**, and as fitness has a physiological basis it is evident that evolutionary ecology must have close links to physiological ecology and population ecology.

Although there is still debate on the units of selection, it is clear from an ecological point of view that selection must act through phenotypes. Consequently, evolutionary ecology must look at living organisms and their environment. Despite the extreme usefulness of modern molecular tools in ecology, for example, for the identification of genotypes and phylogenetic relationships, molecules cannot be subjects for ecology. While "molecular evolution" is easy to define as the prerequisite of evolutionary change in organisms, the term "molecular ecology" does not mean the ecology of molecules. Rather, it defines the study of the population biology, ecology and evolution of organisms using techniques developed in molecular biology laboratories (see Hewitt, Chapter 14). In this chapter, I will outline some important concepts of present-day evolutionary ecology. I cannot review the field completely and I will use examples exclusively from freshwater ecosystems as this is my area of expertise.

## Ultimate factors

The main goal of evolutionary ecology is the search for ultimate factors. While classical ecology was more interested in *how* the expression of traits is controlled by environmental (i.e. proximate) factors, evolutionary ecology asks *why* organisms display certain traits, that is, what is the adaptive value of a trait (ultimate factors). There is a fundamental difference between ecosystem and community ecology on the one hand, and population ecology on the other. Neither an ecosystem nor a community have a common genome, hence ultimate factors for community features can only be derived indirectly from responses of populations to selection pressures. I will illustrate this concept with an example.

**Succession** is a key concept in ecology. In temperate lakes, due to the short life cycles of most pelagic organisms, which are much shorter than the seasonal cycle, seasonal **autogenic succession** is a major community property. In many of these lakes, the composition of the **zooplankton** differs dramatically between summer and winter. The summer months are dominated by cladocerans (waterfleas), calanoid copepods and few, small cyclopoid copepods, while a few species of cyclopoids can make up 90 percent of the zooplankton during winter and early spring (Fig. 10.1). This species shift changes the whole food-web structure, as cladocerans are mostly filter-feeding herbivores which also feed on bacteria, calanoids are more selective herbivores and omnivores, but cyclopoids have an ontogenetic shift in their feeding habits. Their **nauplii** start as herbivores, **copepodites** become gradually more omnivorous, and the adults tend to be carnivorous.

The seasonal change in community structure is mainly a consequence of a unique life cycle of some cyclopoids (e.g. *Cyclops vicinus*). They mature in late winter and produce eggs and nauplii that develop into copepodites. In the fourth (of five) copepodite larval stage (C4), however, the copepodites disappear from the water column. They swim to the bottom of the lake, bury themselves in the sediment and enter into a **diapause** stage that lasts for the whole summer. They reappear from the sediment only in late autumn to complete their life cycle. Hence, cyclopoids are present in the lake during summer, but they are not active. The onset of the summer diapause is controlled by day length. Day length is the proximate factor as it determines *how* the diapause is initiated. However, it does not explain the ultimate factor, *why* the cyclopoids display such a strange life cycle. At first sight, it might

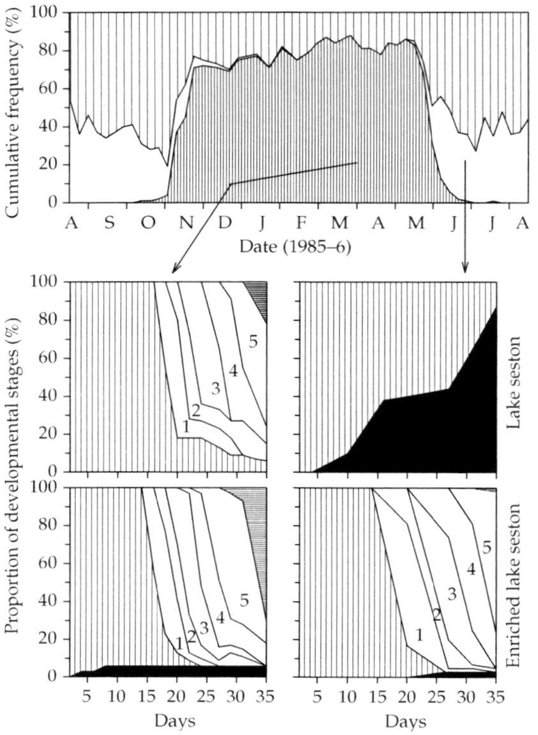


**Figure 10.1** Upper panel: seasonal composition of the zooplankton in a eutrophic north German lake (Schierensee). Densely hatched: four species of *Cyclops* (Copepoda), including *Cyclops vicinus* (see later); white: cladocerans (mostly *Daphnia*; widely hatched: other copepod species. Lower set of panels: an experiment testing the quality of natural seston taken from the lake at different times of the year for the development of *Cyclops vicinus* nauplii. The individual panels show the development of freshly hatched nauplii (vertically hatched) through five copepodite stages (1–5) to adults (horizontally hatched). Black areas indicate the proportion of dead animals. Arrows indicate the time of the experiment. Daily renewed original lake seston was used as food in the upper panels. The seston was enriched with laboratory-grown flagellates in the lower panels. Note that no nauplii developed into the C1 stage and most of them died with lake seston taken when daphnids were abundant in the lake, but development was possible when the seston was artificially enriched (from Santer and Lampert 1995).

be more profitable if they stayed active during summer as they could have multiple generations per season due to faster development in the warm water and rich food supply for the predatory adults, resulting in higher egg production. In search of the ultimate factor that prevents the cyclopoids from exploiting summer conditions, we can test some hypotheses:

**1.** The copepods cannot tolerate high water temperatures due to physiological constraints.
**2.** The life cycle is a historic relict from ancient times when these species lived in temporary ponds that dried out during summer.
**3.** The copepods are susceptible to fish predation. They avoid periods when fish are most active, that is, when young-of-the-year fish are abundant.
**4.** Herbivorous nauplii go through a food bottleneck when they compete with other, more efficient grazers (like *Daphnia*) for common **phytoplankton** resources.

It is possible to isolate diapausing C4-copepodites from the sediment, wake them up, and let them continue their development under short daylight periods, therefore, one can maintain the copepods in the laboratory year round. They survive and grow under summer temperatures, which falsifies the first hypothesis. The second hypothesis cannot be falsified experimentally, but it is unlikely to be true as there is variability in the timing of diapause and a small fraction of the copepodites do not go into diapause at all. Without stabilizing selection, the diapause behavior would probably disappear rather quickly. Avoidance of fish predation may play a role, but fish usually prefer large cladocerans, particularly *Daphnia*, over cyclopoids, which would give the cyclopoids an advantage during summer.

Hypothesis 4 has been tested experimentally by Santer and Lampert (1995) rearing cyclopoid nauplii in the laboratory at different times of the year. These nauplii were offered natural **seston** collected daily from a **eutrophic lake** at the same temperature (12 °C), but at times when they were either present or absent in the lake. In parallel, the natural seston was enriched with laboratory grown **flagellates**. At times, when nauplii were in the lake (early spring), natural seston was sufficient to promote naupliar growth and development. The enrichment with flagellates had no additional effect (Fig. 10.1). After the copepodites had entered diapause, however, nauplii starved to death with natural seston, but grew well in the enriched treatments. This result suggests that food quality and quantity inhibits naupliar development in the field. The poor food situation in the lake was caused by the spring development of filter-feeding *Daphnia* grazing down phytoplankton and creating a **clear-water phase**. Cyclopoid nauplii have a high food threshold for maintenance compared to *Daphnia*, and they are easily outcompeted by daphnids monopolizing the resources. Nauplii produced while daphnids are abundant would have no chance of survival. Consequently, the diapause of preadult cyclopoids can be interpreted as a mechanism to avoid periods of competitive exclusion of the nauplii.

Having identified a possible ultimate factor for the summer diapause in one lake, we can make more general, testable predictions. For example, as food conditions are not perfectly predictable, a fraction of the total population of copepodites, depending on the actual food conditions, should avoid entering diapause. This prediction has, in fact, been borne out (B. Santer and A.-M. Hansen, unpublished). Finally, in highly eutrophic systems where food is always abundant, cyclopoids should cease summer diapause. This has been observed in shallow, hypertrophic Danish lakes.

## Phenotypic plasticity: reaction norms

The basis of natural selection is the performance of different phenotypes in a particular environment. Phenotypic variability may be caused by genotypic variability, but even a single genotype can produce different phenotypes in different environments, which is called **phenotypic plasticity** (Stearns 1992). The reaction to a particular environment may reflect the developmental and physiological program expressed by the genotype to produce a specific phenotype during ontogeny. It is called a **developmental reaction norm** (Schlichting and Pigliucci 1998). The concept of the **reaction norm** (Stearns 1989) is somewhat broader, as it considers not only the development but any kind of reaction of a phenotype to a changing environment. The reaction norm is a property of the genotype and may be subject to natural selection. For example, in a rapidly fluctuating environment it may be advantageous to buffer the metabolism and show an integrated response, while in a slowly changing environment it may be better to closely track the environmental change. Hence, an organism's reaction norm should be related to fitness.

Reaction norms are graphical or other representations of a genotype's phenotype value across environments. Every genotype has a particular reaction norm, thus, a population can be characterized by a set of lines (Nager *et al.* 2000). Phenotypic plasticity implies a non-flat reaction norm. If genotypes vary with respect to their phenotype under particular environmental conditions, but react identically to a change in conditions, all reaction norms will be parallel, that is, the change will have no effect on the differential fitness of the genotypes. However, if the genotypes react differently, reaction norms will be non-parallel and the relative fitness of genotypes will change with the changing environment. Statistically, this will result in genotype by environment interactions. The analysis of variation within and among populations is relevant for predicting if and how a shift might occur.

The concepts of reaction norms and genotype by environment interactions illustrate the changing views of ecology. For a century, physiological ecologists have studied the performance of organisms under differing environmental conditions, asking how they respond, for example, to varying temperatures. They tried to identify characteristics of a species that could be compared to a standard such as Krogh's normal curve of metabolism. Individual variation was considered disturbing and was eliminated by averaging many measurements. However, evolutionary ecologists are particularly interested in the individual differences as they ask how genotype by environment interactions evolve. I will again use an example from aquatic ecology to illustrate the application of this concept.

In a theoretical context, reaction norms are mostly represented by straight lines connecting two points of performance in an environmental gradient. This is a gross simplification as reaction norms may be nonlinear. A typical example of a nonlinear response is the response to temperature in ectotherms. Fitness-related performance of ectotherms in a temperature gradient usually follows a characteristic, asymmetrically skewed reaction norm. At low temperatures, it rises linearly or geometrically until it reaches a maximum at what is assumed to be the optimal temperature (Gabriel and Lynch 1992). At higher temperature, it declines rapidly.

Temperature is an important abiotic factor influencing ecological processes. The current trend of global temperature change is of great concern and the question arises how field populations will react to changes in temperature conditions. It has been shown that *Drosophila* (Huey *et al.* 1991) and bacteria (Lenski and Bennett 1993) show rapid evolutionary responses in thermal sensitivity to laboratory selection, thus, moderate temperature changes may be "buffered" by shifts in genotypic composition in field populations. There are two possibilities how such shifts can occur: (1) a dominance shift of genotypes within a population or (2) immigration of genotypes from populations adapted to different climatic conditions.

We measured temperature reaction norms in the widespread cyclic parthenogen *Daphnia magna*, a waterflea that lives in ponds and shallow lakes with low fish predation (Mitchell and Lampert 2000). Carvalho (1987) had reported differences in temperature responses (survival and fecundity) for **clones** isolated from a permanent pond during different seasons. He concluded that there was genetic adaptation to ambient temperatures as the preferred temperature ranges corresponded to those of the season at which theses clones were most abundant. In order to analyze within population genetic variance and between population genetic differentiation, we sampled eight populations of *D. magna* across Europe from Finland and Russia in the north to southern Spain and Sicily in the south, covering a wide range of climatic conditions. Cultures of seven of the populations were started from **resting eggs** collected in the field and hatched in the laboratory. The eighth (Sicily) population was started from a random sample of active **parthenogenetic** females. From each population, we randomly selected eight individuals that were used to raise **clonal** lineages after screening them for allozyme variation and assuring that they belonged to distinct electrophoretic genotypes. Three replicates each of the 8 × 8 clones were then cultured individually at six temperatures (17–32 °C) and high food levels. We measured the juvenile growth rate of the daphnids as it is closely correlated with the intrinsic rate of population increase, a good fitness parameter (Lampert and Trubetskova 1996), after temperature acclimation of the clones for at least two generations to remove **maternal effects**. In addition, two populations from ponds in Northern Germany were studied extensively over two seasons.

Juvenile growth rate reaction norms to temperature of 26 clones of *D. magna* collected at different seasons from a pond (Lebrade) in Northern Germany are plotted in Fig. 10.2(a). They all exhibit the typical asymmetric shape with a shallow increase and a sharp decrease after the maximum.

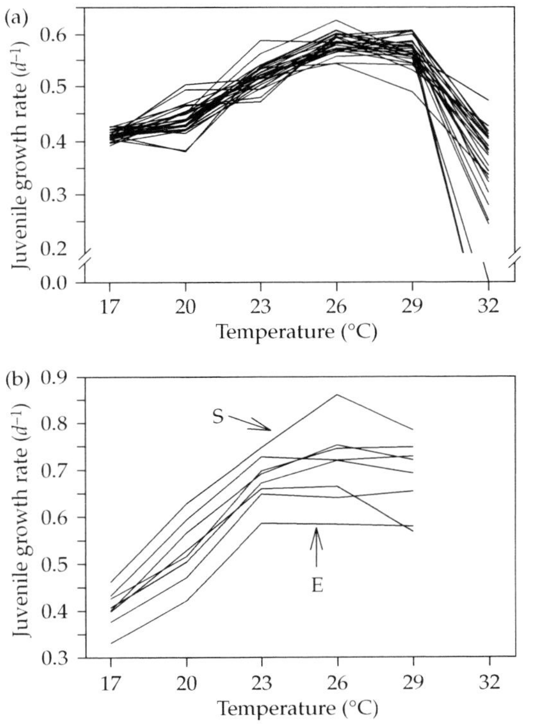


**Figure 10.2** Temperature reaction norms of *Daphnia magna*. Juvenile growth rate is a good proxy for fitness. (a) reaction norms of individual clones isolated from a pond in northern Germany at different seasons of one year (S.E. Mitchell *et al.*, unpublished; data from J. Halves). (b) mean reaction norms of populations (eight clones per population) originating from locations along a latitudinal gradient across Europe. The extremes are populations from England (E) and southern Spain (S) (from Mitchell and Lampert 2000).

The optimal temperature for juvenile growth is approximately 26 °C for all clones, which is rather high considering that their habitat rarely reaches such high temperatures. There were no significant differences between groups of clones sampled during different seasons, and no correlation between electrophoretic genotypes and reaction norms were detected. Although the reaction norms look amazingly similar there are significant effects of temperature and clone on juvenile growth rate and clone by temperature (genotype × environment) interactions. While most reaction norms are parallel below the maximum, the differences increase above. Looking at this bundle of reaction norms we would not expect much effect on genetic composition of a small increase below the maximum, but strong effects above it.

Surprisingly, we find similar results when we compare populations from very different climatic conditions. Although they differ in absolute growth rates, the mean reaction norms of the eight European populations of *D. magna* are almost parallel up to 29 °C (Fig. 10.2b). Unfortunately, the 32 °C treatments could not be included due to high mortality. The grand mean growth rate differs significantly among populations, but there is no significant population × temperature interaction. Within populations (among clones), however, reaction norms differ as there is a significant clone × temperature interaction.

Contrary to Carvalho (1987) who studied a permanent, overwintering population of *D. magna*, we did not find seasonal differences in reaction norms. We also did not find geographical differences. The reason may be that all these populations were intermittent, that is, they went through a phase of bisexuality, produced resting eggs, and started a new population from the **resting egg bank** in spring. Populations of **cyclical parthenogens** may be able to better track their environment when they undergo long periods of parthenogenetic (asexual) reproduction and presumably clonal selection (see De Meester *et al.*, Chapter 11). Sexual reproduction, when it occurs, may disrupt co-adapted gene complexes, and thus disrupt any genotype × environment interactions. Although temperature has a strong influence on the growth rate, the observed geographical differences among populations (highest growth rates in Southern Spain) are probably not due to adaptation for increased growth rate at higher temperatures. Rather, the tendency for higher growth rates in southern populations seems to reflect adaptation to some other factor that correlates with latitude. The parallel reaction norms suggest that this factor acts at all temperatures. In fact, the temperature ranges experienced by the geographically distinct populations are not so different, as the daphnids appear at different seasons, for example, in early spring in Spain, but in summer in Finland.

The conclusion from these two experiments is that moderate changes in water temperature may not have a direct effect on *D. magna*'s geographical distribution. Within population genetic variability and phenotypic plasticity seem sufficient to compensate for the temperature effect. Immigration from other locally adapted populations will not be important as they have very similar reaction norms. This does not preclude possible changes due to indirect effects through changing food resources or predator development. The example shows how ecology benefits from the inclusion of evolutionary ideas, but also that the interpretation of evolutionary results needs a strong input from ecology.

## Phenotypic plasticity: adaptive adjustments

The predictability of a habitat is an important ecological factor. If environmental factors change in a predictable pattern (e.g. in an annual cycle), genotypes can be well adapted to this pattern. The summer diapause of cyclopoids is a good example. Although abiotic factors in a lake are rather predictable (Lampert 1987), biotic factors may change irregularly and unpredictably. For example, resources and the abundance of predators may be strongly influenced by climatic events. According to life-history theory, rapid, unpredictable changes of the environment should favor the selection of phenotypic flexibility rather than of constitutive traits that do not change. By reacting phenotypically to environmental changes, organisms can be optimally adapted to various environments in contrast to a generalist genotype that would be suboptimally adapted under all conditions (Gabriel and Lynch 1992).

Adaptive phenotypic plasticity has been demonstrated frequently in freshwater organisms, both for effective resource utilization (Lampert and Brendelberger 1996) and predator defense (for review see Tollrian and Harvell 1999). As predation is a very strong selective force in freshwater communities, the most striking and best analyzed examples are related to predator defense. Predation in fresh waters is strongly size dependent, which has serious consequences for community structure. There are two groups of predators that differ considerably in their prey selection. Most vertebrate predators (fish, salamanders) are visual hunters, hence they select for the most conspicuous (i.e. the largest) prey, except fish larvae and very small fry that may be gape limited. In contrast, most invertebrate predators detect their prey by mechano- or chemosensors. They have difficulties handling prey and are gape limited, and thus, they select small prey. The defense strategy against fish is to be small and invisible while large size and inhibitory appendages protect against typical invertebrate predators. Without involving phenotypic plasticity, Taylor and Gabriel (1992) formulated a physiological model to describe the optimal life history for indeterminately growing *Daphnia*. In the presence of a fish predator, daphnids must invest a large part of their available resources into reproduction early in life. As resources invested into offspring are no longer available for growth, the daphnids must remain small, but they release at least one clutch of offspring before becoming vulnerable to size selective mortality. In contrast, if they are under threat of invertebrate predation, daphnids must delay reproduction and invest everything into growth until they grow too large for the predator. They will then reproduce later, but being large they can produce more offspring per clutch.

These types of life histories have been described for cladocerans previously (Lynch 1980), but it has only recently been discovered that they can be expressed by a single genotype in response to a particular predator (Tollrian and Dodson 1999). Such adaptive phenotypic reactions are called **inducible defenses** (see De Meester *et al.*, Chapter 11). As well as unpredictable predation pressure, there are other prerequisites for the evolution of inducible defenses: (1) the defense must have a cost associated, that is, it must pay off not to have it when no predators are around, (2) the defense must be effective to offset the costs, (3) the prey must have a reliable cue about the presence and activity of the predator before it is eaten. It has been shown that this is mostly a chemical signal (**kairomone**). Numerous inducible defenses regarding morphology, life history and behavior, have been described for freshwater organisms (e.g. ciliates, rotifers, cladocerans, snails, fish, and algae) (see articles in Tollrian and Harvell 1999).

Stibor and Lüning (1994) showed how *Daphnia hyalina* change their life history in response to fish and invertebrate predators. They cultured neonate daphnids under identical conditions in filtered water that had hosted either small fish, midge (*Chaoborus*) larvae or a predatory aquatic bug (*Notonecta*). They then measured the allocation of energy to growth and reproduction. As predicted by Taylor and Gabriel's (1992) model, in the presence of fish scent daphnids began allocating body mass to reproduction when they were smaller than controls without a chemical cue (Fig. 10.3). In contrast, *Chaoborus* treated daphnids began allocating resources to reproduction when they were larger than the controls. However, when they received a chemical signal from *Notonecta*, the second invertebrate predator, daphnids reacted as in the presence of fish: They matured at a smaller size. Although this seems unreasonable as both predators are insects, it makes sense when one compares the feeding strategies of the two. *Chaoborus* is a gape limited predator that must swallow its prey whole, while *Notonecta* holds the prey with its legs, pinches it, and sucks the contents out. It is not gape limited and selects large prey that are more profitable when handling time is important. Though for different reasons, the effects of fish and *Notonecta* are the same. Consequently, the phenotypic response of *Daphnia* is identical despite the very likely different chemical signals involved.

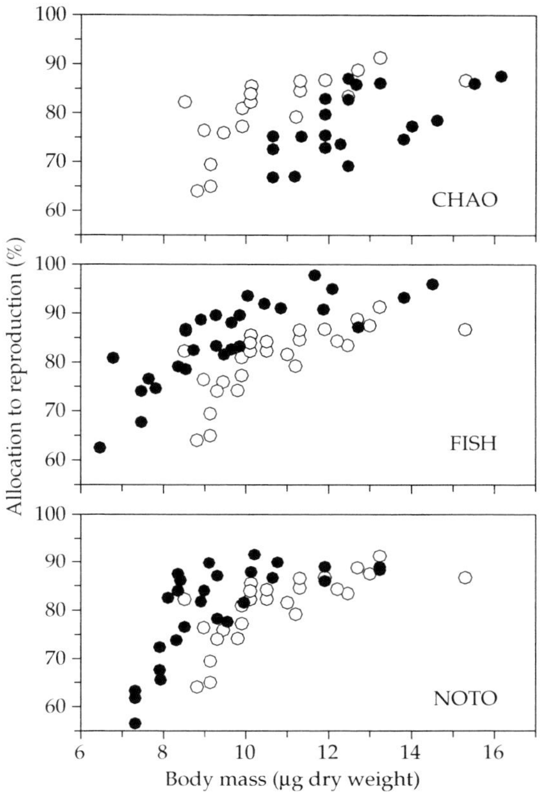


**Figure 10.3** Phenotypic plasticity in the allocation of energy of *Daphnia hyalina* to reproduction in response to a chemical (kairomone) signalling the presence of a predator. Open circles: controls. Full circles: treatments with predator kairomones from *Chaoborus* larvae (CHAO), fish and *Notonecta* (NOTO). *Chaoborus* larvae select small prey, while fish and *Notonecta* prefer large prey (from Stibor and Lüning 1994).

Both constitutive defensive traits and induced defenses have the same ultimate causes, but it depends on the predictability of the ultimate factor relative to the life span of the organism whether selection favors a trait or the plasticity of a trait, that is, whether the trait is genotypically or phenotypically controlled. The spines of a porcupine and the induced spines of a rotifer represent two solutions to the same problem. In any case, the evolution of defenses is an important feedback in ecological terms that facilitates the coexistence of predator and prey in a community.

## Egg banks: records of microevolution

Ecological and evolutionary processes work on different time scales, but there is some overlap. Beyond time scales relevant for phenotypic changes, depending on generation times, more and more evidence has accumulated that evolutionary change can be rapid and can have hidden effects on the interactions in communities (J.N. Thompson 1998). Although rapid microevolutionary change may be a transient phenomenon that does not necessarily result in macroevolutionary patterns, as it is mainly a result of fluctuating selection, it is the normal mode of evolutionary change and it is not only of ecological interest but also a topic of evolutionary ecology. Many of the examples listed by J.N. Thompson (1998) concern evolutionary changes following the invasion of new species into communities. The reason for this may be that invasive species are usually considered a threat, and therefore invaded communities are monitored over longer periods, so that evolutionary changes can be detected. Normally, however, they will not be detected as the environmental change will only be noticed in retrospect when prechange genotypes are no longer available to be tested.

The recent discovery of egg banks in lake sediments, analogous to seed banks in the soil, offers a possibility to observe microevolutionary processes in retrospect. Various aquatic organisms produce resting stages that can survive dormant in lake sediments for many years. The oldest resting eggs hatched so far, from the copepod *Diaptomus sanguineus*, were 330 years old (Hairston *et al.* 1995). These egg banks have two important functions. First, like seed banks, they act as an insurance against catastrophic events as the sediments are a source for recolonization after a severe disturbance. Second, if resting eggs from deeper sediment layers get a cue to hatch, they re-introduce long-dormant genotypes back into the population and counteract selection (Hairston 1996).

Cyclically parthenogenetic *Daphnia* switch to bisexual reproduction under unfavorable conditions (see De Meester *et al.*, Chapter 11). The product of bisexual reproduction are two **diapausing eggs** enclosed in a resistant case (**ephippium**) that can settle to the sediment, get buried and survive for decades (Fig. 10.4). Only resting eggs in surface sediments are likely to get the right cue and hatch, hence the deeper sediment layers contain older eggs and information about past genotypes. For example, Cousyn *et al.* (2001) found rapid local adaptation of *Daphnia magna* to changing fish predation pressure reflected in the phototactic responses of hatchlings from various sediment depths, despite the absence of genetic differentiation at the level of neutral genetic markers (**microsatellites**).

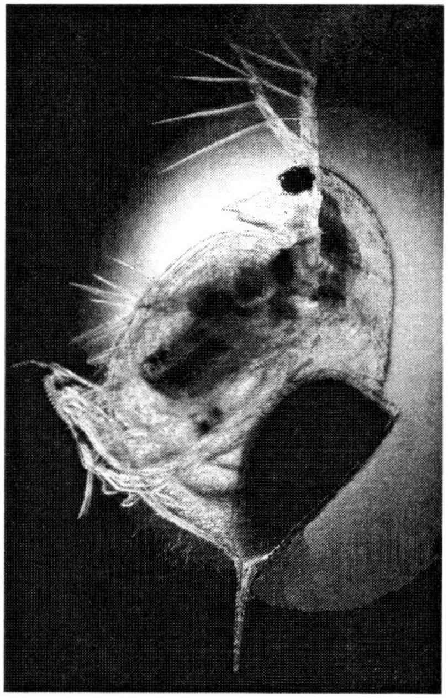

**Figure 10.4** *Daphnia* shedding a saddle-like carapace structure (ephippium) that contains two bisexually produced resting eggs. The ephippium is resistant against harsh environmental conditions (desiccation, freezing, anoxia). It can sink to the sediments of a lake and survive for decades before the resting eggs hatch (by permission of V. Alekseev).

Lake Constance in the south of Germany, the largest German lake (500 km$^2$), has undergone slow, but dramatic changes during its recent history. Beginning in the 1960s, the **oligotrophic lake** showed signs of eutrophication due to phosphorus enrichment from sewage inputs. The phosphorus content increased exponentially, algal abundances increased and water transparency decreased. The species composition of phytoplankton, zooplankton, and fish changed until the lake was considered eutrophic around 1980. In the succeeding years, the trend was reverted with large investments into sewage treatment, and the restoration of water quality was successful. The phosphorus status of 1960 was regained in the year 2000 and changes to the species composition of phytoplankton followed this development with a time lag. *Daphnia galeata* was an important component of the zooplankton of the lake during this whole period and, thus, experienced strong changes in its food conditions from an oligotrophic to a eutrophic lake and back.

Due to its great depth (max. 250 m), Lake Constance has clearly laminated, undisturbed sediments. The age of individual layers and of the resting eggs therein can be determined easily. Weider *et al.* (1997) isolated ephippia from periods before the eutrophication, at peak eutrophication and after the restoration. They were able to hatch them and study the genotypic composition of the individual populations by allozyme electrophoresis. They discovered a clear correlation of certain alleles with the phosphorus content in the lake (an indicator of the trophic state) and concluded that microevolutionary changes had occurred in conjunction with ecosystem changes.

As the resting eggs had been hatched, clones could be established that had been living in the lake during the various phases. It was then possible to measure the fitness of these historic clones under simulated conditions of a eutrophic lake. Hairston *et al.* (1999) measured the growth performance of ten clones each from three periods (before, peak, past eutrophication) in the presence of a toxic cyanobacterium (*Microcystis*) that had been isolated from Lake Constance during 1972. The appearance of such cyanobacteria is typical for eutrophic lakes. The strain of *Microcystis* had been found to inhibit the feeding rate and growth of *Daphnia* when present in their food at a proportion of more than 10 percent.

Neonates of each clone were raised in a flow-through system either in a pure suspension of the green alga *Scenedesmus* (good food) or a mixture of 80 percent *Scenedesmus* and 20 percent *Microcystis* in terms of carbon (poor food). Growth rates in both food types were measured and the inhibition of the growth rate was calculated as a measure of the effect of the cyanobacteria. There was a significant difference in mean growth rate inhibition between clones from the 1960s (beginning of eutrophication) and those from 1978/80 (peak eutrophication). The older clones showed higher depressions of the growth rate (*c*.35 percent) compared to the younger ones (*c*.20 percent), that is, the younger clones were better adapted to eutrophic conditions. Mean growth rate inhibition did not differ between peak and post eutrophication clones.

The individual reaction norms for food quality (Fig. 10.5) revealed the mechanism of evolutionary change. A large diversity of genotypes existed in the lake during the pre-eutrophication period. They differed with respect to growth rate as well as to the steepness of the reaction norm, that is, the phenotypic plasticity in response to toxic cyanobacteria. Slow-growing and particularly phenotypically plastic genotypes disappeared from the population during eutrophication. These genotypes seem to reappear during the recovery of the lake, although the differences were not yet significant. It is interesting to note that the mean growth rates in good food during the pre-eutrophication period and in poor food during the peak-eutrophication period (i.e. the realized rates under these conditions) are nearly identical. One might, therefore, ask why selection did not favor fast-growing genotypes before eutrophication. There are evidently other selection factors that are important. One possibility is suggested by the fact that fast-growing genotypes on average mature at a larger size than slow-growing genotypes. Size-dependent predation by visually hunting fish would eliminate genotypes with a large size at maturation. During peak-eutrophication, when realized growth rates were low, selection would not have acted to reduce size at maturation, that is, unlimited growth rates could increase. A second possibility may be a trade-off between high growth rates when food is abundant and poor performance at low food concentrations. The enrichment with phosphorus resulted not only in a shift in species composition of phytoplankton, but also in increased absolute amounts of small edible algae. Hence, starvation conditions with low food abundances no longer prevailed and

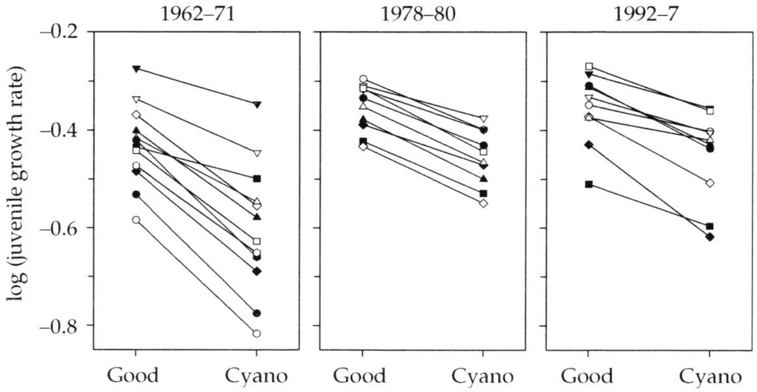


**Figure 10.5** Reaction norms to food quality of *Daphnia galeata* hatched from resting eggs isolated from aged sediments of Lake Constance. Good food is a pure suspension of the green alga *Scenedesmus obliquus*. Poor food (cyano) contains 20 percent of a toxic cyanobacterium (*Microcystis*) isolated from the lake. Years indicate the time span when the resting eggs were produced and buried in the sediment. During this time Lake Constance went through a phase of eutrophication by anthropogenic effects and restoration. 1962–71 was the pre-eutrophication period 1978–80 was the peak of the eutrophication. 1992–7 represents the successful lake restoration period (from Hairston *et al.* 2001).

selection could no longer favor performance at low food levels. The latter possibility has not yet been explored experimentally.

Although microevolutionary changes have frequently been reported, they have rarely been documented in terms of fitness. Preserved genotypes in egg banks can be helpful in elucidating genetic properties of past populations and make shifts over time visible. This is another area where insights emerge from the combination of evolutionary biology looking at genetic variability and changes in genotypic distributions and ecology identifying the selective forces that drive the process.

## Feedbacks and effects at the ecosystem level

In an ecosystem, organisms interact with each other and modify the abiotic conditions. They are sometimes called ecosystem engineers. Evolutionary adaptations in one species may, therefore, have implications on other species and ecosystem attributes (J.N. Thompson 1998) that can be viewed as feedbacks. I will elaborate on this chain of reasoning using **diel vertical migration** (DVM) of zooplankton, one of the best studied ecological phenomena in lakes.

Many zooplankton species display a pronounced diel migration behavior in the pelagic zone. The normal pattern is that they leave the surface waters (**epilimnion**) at dawn and spend the day in the dark, cold bottom waters (**hypolimnion**). They return to the surface after sunset to feed during the night. The phenomenon is long known both in freshwater and marine environments, but until the 1970s research focused on the neurophysiological basis of the behavior (photo- and geotaxis) and the proximate factors of its control. The velocity of light change in the morning and in the evening was found to be the trigger initiating the migration. It was only after the inclusion of evolutionary ideas into ecology that interest shifted to the question *why* zooplankton show this behavior, that is, what is the ultimate factor driving it? At first sight rhythmic migrations should be disadvantageous for herbivorous zooplankton. Primary production is located near the surface where the light is, hence that is were the food is located, and more food results in more offspring production. Moreover, development is faster in the warm surface waters, especially for animals carrying their eggs. Going down in the dark, cold hypolimnion must result in fewer offspring and slower development, which is a fitness disadvantage. A cheater, not participating in the mass migration and staying near the surface would have a higher fitness than a migratory genotype, hence the migration would not be an **evolutionarily stable strategy**. However, it evidently is a stable strategy as diel vertical migration is so widespread.

Many hypotheses have been advanced and tested in order to solve this paradox. They fall into two categories:

**1.** DVM produces a metabolic or demographic advantage that outweighs the disadvantage of poor food and low temperature. Regularly changing conditions encountered during migration lead to a more efficient energy use. Higher fitness is gained through enhanced offspring production.
**2.** DVM is a strategy to avoid mortality risk from visually hunting predators (fish) that depend on light. Fitness is increased through reduced mortality.

Note the two different assumptions of these hypotheses: The first supposes an advantage of migration, the second ones assumes a reduced disadvantage compared to staying near the surface.

Field and laboratory tests have not provided support for the "metabolic advantage" hypothesis (Lampert 1993). Even a slight energy bonus through more efficient energy use could not compensate the negative demographic effects of low temperature. On the other hand, there is now much evidence for the "predator avoidance" hypothesis, from which a number of new testable predictions can be derived:

**1.** Zooplankton must migrate upwards in the evening and downwards in the morning to avoid light. This is, in fact, the normal behavior. Rare cases of "reversed" migrations can be explained by indirect effects, e.g. the "normal" migration pattern of invertebrate predators.
**2.** DVM should be most pronounced for the most conspicuous (e.g. large) zooplankton (most endangered by visual hunters). This has been reported frequently. For example, juveniles of *Daphnia* show only small migration amplitudes while adults, in particular egg-bearing females, migrate strongly.
**3.** The amplitude of migration should depend on the abundance and activity of planktivorous fish. There is now evidence that the amplitude of migration varies with the year-class strength of fish, varies seasonally with fish activity and is lacking in fishless lakes (for review see Lampert 1993). Gliwicz (1986) reported varying migration amplitudes of the copepod *Cyclops abyssorum* in mountain lakes that had been stocked with fish for known periods of time. *Cyclops* in fishless lakes did not migrate. The first evidence for migration was found in lakes that contained fish for 25 years, and a strong migration pattern occurred in lakes stocked 1000 years ago. Gliwicz (1986) interpreted these results as evidence for rapid local adaptation. This interpretation became questionable after the discovery that DVM is not a fixed behavior but is an induced response to chemical cues indicating the presence of fish. Zooplankton react to the light, but they do this only if a fish kairomone signals danger. Thereby they avoid the costly migration when there is no need for it. The first large scale proof for the induction came from experiments in 12 m high indoor water columns (Plön Plankton Towers). Loose (1993) studied the vertical distribution of *Daphnia hyalina* in the stratified columns that were connected by tubing to small external aquaria. Water from the epilimnion was cycled through these aquaria. The daphnids did not migrate in the control tower (empty aquarium), but started migrating immediately if the aquarium contained a single fish. Again, large individuals migrated deeper than small ones. The only way the daphnids could detect the fish in the outside aquarium was by a chemical cue. Although this discovery was a strong argument in support of the predator avoidance hypothesis, it weakened the field results of Gliwicz (1986) as *Cyclops* in the fishless lakes could not detect fish chemicals, hence they would not migrate even if they had the genetic potential.

A reduction of mortality by planktivorous fish can be gained in different ways, for example, by being less conspicuous (smaller) or by DVM. One should, therefore, expect to find genetic variability for both strategies. This has been demonstrated in the same Plankton Tower system by De Meester *et al.* (1995). They collected various clones of *Daphnia galeata* from a lake and measured several of their life-history characteristics. They then selected three electrophoretically different clones to inoculate the Towers with equal amounts of each clone. In the presence of a fish kairomone, the clones displayed different DVM strategies. The clone with the largest body size separated clearly into size classes, the large adults showing the largest migration amplitudes. Migration in the clone with the smallest body size was less pronounced. The third clone, intermediate in size, did not exhibit significant DVM. During the first four weeks of the experiment, in the physical absence of predators, the daphnids built up large populations, but all three clones coexisted with fluctuations in their relative abundance. Then both towers received some free-ranging small fish that would prey on the daphnids. As a result, the migration behavior was enhanced and the *Daphnia* population densities decreased. Three weeks later, the nonmigratory clone was eliminated completely, but the two other clones coexisted at a 1 : 1 ratio. Evidently, they had the same fitness under the experimental conditions, but they gained this fitness in different ways. One clone was larger (i.e. would have more offspring) and migrated deeper, thus, incurring higher costs from migration. The other clone stayed smaller, but migrated to a lesser extent.

So far we have looked at DVM only from the point of view of zooplankton. However, DVM has far reaching consequences for the ecosystem. Being rather unselective filter-feeders, *Daphnia* can monopolize algal resources and have a strong impact on the phytoplankton. If the daphnids migrate to the hypolimnion they cannot graze on phytoplankton during the day and phytoplankton can grow unimpeded while light is available. They will only be harvested at night. Large diel fluctuations of grazing rates correlated with vertical shifts in zooplankton biomass have been measured *in situ*. A trophic cascading effect results from the zooplankton's adaptation to predator pressure.

Even if there is no threat of planktivores, zooplankton may adjust their vertical distribution to optimize fitness. During summer, many lakes do not exhibit the highest algal concentrations near the surface but in deeper waters, provided light can penetrate deep enough. They show deep-water chlorophyll maxima. Herbivorous zooplankton then face a dilemma. They can either stay in the deep waters where food is abundant but temperature is low, or they can use the warm surface water for faster development but produce fewer eggs due to resource limitation. Under such conditions, daphnids should allocate the time spent at different depths so that they gain maximum fitness. They should spend part of the time feeding in the cold and then return to the warmth. Note that this would affect the vertical distribution, but would not induce rhythmic DVM as it is independent of light.

A model of daphnid growth in various temperature–food conditions predicts that the proportion of the population found in deep waters feeding at any time must be larger if the vertical temperature gradient is smaller. Again the Tower system was used to test this prediction. In fact, the proportion of daphnids in the hypolimnion varied between 80 percent at a very shallow temperature gradient (20–18 °C) and 40 percent at a steep gradient (20–10 °C), which indicated that the daphnids were distributed to optimize their fitness with respect to offspring production as long as they were not driven down by predators (Lampert *et al.* 2003).

Optimizing the fitness of individual *Daphnia* causes an effect on the ecosystem level. Normally, phytoplankton would be distributed across the epilimnion and their concentration would decrease below the thermocline. Under such conditions, in the absence of fish, daphnids will permanently graze in the epilimnion and not go down. Hence, grazing pressure (i.e. algal mortality) will be high in the epilimnion but not in the cold water. With increasing transparency due to grazing (clear-water phase) light can penetrate deeper and algae near the thermocline can grow better and profit from the increased hypolimnetic nutrient concentrations. The result will be a deep-water chlorophyll maximum (Fig. 10.6). Experiments in the Plankton Towers have demonstrated this effect. Only when the food resources in the epilimnion become too low to be profitable to the zooplankton, will *Daphnia* start exploiting the deeper waters (W. Lampert, unpublished). However, the deep-water algal maximum may persist for a long time as the filtering rate of the daphnids will be much lower in the hypolimnion than in the epilimnion due to the low temperature. There will be enough grazing in the depleted epilimnion to maintain low

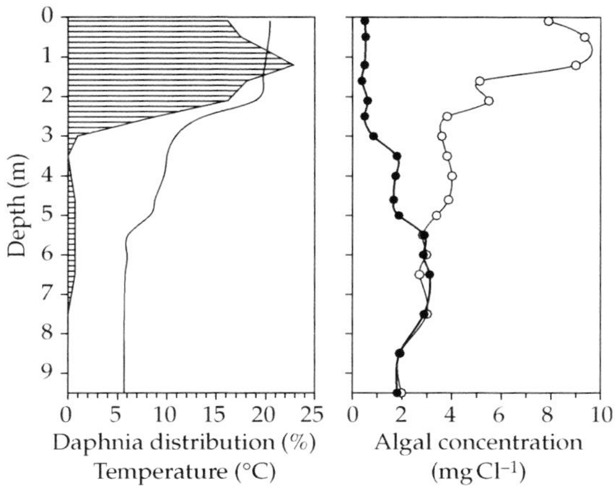


**Figure 10.6** Experimental study on the effect of grazing by *Daphnia hyalina* on the vertical distribution of algae (*Scenedesmus obliquus*) in the Plön Plankton Tower system. Left panel: vertical distribution of *Daphnia* (hatched area) in relation to temperature (solid line). Right panel: vertical distribution of algae after eight days of grazing by *Daphnia* (full circles) and in controls without grazing (open circles).

algal abundances, even if daphnids spend only part of their time there (see above).

The distribution of algal chlorophyll in a lake in space and time is a typical ecosystem property. It has not evolved as a system feature but is rather what would be called an "emergent property." However, it can occur as a consequence of the behavior of grazers choosing their optimal habitat. This is an example of an ecosystem property resulting from individual fitness optimization. I have developed this line of reasoning to demonstrate how principles of evolutionary ecology can be applied to ecosystem ecology.

## Conclusions

I believe that bridging the gap between ecosystem ecology and evolutionary ecology is the great challenge of ecology in the future. The title of this volume *"Evolution: from molecules to ecosystems"* is still rather ambitious. An ecosystem clearly cannot evolve in the Darwinian sense. It is neither a defined entity nor does it have a common genome as the basis for evolution. There is no room for a superorganism concept in ecosystem ecology although the historical conflict between different schools of thought is still ongoing (Hagen 1992). I do not see a reason for this conflict. Although it will be impracticable to explain the structure of an ecosystem by natural selection acting on every single component, we should still be able to define the major selective forces, predict the response of the organisms and the consequences for community assemblage. The role of rapid evolution for the occurrence of "emergent" ecosystem properties like stability and elasticity needs particular attention (J.N. Thompson 1998). Considering that ecosystems themselves cannot evolve, but can be modified as evolution takes place within the ecosystem frame, justifies the title of this book.

Many members of my research group have contributed to the examples I have used in this chapter. There is no space to list all the names, but I want to thank them all. I am particularly grateful to Victor Alekseev, Nelson G. Hairston, Jr, Suzanne Mitchell, Barbara Santer, and Herwig Stibor for permission to use their material. Manuel Serra, Lawrence J. Weider and Colleen Jamieson provided extremely helpful comments on an earlier draft of this article.

CHAPTER 11

# Evolutionary and ecological genetics of cyclical parthenogens

Luc De Meester, Africa Gómez, and Jean-Christophe Simon

**Cyclical parthenogenesis** (CP) is defined as the more or less regular alternation of sexual and **parthenogenetic** reproduction (Bell 1982; Hughes 1989). The terminology is, to some extent, unfortunate, as the alternation between sexual and parthenogenetic reproduction is not necessarily genetically determined in the life cycle, but responds largely to environmental conditions, the resulting pattern often being far from "cyclical." In general, parthenogenetic reproduction is sustained as long as environmental conditions remain favorable, whereas sexual reproduction is elicited by cues anticipating unfavorable conditions. Cladocerans, rotifers and aphids are the best-known representatives of cyclical parthenogens (CPs) (Box 11.1). With their capacity to combine sexual and asexual reproduction, many plants, cnidarians, bryozoans, and protists share a common characteristic with CPs, as they enjoy the benefits of both sexual and **clonal reproduction** (Lynch and Gabriel 1983). However, in CPs *sensu stricto*, the asexual reproductive phase involves the development of unfertilized eggs (i.e. they are parthenogenetic as opposed to reproducing by fission and budding), the asexually generated **ramets** are most often physiologically unconnected, and there is an alternation of a parthenogenetic

**Box 11.1 Cyclically parthenogenetic organisms**

Cyclical parthenogens encompass approximately 15 000 animal species that belong to at least six major and highly divergent **clades**, indicating a polyphyletic origin for this life cycle. Although the most widely known CPs are cladocerans (four orders of crustaceans with over 450 spp.), rotifers (over 2000 spp. belonging to the class Monogononta of the phylum Rotifera) and aphids (over 4000 spp. of the superfamilies Aphidoidea and Phylloxeridoidea), they are not the most speciose cyclically parthenogenetic groups. There are over 5000 parasitic spp. of cyclically parthenogenetic digenean trematodes, while the gall wasps of the family Cynipidae include over 2000 cyclically parthenogenetic spp., mainly parasitic on *Quercus* trees. The Cynipidae have a single, genetically determined parthenogenetic generation between sexual cycles. There are 20 cyclically parthenogenetic spp. of Cecidomid dipterans and a single cyclically parthenogenetic coleopteran species, *Micromalthus debilis*. In addition, some parasitic nematodes could be considered CPs as they combine **apomixis** with sexual reproduction in the same life cycle (Fisher and Viney 1998).

Cladocerans and rotifers are aquatic organisms that mainly inhabit inland waters, whereas aphids are small homopteran insects that feed on the phloem sap of plants. All three groups rely on a fast population growth rate and the production of resting stages to cope with the unpredictability of their habitat.

phase with a recurrent phase of sexual reproduction. Yet, the distinction between CP and other life cycles that combine sexual and asexual reproduction is somehow arbitrary (Hebert 1987).

The alternation of parthenogenetic and sexual reproduction has a strong influence on the genetic structure and evolution of cyclically parthenogenetic populations. The recurrent phase of sexual recombination releases hidden genetic variance on which natural selection can act, whereas selection during the phase of clonal reproduction is very efficient as it acts on both additive and interaction components of genetic variation (Lynch and Gabriel 1983; Deng and Lynch 1996). Moreover, CPs often show high population growth rates during the parthenogenetic phase. Both theoretical and empirical studies indicate that populations with regular bouts of sexuality do not suffer from the disadvantages associated with obligate asexuality, such as accumulation of deleterious mutations and reduced genetic diversity (Lynch and Gabriel 1983). Given these advantages, it is striking that CP is so rare in animals (Box 11.1). Constraints related to the coordinated formation of eggs through both meiosis and mitosis have been advanced to explain the paucity of cyclically parthenogenetic taxa (Lynch and Gabriel 1983). In addition, theoretical models suggest an intrinsic instability of the coexistence of the sexual and the asexual phases of the life cycle (Burt 2000). According to Hebert (1987), a reliable mechanism of environmental sex determination and the production of two egg types have been central for the adoption of CP.

Because **clonal** lineages of CPs can be maintained in the laboratory, a straightforward experimental approach can be used to partition environmental and genetic variation in populations. In addition, standard breeding studies are also possible. For this reason, cladocerans (*Daphnia*), rotifers (*Brachionus*) and aphids (*Acyrthosiphon, Rhopalosiphum, Sitobion*) are widely used model organisms for ecological and evolutionary research. Although allozyme studies have yielded important insights into the population genetic structure of CPs, these markers often exhibit low polymorphism. The recent advent of **PCR**-assisted molecular techniques has revolutionized our understanding of the genetic structure and **phylogeography** of natural populations. For example, the use of **microsatellites** allows studies of **resting egg banks** and enables discrimination of clonal individuals in field and experimental populations.

Here, we first describe the life cycle of cladocerans, rotifers, and aphids. Then, we address six topics in which major progress has been made in the last decade, emphasizing the similarities and differences among the three groups: (1) mating patterns and mate recognition in CPs; (2) variation in reproductive biology and loss of sexual reproduction; (3) evolutionary implications of resting egg banks; (4) the structure of genetic variation within and among populations; (5) adaptive polymorphism and local **adaptation**; and (6) **cryptic species** and hybridization. Due to space constraints, we do not provide a full account of the literature on each of the three groups, but rather we focus on emergent patterns that illustrate the importance of subtle differences in life cycle with respect to evolutionary dynamics, and we try to identify future avenues of research for each group. Therefore, we mainly cite recent studies and reviews and urge interested readers to consult elsewhere for a complete overview of the field (e.g. Hebert 1987; Hughes 1989; Carvalho 1994; De Meester 1996; Hales *et al.* 1997).

## The cyclically parthenogenetic life cycle

The cyclically parthenogenetic life cycles of cladocerans, rotifers and aphids are variants on a common theme (Box 11.2). Their parthenogenetic reproduction is **apomictic** (i.e. clonal; Hebert 1987; Hales *et al.* 1997). The rate of population increase during the parthenogenetic phase is very high in all three groups, because of their short generation time (*c.*2 days in rotifers, 6–7 days in aphids, and 9–11 days in *Daphnia*, at 20 °C) and their high fecundity (a single parthenogenetic aphid may produce 50–100 offspring and a single clutch of cladocerans may contain more than 100 eggs). This affords CPs a key ecological feature: the rapid exploitation of ephemeral resources. Sexual reproduction in all three groups is triggered by environmental cues and results in the release of genetic variance as well as influencing demography of the

## Box 11.2 The cyclically parthenogenetic life cycle of cladocerans, rotifers, and aphids

*Cladocerans*

As long as environmental conditions remain favorable, cladocerans (e.g. *Daphnia*) reproduce by amictic **parthenogenesis** (De Meester 1996; Fig. 11.1a). Populations are dominated by diploid parthenogenetic females that belong to coexisting clonal lineages. The relative abundance of these clones reflects their relative success in the habitat. Sexual reproduction is induced by unfavorable conditions or by stimuli that are associated with the onset of unfavorable conditions. These stimuli include high population densities, food shortage, the presence of predators, a shortening of the photoperiod, and changes in temperature. Males are produced parthenogenetically and are, thus, genetically identical to their mothers. The same female can produce both meiotic and ameiotic eggs. The sexual eggs are **resting eggs** that are deposited in a protective envelope (**ephippium**) formed by the carapace of the mother in most taxa (Fig. 11.2a). These eggs undergo an embryonic **diapause** after a few divisions and are thus actually embryos in an arrested state of development. They remain viable for a long time and are resistant to drying and freezing, while the ephippium protects them against mechanical damage and digestive enzymes of fish and birds (De Meester *et al.* 2002). The production of resistant resting eggs enables populations to recolonize the local habitat when the environment becomes favorable again and promotes dispersal to other habitats. Resting eggs hatch into parthenogenetic females in response to stimuli associated with favorable conditions.

*Monogonont rotifers*

The basic monogonont rotifer life cycle resembles that of cladocerans with the main difference that **haplodiploidy** is the mechanism of sex determination (Fig. 11.1(b); Nogrady *et al.* 1993). Parthenogenetic females usually produce eggs, but some species are viviparous (e.g. *Asplanchna*). In contrast to cladocerans, there are two types of females, sexual (**mictic**) and parthenogenetic, in most rotifer species. They usually co-occur in the population, and are morphologically identical but differ in life history traits. However, some genera (*Conochilus, Sinanterina, Asplanchna*) have **amphoteric** females, with the ability to reproduce both sexually and parthenogenetically. Cues that induce sexual females vary according to species and genotype, but often include crowding, photoperiod, and dietary compounds (Nogrady *et al.* 1993). Sexual females produce meiotic eggs that develop into haploid males if unfertilized. Fertilized oocytes develop into resting eggs (Fig. 11.2b). Rotifer males are smaller than females and lack a digestive system. At least some species produce two kinds of resting eggs: sexual and "pseudosexual" eggs. The latter are produced amictically in the absence of males and enter an obligatory brief diapause before resuming development and hatching. When cultured under low food conditions, *Synchaeta pectinata* exhibits a diversified bet-hedging strategy by producing various mixtures of diapausing and **subitaneous eggs** (Gilbert and Schreiber 1998). The resting eggs of monogonont rotifers, also called cysts or dormant eggs, are actually embryos in an arrested state of development. They are protected by a multilayered shell and can survive desiccation and other unfavorable conditions. After receiving appropriate stimuli, resting eggs hatch into parthenogenetic females that start another cycle (Nogrady *et al.* 1993).

*Aphids*

Aphids typically display an annual life cycle with one (some Phylloxeridae and arctic Aphididae), but often more than 20, apomictic parthenogenetic generations per year followed by a single sexual generation that overwinters as eggs (Fig. 11.1(c); Dixon 1998). Aphids with a two-year life cycle can be found in the tribes Fordini and Hormaphidini (Aphidoidea), and in the families Adelgidae and Phyloxeridae (Phylloxeridoidea). Approximately 10 percent of aphid species are host-alternating and spend autumn, winter and spring on a woody host, and summer usually on herbaceous plants. Non host-alternating aphids typically complete their entire life cycle on one or a few plant species of the same genus or family. They feed on the phloem sap of plants. This food is rich in sugars, but poor in some essential amino acids, and obligate bacterial symbionts supply their host with deficient amino acids (see van Ham *et al.*, Chapter 9). Population growth rates are maximized through viviparity and telescoping of generations in many aphids. At birth, aphids already have embryos developing in their gonads and their oldest embryos have also started to develop their own gonads. This telescoping of generations is characteristic of aphids and results in very high rates of increase, which can lead to well-known aphid outbreaks. In conjunction with a broad range of maternal and grand-maternal effects, it also enables aphids to anticipate seasonal changes in their resources (Dixon 1998). Sexual reproduction occurs in autumn and is induced by decreasing day length and temperature. Various degrees of investment in sex can be found in the same species or even in the same population of aphids (Simon *et al.* 2002).

population through allocation of resources to male production (see Serra *et al.*, Chapter 12). In rotifers and cladocerans, the dormant sexual stages are crucial to dispersal and recolonization of the habitat, whereas dispersal in aphids relies on winged adults.

As there is a trade-off between investment in sexual and asexual reproduction in CPs, there is strong selection for a subtle adjustment between the allocation of energy to each type of reproduction and environmental conditions (for the evolution of the timing of sex in rotifers see Serra *et al.*, Chapter 12). In temporary habitats, the **fitness** of a **clone** can best be measured as the amount and quality of the sexual **resting eggs** produced at the end of the growing season. Thus, life-history strategies of parthenogenetic females in temporary habitats should maximize resting egg production. Assuming that population size increases during the growing season, the later a clone switches to sex, the greater its potential fitness because more resting eggs will be produced. This strategy is, however, constrained by the need to switch to sexual reproduction before the arrival of unfavorable conditions. Therefore, we expect selection to favor strategies that predict the onset of unfavorable conditions combined with risk spreading. In aquatic CPs, investment in sexual reproduction is higher in temporary and small habitats than in large habitats allowing permanent population persistence throughout the year (Hebert 1987). In aphids, sexual reproduction results in the production of cold-resistant eggs that are the only form to survive cold winters, and allows proper synchronization with host plant phenology (Rispe *et al.* 1998).

## Mating patterns and mate recognition in cyclical parthenogens

In cyclically parthenogenetic populations that are characterized by regular bouts of sexual reproduction, genotypic frequencies are often in Hardy–Weinberg equilibrium. This has been interpreted as an indication that mating is random. Although the observation of Hardy–Weinberg equilibrium indeed suggests that assortative mating does not have a big impact on most loci and does not lead to inbreeding, one must be careful concluding that it implies no assortative mating whatsoever. Indeed, in populations that harbor a high diversity of clones, assortative mating with respect to specific phenotypic traits need not be reflected at the level of **neutral markers**. Some assortative mating has been inferred by studies on seasonal **succession** of clones (De Meester 1996).

Cyclical parthenogens face the risk that mating among individuals of a clone, genetically equivalent to selfing, may be common in populations that are characterized by a dominance of a few clones (e.g. in small populations). Strong inbreeding depression has been demonstrated in cladocerans (50 percent less survival until maturity of selfed compared to outbred offspring and reduced performance in other life-history traits; review by De Meester and Vanoverbeke 1999), rotifers (Birky 1967), and aphids (Dedryver *et al.* 1998; Rispe *et al.* 1999). Such strong inbreeding depression suggests that selfing or severe inbreeding is not common in populations of CPs. This could be due to random mating in populations harboring a high number of clones, or to the evolution of specific inbreeding-avoidance mechanisms. In aphids, clonal selfing may be strongly reduced

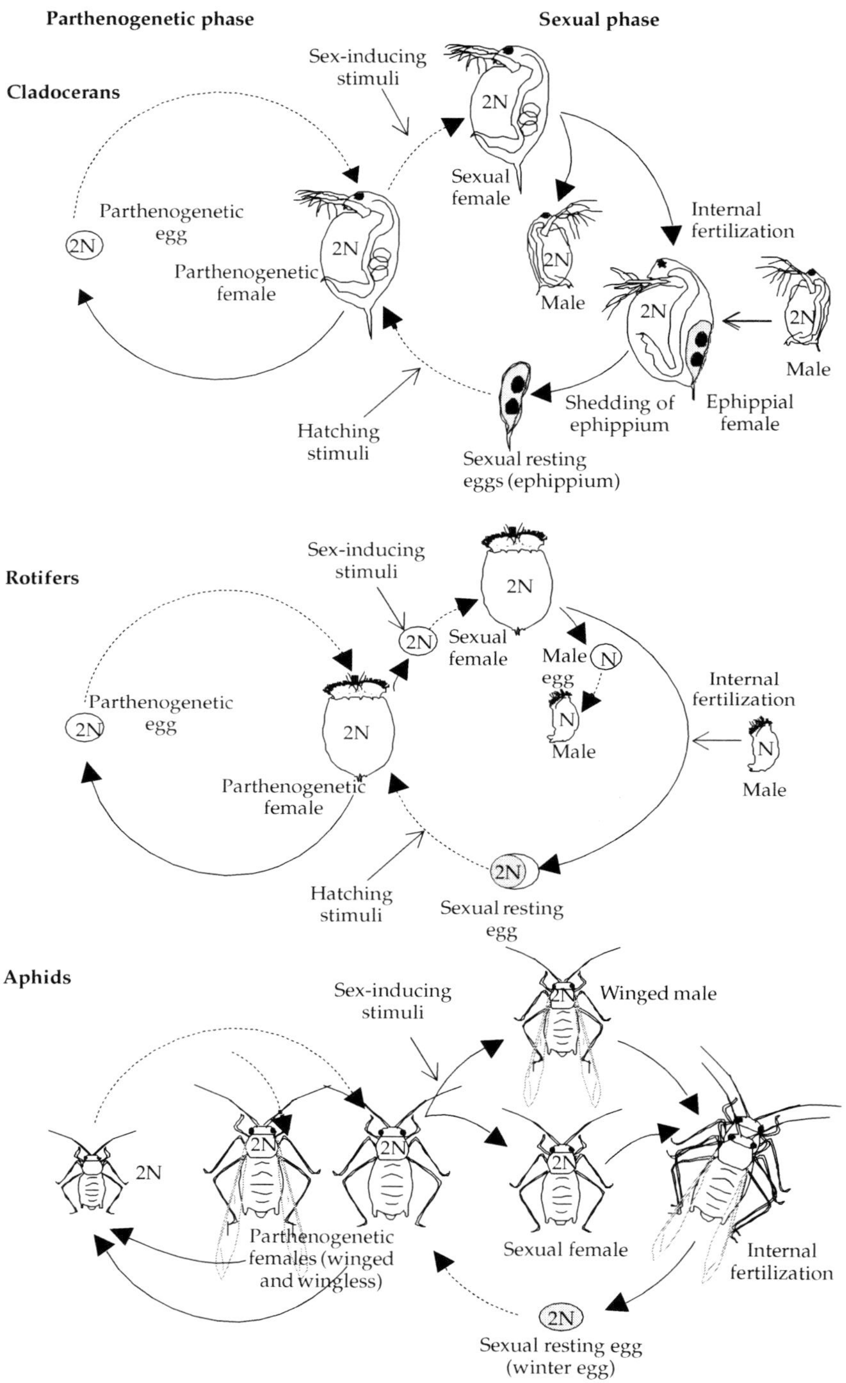


**Figure 11.1** Schematic life cycle of three groups of cyclical parthenogens: (a) cladocerans (*Daphnia*); (b) monogonont rotifers (*Brachionus*); (c) aphids. Continuous lines indicate reproductive events and dotted lines indicate developmental events (for further details see Box 11.2).

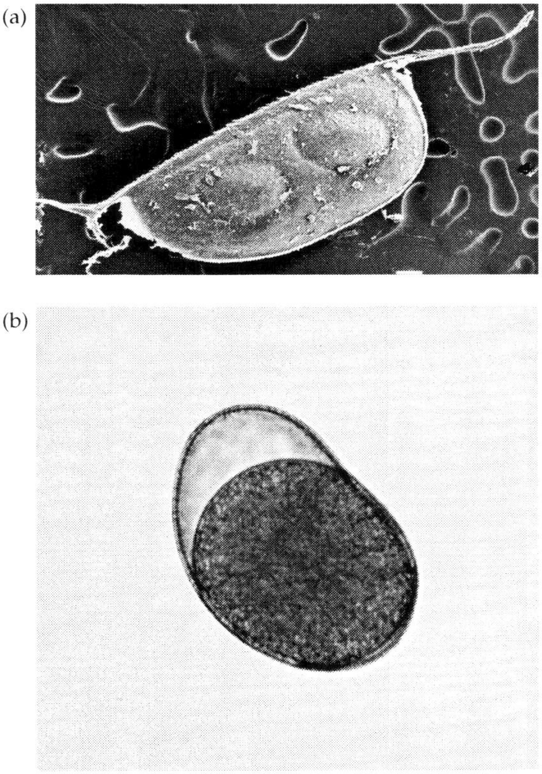

**Figure 11.2** (a) Ephippium with two resting eggs of the cladoceran *Daphnia magna* (SEM photograph L.D.M.); (b) resting egg of the rotifer *Brachionus* (photograph by R. Ortells).

by the fact that most sexuals are winged and that host-alternating species have an obligatory winged phase linked to sexual reproduction, which takes place on a different host plant than the clonal phase (Dixon 1998). In cladocerans, De Meester and Vanoverbeke (1999) have provided evidence that an uncoupling at the clonal level of the responses to stimuli inducing males and sexual eggs may result in a substantial reduction of the occurrence of clonal selfing. Innes and Dunbrack (1993) observed that some *Daphnia* clones do not produce males at all, and interpreted this as an inbreeding avoidance mechanism. A similar pattern with clones producing mostly sexual females was also found in aphids (Rispe *et al.* 1999). Finally, De Meester (1992) and Brewer (1998) showed that habitat selection (vertical distribution) of male *Daphnia* is different and less genotype-specific than that of female *Daphnia*, a pattern that may also result in a lower probability of intraclonal mating.

Whereas mating behavior has only been sporadically studied in cladocerans (Forró 1997) and aphids (Guldemond and Dixon 1994), rotifer mating behavior is relatively well known, especially in the genus *Brachionus* (Snell 1998). Mate recognition in rotifers depends on contact chemoreception of a sex **pheromone** (a surface glycoprotein in *B. plicatilis*) on the female's body surface (Snell 1998). Rotifer sex pheromones seem to be species-specific (Gómez and Serra 1995).

The performance of cross-mating tests to assess reproductive isolation in cladocerans suffers from the fact that hybridization among taxa seems to be widespread, even among distantly related species (Colbourne and Hebert 1996). Whereas Crease and Hebert (1983) did not find evidence for the presence of sex pheromones in *Daphnia magna*, Carmona and Snell (1995) reported the presence of glycoproteins on the ovaries of *D. obtusa* and *Ceriodaphnia dubia*, and hypothesized that these molecules might be involved in contact recognition of males, as they are in rotifers. In aphids, mating females attract males by a sex pheromone. There is a wide range of variation in the chemical composition of sex pheromones and their patterns of emission in aphids (Guldemond and Dixon 1994).

Given the overall evolutionary importance of mating patterns, experiments are needed in all three groups to determine whether mate choice occurs, and to what extent there is a decline in fertilization success with genetic distance.

## Evolutionary implications of diapausing eggs and resting egg banks in cyclical parthenogens

Aphids, rotifers, and cladocerans all produce **diapausing eggs**, but rotifer and cladoceran populations build up large resting egg banks due to the ability of the eggs to survive over several growth seasons, which has not been reported for aphids. Moreover, the resting stages of rotifers and cladocerans are the main dispersing agents, while this is not the case in aphids. Although the production of diapausing eggs in all three groups can be viewed as a strategy to cope with seasonality as well as with the unpredictability of the habitat, the latter aspect is apparently less important in aphids.

Most aphids overwinter as diapausing eggs that contain anti-freeze (Blackman 1987). The diapausing eggs are deposited on their host plants. Heritable differences in diapause length exist among lineages of aphids but hatching time in tree-dwelling species may also depend on bud burst phenology of their host-plants (Dedryver *et al.* 1998). Some species of aphids also include hibernating and aestivating forms that arrest development or reproduction at various stages depending on species (Dixon 1998). However, there is no evidence for diapause lasting for more than one year (Blackman 1987).

As rotifer and cladoceran resting eggs can remain viable for many years (up to decades at least), they accumulate in the lacustrine sediments, forming resting egg banks (Brendonck *et al.* 1998). Resting stages allow for dispersal to other habitats as well as for dispersal in time, though depending on local conditions such dispersal may not necessarily result in **gene flow**. The latter may increase genetic diversity in the population in the case of fluctuating selection, while slowing down evolution in the case of **directional selection** (Hairston and De Stasio 1988). As resting egg banks contain eggs produced over several years, this effectively results in overlap of generations similar to that of long-lived **iteroparous** species. The evolutionary and ecological importance of resting egg banks resides to a large extent in their staggering sizes. Reported resting egg bank densities in both rotifers and cladocerans range from $1 \times 10^4$ up to almost $1 \times 10^7$ eggs $m^{-2}$ (Snell *et al.* 1983; Marcus *et al.* 1994; Cousyn and De Meester 1998). The large size of these egg banks, estimated to contain billions of eggs in a moderately sized pond, may effectively buffer against population bottlenecks and the genetic impact of immigrants (De Meester *et al.* 2002). One can safely assume that most rotifer and cladoceran populations in intermediate or large habitats start the growing season with a prodigious clonal diversity ($\gg 10^6$ genotypes). Resting eggs are a key factor in the evolutionary success of rotifers and cladocerans, because of their ability to survive harsh conditions, the increased dispersal potential and the contribution to population genetic variability (King 1980). In addition to their effect on population genetic structure, resting egg banks potentially have important ecological effects. They strongly reduce local extinction rates, potentially alter the outcome of competition and predation, and increase the likelihood of clonal coexistence through the storage effect (Cáceres 1997). In addition to these ecological and evolutionary consequences, resting egg banks can also be used to reconstruct microevolution in the recent past (see Lampert, Chapter 10). Field studies of resting egg viability, dynamics (e.g. average time spent in the egg pool, proportion of eggs hatching every year; see DeStasio 1989) and **dormancy** strategies, as well as theoretical studies exploring these issues are needed in both cladocerans and rotifers in order to understand the evolutionary role of resting egg banks.

## Patterns of loss of sexual reproduction

Transitions to obligate asexual reproduction are quite frequent in CPs, leading to the coexistence of cyclical and obligatory parthenogenetic populations. Research into the patterns of loss of sexual reproduction in cyclically parthenogenetic lineages can shed light on the balance between the advantages and costs of sexual reproduction in relation to environmental conditions. Morphological and molecular phylogenetic analyses show that CP is the ancestral state in cladocerans and aphids (Taylor *et al.* 1999; Von Dohlen and Moran 2000). The loss of sex in the life cycle of cladocerans and aphids is, therefore, a derived character, and apparently arose independently on several occasions (Moran 1992; Colbourne and Hebert 1996; Delmotte *et al.* 2001).

Cladocerans show a pronounced pattern of geographic **parthenogenesis** (Lynch 1984a): populations occurring in temperate areas are CPs, whereas many taxa found further north are **obligate parthenogens** (OPs) and are often **polyploids** (Colbourne and Hebert 1996). Weider and co-workers (1999a,b) have compiled an impressive data set on the occurrence and distribution of obligatory parthenogenetic taxa and their genetic diversity around the circumarctic region. The high genetic diversity in obligatory parthenogenetic *Daphnia* reflects their polyphyletic origin, as many asexual clones are independently derived from the ancestral cyclically parthenogenetic populations (Innes and Hebert 1988). In *Daphnia*, **obligate parthenogenesis** (OP) is caused by a meiosis suppressor gene that can spread through the population because many obligatory parthenogenetic lines retain the ability to produce males. When males of obligatory parthenogenetic clones mate with cyclically parthenogenetic females, half of the offspring clones are obligatory parthenogenetic (Innes and Hebert 1988; Crease *et al.* 1989). This process makes the boundary between obligatory and cyclically parthenogenetic populations potentially very dynamic.

Although a whole class of rotifers, Bdelloidea, comprises ancient obligatory asexual species (Mark Welch and Meselson 2000), OPs have not so far been described in monogonont rotifers. However, males are rare in some common species, or have never been reported in some localities (e.g. *Keratella cochlearis, Brachionus calyciflorus*; Bell 1982). The fact that sex is seemingly lost after long-term culture of CP clones in the laboratory (A. Gómez, pers. observation) suggests that the loss of the sexual phase might be more common than thought in rotifers.

The loss of the sexual phase in aphids is frequent and has important ecological and evolutionary consequences (Simon *et al.* 2002). The obligatory parthenogenetic species (*c.*3 percent of the species) are scattered among the different aphid taxa, except for the tribe Tramini (Lachnidae) that is completely asexual and overwinters in ant nests (Moran 1992). Partial loss of sexual reproduction is common: 57 percent of host alternating (over 137 species) and 33 percent of non host-alternating (over 135 species) aphids show coexisting sexual and asexual populations (Moran 1992). Because the production of sexuals mainly relies on photoperiodic cues, a change in latitude can potentially result in the loss of sex with no genetic change (Moran 1992). However, in most species studied so far, OPs were found not to respond at all to sex-inducing cues as provided in the laboratory, and were found to be genetically distinct from their cyclically parthenogenetic sister lineages (Simon *et al.* 1999a). Because it is difficult to breed aphids in the laboratory, few attempts have been made to elucidate the inheritance of the transition to OP, although some studies suggest that a single gene may be responsible for breeding system variation in certain species (Simon *et al.* 1999a). Males from OPs can fertilize cyclically parthenogenetic females and transmit allele(s) suppressing the formation of sexual females, leading to the production of new obligatory parthenogenetic lineages (Simon *et al.* 1999a). Gene flow between cyclical and obligatory parthenogenetic lineages mediated by male-producing OPs, thus, occurs both in *Daphnia* and in aphids, with the potential for asexuality genes to spread in a contagious fashion and rapidly convert cyclical into obligatory parthenogenetic lineages (Innes and Hebert 1988; Simon *et al.* 2002). However, little is known about the genes involved in the loss of sex, which may differ between the two taxa (meiosis suppressor genes in *Daphnia*, Innes and Hebert 1988; periodicity genes or genes regulating hormonal expression in aphids, Simon *et al.* 1999a). The recurrent generation of new asexual lineages may be essential for the long-term maintenance of obligate parthenogenetic reproduction. The phylogenetic distribution of OPs among the aphids indicates that they are usually evolutionary dead ends and that most of them have closely related cyclically parthenogenetic counterparts. Some OPs are considered to be distinct species, but they can easily be placed within genera and species that have retained sex. Only a few OPs appear to be sufficiently ancient to have resulted in substantial genetic divergence (Moran 1992). But even in such putatively ancient asexuals, recent assessment of allelic divergence actually suggests a much more recent origin than previously thought (Normark 1999; Delmotte *et al.* 2003).

An important difference between aphids and cladocerans is that obligatory parthenogenetic lineages in aphids do not produce resting eggs. Rispe *et al.* (1998) argue that the maintenance of sexual reproduction in many aphids may be linked to an ecological function: the resistance to frost of sexually produced diapausing eggs. As a result, OPs should occur in warm regions, and CPs in colder regions, which is often the case (Dixon 1998; Rispe *et al.* 1998), and which is the opposite pattern to that found in cladocerans. This ecological correlate of sexual reproduction has also been invoked in rotifers (King 1980). It is noteworthy that cladocerans have circumvented this constraint, as OPs produce their resting eggs parthenogenetically (Hebert 1987). Although some rotifer species likewise are capable of producing short-term diapausing eggs in the absence of males (Gilbert and Schreiber 1998), the production of true resting eggs seems to require sexual reproduction. Interestingly, these species also produce sexual resting eggs, and are thus not obligatory parthenogenetic.

Through examination of the genomic differences between CPs and their obligatory parthenogenetic relatives, one can evaluate whether asexual lineages can avoid the accumulation of deleterious mutations (**Müller's ratchet**), improving their chance to persist over long periods of time. Sequence analysis suggested a significant acceleration of deleterious mutation accumulation in the asexual aphid species *Tuberolachnus salignus* (Normark and Moran 2000). Sullender and Crease (2001) recently investigated the activity of a **transposable element** in *Daphnia pulex*. While the transposable element was probably active in cyclically parthenogenetic populations, it had largely been inactivated in obligatory asexuals, allowing them to avoid deleterious genome changes associated with **transposition** (see Fontdevila, Chapter 16). Interestingly, the ancient asexual Bdelloid rotifers similarly appear to have lost their **retrotransposons**, in contrast to the cyclically parthenogenetic Monogonont rotifers (Arkhipova and Meselson 2000).

## Genetic diversity in local populations

Genetic diversity in cyclically parthenogenetic populations can be assessed at both the allele and the multilocus genotype level, depending on whether one focuses on the sexual offspring or the parthenogenetically reproducing population of females, respectively. There are numerous studies documenting genetic diversity in cladocerans, rotifers, and aphids, and several reviews have been published (cladocerans: Hebert 1987; Carvalho 1994; De Meester 1996; rotifers: Gómez *et al.* 1995; Gómez and Carvalho 2000; aphids: Hales *et al.* 1997). The genetic structure of cyclically parthenogenetic organisms is largely determined by the length of the growing season, as it determines the number of parthenogenetic generations between sexual phases (Pfrender and Lynch 2000). Selection among clones can be substantial during the parthenogenetic phase, such that populations may exhibit a clonal structure at the end of the growth period, characterized by lower genetic diversity and deviations from Hardy–Weinberg equilibrium. Reduction of genetic diversity during the course of the growing season has been observed in aphid (Rhomberg *et al.* 1985; Sunnucks *et al.* 1997), rotifer (Gómez *et al.* 2000), and cladoceran populations (Lynch 1984b). Low clonal diversity and strong deviations from Hardy–Weinberg equilibrium are often observed in **zooplankton** populations inhabiting small permanent ponds (cladocerans: Hebert 1987; rotifers: Ortells *et al.*, unpublished). This is also the case for aphids living in isolated small patches of host plants (Rhomberg *et al.* 1985; Hebert *et al.* 1991), although the existence of winged forms combined with a more continuous habitat makes it difficult to delimit local populations. The structure of populations inhabiting intermittent habitats (e.g. prone to drought for cladocerans and rotifers or to frost for aphids) is essentially sexual, and these populations are characterized by high clonal diversity, stable allele frequencies, and no deviations from Hardy–Weinberg equilibrium (cladocerans: Hebert 1987; rotifers: Gómez *et al.* 1995; Gómez and Carvalho 2000; aphids: Simon and Hebert 1995; Sunnucks *et al.* 1997).

## Dispersal ability and genetic structure at the regional scale

Although the extent of dispersal in aphids is controversial due to the difficulties in directly documenting their movements, the occurrence of winged forms substantially increases the potential for dispersal and gene flow in most species. Studies using high resolution genetic markers have indeed demonstrated long range and intense dispersal in at least some species (Simon *et al.* 1999b; Sunnucks *et al.* 1997). The production of winged forms is often induced by crowding, but other biotic (host plant quality, natural enemies) and abiotic (light, temperature) as well as genetic and **maternal effects** may also be important (Dixon 1998). The life cycle has a strong influence on the dispersal of sexual forms in aphids. In host-alternating species, sexual reproduction is associated with an obligate host shift, while non-host-alternating species may remain on the same plant. In the former species, both **presexual forms** and males are winged, whereas most aphid species that do not alternate hosts have wingless males.

While many aphids are winged, and therefore able to disperse actively, rotifers, and cladocerans rely on passive dispersal by resting eggs. The diapausing and resistant features of resting eggs allow for long-distance transport through wind, water or animal vectors such as waterfowl or aquatic insects. Given the relatively high dispersal capacity in all three groups, it is striking that pronounced changes in genotype composition have often been observed even among neighboring populations, for both neutral markers and for ecologically relevant traits (aphids: Via 1999; Dedryver *et al.* 2001; rotifers: Gómez *et al.* 1995; cladocerans: reviewed by De Meester 1996; Morgan *et al.* 2001). De Meester *et al.* (2002) discuss this apparent paradox between high dispersal capacity and low levels of gene flow. In very small populations harboring a small number of clones, part of the genetic differentiation among populations for neutral markers can be explained by a **hitchhiking** effect of neutral alleles with temporarily successful clones (Rhomberg *et al.* 1985; Vanoverbeke and De Meester 1997). Irrespective of this process, however, rapid monopolization of resources by the first colonist of a habitat, due to the fast population growth rate, the production of a resting egg bank and fast local adaptation, results in very strong priority effects in CPs. As such, the observed differences among populations often reflect long-lasting **founder effects** with respect to neutral markers (Boileau *et al.* 1992; De Meester *et al.* 2002). In aphids, despite high potential gene flow generated by intense dispersal, there is evidence for regional differentiation with climate and host plant being the main structuring factors (Sunnucks *et al.* 1997; Simon *et al.* 1999b; Via 1999; Dedryver *et al.* 2001), illustrating that the balance between selection and migration is very important in determining the extent of regional differentiation.

## Genetic polymorphism, phenotypic plasticity, and local adaptation in cyclical parthenogens

The cyclically parthenogenetic life cycle should allow rapid local genetic adaptation, as it combines effective selection on the whole genetic component of variation during the parthenogenetic phase with the release of hidden genetic variation through recurrent sexual recombination (Lynch and Gabriel 1983; De Meester 1996; De Meester *et al.* 2002). In many cases, studies of ecologically relevant traits have reported patterns of local adaptation in cladocerans (Ebert 1994; De Meester 1996; Boersma *et al.* 1998; Declerck *et al.* 2001) and aphids (Sunnucks *et al.* 1997; Via 1999), while so far little has been done in this field using rotifers.

In aphids, host-plant specialization is a major component of local adaptation. Considerable intraspecific genetic variation exists for host use in aphids, with different clones showing their highest fitness on different host plant species, and there is growing evidence from studies using molecular markers for population subdivision according to host plant in many aphid species (Sunnucks *et al.* 1997; Via 1999; Hawthorne and Via 2001). The interactions between aphids and their natural enemies can also lead to local adaptation. Recent work has shown potential for coevolution in the *Acyrthosiphon pisum* (pea aphid)/*Aphidius ervi* (parasitoid) system by documenting genetic variation for both the ability to infect in parasitoid populations and the susceptibility to infection of aphid clones, and by providing evidence for specific and reciprocal genetic interactions (Henter and Via 1995). There is also good evidence for parasite–host coevolution in the *Daphnia*–microparasite system, with reports of genotype-dependent susceptibility to specific parasite strains (Carius *et al.* 2001) and patterns of local adaptation of parasite and host (Ebert 1994).

In their interactions with predators, aphids, cladocerans, and rotifers exhibit striking predator-induced

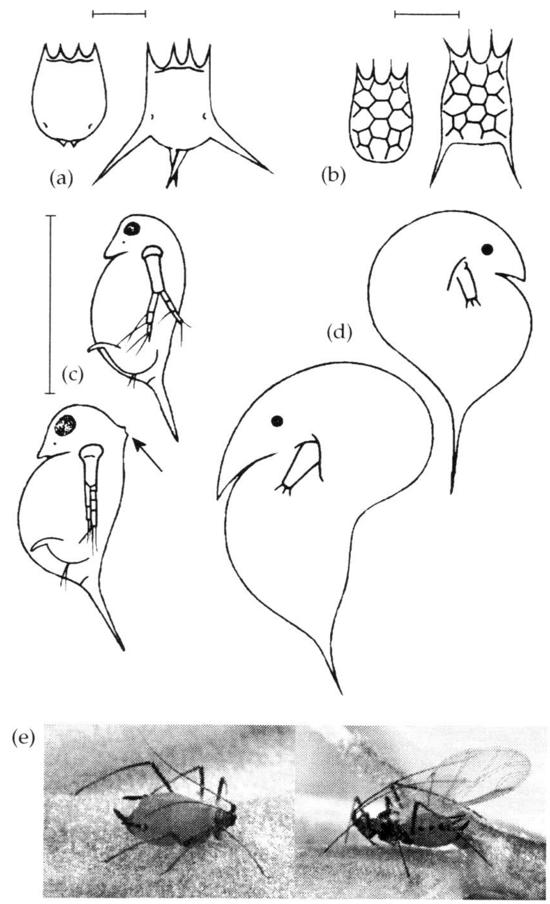


**Figure 11.3** Phenotypic plasticity and polyphenism in cyclical parthenogens. Control and predator-induced phenotypes of clones of (a) the rotifer *Brachionus calyciflorus* (predator: the rotifer *Asplanchna*), (b) the rotifer *Keratella quadrata*, (c) the cladoceran *Daphnia pulex* (arrow points to neckteeth induced by the presence of the phantom midge larvae, *Chaoborus*), and (d) the cladoceran *Daphnia carinata* (induction by water beetles) (from Lampert and Sommer 1997, reproduced with permission). (e) Wingless and winged forms of the aphid *Megoura viciae* (copyright Bernard Chaubet).

defenses (Fig. 11.3(a–d); see also Lampert, Chapter 10). Many traits show adaptive and inducible shifts in trait values in response to predators, including behavioral (diel vertical and horizontal migration, alertness, swarming), morphological (neck teeth, helmet development, tail spine size, size at maturity, eye diameter, wings) and life history traits (development time, size, and number of eggs) (Tollrian and Harvell 1999; Weisser *et al.* 1999). Studies involving different clones and traits have revealed that most clones show a significant response to predator presence in a different set of traits (Boersma *et al.* 1998), which may often result in similar fitness in the presence of a predator (De Meester *et al.* 1995). Also, interactions are complex, as they often involve trade-offs among defenses against multiple antagonists (Decaestecker *et al.* 2002). Spine induction by some rotifers (*Brachionus, Filinia, Keratella*) in the presence of copepods and predatory rotifers (e.g. *Asplanchna*) has been shown to deter predation significantly. The complex phenotypic changes in morphology associated with diet and cannibalism in *Asplanchna* have been studied in detail (see Gilbert 1980; Nogrady *et al.* 1993).

The case of predator-induced defenses underscores the large phenotypic plasticity found in most CPs. In aphids, **polyphenism** (i.e. the occurrence of multiple, often morphologically distinct phenotypes that occur during the life cycle of a clone) is one of the most striking examples of this phenotypic plasticity (Fig. 11.3e). Polyphenism is probably more extensive in aphids than in any other invertebrate group (Moran 1992). Up to 10 different phenotypes or morphs can be produced during the annual life cycle of an aphid clone (Dixon 1998). They can differ in various features including morphology (winged, wingless), reproductive mode (sexual, parthenogenetic), physiology, life-history traits (development, fecundity, survival), behavior (dispersal, defense), and host utilization (summer and winter host in host-alternating species) (Dixon 1998). A spectacular example is the existence of soldiers with enlarged prehensile forelimbs and frontal horns (Stern and Foster 1998). The soldiers are sterile and show defensive behavior to protect the colony against predator attacks. The high genetic relatedness among clonal individuals of the same colony may indeed lead to the emergence of altruistic behavior (Hales *et al.* 1997). The occurrence of mictic and **amictic** females in rotifers is similar to the polyphenism reported for aphids in that different eggs either develop into sexual or asexual females. Polyphenism has also been described in the rotifer genus *Asplanchna*, which shows different morphotypes in the same clone in response to cannibalism (Gilbert 1980).

The above examples show that CPs combine extensive phenotypic plasticity with high evolutionary flexibility in terms of genetic responses to changes in ecological conditions. Such a pattern arises from the combined effects of long-lived clones with short individual generation times. The key to the high levels of phenotypic plasticity is that individual genotypes (**genets**) often live for a long time (for one to many years). As a result, the environment becomes fine-grained at the level of the genotype, thus, resulting in selection for phenotypic plasticity that tracks seasonal changes in environmental conditions. On the other hand, individuals have short generation times and high reproductive capacity, which creates conditions for strong clonal selection and rapid shifts in genotypic composition, especially because selection also acts on the interactive genetic component of variation. Added to such effects is the extensive genetic diversity that is created during sexual reproduction and stored in the resting egg bank. One can, thus, predict that CPs are likely to adapt rapidly to local conditions (e.g. Cousyn *et al.* 2001) through selection for genotypes exhibiting an adaptive combination of phenotypic plasticity responses. Clones of CPs can, therefore, be thought of as extreme generalists with respect to their local environment, but specialists with respect to a wider regional setting.

## Cryptic species and hybridization

As CP promotes long-lasting founder effects as well as rapid local adaptation, it may promote provincialism in **phylogeographic** distributions and, eventually, **allopatric speciation**. This has been reported for both cladocerans (Hebert and Wilson 1994; Hebert and Finston 1996; De Meester *et al.* 2002) and rotifers (Gómez *et al.* 2000). Molecular phylogenetic analyses in these two groups have revealed that a remarkable stasis in morphology hides the occurrence of ancient cryptic species complexes (Colbourne *et al.* 1997). These studies indicate that some taxa currently included in the same morphological species have evolved separately for millions of years. For instance, molecular phylogenetics has revealed the existence of ancient cryptic species in the rotifer *Brachionus plicatilis*, previously regarded as a cosmopolitan species. Nuclear and mitochondrial sequence variation revealed that at least nine cryptic species exist, some of them occurring in **sympatry** and showing a remarkably wide geographical distribution (Gómez *et al.* 2002). In *Daphnia*, sister species often differ in their habitats (e.g. ponds versus lakes), which is suggestive of speciation via habitat transitions (Taylor *et al.* 1998).

In cladocerans, taxonomy is further confounded by the abundant occurrence of natural hybrids. Hybridization occurs between *Daphnia* taxa separated for up to 9–50 million years, which indicates that reproductive incompatibility evolves very slowly (Colbourne and Hebert 1996; Schwenk *et al.* 2000). Intriguingly, hybrids are often very abundant and may dominate entire populations (Schwenk and Spaak 1995). The abundance of hybrids in many lakes may be related to the fact that they have advantageous combinations of parental traits (Schwenk and Spaak 1995), but how these combinations originate needs further investigation. *Daphnia* hybrids inhabiting permanent lakes do not face the problems associated with sexual reproduction. As they may survive for many generations, the abundance of hybrid clones may merely reflect their success in clonal competition rather than common hybridization events. The age of hybrid lines and the frequency of hybridization events are issues that need further attention. Analysis of nuclear and mitochondrial DNA markers has revealed that hybridization can be asymmetric, one of the parental species always being the mother (Schwenk and Spaak 1995). Such effects may be due to the timing of sexual stages, but may also reflect selection on the hybrids themselves. The occurrence of interspecific hybrids allows for the possibility of **introgression** and **reticulate evolution** (see Fontdevila, Chapter 16). At least one cladoceran species, *Daphnia mendotae*, has been shown to be of hybrid origin (Taylor *et al.* 1998; Schwenk *et al.* 2000).

In aphids, there are many complexes of closely related forms that share a host on which sexual reproduction occurs and are able to hybridize. Although natural hybridization may be reduced by mate recognition systems, it has been documented in several species (Guldemond and Dixon 1994; Sunnucks *et al.* 1997). In some cases, it seems that hybridization could lead to the emergence of asexual lines of aphids (Delmotte *et al.*, in press). In addition, there is growing evidence that **sympatric speciation** may occur following host shifts (Guldemond and Dixon 1994; Via 1999), a phenomenon particularly well studied in the pea aphid, *Acyrthosiphon pisum*, for which subspecies or host races are known. Reproductive isolation between **sympatric** races of pea aphids specialized either on clover or alfalfa has been demonstrated from habitat choice exerted by winged aphids and allozyme analysis suggesting reduced gene flow (Via 1999). The mechanisms involved in host shift are not completely understood but presumably result from changes in host recognition or differential reproductive performance (Hawthorne and Via 2001). Besides host shift, reproductive isolation and speciation in aphids may also result from the loss of sexual reproduction leading to abrupt isolation of the newly formed asexual lineage from the sexual gene pool (Simon *et al.* 1999a). Self-fertilization and **allochronic isolation** (due to temporal differences in the production of sexuals between aphid taxa) have also been invoked as possible mechanisms for reproductive isolation (Guldemond and Dixon 1994).

## Concluding remarks

Cyclical parthenogenesis, with its combination of sexual and asexual reproduction, enables aphids, cladocerans, and rotifers to make effective use of ephemeral resources while, at the same time, maintaining extensive potential for adaptive evolution. Cyclical parthenogens are among the most abundant organisms in continental aquatic habitats, and also constitute one of the major pests on cultivated plants. Resistance to insecticides, host plant switches and the evolution of predator-induced defenses illustrate the evolutionary potential of CPs. In all three groups, the sexual phase is associated with the production of resistant **propagules**. The major differences in genetic structure within and among populations between aphids on the one hand and rotifers and cladocerans on the other seem to arise from differences in dispersal ability and the capacity for long-term dormancy in the sexually produced eggs of the latter two groups, which allows them to cope with temporal unpredictability. The resting eggs of rotifers and cladocerans accumulate in lake and pond sediments, influencing long-term adaptive potential as well as various population and community level ecological characteristics. The sexual phase of the life cycle has apparently been lost repeatedly in these groups of CPs. Loss of the sexual phase has important implications for the evolutionary fate of the resulting lineages, and complicates the efforts of researchers trying to elucidate the population structure of CPs. Cyclical parthenogens display wide phenotypic plasticity, both behavioral and morphological, often of an adaptive nature, and they exhibit striking patterns of local adaptation. Finally, their evolutionary peculiarities have played an important role in shaping deep phylogeographic structures and allopatric speciation accompanied with little morphological change.

We thank the organizers of the symposium in Valencia for inviting us to attend this highly interesting meeting. L.D.M. acknowledges financial support from the K.U. Leuven Research Fund (grant 0T/00/14). We thank Gary Carvalho, Manuel Serra, Roger Hughes, and two anonymous referees for constructive criticism of previous versions of this manuscript.

CHAPTER 12

# The timing of sex in cyclically parthenogenetic rotifers

**Manuel Serra, Terry W. Snell, and Charles E. King**

**Cyclical parthenogenesis** is a mode of reproduction that combines **parthenogenesis** with episodic sexual recombination. It has evolved independently several times, but is largely restricted to parasitic trematodes, gall wasps, gall midges, aphids, and two groups of **zooplankters**: cladocerans and monogonont rotifers (Bell 1982; see also De Meester *et al.*, Chapter 11). Cyclical parthenogenesis can be regarded as an optimal combination of a demographic phase during which there is rapid clonal propagation, and a sexual phase during which meiosis and recombination create new genotypic variation. Each of these phases has its own evolutionary and ecological advantages and costs.

Cyclical parthenogens provide a group of model organisms for examining several important aspects of the evolution of sexual reproduction. First, the consequences of sexual and asexual reproduction on the evolutionary dynamics of **adaptation** can be analyzed in a single species (De Meester *et al.*, Chapter 11). Second, because sex is facultative in cyclical parthenogens, the level of sexual reproduction can be correlated with environmental factors and habitat characteristics. The costs of sex can be easily measured and incorporated into models accounting for different patterns of sexual reproduction. The optimal allocation of resources into sexual and asexual offspring can be analyzed with similar methods to those used in the study of other life-history trade-offs. Third, the effects of "bad choices" regarding when to switch from asexual to sexual reproduction are readily investigated in this group, as well as the effects of environmental changes, including those from anthropogenic sources, on sexual and asexual cycles.

Rotifers make up a phylum composed of taxa that differ greatly in their mode of reproduction: Seisonids are obligate sexuals, bdelloids are obligate parthenogens, and monogononts are cyclical parthenogens. Monogonont rotifers live in seasonal and ephemerally favorable aquatic habitats and differ in their seasonal distribution and their residence time in the plankton. In rotifers, asexual reproduction by **apomixis** produces **clonal** females (Fig. 11.1(b) in De Meester *et al.*, Chapter 11). Multiple generations of asexual reproduction are punctuated with occasional bouts of male production and sexual recombination. The product of sexual reproduction in monogononts is a thick shelled **resting egg** that typically has an extended period of **dormancy** before hatching. Monogonont rotifer populations in temperate climates are re-established each year by hatching from a **resting egg bank** in the sediments (Pourriot and Snell 1983). After a phase of clonal propagation in the plankton, sex is triggered by environmental factors such as population density (e.g. Snell and Boyer 1988; Carmona *et al.* 1995). Sexual reproduction starts when asexual females parthenogenetically produce both sexual and asexual daughters. Sexual females then produce haploid eggs that develop into either haploid males or, if fertilized, into resting eggs. Haploid egg fertilization is only possible if the sexual female is inseminated within a few hours of birth (Snell and Childress 1987). If a sexual female is not inseminated while young, she will only produce males. Sexual, resting eggs are able to survive adverse conditions and may remain viable in the sediment for decades (Marcus *et al.* 1994; see also Lampert, Chapter 10). Resting eggs are likely to be the major mode of geographic dispersal.

In this paper, we review a considerable amount of information from modeling, laboratory, and fieldwork on the rotifer life cycle. Intensive research has been performed on a rotifer genus (*Brachionus*), and we will focus mainly on this research. Our aim is to get insights into the evolutionary forces shaping the timing of the asexual and sexual phases of reproduction in monogonont rotifers. We address the causes of the maintenance of sex in rotifers and its optimal timing. Finally, we consider the factors operating on the reproductive process in a variety of habitats and relate the effects of environmental change to the probability of population extinction.

## Evolutionary hypotheses for the origin and maintenance of sex in rotifers

The field of population biology has been deeply concerned—some might say obsessed—with the question "why sex?" (Bell 1982). It is easy to understand the reason for this concern. Assuming that a typical asexual female is able to produce the same number of offspring as a typical sexual female, and further assuming that offspring **fitness** is independent of mode of reproduction, an asexual female will have twice as many daughters as a sexual female. That is, since only females can produce eggs, asexual reproduction will have a twofold advantage over sexual reproduction. With this huge fitness advantage, any mutation that leads to a substitution of parthenogenesis for sexual reproduction would be expected to spread rapidly through the population. The twofold cost of sex can be stated as the cost of producing males or, if we focus on the parent–offspring relationship, as the 50 percent loss in the proportion of genes shared by a female and her sexual offspring. Given that parthenogenesis has such a massive fitness advantage over sexual reproduction, why is it that almost all species reproduce sexually?

The literature is replete with suggestions on how the twofold cost of sex is paid. More than 20 theoretical explanations attempt to account for the short-term advantage of sex (see West *et al.* 1999). An in-depth review of these explanations is beyond the scope of the present chapter. Instead we will describe them briefly, focusing on aspects that have particular relevance to understanding the life history of rotifers.

A number of hypotheses on the maintenance of sexual reproduction are based on fitness advantages of behaviors involved with either sexual selection or male parental care. For example, **sexual selection**—that is, female choice of mates or male–male competition for mates—may lead to selective incorporation of high fitness paternal genes. Alternatively, the cost of sex may be reduced if male participation in parental care permits females to enlarge their number of progeny. Both of these hypotheses have greater relevance to organisms with a more sophisticated behavioral repertoire than rotifers. Despite the fact that rotifer mating behavior has been reasonably well studied, no evidence of either male rejection by females or parental care by males has been reported. A third category of hypotheses is based on the fact that sexual recombination produces genetically variable offspring. It is presumed that parthenogenetic species have limited evolutionary potential to adapt to environmental heterogeneity, or that they are less able than sexual forms to take advantage of rare, independent mutations that can recombine sexually to produce high-fitness individuals. Perhaps sex functions to purge deleterious mutations (as claimed by **Müller's ratchet** and the **mutational deterministic hypothesis**; see García-Dorado *et al.*, Chapter 3), or to facilitate rapid evolution in a host–parasite "arms race" (the **"Red Queen" hypothesis**). These, as well as other mechanisms, might interact in a synergistic fashion to maintain sex (West *et al.* 1999).

Müller's ratchet envisions an obligate asexual organism in which there exists a class of individuals that has the smallest number of deleterious mutations in the population. Assuming that fitness is inversely related to the number of mutations, this class represents the group with the highest fitness in the population. Note that groups of any size, including the class with the highest fitness, are subject to loss from the population because of stochastic processes. Under Müller's ratchet, if the class with the highest fitness is lost, the group with the next smaller number of mutations then becomes the class with the highest fitness. In other words, under this model, fitness must always deteriorate.

At first glance it would seem that the rotifers would provide perfect material to test the ideas incorporated in Müller's ratchet. Bdelloid rotifers lack sexual reproduction and might be expected to undergo a stochastic deterioration, whereas the periodical sexual reproduction of monogononts should allow them to purge part of their **mutational load**. The huge size of many rotifer populations makes it unlikely that whole fitness classes will be lost by chance. In fact, even though beneficial mutations tend to have very low probabilities (approximately $10^3$ times lower than rates of harmful mutations), their occurrence is much more likely in most populations than is the loss of an entire fitness class due to stochastic events. It is not surprising that Mark Welch and Meselson (2001) found that nucleotide **substitution rates** in monogononts are similar to those of the asexual bdelloids. The authors, therefore, concluded that mutational load is not a selective pressure for the maintenance of sexual reproduction in monogononts. While exclusion during recombination is one means by which deleterious mutations can be eliminated, monogonont rotifers have a life-history stage that may also contribute to the same purgative effect. The haplodiploid mechanism of sex determination in monogononts suggests that male haploidy may function to reduce mutational load by exposing deleterious recessive mutations to selection. Males are typically smaller and morphologically and physiologically less complex than females of the same species. For instance, males lack a functional digestive system. It is however reasonable to speculate that they still express an important part of their genome (e.g. constitutive genes for energy metabolism). Therefore, the estimation of the mutational load is expected to reflect, at least in part, whether or not the analyzed gene is expressed in males.

The search for evolutionary advantages of sex has not been accompanied by a corresponding effort to look for fitness benefits of asexual reproduction (Brookfield 1999). Several features of parthenogenesis besides the twofold argument would facilitate its evolution from obligate sexual ancestors. A very powerful advantage of parthenogenesis is that it enables clonal propagation without breaking up maternal gene combinations. Clonal genotypes with the highest state of adaptation to an environment will be passed on without dilution by recombinational load since the entire genome is the unit of selection. In cyclical parthenogens a slippage in directional evolution may be associated with sexual reproduction (Lynch and Deng 1994). Ultimately, therefore, parthenogenesis leads to the production of a large number of high-quality females that are adapted to their environment.

In order for reproduction by clonal propagation to be advantageous, maternal and offspring environments must be similar so that the fitness value of particular gene combinations remains rather constant. In rotifers, sex is associated with dormancy, and thus with dispersal in time and space. This is a quite general pattern in the living world, from plants to animals: if sexual and asexual reproduction are combined in a single life cycle, then sex tends to occur when there is a low correlation between parental and offspring environments. It is under these conditions that high genetic diversity may play a major role in facilitating adaptation. However, in rotifers it is not obvious that environmental correlations between parents and their sexual offspring must be particularly low, as we have little information on the environmental cues that trigger hatching of resting eggs and few measures of the intensity of clonal selection that must occur among the post-hatching offspring.

Because of space limitations, we have not dealt with the question of how rotifers adapt to changes in their environment through the generation of new genetic variation. This process must involve, at least in part, the recombinant resting eggs produced by sexual reproduction. However, theoretical models also emphasize that recombination is not needed to produce new favorable combinations of alleles. Instead, given the huge population sizes of rotifers in even moderate-sized ponds, mutation is fully capable of generating high levels of new variation (King 1980). Moreover, clonal selection can rapidly amplify favorable novel mutations and combinations of mutations rendering them far less likely to be lost by random **genetic drift** than are newly formed genotypes produced by sexual recombination (King and Schonfeld 2001). Results from these models suggest that sex in rotifers may have relatively little value as a means of generating novel genotypes.

Historical and developmental constraints may be relevant for the maintenance of sex in rotifers. Estimations of population densities in the field suggest that most rotifers lack the capacity to survive the adverse periods that occur during parts of a typical annual cycle in temperate zone lakes. This implies that long-term fitness in these cyclical parthenogens depends upon the size and quality of the resting egg pool. Since resting eggs are produced sexually, the critical functional role of dormancy may by itself be a strong enough selective factor to maintain sex in monogononts. A similar argument has been developed for the maintenance of sex in cyclically parthenogenetic aphids (see De Meester *et al.*, Chapter 11). In contrast to monogononts, bdelloid rotifers lack both sex and resting eggs. Instead, **anhydrobiosis** is used to survive periods in which the environment is unfavorable. Asexual resting eggs are produced by some cladocerans and in a monogonont rotifer species. These cases underscore the function of the resting egg as a life-history stage promoting survival through adverse periods. However, they are clearly exceptional even though they do indicate that it is possible to decouple sex and resting egg production. The fact that most species have not done so may reflect historical constraints, or provide clues to the function of sex in cyclically parthenogenetic rotifers.

## Empirical studies on the timing of sex in rotifers

Theoretical studies have demonstrated the importance of habitat predictability, variation in resource availability, and demographic parameters in determining the timing of rotifer sex (Serra and King 1999). Two parameters have been useful for characterizing the timing of sex: the **sexual reproduction threshold** (i.e. the population density at which sexual reproduction is initiated, this determining when sex occurs in a population growth cycle), and the **sexual reproduction ratio**, which indicates the proportion of females that are sexual.

One of the most consistent observations about sexual reproduction is its association with population density. Over the past 70 years researchers studying a variety of rotifer species in laboratory and natural environments have reported a positive relationship between sexual reproduction and population density (e.g. Snell and Boyer 1988; Carmona *et al.* 1995). In an intensively studied genus (*Brachionus*), sexual reproduction appears to be triggered when population density exceeds a species-specific or population-specific threshold and is switched off when population density falls below that threshold. This density response can be modified by a variety of environmental factors such as food quantity (Snell and Boyer 1988). Environmental factors such as temperature, salinity (see Pourriot and Snell 1983), and free ammonia (Snell and Boyer 1988) can suppress sexual reproduction even when population density thresholds are exceeded. Vitamin E and long photoperiods are primary environmental triggers in other genera (*Asplanchna* and *Notomata*, respectively; see Gilbert 1993). Several authors have estimated the sexual reproduction threshold for natural and laboratory rotifer populations. For natural populations of *Asplanchna*, sexual reproduction thresholds were estimated to range from 2.3 to 100 females/l (Snell *et al.* 1999). Carmona *et al.* (1995) estimated sexual reproduction thresholds of 6.6 and 23 females/l for congeneric populations of *Brachionus*. Snell and Boyer (1988) recorded a sexual reproduction threshold of 147 females/l in laboratory populations of *B. plicatilis*.

A second parameter of the sexual reproduction pattern is the sexual reproduction ratio, that is, the proportion of sexual daughters produced by a female when the sexual reproduction threshold is exceeded. A number of studies have estimated sexual reproduction ratios in laboratory and, occasionally, in natural rotifer populations (Snell *et al.* 1999). For laboratory populations of *Asplanchna* a range of 8–69 percent sexual daughters has been reported after sex has been initiated. Estimates for *Brachionus* range from 1–89 percent, but rarely are higher than 50 percent. Carmona *et al.* (1995) recorded sexual reproduction ratios over 17 months in natural populations of two **sympatric** *Brachionus* species. For *B. plicatilis*, on 67 percent of the tri-weekly sampling dates sexual reproduction ratios were higher than 5 percent and fluctuated around 10 percent. In contrast, *B. rotundiformis* sexual reproduction ratios exceeded 5 percent on only 20 percent of the sampling dates, but then the sexual reproduction ratios were higher than those observed in *B. plicatilis*. These authors describe the pattern of sexual reproduction of *B. plicatilis* as continuous, in contrast to the periodic pattern of *B. rotundiformis*. Although sexual reproduction in both species was correlated with population density, the density threshold in *B. rotundiformis* was approximately three times higher than in *B. plicatilis*, which seems to account for the divergence in sexual reproduction patterns.

## Optimal timing of sexual reproduction and resting egg production

The optimal combination of sexual and asexual reproduction in rotifers can be conceptualized as a trade-off between the short-term advantages of asexual reproduction and the advantage of sex. Asexual reproduction facilitates fast colonization when habitats become suitable. Sexual reproduction is associated with dormancy and survival during adverse periods. One answer to the question of when sexual reproduction should be initiated is based on the role of resting eggs in allowing survival during adverse periods: sex should occur when the habitat becomes unsuitable. This view will be termed the **habitat deterioration hypothesis** (Table 12.1).

Alternative hypotheses have been proposed. Gilbert (1974) suggested that sexual reproduction is optimally induced at high population density because male–female encounters are more likely. If sexual reproduction occurs at low population density most sexual females would be male producing, but males would still be rare, and resting egg production would be low. In their model simulating the rotifer "male–female encounter hypothesis," Snell and Garman (1986) showed that the probability of a female being fertilized increases almost linearly when male density increases from 50 to 700 individuals/l. Using branching process analysis, Muenchow (1978) developed a stochastic version of the **male–female encounter hypothesis** (Table 12.1). He suggested that sexual reproduction

**Table 12.1** Hypotheses on the optimal pattern of sexual reproduction in rotifers

| Hypothesis | Rationale | Predicted relationship between sexual reproduction and population density |
|---|---|---|
| Habitat deterioration | Sex initiation occurs when population growth is no longer possible | *Indirect*: high density causes habitat deterioration, which in turn induces sex<br>*Spurious*: habitat deterioration (population decline) produces a peak in the population density time series; habitat deterioration induces sex |
| Male–female encounter | Sex initiation occurs at high population density because male–female encounters are then more likely | *Direct*: high density is the selective condition for inducing sex |
| Resource-demanding | Sex occurs in good environmental conditions because sex (male and resting-egg production) is resource-demanding | *Spurious*: good conditions result in high density; good conditions induce sex |

should occur at high densities as a way to decrease the variance associated with demographic stochasticity, which is highest at low abundances. Note that this conclusion holds whether the fertilization rate is density-dependent or not. High variance in male–female encounter rates would decrease the among-year geometric mean of encounters, the most relevant fitness measure in this case.

Sexual females are sometimes common during the early, exponential phase of population growth in mass cultures, when resources are abundant. Several authors (e.g. Gilbert 1980; Snell and Boyer 1988) noted that sex is resource demanding, because a higher amount of resources seems to be allocated to resting eggs than to asexual eggs. The threshold food concentration for sexual reproduction is 10 times higher than that for asexual reproduction in *B. plicatilis* (Snell and Boyer 1988). Life-table experiments showed that an asexual female has much higher fecundity than a fertilized sexual female (see Serra and Carmona 1993).

There is also evidence for a higher lipid level in resting eggs than in asexual eggs (Gilbert 1980). In addition, sexual reproduction also allocates resources to males. From these observations, it follows that sexual reproduction is more resource demanding than asexual reproduction. Accordingly, Snell and Boyer (1988) proposed that sexual reproduction should occur under good environmental conditions and be suppressed when the physical or chemical environment or food resources are less optimal. This we will term the **resource-demanding hypothesis** (Table 12.1). Other authors have reached similar conclusions from experiments on the effects of environmental factors on resting egg production in rotifer mass cultures (Hagiwara *et al.* 1988). The role of favorable food conditions was also used by Gilbert (1980) to explain the relationship between a vitamin-E rich diet, large *Asplanchna* morphotypes, and sexual reproduction.

Unfortunately, field observations and laboratory experiments have not provided an unambiguous test to discriminate between these hypotheses. The common association of sexual reproduction with high population density is compatible with all of the above hypotheses (Table 12.1). Under the male–female encounter hypothesis, population density would provide a direct selective advantage for sex initiation. However, if habitat deterioration gives a selective advantage to sex initiation, the correlation with high population density could be explained in two ways: (1) high density is expected to cause habitat deterioration due to resource depletion, and (2) peaks in the population density time series are expected just before habitat deterioration. Associations between sexual reproduction and population density are also predicted by the resource-demanding hypothesis as both are concomitant effects of good environmental conditions. Finally, peaks in a field population density time series could be associated with sex initiation because sexual reproduction would cause a decrease in population density due to the shift from **subitaneous egg** to resting egg production. This is especially true if sexual reproduction occurs at high rates.

## Theoretical analysis of the optimal patterns of sexual reproduction

R. A. Fisher's brilliant insight establishing sex allocation theory demonstrated that the optimal sex ratio in a population is 50 percent males : 50 percent females. Since each offspring in a sexual population has one male and one female parent, if there is an excess of males in the population the average female makes a greater reproductive contribution to the next generation than the average male, hence the production of female offspring would be favored. The opposite is true if there is an excess of females. This situation creates a frequency dependent selection that is expected to produce an equal number of male and female progeny. This argument is based on the assumption that it is equally expensive to produce male and female offspring. If this assumption is not true, Fisher's result can be restated in the form: the resources allocated by parents to male and female offspring production must be equal (e.g. Ridley 1996).

Rotifer males are smaller and morphologically simpler and, therefore, energetically less expensive to produce than females. They develop from unfertilized haploid eggs produced by sexual females. Recently, Aparici *et al.* (1998) have extended

Fisher's theory to rotifers. In many systems it is nearly impossible to accurately quantify the amount of resources devoted to producing each sex under natural conditions. However, sexual rotifer females produce either males or resting eggs, but not both. Hence, if there is to be equal investment of resources into male and female offspring, half of the sexual females must produce resting eggs while the other half must produce haploid males. This could be achieved if selection acts to alter the threshold age that marks the latest age at which females can be fertilized. A female that has not been fertilized by the threshold age can only produce male offspring in the future. Thus, if the threshold age is shifted to an earlier point in the life history, male production will be favored. Similarly, if it is shifted later, there will be fewer males produced. The threshold age at which the fertilization probability under a given set of environmental conditions is 1/2, is therefore an **evolutionary stable strategy**. This means that maximization of resting egg production is subject to the constraints imposed by sex allocation as well as the basic physiological and genetic constraints affecting egg formation.

In rotifers, there is an interaction between population density and the threshold age for the loss of fertilization potential. As shown by Snell and Garman (1986), fertilization is more likely at high densities. Sexual reproduction at high densities is expected to be associated with short threshold age of fertilization, so that a compensation maintaining the average fertilization probability at 1/2 occurs. In fact, this probability could also be achieved by initiating sexual reproduction at an appropriate density. However, regardless of what trait would adjust the fertilization rate—either the population density at which sexual reproduction is initiated or the threshold age of fertilization is actually selected for—effective fertilization rate is expected to be 1/2, and consequently density-independent when averaged over the realized range of densities. This conclusion seems to militate against the male–female encounter hypothesis in the sense that higher density is not expected to cause higher proportions of resting egg producing sexual females.

Snell (1987) viewed the problem of optimization of the sexual reproduction pattern as a trade-off between current sexual reproduction and current parthenogenetic growth, and consequently between current and future production of sexual females within a population growth cycle. When the sexual reproduction ratio is low, most females are producing asexual daughters and contributing simultaneously to both present population growth and the future number of sexual females. These trade-offs in the population growth cycle are similar to (1) the trade-off between reproduction (analogous to sexual reproduction) and somatic growth (analogous to parthenogenetic growth) and (2) the trade-off between present and future reproduction. Both of these trade-offs are well known in life-history theory. This view stresses that rotifers have discrete sexual generations whose fitness is the net recruitment rate of resting eggs, that is, the number of resting eggs produced during the growing season.

As a synthesis of the previous and their own theoretical work, Serra and King (1999) proposed a relationship between habitat features and optimal patterns of sexual reproduction (those maximizing resting egg production), which we summarize in Table 12.2. Habitats can be classified following two criteria: their predictability regarding the end of the growth period and the importance of competition (negligible or not). Table 12.2 implicitly assumes that, if competition is important, population density will achieve a quasi-stationary phase, and that the yearly periods of colonization and extinction of the water column population are relatively short and can be neglected. Hence, the optimal pattern in the stationary phase would be the optimal pattern for the whole growth period, and it does not make a difference whether its end is predictable or not. Consequently, three optimal sexual reproduction patterns are expected for four habitat types.

Optimal sexual reproduction patterns in predictable, noncompetitive habitats were studied by Serra and Carmona (1993), who assumed time-dependent, density-independent growth rates. Potential—without sex—population growth rate was assumed to vary from zero at the onset of colonization, to a maximum, then to zero (end of habitat suitability), and then to negative values. Such shifts might occur in habitats with an ephemeral

**Table 12.2** A demographic habitat classification for optimal patterns of sexual reproduction in rotifers (after Serra and King 1999)

| Population growth | Sexual reproduction feature | Length of population growth period | |
|---|---|---|---|
| | | Predictable | Unpredictable |
| Density independent | Sex initiation | Just before the habitat becomes unsuitable; population density relatively high | As soon as habitat deterioration becomes possible; population density relatively low |
| | Ratio | High (close to 1); sexual reproduction is ephemeral | Intermediate; larger for long habitat durations sexual reproduction is extended |
| Density dependent[a] | Sex initiation | When the population density approaches the carrying capacity (high density) | |
| | Ratio | Intermediate; the better the habitat, the higher the sexual reproduction ratio; sexual reproduction is extended | |

[a] Density-dependent growth is associated with a long stationary population growth phase. Consequently, it does not make a difference whether the end of the growth period is predictable or not, since the sexual reproduction pattern occurring in the stationary phase dominates over the patterns in the declining phase.

seasonal growth, or with poor conditions so that population densities are low and competition can be neglected. Serra and Carmona (1993) found that a **big-bang strategy** (all reproduction being first parthenogenetic, then sexual) should evolve, a result analogous to that commonly found using life-history models for the allocation of resources into somatic growth and reproduction (all resources first being allocated into growth, then into reproduction). Under these conditions sex should be initiated a short time before the habitat becomes unsuitable, and simultaneously the sexual reproduction ratio should become one. In this way, the maximum possible population density is achieved, providing the maximum number of mothers. Therefore, sexual reproduction is expected to be punctuated, with an ephemeral period of sex just before the habitat becomes unsuitable. This theoretical finding is in agreement with the habitat deterioration hypothesis.

Optimal patterns of sexual reproduction in a competitive habitat were studied by Snell (1987) who conducted simulations using a **discrete growth model** with competition to describe population dynamics. A periodic pattern consisting of several parthenogenetic generations followed by a mixture of sexual and asexual females was assumed, and the effects of both the frequency of the sexual phase and the proportions of sexual females were explored. He found that resting egg production was maximized by higher allocations to sexual daughters at higher intrinsic growth rates. Serra and King (1999) reached similar conclusions using a **continuous growth model** with two state variables (asexual and sexual female densities). Intraspecific competition was assumed to decrease linearly the birth rate. These authors analytically found that the optimal sexual reproduction ratio, if constant, was $1 - (b/q)^{1/2}$, where $b$ is the intrinsic birth rate (without competition) and $q$ is the mortality rate. This result was interpreted as being due to lower relative costs of sex in good conditions, and can be related to the resource-demand hypothesis, since it provides an optimizing argument to the finding of high sexual reproduction levels in good environments. Using computer simulation,

Serra and King (1999) also found that an intermediate strategy—that is, simultaneous allocation into asexual and sexual offspring—was better than a big-bang strategy. In summary, intermediate sexual reproduction patterns and extended periods of sexual reproduction are expected if rotifer populations grow density-dependently.

The optimal pattern of sexual reproduction when the length of the population growth period is uncertain and cannot be predicted by the rotifers was discussed by Serra and King (1999; Table 12.2). They used an analogy between individual resource allocation to reproduction and somatic growth and the allocation of resources to sexual and asexual daughters. Then, by analogy with the theory of life history optimization in randomly varying habitats, they proposed that a bet-hedging strategy should be optimal. An intermediate sexual reproduction ratio is a way to produce at least a few resting eggs during bad years while not stopping parthenogenetic growth during good years. This would imply an extended period of sexual reproduction. Evidence in support of this expectation was found in a field study of two sympatric populations of congeneric species with different seasonal distributions (Carmona *et al.* 1995). The winter species—supposedly living in a less predictable habitat—showed sexual reproduction almost continually, in contrast to the pattern found in the summer species.

In randomly varying habitats, early sex seems to be optimal. However, early sex, despite occurring at intermediate ratios, implies a cost for population growth that is necessary in order to protect against unpredictable years with short growth periods leading to resting egg failure. An alternative protection against total failure of sexual reproduction in a given year is the resting egg bank found in the sediments. It is well known that rotifer resting eggs can remain viable for decades (Marcus *et al.* 1994). Not all of the resting eggs that are produced in one year hatch when good conditions occur. As a consequence of this bet-hedging tactic, the risks of delayed sex may be rewarded by an increased output of resting eggs when the bet is won, and extinction is avoided because of the resting egg bank when the bet is lost. These arguments have been accounted for in recent developments of the dormancy theory, in which both optimal production of resting stages and optimal rates of dormancy have been analyzed simultaneously (Spencer *et al.* 2001).

## The impact of environmental change on the timing and proportion of sex

In monogonont rotifers, environmental change can have a stronger effect on sexual reproduction than on asexual reproduction. This is because sexual reproduction takes longer to complete than asexual reproduction and requires more resources. Sexual reproduction consists of several steps in a cascade of events from sex initiation to resting egg production (Table 12.3). The entire process spans three generations and takes several days to complete. Laboratory experiments with *Brachionus calyciflorus* at 25 °C demonstrate that under optimal conditions sexual reproduction takes about 4 days to form resting eggs (Preston *et al.* 2000). If environmental conditions deteriorate during this period, resting egg formation could fail (Serra and Carmona 1993). A switch from asexual to sexual reproduction, therefore, is a commitment that must culminate in resting egg formation; otherwise the original parental females will die without leaving progeny.

In addition to the time required for individual females to switch their reproductive physiology to resting egg production, time is required for a rotifer population to reach a density sufficient to trigger sex. When a population is re-established each year, newly hatched rotifers from the resting egg bank reproduce asexually, growing to considerable densities because resources are initially abundant. Sex is triggered as populations reach densities of 1–100/l. It takes rotifer populations several days to grow from the low densities following resting egg hatching to the density threshold that triggers sex. As an example, at low population densities and in the absence of competition, the instantaneous population growth rate of *Brachionus* can approach 1 $d^{-1}$ (Snell and Serra 2000). With such a growth rate, it would take a rotifer population 9–10 days to grow from a resting egg hatching density (e.g. 0.01/l) to a density sufficient to trigger sex (100/l). Adding to this the time required for females to form resting

**Table 12.3** Life cycle stages needed for resting egg production, and fitness losses associated with environmental stressors

| Stage | Environmental stressor | Fitness loss |
|---|---|---|
| Parthenogenetic reproduction to sexual reproduction threshold | 1. Suboptimal temperature, salinity, food level, etc.; pollutants, endocrine disruptors | 1. Slow population growth due to reduced birth rate and/or survival rate |
| | 2. Episodic catastrophes truncate population density before threshold | 2. Insufficient time for population growth to reach sexual reproduction threshold |
| Sex initiation | 1. Pollutants, endocrine disruptors | 1. Interference with chemical signals of population density |
| | 2. Episodic catastrophes truncate sexual reproduction development | 2. Insufficient time for transition from asexual to sexual females |
| | 3. Suboptimal environment; pollutants, endocrine disruptors | 3. Reduced sexual reproduction ratio (production of sexual daughters) |
| Fertilization | 1. Pollutants reduce swimming speed | 1. Low male–female encounter probability |
| | 2. Pollutants, endocrine disruptors | 2. Low recognition probability |
| | 3. Poor quality diet, pollutants, endocrine disruptors | 3. Low male fertility |
| Resting egg production and quality | 1. Poor quality diet, pollutants, endocrine disruptors | 1. Low fertilized female fecundity and low hatching percentage |

eggs after receiving a stimulus to initiate sexual reproduction, we get a minimum total of about 2 weeks required for resting egg formation.

Some evidence indicates that sexual reproduction also requires special resources. Rapid asexual reproduction was observed when *Brachionus plicatilis* was fed the alga *Chlorella* (Snell and Hoff 1987). However, when sex was triggered in these populations, resting egg production failed. The source of this failure was traced to male fertility. Adding the alga *Tetraselmis* to the diet of the maternal sexual females restored fertility to their sons. These data suggest that as yet unidentified nutrients are required for male fertility, but not for female fertility. Food limitation also appears to differentially affect sexual and asexual females so that sexual females have a higher threshold food concentration (Snell and Boyer 1988).

The time and resources required for sexual reproduction constrain its occurrence in a growing season and determine the optimal allocation to asexual and sexual reproduction. Therefore, an important question is, how does environmental change alter the time and resources required for sexual reproduction in rotifers? If the time necessary for sexual reproduction is lengthened or the resources available for sexual reproduction are reduced, resting egg production could be seriously diminished (Table 12.3). The size and stability of the resting egg bank determines the long-term persistence of monogonont rotifer populations (Snell and Serra 2000).

The effects of a variety of environmental changes on sexual reproduction can be predicted. For example, Snell and Serra (2000) simulated the effect of increases in the frequency and severity of catastrophic population crashes on resting egg bank dynamics and the probability of population extinction. Catastrophes could result from events like an early period of cold weather, rapid oxygen depletion, a sudden bloom of cyanobacteria, or acute exposure to toxicants in runoff or effluent. Catastrophes occurring once every 5 years in the Snell–Serra model had little effect on the probability of population extinction over 100 years. However, a rotifer population experiencing three catastrophic crashes per year is virtually certain to go extinct within 100 years. These results stress the buffering effect of the resting egg bank, but also how that effect can be altered. A series of several bad years in a row could tip the balance toward long-term depletion of the resting bank and send a population towards extinction. Natural rotifer populations are often food limited, even in nutrient-rich systems (Merriman and Kirk 2000). Resource competition is probably common (Grover 1997). If rotifer species typically exist in marginal resource conditions, environmental change resulting from climate cycles of increased or decreased temperature or drying could further reduce opportunities for resting egg production. Several rotifer species are susceptible to cyanobacteria toxins, which can alter rotifer abundance and species composition (Gilbert 1996). Environmental changes increasing the frequency or intensity of cyanobacterial blooms would almost certainly reduce opportunities for rotifers to produce resting eggs.

Humans have accelerated the rate of environmental change in many ecosystems. In aquatic communities, human activities have increased the mean annual temperature of surface waters and the frequency of extreme temperature events (Moore *et al.* 1996). Pesticide exposures alter the composition of zooplankton communities (Hanazato 1991) and represent yet another source of catastrophic mortality for rotifer populations. For many temperate rotifer populations, resting egg production is usually highest in spring and summer when exposures to pollutants in runoff and effluents are highest. Small reductions in population growth rate from toxicant exposures can seriously reduce resting egg production. Reductions in growth rate of only 20 percent caused resting egg bank size to decline over 100 years (Snell and Serra 2000). When assessing the potential negative effects of these changes, testing for the effects on asexual reproduction is not enough. The effect of environmental change on sexual reproduction is probably more important for determining long-term population survival. The resting egg bank can buffer transient negative effects, but sexual reproduction is expected to be more sensitive to environmental changes. Moreover, the evolutionary adjustment of sexual reproduction to environmental challenges is expected to be more difficult, due to its higher complexity.

The reliance of sexual reproduction on chemical communication may further make it susceptible to pollutants. Sexual reproduction requires chemical signals about population density, male recognition of conspecific females through contact chemoreception, and sperm–egg fusion in the pseudocoelom (Snell 1998). Chemical communication systems are vulnerable to interference from pollutants, but these effects are currently poorly understood in aquatic systems. A new class of pollutants gaining attention in invertebrates is endocrine disruptors. Compounds known to have endocrine disrupting activity in other animals have effects in rotifers (Preston *et al.* 2000). The effect of some vertebrate endocrine chemicals on rotifer reproduction has also been demonstrated (Gallardo *et al.* 1997).

Perhaps a more direct effect of pollutants is on rotifer swimming speed. The probability of encounter between male and female rotifers is critical for determining the proportion of sexual females that get fertilized (Snell and Garman 1986). A key parameter determining encounter probabilities is male and female swimming speed. Exposures to 110–330 $\mu$g pentachlorophenol/l for 30 min reduced female swimming speed 20–30 percent in four of seven rotifer species (Preston *et al.* 1999).

In summary, environmental change poses one of the biggest challenges to the long-term persistence of rotifer populations. Sexual reproduction seems to be the point in the rotifer life cycle that is most vulnerable to disruption. A variety of anthropogenic activities lengthen the time required for sexual reproduction, reduce resource availability, or truncate population growth cycles, diminishing resting egg formation. Reductions in deposition into resting egg banks can tip the balance towards population extinction.

## Final remarks

To integrate some of the above points, let us consider an idealized population starting from a single resting egg that hatches to produce a "stem female." The stem female and her descendants, for at least several generations, will all be asexual females. During this phase the population increases in size through **clonal reproduction**. Eventually, an environmental signal will trigger the production of sexual females, but as discussed above, not all females respond to this signal. As we know from both theory and observation, the sexual reproduction ratio will be intermediate in many conditions. This ratio commits a part of the population to sexual reproduction while the other part continues to reproduce asexually as might be expected under a "bet-hedging" scenario. Note that half of the sexual females are expected to produce males while the other half produce resting eggs (Aparici *et al.* 1998). If the asexual fraction of the population that is engaging in the bet-hedging strategy becomes divided at a later time into sexual and asexual fractions, the overall production of resting eggs will increase.

This simplified clonal growth cycle suggests an interpretation of the "cost of sex" concept from the point of view of a particular **clone**. The parthenogenetic phase would be a way to produce a larger number of sexual females, and eventually a higher resting egg output. However, given that each resting egg produced needs half of the sexual females to be male producing, the cost of sex for the whole cycle (from resting egg to resting egg) is 1/2 in the sense that twice as many resting eggs could be produced if they could be formed asexually.

Population biologists studying rotifers are nearing an understanding of the key evolutionary processes shaping the life history of these fascinating cyclical parthenogens. We note with more than passing interest that the analyses presented here are critically dependent on approximately equal measures of results from field, laboratory, and theoretical studies. This is precisely the triad of "modern" population studies that Thomas Park called for well over 50 years ago and that remains as urgently needed today as it was then.

This work was partially supported by a grant (BOS2000-1451) from the Ministry of Science and Technology (Spain). We thank M. J. Carmona for her suggestions on the manuscript. We also thank J. J. Gilbert and two anonymous reviewers for their useful comments and criticism.

CHAPTER 13

# From ecosystems to molecules: cascading effects of habitat persistence on dispersal strategies and the genetic structure of populations

**Robert F. Denno and Merrill A. Peterson**

Knowledge of the evolutionary links between ecosystem-level properties and variation at the genetic level predates even the discovery of the molecular basis of genes. Much of the research in this arena has been motivated by observations that organisms possess attributes that enhance their performance in their local ecosystems. Early experiments provided convincing evidence such presumed **adaptations** to the environment often have a genetic basis, by demonstrating heritable ecotypic variation among populations (e.g. Clausen *et al.* 1940). More recent studies have shown that variation at the level of specific genes or gene products is clearly the result of adaptation to ecosystem properties (e.g. Koehn and Hilbish 1987).

Although the link between local adaptation and ecosystem properties is intuitively appealing, it is important not to overlook less obvious, but perhaps equally important evolutionary effects of ecosystems that may involve processes other than natural selection. One such process is **gene flow**, or the movement of genes among populations. Gene flow, which occurs via the movement of either individuals or gametes, tends to homogenize the allele frequencies of populations. As such, gene flow may counter the effects of geographic variation in natural selection and/or minimize the degree to which **genetic drift** can lead to differentiation among populations (Slatkin 1985). Through these effects, gene flow can limit local adaptation (Storfer 1999) interfere with coevolutionary processes (Lively 1999), and restrict the conditions under which speciation can occur (Servedio and Kirkpatrick 1997). Thus, if ecosystem-level processes do indeed influence gene flow, their evolutionary effects may be far-reaching.

If we are to understand how ecosystem-wide properties might influence gene flow, we must first examine the factors that influence the dispersal strategies of organisms, because it is via dispersal that gene flow occurs. Decades of research have revealed that the evolution of dispersal strategies can be molded by a wide range of ecological processes. For example, at the population level, dispersal away from areas of high competitor density can minimize intraspecific competition (Denno and Peterson 1995). In addition, dispersal away from natal sites may enhance **fitness** by reducing the probability of inbreeding (Perrin and Goudet 2001). At the community level, dispersal may allow individuals to escape interspecific competitors or natural enemies, or find organisms that serve as prey or hosts (Ronce *et al.* 2001). Dispersal strategies may also be influenced by ecosystem-level phenomena. For example, the persistence of habitat patches may vary considerably among ecosystems. In relatively ephemeral habitats, the advantages of dispersal are clear, as long-term persistence in a region is not possible without dispersal to new habitat patches (Southwood 1962). In contrast, in relatively persistent habitats, the advantages of dispersal are diminished, since long-term persistence can be achieved without colonizing new patches. In such persistent habitats, selection may favor a reduction in the tendency to disperse if dispersal is risky or energetically costly (Roff 1990).

Because levels of gene flow are presumably influenced by dispersal strategies, any of the mentioned ecological processes might have cascading effects that ultimately influence the genetic differentiation of populations. In keeping with the overarching theme of this volume, we will restrict our focus in this chapter to the link between ecosystem-level processes and population-genetic differentiation. Making a strong case for such cascading effects of ecosystem processes is a difficult multi-step process. First, one must document the existence of among-population variation in dispersal strategies. Next, the variation in dispersal strategies must be attributable to ecosystem-level processes. Finally, it must be established that the dispersal strategies have influenced gene flow and/or population-genetic differentiation.

Documenting that dispersal strategies vary among populations is a challenge for most species, in which the only means of quantifying dispersal parameters for a given population is through labor-intensive efforts such as mark–release–recapture. The prospect of conducting such studies in multiple populations to determine if populations differ in dispersal strategies is daunting at best. Species with an obvious dispersal polymorphism provide a welcome alternative, because rather than assessing actual dispersal distances, one can use the proportion of individuals in a population that are of the dispersive morph as an index of the dispersal strategy of that population (Denno *et al.* 1991). Examples of dispersal-polymorphic species are numerous, and include insects with both flightless short-winged and flight-capable long-winged forms (Roff 1990), plants with fruits or seeds that differ in their potential for dispersal (Parciak 2002), and mammals with distinct dispersive forms (O'Riain *et al.* 1996). Using such species, several researchers have indeed found that dispersal strategies vary among populations (e.g. Roff 1990; Parciak 2002).

Relating variation in dispersal strategies to ecosystem-wide phenomena is similarly challenging. For one, variation in dispersal strategies may be greater among species than within species, necessitating interspecific comparisons of the dispersal strategies of organisms occupying different ecosystems. Such comparisons may often be confounded by **phylogenetic nonindependence** (Wagner and Liebherr 1992). Furthermore, because many factors can influence dispersal strategies (Ronce *et al.* 2001), it is difficult to make a strong case that a particular ecosystem-level factor has shaped the dispersal strategy of a given population. To do so requires a detailed understanding of the population- and community-level factors that may also influence the dispersal strategies of a particular species.

The final difficulty in demonstrating the cascading evolutionary effects of ecosystem-level processes is in establishing a link between dispersal strategies and either gene flow or the genetic structure of populations. Although several methods of estimating gene flow have been developed (Rousset 2001), the most commonly used estimates are based on Wright's (1951) **fixation index** ($F_{ST}$). This measure of population-genetic structure describes the degree to which variation in allele frequencies in a region can be attributed to differences among local populations. Wright showed that one could estimate the number of migrants exchanged between populations ($N_e m$), using the equation $F_{ST} \approx 1/(4N_e m + 1)$. Unfortunately, the quantity $N_e m$ may vary for two very different reasons: either the effective population size, $N_e$, may vary, or the fraction of individuals moving among populations, $m$, may vary. Of these, only $m$ would be expected to vary with dispersal. However, because it is impossible to disentangle the two factors, it is difficult to rigorously attribute variation in $N_e m$ to variation in dispersal (Whitlock and McCauley 1999). Furthermore, many of the assumptions of Wright's models may not be realistic for natural populations, reducing the likelihood that $F_{ST}$-based measures of gene flow will be accurate (reviewed in Whitlock and McCauley 1999). Nonetheless, the fact that such gene flow estimates are generally consistent with independent measures of dispersal for a wide diversity of organisms (Govindaraju 1988; Peterson and Denno 1998; Bohonak 1999) is encouraging, and suggests that this approach for estimating gene flow and population-genetic structure is informative.

Taking the above difficulties together, it is clear why it is such a challenge to demonstrate that ecosystem-level processes can shape the evolution of populations, via effects on levels of gene flow. In this chapter, we describe the results of a long-term research program on the evolutionary ecology of dispersal strategies in saltmarsh-inhabiting planthoppers in the genus *Prokelisia* (Hemiptera: Delphacidae). The unique nature of this system has allowed us to overcome most of the difficulties outlined above, enabling us to make a strong case for the cascading evolutionary effects of ecosystem-level processes on the genetic structure of populations. In this chapter, we describe research showing that dispersal strategies vary among populations, and that this variation has been shaped by habitat persistence. Subsequently, we demonstrate that these dispersal strategies in turn have ultimately influenced the genetic structure of *Prokelisia* populations.

## Flight polymorphism, habitat persistence, and the incidence of dispersal

### Distribution, host plants, and habitats of *Prokelisia* planthoppers

The most abundant herbivorous insects on North American tidal marshes are *Prokelisia* planthoppers that feed and develop exclusively on the phloem sap of *Spartina* cordgrasses. Two species, *P. marginata* and *P. dolus*, occur **sympatrically** along much of the Atlantic and Gulf coasts, where they exploit *Spartina alterniflora* (Denno *et al.* 1996). On the Pacific coast, the two planthoppers are **allopatric** with *P. marginata* restricted to northern California, and *P. dolus* occurring in the isolated marshes of southern and Baja California (Denno *et al.* 1996).

*Spartina* dominates the vegetation of these marshes, often forming extensive pure stands within the intertidal zone (Denno *et al.* 1996). Within this zone, *Spartina* ranges from approximately mean high water level (the mean elevation of high tide) to elevations as much as 2 m below that level. Along this elevational gradient, the structure and growth dynamics of *Spartina* vary, but the vegetation can be conveniently divided into two broad habitat types. Specifically, meadow vegetation on the high marsh consisting of short-form plants grades abruptly into creekside vegetation in the low marsh, comprised of taller and more robust plants.

The primary habitat where reproduction and development occur differs for the two *Prokelisia* species. *P. marginata* develops primarily on low marsh *Spartina* whereas *P. dolus* occurs mostly on high-marsh vegetation (Denno *et al.* 1996). Along the central Atlantic coast, both planthoppers are **trivoltine** (three generations per year) and adults are absent from the marsh during winter, whereas on the Gulf coast both species are **multivoltine** (many generations per year) and reproduce year-round (Denno *et al.* 1996). On the Pacific coast, *P. marginata* is trivoltine and adults are absent during winter. In southern California and Mexico, adults of *P. dolus* are present throughout the year and the species breeds continuously (Denno *et al.* 1996).

Along the mid-Atlantic coast, both species overwinter on the high marsh as active nymphs nestled in leaf litter, as does *P. marginata* on northern California marshes (Denno *et al.* 1996). In contrast, nymphs do not survive the winter in low-marsh habitats in either mid-Atlantic or northern California populations (Denno *et al.* 1996). For *P. marginata*, density-dependent dispersal results in interhabitat movements between overwintering habitats on the high marsh and summer sites for development in the low marsh (Denno and Peterson 1995; Denno *et al.* 1996). Low-marsh *Spartina* is much more nutritious than high-marsh vegetation during much of the year, resulting in higher performance of *P. marginata* on low-marsh plants (Denno and Peterson 2000). In contrast, *P. dolus* both overwinters and develops on high-marsh *Spartina* (Denno *et al.* 1996). Because females of both species insert their eggs only in living leaf blades (Denno 1994), eggs are able to persist only in habitats with living plants. All stages of *Prokelisia* planthoppers, however, are able to withstand short-term inundation by tides (Denno *et al.* 1996).

## Wing polymorphism and flight capability

Adults of both *Prokelisia* species are wing dimorphic, with both fully winged adults (**macropters**) and adults with reduced wings (**brachypters**) occurring in most populations (Fig. 13.1; Denno 1994). Macropters are capable of long-distance flight (>30 km), whereas the movement of brachypters is restricted to walking or jumping over only short distances of up to several meters (Denno 1994; Denno *et al.* 1996). Female brachypters reproduce at a significantly earlier age and are more fecund than macropters, illustrating the trade-off that exists between dispersal ability and reproductive effort (Denno *et al.* 1989; Denno 1994). Similarly, macropterous males sire fewer offspring than their brachypterous counterparts (Langellotto *et al.* 2000).

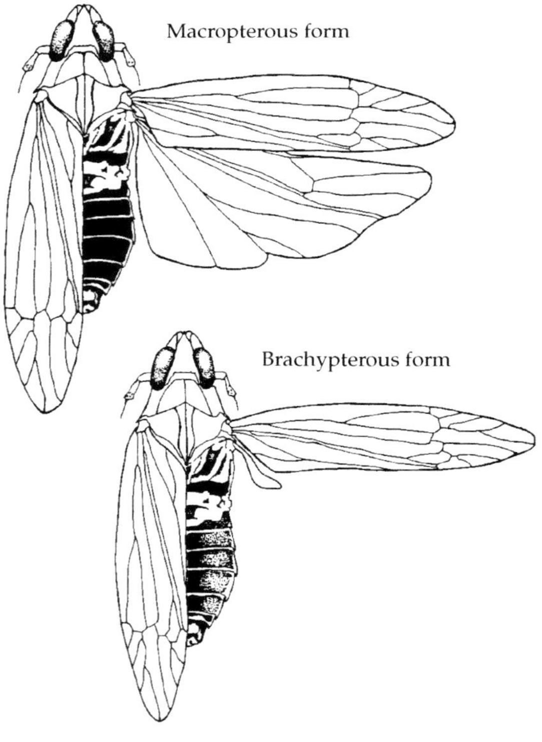


**Figure 13.1** Macropterous (flight-capable) and brachypterous (flightless) forms of *P. marginata*.

Wing form in *Prokelisia* planthoppers is determined by a developmental switch that is sensitive to environmental cues (Denno *et al.* 1991). The sensitivity of the switch is under **polygenic** control (Denno 1994). Population density is, by far, the most significant of the environmental factors known to affect wing form in *Prokelisia* planthoppers. Specifically, the production of macropters is positively density dependent in both species (Denno and Roderick 1992). Thus, wing-form is genetically determined, but can be modified to some extent by environmental factors such as population density.

## Geographic variation in the incidence of dispersal and its genetic basis

There is significant geographic variation in the dispersal capability (percentage macroptery) of both *P. marginata* and *P. dolus*. The most flight-capable populations of *P. marginata* occur along the Atlantic coast (mean ± SE = 92 ± 2 percent macroptery), followed by Pacific populations (29 ± 4 percent), and then Gulf populations (17 ± 5 percent) (Denno *et al.* 1996). Overall, the dispersal capability of *P. dolus* is much less than that for *P. marginata*, with levels of macroptery significantly higher in both Atlantic (8 ± 2 percent) and Gulf coast populations (6 ± 2 percent) than in Pacific coast populations (0.2 ± 0.1 percent) (Denno *et al.* 1996).

Within coastal regions, levels of macroptery remain rather constant, but there can be occasional exceptions. Higher-than-average levels of macroptery have been observed in several Gulf coast populations of *P. marginata* and *P. dolus* as well as in a few mid-Atlantic populations of *P. dolus*. In these exceptional cases, high levels of macroptery are usually associated with outbreak densities and density-dependent dispersal (Denno *et al.* 1996).

To ensure that coastal differences in macroptery are genetically based and are not due in large part to differences in environment (e.g. population density), Atlantic coast and Gulf coast populations of the two

*Prokelesia* species were raised under the same range of densities (Denno *et al.* 1996). When this was done, Atlantic and Gulf coast populations of both species differed dramatically in their tendency to produce macropters (Denno *et al.* 1996). For the two populations of *P. marginata*, more adults molted into macropters as rearing density increased, but the overall incidence of macroptery was significantly higher in the Atlantic population (84 percent) than in the Gulf coast population (38 percent). Likewise, macroptery increased with crowding in both populations of *P. dolus*, but this response was significantly higher for the Atlantic coast (49 percent) than the Gulf coast population (8 percent). These data suggest that selection for dispersal capability has acted differentially between the Atlantic and Gulf coast populations, resulting in a greater production of macropters at a given density in the Atlantic populations of both *Prokelisia* species. Notably, these density-wing form responses paralleled the incidence of macroptery observed in the field. Thus, inter-coastal variation in macroptery observed in field populations of the *Prokelisia* planthoppers likely reflects genetically-based differences in dispersal capability.

## Habitat persistence

For *Prokelisia* planthoppers, habitat persistence is determined in large part by three interacting factors: winter severity, tidal disturbance, and marsh elevation (Denno *et al.* 1996). Due to tremendous regional differences in the destruction and disturbance of the low-marsh habitat during winter, it is likely that geographic variation in habitat persistence is much greater for low-marsh inhabitants such as *P. marginata* than for high-marsh occupants like *P. dolus*. In particular, in regions where low-marsh habitats are destroyed or disturbed during winter (Atlantic and north Pacific coasts), planthoppers should fail to persist, and exploitation of these habitats should depend on annual recolonization. In contrast, planthopper persistence should be higher and recolonization less critical for the exploitation of either low-marsh habitats in regions with minimal tidal and winter disturbance (Gulf coast), or high-marsh habitats, which are seldom disturbed in any region.

To verify these expectations, geographic variation in habitat persistence was assessed in the two planthopper species by examining how well each species endures the winter in its primary habitat for development on both Atlantic and Gulf coasts (Denno *et al.* 1996). Patterns of habitat occupancy during the warm season confirm that regardless of region, the proportion of the *P. marginata* population inhabiting the low marsh is much higher than that for *P. dolus* (Fig. 13.2). This pattern corroborates that *P. marginata* develops mostly in low-marsh habitats and that *P. dolus* resides primarily on the high marsh during the growing season. For *P. marginata*, the proportion of the population in the low marsh during

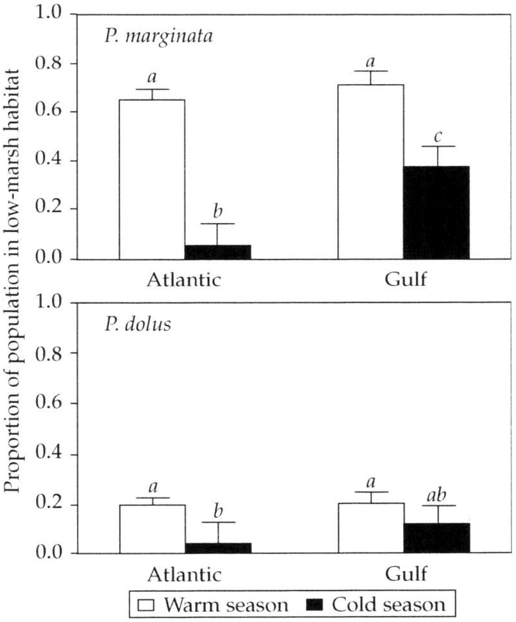


**Figure 13.2** Warm and cold seasonal patterns of low-marsh occupancy (proportion of population in low marsh compared to high marsh) by *Prokelisia marginata* and *P. dolus* along the Atlantic and Gulf coasts. Along the Atlantic, where low marsh habitats are destroyed during winter, a much smaller proportion of *P. marginata* individuals remain during winter in that habitat than along the Gulf coast, where low-marsh habitats are less disturbed during winter. During the warm season, most individuals of *P. marginata* occupy the low marsh on both coasts. In contrast, the proportion of the *P. dolus* population in the low marsh remains low during both seasons in both regions. Means with different letters are significantly different ($P < 0.05$) (modified from Denno *et al.* 1996).

the warm season is high in both Atlantic (64 percent) and Gulf coast regions (72 percent) (Fig. 13.2). During the cold season, however, the proportion of the population remaining in the low-marsh plummets to 7 percent on Atlantic marshes, but drops to only 40 percent in Gulf coast marshes. These data demonstrate that *P. marginata* is better able to endure winter conditions on the low marsh along the Gulf than it is along the Atlantic coast. In contrast, *P. dolus* shows only a slight shift of the population from the high marsh to the low marsh during the warm season along the Atlantic coast, and exhibits no such shift along the Gulf coast (Fig. 13.2). Thus, regardless of region, most (>80 percent) of the *P. dolus* population resides on the high marsh where it persists well throughout winter (Fig. 13.2). Collectively, these data suggest that low-marsh habitats along most of the Atlantic coast are temporary and that planthoppers fail to persist there during harsh winters, as is also the case on the north Pacific coast (Denno *et al.* 1996). Thus, the low-marsh habitat is far more stable along the Gulf coast, where tidal disturbance is less, winters are equitable, and planthoppers persist. Planthoppers also persist well in high marsh habitats in all regions due to reduced tidal disturbance and ice shearing during winter.

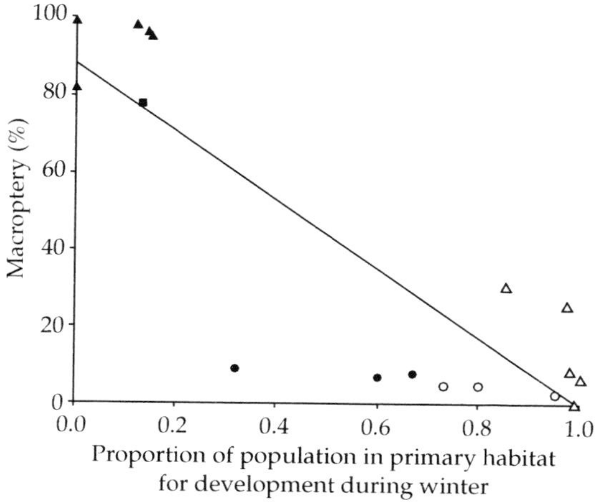


**Figure 13.3** Relationship between the level of macroptery (%) in populations of *Prokelisia* planthoppers and the proportion of each species' population spending the winter in the primary habitat for development ($y^{\text{macroptery}} = 89.8 - 90.8x$, $r^2 = 0.75$, $P < 0.001$). Populations which persist poorly through winter in their primary habitat exhibit high levels of dispersal capability. *P. marginata* populations: Atlantic coast (▲), Gulf coast (•), and Pacific coast (■). *P. dolus* populations: Atlantic coast (△) and Gulf coast (○) (modified from Denno *et al.* 1996).

## Relationship between habitat persistence and the incidence of dispersal

At a regional spatial scale, the incidence of macroptery in planthopper populations is inversely related to habitat persistence, indexed as the proportion of each species' population able to endure through winter in its primary habitat for development (Fig. 13.3; Denno *et al.* 1996). At Atlantic and Pacific locations, a small proportion (<20 percent) of the *P. marginata* population remains through winter in the primary habitat for development (low marsh), and as expected, associated levels of macroptery during the growing season are high (>75 percent). Along the Gulf coast, however, a higher proportion (30–70 percent) of the *P. marginata* population remains through winter in the low marsh, and associated levels of macroptery are relatively low (<15 percent). For *P. dolus*, the proportion of individuals remaining through winter in their primary habitat for development (high marsh) is high (>70 percent) at all locations along the Atlantic coast and accordingly, levels of macroptery are low (<35 percent). These results suggest that along the Atlantic and Pacific coasts, large-scale dispersal is necessary for *P. marginata* to recolonize its preferred low marsh habitat, because in those regions, that habitat is unsuitable for winter survival. The low levels of macroptery in most Gulf coast populations of *P. marginata* are possible because this species can remain on the low marsh year-round in this equitable region. Because *P. dolus* can remain throughout the year in its primary habitat for development on the less-disturbed high marsh, levels of dispersal in all populations of this species are low.

## Incidence of dispersal, levels of gene flow, and the genetic differentiation of populations

With the link between habitat persistence and the incidence of dispersal in planthopper populations established at the regional scale, we now proceed to examine the evidence that such differences in dispersal influence gene flow and the genetic structuring of populations. We have used allozyme data to determine if levels of gene flow and genetic differentiation among populations vary with dispersal ability (Peterson and Denno 1997). Two hypotheses were tested. The first hypothesis was that overall levels of population-genetic differentiation would reflect inter-coastal variation in dispersal capability. A second expectation was that genetic **isolation by distance** (a decline in gene flow with increasing geographic distance) would be more pronounced in regions where the incidence of macroptery is low (Gulf coast) than in regions where macroptery levels are high (Atlantic coast).

To assess allozyme variation among *Prokelisia* populations, planthopper samples were taken from marshes along Atlantic and Gulf coasts at approximately 175-km intervals (see Peterson and Denno 1997). In all, 27 locations were sampled, 16 locations along the Atlantic coast and 11 along the Gulf coast. For each sample location and planthopper species, 36 individuals were assayed for variation in seven polymorphic enzymes (see Peterson and Denno 1997). Population-genetic structure was estimated by calculating values of $G_{ST}$ (Nei 1973) for each species on each coast. Because $G_{ST}$ is an estimator of $F_{ST}$, we also used the values of $G_{ST}$ to estimate gene flow ($N_e m$), using Wright's (1951) equation.

## Dispersal strategies and gene flow

The hypothesis that gene flow declines more rapidly with geographic distance among sedentary (Gulf coast) populations than among mobile (Atlantic coast) populations was tested by comparing patterns of genetic isolation by distance within both the Atlantic and Gulf coast regions for each planthopper species (Peterson and Denno 1997). For this analysis, gene flow ($N_e m$) was estimated for all pairwise combinations of populations in each region, and the significance of the relationship between gene flow and the pairwise geographic distances among populations (the isolation by distance slope) was assessed. A steeper negative slope (or greater genetic isolation by distance) was predicted for sedentary populations (Gulf coast) than for more dispersive populations (Atlantic coast).

For the sedentary *Prokelisia dolus*, there was a significant negative relationship between gene flow ($N_e m$) and distance among populations on both coasts, especially on the Gulf where levels of macroptery are very low (6 percent) (Fig. 13.4). By contrast, populations of the more mobile *P. marginata* showed little evidence for genetic isolation by distance on either coast (slopes not significantly different from zero). When regional isolation-by-distance slopes were plotted against regional means of macroptery (percent) for each species, the most negative slopes were associated with sedentary populations of *P. dolus* on the Gulf coast, and near-zero slopes were characteristic of the most mobile populations of *P. marginata* along the Atlantic seaboard. However, with only four data points, the trend was not statistically significant (Peterson and Denno 1997). Nonetheless, these results are consistent with those from an extensive review of isolation-by-distance relationships for phytophagous insects (Peterson and Denno 1998). This review found that isolation-by-distance slopes were steepest for moderately mobile species such as *P. dolus* and most shallow for mobile taxa like *P. marginata*. Thus, our review based on interspecific comparisons (Peterson and Denno 1998) and our intraspecific assessment (Peterson and Denno 1997) confirm that species with moderate dispersal capabilities exhibit the greatest isolation by distance. By contrast, isolation by distance is weak in highly mobile species such as *P. marginata*, and is likely the result of the homogenizing effects of gene flow. In very sedentary species (e.g. scale insects), very limited gene flow will allow nearly all populations to diverge, resulting in a shallower isolation by distance slope (Peterson and Denno 1998).

## Dispersal strategies and the genetic structure of populations

To test the hypothesis that intraspecific variation in dispersal strategies influences overall levels of population-genetic subdivision, we compared the genetic subdivision of Atlantic coast populations of

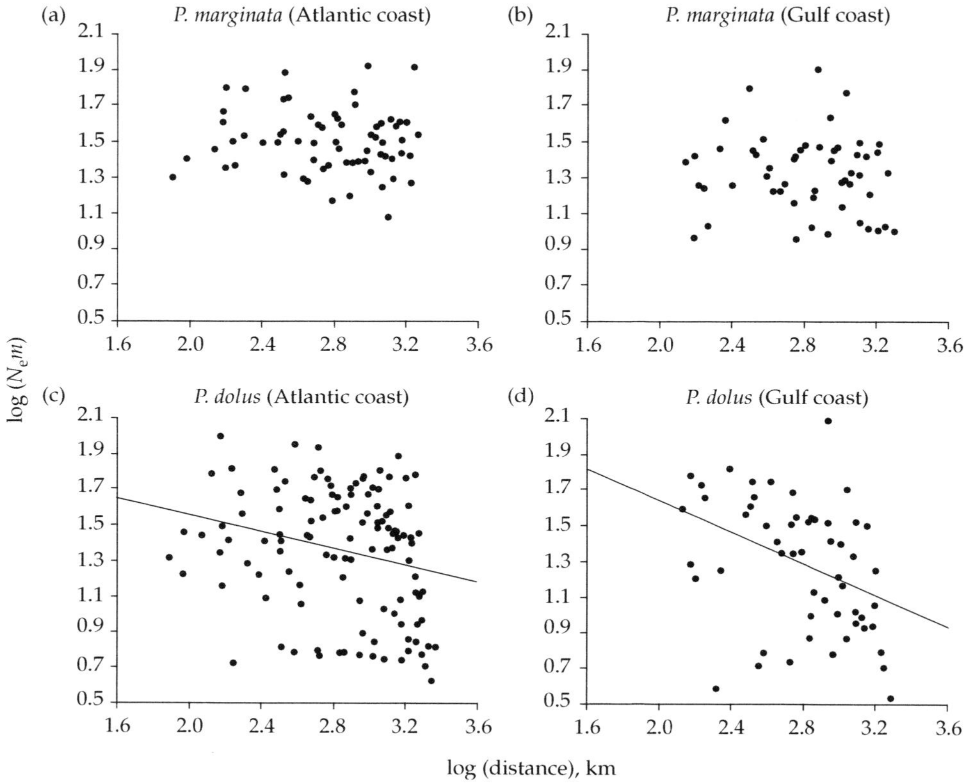


**Figure 13.4** The relationship between log-transformed gene flow estimates ($N_e m$) and geographic distance for all pairwise combinations of populations of (a) *P. marginata* along the Atlantic coast ($P = 0.57$), (b) *P. marginata* along the Gulf coast ($P = 0.47$), (c) *P. dolus* along the Atlantic coast ($y = 2.011 - 0.231(x)$, $r^2_{adj} = 0.05$, $P < 0.05$), and (d) *P. dolus* along the Gulf coast ($y = 2.528 - 0.444(x)$, $r^2_{adj} = 0.14$, $P < 0.01$). The significance of slopes was determined using Mantel's randomization test, followed by OLS regression to determine the regression coefficients (modified from Peterson and Denno 1997).

each species with that among Gulf coast populations (Peterson and Denno 1997). For both species, Atlantic coast populations are characterized by higher levels of macroptery, particularly in *P. marginata*, and thus we predicted that populations there would be less subdivided (lower $G_{ST}$) than along the Gulf coast. Indeed, populations of *P. marginata* along the Gulf coast exhibited significantly greater genetic subdivision ($G_{ST} = 0.023 \pm 0.0003$; mean $\pm$ SE) than along the Atlantic coast ($G_{ST} = 0.015 \pm 0.0006$). Gulf coast populations of *P. dolus* were slightly more genetically subdivided ($G_{ST} = 0.033 \pm 0.0007$) than were the Atlantic coast populations of this species ($G_{ST} = 0.028 \pm 0.0015$) (Peterson and Denno 1997). Thus, by measuring the relationship between regional variation in planthopper dispersal capability and the genetic differentiation of populations, it was demonstrated for the first time that intraspecific variation in dispersal strategies influences the genetic structure of populations.

## Summary

The objective of the research reviewed in this chapter was first to elucidate how ecosystem-level phenomena (habitat persistence) can shape the dispersal strategies of populations, and then to demonstrate the consequences of those strategies for gene flow and the genetic differentiation of populations. Demonstrating such cascading evolutionary effects of processes occuring at the ecosystem-level has been problematic for three main reasons. First, distinguishing migrants from nondispersers is unrealistic for most taxa because they are morphologically indistinguishable (but see Denno *et al.* 1991; O'Raian *et al.* 1996; Parciak 2002). Thus, for most taxa, determining the degree to which dispersal strategies vary among populations requires intensive research in each of many populations. Second, determining the ecological factors driving the evolution of dispersal strategies of a particular species is only possible with a detailed understanding of the population biology of that species. Third, most attempts to understand the evolutionary causes and consequences of dispersal strategies have relied on interspecific comparisons that may suffer from phylogenetic nonindependence (Wagner and Liebherr 1992).

The *Prokelisia* planthoppers described in this review have provided a unique opportunity both to examine the evolutionary consequences of geographic variation in dispersal ability, and to explore the ecological factors that maintain this variation. Not only do these wing-dimorphic insects allow for the easy recognition of flight-capable individuals in a population (Denno 1994), but they also can exhibit extreme intraspecific variation in dispersal ability across populations (Denno 1994; Denno *et al.* 1996). This combination of features has allowed us for the first time to rigorously identify those factors that promote a particular dispersal strategy and then discover the consequences of that strategy on population-genetic structure. This chapter has provided us with an opportunity to coalesce several studies on *Prokelisia* planthoppers to demonstrate these links between population ecology and genetics.

Through our research on this system, we have determined that habitat persistence explains much of the variation in the regional dispersal strategies of *Prokelisia* planthoppers, a finding that is supported intraspecifically by *Prokelisia* planthoppers (Denno *et al.* 1996), and interspecifically by phytophagous insects at large (Roff 1990; Denno *et al.* 1991). Furthermore, we have found that regional variation in dispersal strategies has importance population-genetic consequences in these insects. Most compelling is our finding that in *P. marginata*, population-genetic subdivision was greater along the Gulf coast than along the Atlantic coast, mirroring regional variation in dispersal strategies. Thus, these planthoppers enabled us to answer questions that have remained intractable for many systems. In so doing, they have opened the doors for the study of dispersal, not only by facilitating the investigation of ecosystem-level processes that promote it (Denno *et al.* 1991, 1996, 2000), but also by easing the study of the cascading effects of these processes on the genetic structuring of populations (Peterson and Denno 1997, 1998).

We argue that our results are not specific to wing-dimorphic planthoppers, but likely reflect the far reaching evolutionary effects that habitat persistence has for many taxa. To be sure of this generality, we clearly need more examples. Further studies with other species exhibiting dispersal polymorphisms would be an obvious place to start, but **phylogenetically independent contrasts** between species would also be informative.

Now that the cascading evolutionary effects of habitat persistence have been documented, it is important that we assess how far-reaching those effects may be. For example, if local adaptation is compromised by extensive gene flow among populations (e.g. Storfer 1999), one might hypothesize that taxa in persistent habitats should evolve local adaptations more than taxa in temporary habitats. This hypothesis could be tested by comparing (via intraspecific comparisons or phylogenetically independent contrasts) the geographic scale of local adaptation for populations occupying persistent versus temporary habitats. Perhaps more significantly, our findings suggest that ecosystem-level phenomena such as habitat persistence may influence the likelihood of speciation. Generally, the differentiation of populations is regarded as the first step in the speciation process. Since the effects of habitat persistence on dispersal strategies apparently dictate the degree to which populations diverge, speciation rates may be higher for **clades** of taxa that specialize in exploiting persistent habitats, compared to those clades that exploit temporary habitats. Thus, the cascading evolutionary effects of habitat persistence may influence not only microevolutionary processes such as gene flow, population differentiation, and local adaptation, but may also play a role in diversification.

We gratefully acknowledge the University of Valencia for inviting and supporting our participation in the international symposium entitled "Evolution: from molecules to ecosystems," a symposium that commemorated the fifth centennial of the University's existence and launched the Cavanilles Institute for Biodiversity and Evolutionary Biology. In particular, Andrés Moya and Joaquin Baixeras served as superb hosts throughout the symposium held in Valencia 2–4 November, 2000. This chapter benefited from the careful reviews of Enrique Font, Sir Richard Southwood, and two anonymous reviewers. The research reviewed here was supported in part by National Science Foundation Grants DEB-9527846 and DEB-9903601 to RFD. Preparation of this review was supported in part by National Science Foundation Grant DEB-0212652 to MAP.

CHAPTER 14

# Using molecules to understand the distribution of animal and plant diversity

Godfrey M. Hewitt

Many questions in ecology can now be addressed through the use of molecular markers, making **molecular ecology** a thriving multichanneled interface. Using recently available DNA sequence information, a wide spectrum of ecological topics is being investigated. These range from paternity analysis, through population structure, to speciation and systematics; they include all living kingdoms in all realms of the globe, and have application from pest control and animal breeding to conservation and sociobiology. The field is growing rapidly, as witnessed by the number of papers and journals reporting on it. In particular, DNA markers allow the measurement of genetic diversity and an assessment of its geographic distribution from broad range to fine scale. Such genetic variation underlies **adaptation**, functional diversity, and biodiversity at individual, population, and species levels.

## Diversity—ancient and modern

The present biodiversity across the globe has been produced through aeons, and contains phylogeny from ancient to modern times that can be investigated using DNA sequences. Our cooling planet's crust is moved and fractured by magma from the hot interior, and such plate tectonics involve the movement of continents and seas, the formation of mountains, islands and straits, and changes in heat distribution and climate. The continents and seas were very differently distributed some 225 million years ago (mya), when all land was collected as the continent of Pangea. This proceeded to break up and the components drifted to their present positions, producing new oceans and reassociated landmasses. Consequently, deep phylogenies of animal and plant groups reflect these ancient geographic associations and separations. Great mountain ranges like the Alps (36–23 mya), Himalayas (20–5 mya), and Andes (15–6 mya) were produced following the collision of Africa with Europe, of India with Asia and the Nazca Plate with South America. Ocean floor fissures and volcanic hot spots have produced remote islands and archipelagoes such as the Hawaiian Chain (30 mya to now), Galapagos (perhaps 80 mya to now) and Canary Islands (20 mya to now). Such old events have divided species and provided opportunities for colonization; they have molded the middle phylogenies of many genera and families. More recently, the Quaternary period (2 mya to now) has seen the growth of ice caps at the poles with a series of increasingly severe ice ages. This progressive cooling has probably been caused by the redistribution of the landmasses and changes in the oceans that absorb heat from the sun and carry it around the world. Because of the periodic growth of ice sheets, the sea level has fallen to uncover large areas that are presently shallow seas, and joined now separate lands. These more recent changes have affected the distribution of species, and been concerned with the process of speciation. They are responsible for the shallow phylogenies of species, subspecies, and populations.

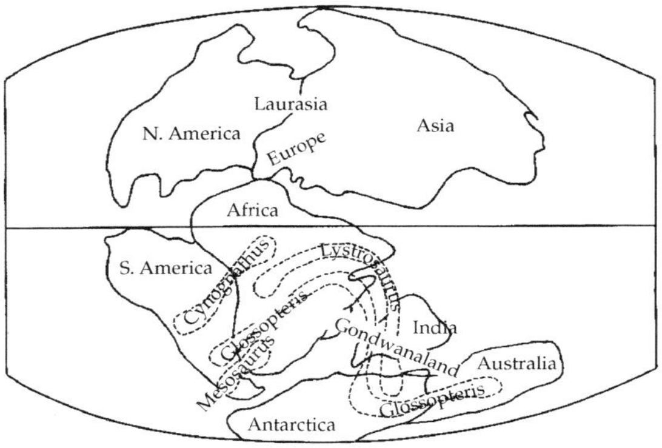


**Figure 14.1** The break-up of Pangea began around 200 mya into Laurasia and Gondwanaland, and the break-up of Gondwanaland around 180 mya into the southern continents through plate tectonic movement. The distribution of some Permian fossil organisms across Gonwanaland is indicated.

The disjunct distribution of ratite birds on various continents provides a classic example of how ancient events can explain current diversity. The break-up of Pangea and then Gondwanaland into the southern continents (200–150 mya) (Fig. 14.1) is thought to have caused the separation of their ancestors, which evolved to produce ostriches in Africa, emus and cassowaries in Australia, rheas in South America, and kiwis and moas in New Zealand. Molecular data confirm that these are indeed a monophyletic group (Sibley and Ahlquist 1990). Many other animals and plants have such disjunct distributions that are probably due to **continental drift**, including marsupials, frogs, lungfishes, midges, grasshoppers, and southern beech trees. Molecular phylogenies can help to elucidate their relationships and origins. For example, early divergence around 180–130 mya to produce the monotreme, marsupial, and placental mammals from a common ancestor is unclear and several molecular studies have been made, including mitochondrial and nuclear DNA and protein sequences (see Zardoya and Meyer, Chapter 18). The **paralogous** copies comprising the β-globin gene clusters offer a promising approach to understanding how diversity at this ancient level evolved (Lee *et al.* 1999). Considering somewhat later events, mitochondrial and nuclear DNA data reject the monophyly of Australian marsupials, and intercontinental dispersal and multiple entries must have produced this fauna (Springer *et al.* 1998).

The major uplifts of the Andes (12–3 mya), as the Nazca Plate subducted beneath the South American Plate, has produced ridges and basins to the east. A pioneering study of the molecular genetic relationships of small mammals in western Amazonia has revealed **phylogeographic** signals that reflect these palaeogeographical features (da Silva and Patton 1998). Some 11 out of 17 taxa show distinct **haplotype clades** upstream and downstream of the central Rio Jurua, coincident with where it crosses the Iquitos Arch. This ridge was produced by the Pliocene Andean foreland dynamics. The extent of mitochondrial DNA (mtDNA) *cyt-b* sequence divergence is from 4–13 percent in these taxa and indicates disjunction some 2–6 mya in the Pliocene. This concordance in time and space clearly indicates that the arch and basin formation have been instrumental in the divergence of these taxa, and probably many more. The genetic diversity seen in other organisms, particularly birds (García-Moreno *et al.* 1999), suggests that the Andean uplifts since the Miocene have been important in generating new species and so increasing biodiversity (Hewitt 2000).

The Hawaiian Islands in the mid-Pacific are the result of a volcanic hot spot, where a plume of magma from the earth's core penetrates the crust, and produces a chain of islands as the plate moves over it. Northwest, to the leeward of the eight extant aerial islands is a long ridge (1500 km) of largely submerged islands that were formed and eroded sequentially over some 30 mya. Because of their isolation and accurate K–Ar dating, these islands have been the prime theatre for studies of colonization and speciation in many organisms (Wagner and Funk 1995). Molecular studies with mitochondrial, chloroplast and nuclear DNA, particularly in *Drosophila*, crickets, spiders, planthoppers, honeycreepers, silverswords, and lobelias, has provided strong evidence for repeated **serial colonization**, adaptation, and speciation as new islands emerged, and has also allowed geologically dated calculation of rates of sequence divergence. In some cases, reverse colonization of older islands can be deduced. Interestingly, the *Drosophila* DNA lineages appear to have diverged from their mainland ancestors between 32 and 10 mya, so initial colonization was of now submerged islands, since the oldest main island of Kauai is only 5 million years old (Fleischer *et al.* 1998).

Other volcanic island systems, like the Galapagos in the Pacific and the Canaries in the Atlantic, have also been used to research questions on the creation of diversity through colonization, adaptation, and speciation (Grant and Grant 1996b; Juan *et al.* 2000). The endemic beetle fauna of the Canary Islands is particularly rich and DNA studies have been used to clarify how this species diversity arose. The southern tip of the oldest island, Fuerteventura, is some 20 million years old and the youngest, El Hierro in the west, less than 1 million years old, but there is still volcanic activity along the whole chain (Fig. 14.2). Like Hawaii, the primary colonization and speciation is as an island emerges, but this is overlaid by back colonization, extinction, and recent colonization in the last 3 million years (Emerson *et al.* 2000). The Galapagos finches also appear to have radiated after their arrival only some 3 mya, as estimated from molecular data. The more complex volcanism of these latter islands is a likely explanation of such colonization pattern differences.

These examples show how the biodiversity of a region is a product of ancient and modern events, and to understand its composition one must take account of evolutionary history from deep in the past. Because the events of the Quaternary are recent, they are more amenable to study, with in general more detailed fossil records, palaeoclimatic information, and DNA lineages. This attracts more study and allows greater interpretation of events and processes; it is currently providing many new explanations as such studies are combined.

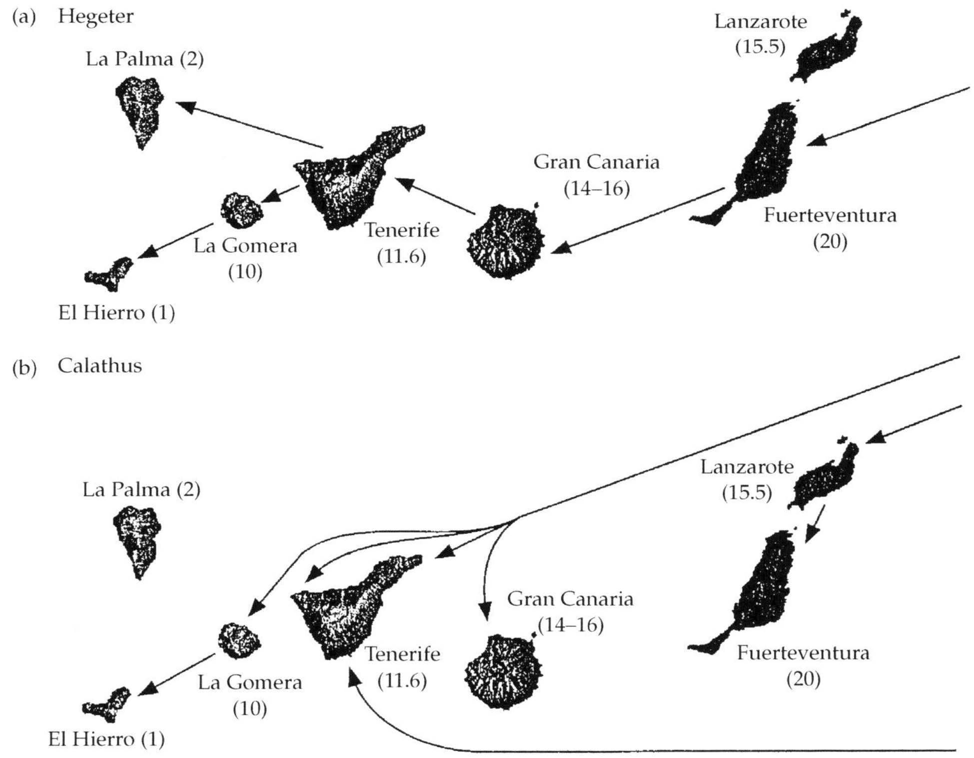


**Figure 14.2** Proposed colonization routes for the beetle genera (a) *Hegeter* and (b) *Calathus* among the Canary Islands as deduced from DNA phylogenies and geological dating of these volcanic islands (see Emerson *et al.* 2000).

## Quaternary ice ages

We are currently in a warm interglacial period, the Holocene, among the ice ages. These increasingly severe climatic oscillations officially became the Pleistocene with the formation of the Arctic ice cap, although the Antarctic ice cap had been growing for some 30 million years. The **Croll–Milankovitch theory** proposed that ice ages are driven by variations in the earth's orbit around the sun (Bennett 1997). Orbital eccentricity has a 100 000-year cycle, axial tilt cycles over 41 000 years and precession due to axial wobble has a 23 000-year cycle; these modify insolation and combine to cause major oscillations in climate. In the early Pleistocene these showed a 41 000-year frequency, while from about 0.9 mya the ice ages have recurred on a 100 000-year cycle. Recent paleoclimatic research has shown that the major cycles comprise millennial scale oscillations of considerable rapidity and magnitude. The clearest evidence for this comes from the deep continuous ice cores in Greenland and the Antarctic, which is supported by fossil pollen and beetle data. Interestingly, some of the first indications of such rapid paleoclimatic changes came from beetle fossil remains, some time before the annually layered snow of the ice cores provided such startling serial temporal resolution (Williams *et al.* 1998).

Recently the **Greenland Ice Core Project** has obtained cores over 2 km deep, which provide physical and biological climatic signals reaching back some four ice ages (400 000 years). The last ice age and previous interglacial (120 000 years) have been analyzed in most detail to reveal some 24 major warm periods lasting thousands of years and many shorter fluctuations. The $\delta^{18}O$ data indicate very rapid warming of the order of 10 °C in a few decades and a slower decline (Alley 2000). Nitrogen–Argon isotope measures suggest that temperature changes were even more extreme—perhaps a 15 °C difference. The most likely candidate to produce such oscillations is the switching on and off of the North Atlantic Deep Water Conveyor bringing warm tropical waters to northern latitudes. Great ice surges from the massive North American Laurentide ice sheet and possibly from the Scandinavian ice sheet seem to upset the thermo–haline balance of the Conveyor.

These North Atlantic events are part of a global interaction, and in other parts of the world ice cores, sediment cores, and animal and plant remains describe related major climatic fluctuations. How these rapid changes were expressed in local climate will be revealed as more data emerge from Asia, the Tropics, and Southern Hemisphere.

## Changes in species distributions

Plant, animal, and microbial life was necessarily greatly affected by these major climatic changes. The fossil species composition found in cores and stratified series from the sea bottom, lake bottom, and on land reflects the organisms' abilities to survive climatic conditions in and around that locality over time. They show the major turnover in species caused by large climatic oscillations within an ice age, and some pollen and beetle records carry signals of the rapid changes that occurred over hundreds of years in northern Europe (Birks and Ammann 2000). Extensive work on pollen cores now provides a particularly detailed record of the presence of plant species over much of Europe and North America from the last ice age, and several long cores reach over three ice ages in Europe. From these networks of cores, species distributions and vegetation composition through time have been reconstructed across Europe and North America (Bennett 1997). Animal fossil records, particularly beetles, complement these, but these are not so comprehensive.

An increasingly clear picture is emerging from these studies of changes in species composition and distribution through the last ice age. The cold climate condensed temperate and tropical regions toward the equator. During the glacial maximum 18 000–24 000 years ago, large ice sheets covered much of North America and North Europe, while North Asia was in deep permafrost with only its mountains glaciated (Fig. 14.3). South of this were tundra and permafrost, and the species now inhabiting these northern latitudes survived to the south. Major mountain ranges in middle latitudes were glaciated and their present species survived at lower altitudes and latitudes. It was generally much drier, with increased desert and savanna and

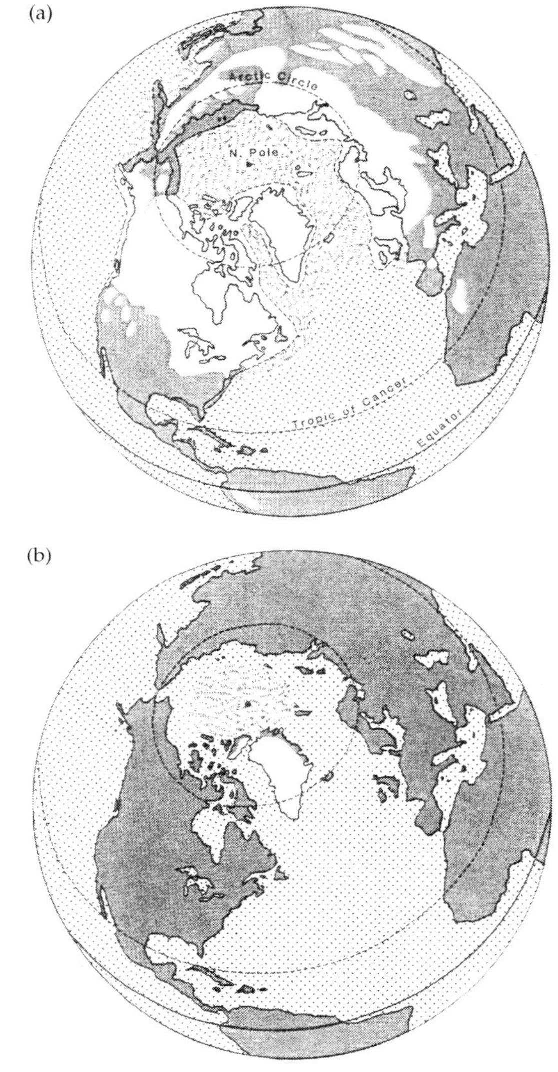


**Figure 14.3** The extent of the ice cap, sea ice and permafrost in the Northern Hemisphere (a) during the last glacial maximum (20 000 years ago) and (b) at present.

fragmented tropical forest. The build up of ice lowered the sea level and several major land bridges emerged, such as the Bering Straits and Sunda Shelf. In higher latitudes species disappeared, while at lower latitudes mountain tree species descended as much as 2000 m. Just where tropical lowland species survived is open to much debate (Colinvaux *et al.* 2000).

The fossil pollen and beetle records show that the response of many species to these sudden climatic changes was also rapid. After the last ice age the climate warmed from around 16 000 years ago and species were able to expand their ranges remarkably quickly. Even large trees in Europe and North America moved north at several hundred meters a year (Bennett 1997), and Mediterranean beetles were found in Britain around 13 000 years ago. Then came the last of the major cold reversals—the Younger Dryas (11 000–10 000 years ago)—when species retreated again. Following this is the present Holocene, when climate warmed again very rapidly and species advanced again into new territories in higher latitudes and higher altitudes.

## Effects on population genomes

It is clear from the fossil record that species distributions have been contracted, expanded, moved and mixed repeatedly by these rapid major climatic shifts. Such changes can be expected to have modified the genetic composition of populations and species in many ways (Hewitt 1993). This is likely to vary among regions with different latitude, topography, and biota. Furthermore, the effects of more recent events may overlay or augment earlier ones. A relatively simple scenario can be envisaged for much of Europe and other northern temperate lands. With a rapid reduction in average temperature of 10–15 °C that persisted for at least a few decades, the northern populations of most species in Europe would not reproduce and go extinct. At the southern, warm edge of the species distribution conditions would also cool and small isolated populations would expand. This would reverse when the climate warmed again.

Ice core, pollen, and beetle data demonstrate a rapid warming after the last ice age and small populations at the northern edge would colonize the now suitable habitat by a series of long distance **founder events** (founding a new colony by a few migrants) with exponential population growth. This, then, effectively excluded much contribution from later migrants (Hewitt 1993). This is called the **leading edge model** producing over time a loss of allele diversity and areas of relative genetic homogeneity, and computer simulation with **leptokurtic dispersal** (more long distance dispersers) demonstrates these dynamics and consequences (Ibrahim *et al.* 1996). Repeated reversals during an expansion, as seen in the climatic record, accumulate these effects. There is increasing evidence for such genetic patterns in many species that colonized previous glacial or tundra regions (e.g. Hewitt 1996, 2000).

This rapid colonization has two other interesting corollary effects. First, once a few long distance founders have colonized and exponentially filled an area, dispersal and **gene flow** from populations behind is limited to logistic growth and so they contribute relatively little genetically—a form of high density blocking. This will tend to restrain most refugial genomes (see later) to the south (Hewitt 1993, 1996). Second, when two separate long distance rapid colonizations first meet, their density will be low and allow some intermingling of founder populations. These will expand to fill the area, but will leave a signal of broad **introgression** of alleles that is difficult to explain on current population dynamics. This **Pioneer** form of leading edge expansion and contact may be contrasted with the **Phalanx** form, where the advance is slow and maintains high population density, and also genetic diversity (Nichols and Hewitt 1994).

Whilst such "southern richness to northern purity" in genetic diversity seems common in north temperate species, some with geographical or biological peculiarities may not show it. Those species with more Boreal, Savannah, and Tropical ranges will also have been affected differently. The mountains of southern Europe, which have allowed altitudinal shifts with climatic changes and so provided **refugia** for many species, will have augmented the pattern of high southern diversity. Such mountain blocks may also act as islands, with the survival and divergence of distinct genomes through climatic oscillations. This may allow speciation (Hewitt 1996), and may also apply to other low latitude mountain systems on other continents, like the Andes, Appalachians, Annamites, and Arusha Highlands (Hewitt 2000).

## Suitable sequences and analyses

To look for and examine these and other possible genetic consequences of ice age range changes requires the development of informative DNA sequences and discriminating analyses. DNA technology has advanced so that sequences of hundreds of nucleotides can be readily obtained from hundreds of individuals across a species range in an average study. Such sequence data can show the genetic relatedness among individuals and the genealogy and divergence of sequences and populations can be determined. In a geographical context this allows the deduction of ancestry, past refugia and postglacial colonization. Such use of genealogical lineages to deduce evolutionary processes is part of the thriving field of **phylogeography** (Avise 2000). It allows us to understand genetic diversity in space and time (Hewitt 2001). A wide range of DNA markers are now in use (see Box 14.1), and one of the fruits of maturing genome projects will be many more (see González-Candelas *et al.*, Chapter 6).

The explosion of DNA data for populations has lead to the development of analytical methods to probe and distill their significance. Allozyme, fingerprint, and banding methods provide allele frequency data only; their alleles cannot be genealogically organized. DNA **haplotypes** (allelic sequence variants) can be ordered into a genealogy and so produce their phylogeny. Combined with their population frequency and geographic distribution, this provides a strong basis for inferences on the species' evolutionary history. The usual phylogeographic approach is to build a phylogeny from haplotype sequences using distance, parsimony, and maximum likelihood methods, and then represent the lineages geographically. There are several further developments that are providing more discriminating analyses (see Box 14.2).

Clearly, each DNA sequence has its own genealogy and they may evolve at different rates. Furthermore, the various methods of analysis probe different aspects of the molecular and spatial history. Consequently, to reconstruct a species phylogeographic history one would ideally like to use a range of sequences (including nuclear, cytoplasmic, sex-linked, autosomal, conserved, neutral, high, and low mutation rate) and apply a suite of pertinent analyses (such as DNA distances, phylogenetic trees, nested clades, networks, mismatch, and specific simulation models). However, this is prohibitively

**Box 14.1 Suitable DNA Markers**

For evolutionary events in the Pleistocene and Holocene fairly rapidly diverging markers are required to better distinguish recent phylogenetic and geographic changes. The variation seen presently in slowly evolving sequences will have arisen over some time, and much of it will be the assortment of ancient mutations.

Mitochondrial DNA (mtDNA) has been very important in animal phylogeographic studies, but much less so in plants where its genetic structure is more complex. Chloroplast DNA (cpDNA) has been favored in plants, but less successfully because of its low divergence rate. mtDNA diverges at some 2 percent/million years on average, while cpDNA is around 0.02–0.1 percent/million years (see Hewitt 2000). Some effort has been put into finding nuclear sequences to complement and augment these cytoplasmic markers, which are generally uniparentally inherited and contain only a tiny fraction of the coding genes.

As far as nuclear DNA is concerned, gene **exons** are conserved and **introns** have not proven as fast evolving as initially expected, only some 0.7 percent/million years on average in mammals. Other noncoding regions are being explored to find suitably variable sequences, and there are several promising candidates.

Noncoding nuclear DNA was used in the European meadow grasshopper, *Chorthippus parallelus* for phylogeography, and the 3′coding region *Antennipedia*-class homeobox gene in the African desert locust *Schistocerca gregaria* (see Hewitt 2001). Both of these sequences have diverged faster than mtDNA, and as fast as the A-T rich control region in the locust. These markers were not easy to find and develop, but improved knowledge of the genome and advanced technology now make the task easier.

**PCR-based hypervariable markers**, like **microsatellites** and **AFLPs** (amplified fragment length polymorphisms), can provide much useful variation for estimating genetic distance and phylogeny within species. However, they need careful analysis due to **homoplasy** and they do not reveal genealogy (see Hewitt and Ibrahim 2001).

**SINEs** (short interspersed nuclear elements) and **transposons** are possibilities, but their structure and dynamics are complex, and they may only be useful in particular cases.

expensive and very time-consuming! Realistically, one needs to carefully identify a specific question, geographic range and evolutionary time scale in making a choice of markers and analyses. For discerning genetic structure produced over the ice ages, a combination of fast evolving mtDNA sequence, noncoding nuclear sequences, and **microsatellites** across the species range should provide most primary information when analyzed with the major methods described.

## Understanding genetic diversity across Europe

Such techniques and approaches are being applied to a variety of plant and animal species around the world, and the studies already accumulated for Europe provide useful exemplars and illustrate some more general principles (more detailed referencing to these may be found in Hewitt 1999, 2000, 2001).

The European grasshopper *Chorthippus parallelus* has been investigated with a range of genetic markers and its biology is quite well understood (Butlin 1998). Its genetic structure and postglacial colonization routes across Europe have been revealed particularly well by a 393-bp fragment of noncoding nuclear DNA analyzed by genetic distance ($K_{st}$ nucleotide differences; see Box 14.2) among populations. There is considerable divergence between Iberian, Italian, Greek, Balkan, and Russian samples, with the rest of northern Europe similar to the Balkan populations. Along with pollen data this strongly indicates ice age refugia in these southern places, with only Balkan populations colonizing the rest of Europe. The reduced genetic diversity in the north is expected from a rapid warming and

## Box 14.2 Recent phylogeographic analytical approaches

**DNA distance phylogeography** uses DNA distance measures among populations, (such as $\gamma_{st}$, $N_{st}$, $K_{st}$) computed from haplotype sequence and frequency, to phylogenetically relate populations across the species range. Their statistical significance can be tested by bootstrap techniques, and in combination with fossil evidence are very useful for phylogeographic inference.

**Nested clade analysis** uses a parsimonious tree of haplotypes and associates them hierarchically into 1-, 2- to *n*-step clades on the mutational changes separating them. Next the geographical association of clades is assessed through the geographical center and variance in their haplotype distances. The statistical significance of the derived measures is assessed with random permutation, and then compared with expectations of geographic genetic structure from gene flow with **isolation by distance**. Departures from these expectations may be attributed to past range fragmentation and expansion.

**Haplotype networks** are produced by joining variant sequences together sequentially by base substitutions, and other single mutations, through 1-, 2- to *n*-step connections until a single full network is formed. Such constructions have evolved from minimum spanning networks, through median and reduced-median networks, to median-joining networks to seek the most parsimonious evolutionary pathway to form the sequences. Such haplotype networks are reticulate, showing alternative pathways and possible homoplasy, which conventional trees do not, and they also show possible haplotypes not yet sampled, mutation hot spots and recombination. These networks have to be combined with the spatial distribution of haplotypes to allow phylogeographic interpretation, but as yet there is no formal method for this.

**Sequence mismatch distribution** plots the frequency in a sample of pairs of sequences against their number of nucleotide differences. Expected distributions are produced by simulation, and comparison with these allows the identification and timing of signals of past expansion and other demographic events. In a well-defined geographic situation they add significantly to phylogeographic interpretation.

**Genetic and demographic simulation** is now possible with powerful desktop computers to explore with breadth how DNA markers evolve in specified molecular, spatial, and demographic conditions over history. This promising approach is being used increasingly, and has already proved useful in testing phylogeographic possibilities in, for example, grasshoppers, locusts, and fishes (Hewitt and Ibrahim 2001).

expansion (Hewitt 1993), and nested clade analysis confirms this signal with large **clade** distances (Hewitt and Ibrahim 2001). The genome divergence among the southern refugial populations is commensurate with their separation over several ice ages, and from these refugia they probably colonized Europe during each warm interstadial. On the basis of chloroplast DNA (cpDNA) the black alder *Alnus glutinosa* shows a very similar genetic structure across Europe, with several disjunct southern genotypes and a major Balkan recolonization.

The European hedgehog *Erinaceus europaeus/concolor* was investigated using a 383-bp *cyt-b* mtDNA sequence, and has a very different pattern of genetic subdivision across Europe, and hence postglacial colonization routes. Three distinct clades dating from the onset of the Pleistocene have colonized Europe from southern refugia in Iberia, Italy, and Balkans with a fourth clade in Turkey and the Levant. There is further Pleistocene division in each of these major clades, and nested clade analysis supports much past fragmentation with some range expansion (Fig. 14.4) (Seddon *et al.* 2001). Mismatch analysis of the western European haplotype clades suggests that several expansions have occurred, with the last postglacial one dominating the north.

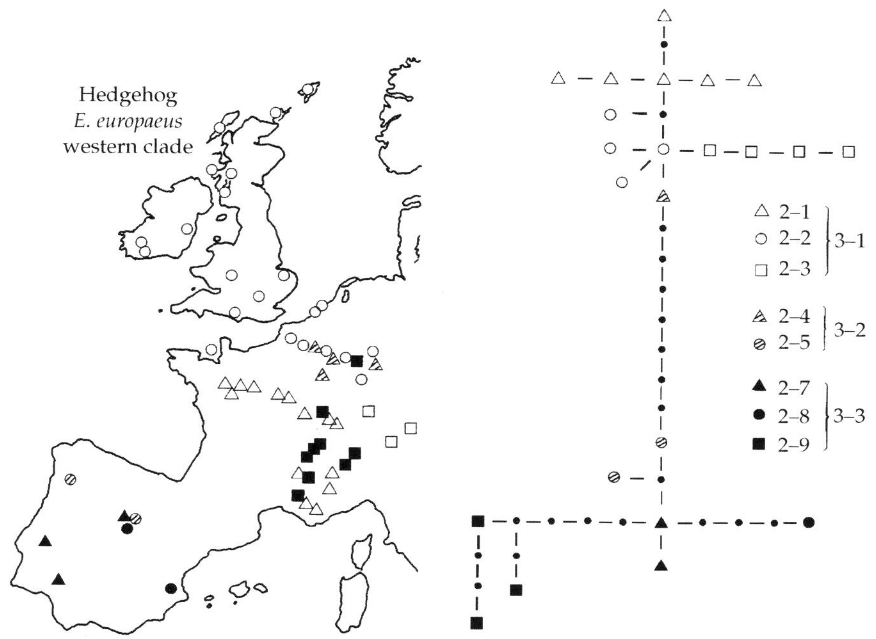


**Figure 14.4** The distribution of mtDNA haplotypes for the westernmost clade of the European hedgehog *Erinaceus europaeus* that is deduced to have colonized northwards from Iberia after the ice age, and a network with nested clades identified (see Seddon *et al.* 2001).

The hedgehog genomes have been diverging in southern Europe for much longer than those of the grasshopper, and have apparently colonized northward as several phalanxes rather than one pioneer expansion. The major deciduous oaks *Quercus robur* and *Q. patraea* share their common cpDNA haplotypes and these demonstrate a refugial and colonization pattern similar to the hedgehog, with genotypes from several southern regions spreading northward.

In Europe the brown bear *Ursus arctos* has two major mtDNA lineages, in the east and west, which represent postglacial colonization from Iberia and around the Black Sea. These two northern colonization routes met in central Scandinavia, while several other lineages remain to the south in Italy and the Balkans. A similar mtDNA lineage pattern across Europe is seen in the shrew *Sorex araneus* group, with *S. araneus* in the east and *S. coronatus* in the west. Most interestingly, several other small mammal species have genomes which meet in central Sweden, which is explained by the contact of western and eastern colonizations when the last of the ice sheet melted around 9000 years ago.

The three refugial colonization patterns, the grasshopper, the hedgehog, and the bear, are repeated in other species for which there are sufficient data (Fig. 14.5) (Hewitt 1999). These reflect the major refugial peninsulae, with the Alps and, to a lesser extent, the Pyrenees as barriers to initial postglacial colonization. Where distinct genomes meet as a result of these postglacial expansions, **hybrid zones** are formed, and there are clusters of these down central Europe, in the Alps, the Pyrenees, and the Balkans. The contact of west and east genomes in many species from the Baltic to the Alps has been termed a **suture zone**, as are the Alps and central Sweden clusters. As more molecular phylogeographies across Europe are produced, it should be possible to augment these paradigm patterns, and a radically different one may be discovered. Indeed, information on freshwater fish is beginning to emerge, for which major rivers like the Danube and Dneiper appear to be important colonization routes.

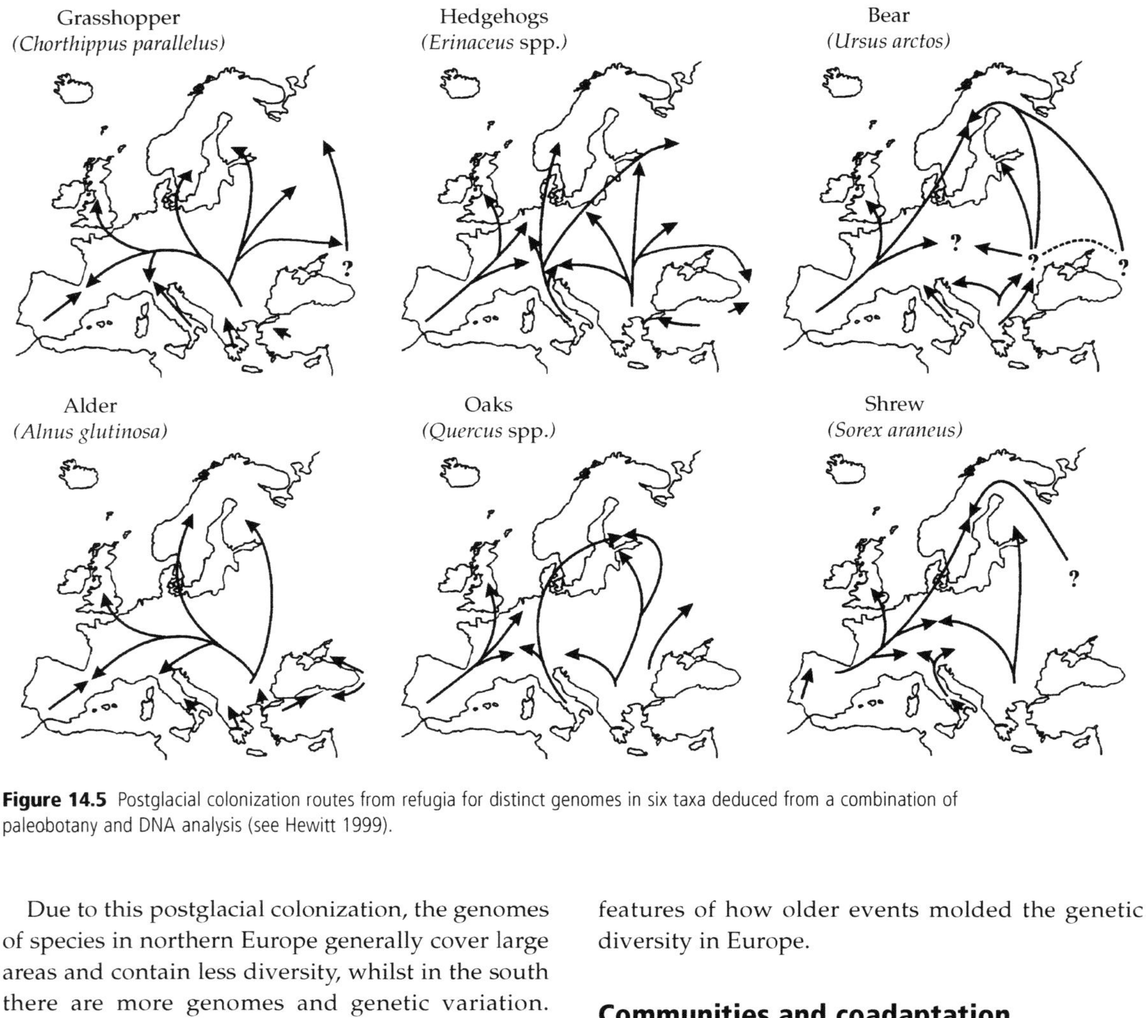


**Figure 14.5** Postglacial colonization routes from refugia for distinct genomes in six taxa deduced from a combination of paleobotany and DNA analysis (see Hewitt 1999).

Due to this postglacial colonization, the genomes of species in northern Europe generally cover large areas and contain less diversity, whilst in the south there are more genomes and genetic variation. These southern populations would seem to have survived several ice ages there, as witnessed by the extent and dissection of their sequence divergence, for example, the grasshopper *C. parallelus* around 0.3–0.5 mya, the hedgehog *E. europaeus/concolor* around 3–6 mya (Table 14.1). The mountain blocks would seem important in this process of divergence and the retention of its products in these more temperate regions (Hewitt 1996). The deeper divergence, as in the hedgehog, has probably been evolving since the Pliocene, and more molecular studies are needed to probe this, like those in fish, beetles, and salamanders (Hewitt 2001). With suitable analyses these should reveal more general features of how older events molded the genetic diversity in Europe.

## Communities and coadaptation

The mixtures of species that form northern temperate communities in Europe, and probably elsewhere, comprise genomes that have survived in various southern refugia (Hewitt 1999). Thus Britain contains oaks and hedgehogs from Iberia, while the grasshoppers and alder came from the Balkans as conditions warmed after the ice age. In Scandinavia there are colonizations by species genomes from refugia in Iberia, Italy, Balkans, and further east, while in central Europe the number of hybrid zones reflects great spatial variation in the genomes of species forming communities. Since the genomes of many of the component species in any northern temperate place have

**Table 14.1** DNA sequence divergence, and hence estimated maximum time of separation in species colonizing Europe after the last ice age

| Organism | mtDNA sequence | Divergence (%) | Maximum age (millions of years) | Refugia |
|---|---|---|---|---|
| *Bombina bombina* (fire bellied toad) | RFLP | 9.4 | 5 | (I) B B |
| *Erinaceus europeaus* (hedgehog) | *cyt-b* | 6–12 | 3–6 | S I B |
| *Triturus cristatus* (crested newt) | RFLP | 4–8 | 2–4 | S I B |
| *Arvicola terrestris* (water vole) | *cyt-b* | 4–7.6 | 2–4 | S I B |
| *Crocidura suaveolens* (white toothed shrew) | *cyt-b* | 3–6.4 | 1.5–3.2 | S (I) B |
| *Mus musculus* (house mouse) | RFLP | 3.4 | 1.7 | W and E |
| *Microtus agrestis* (field vole) | RFLP | 2 | 1 | W and E |
| *Sorex araneus* (red toothed shrew) | *cyt-b* | 1–3.8 | 0.5–2 | S (I) B |
| *Ursos arctos* (brown bear) | control region | 2.7–7 | 0.35–0.85 | S (I) (B) |
| *Chorthippus parallelus* (meadow grasshopper) | 6.7 kb | 0.7–0.9 | 0.3–0.5 | (S) (I) B |

The southern refugia of distinct genomes are shown: S = Iberia, I = Italy, B = Balkans, W = west, E = east. Those not expanding out of their peninsula are in parentheses.

only been together for a few thousand years, any close **coadaptation** must have occurred in that time. Of course, species may have coexisted with different genomes of the other species in their own glacial refugia, and so could be preadapted to some extent. But during the Pleistocene species ranges have been fragmented and expanded many times across Europe along possibly different colonization routes, so they could have been compatriot with several different genomes of other species. Such a varied pulsating spatial evolutionary history makes the understanding of ecological relationships frightfully complex, but it happened and must be incorporated into our thinking.

We may expect the relationship and genetic coadaptation between a parasite and host, or a predator and prey species to be less affected by climatically induced range changes. The host will expand its range from a refugium when conditions suit, and the dependant species will follow closely, so compatriot species genomes should tend to stay together. However, in the case of the spruce bark beetle, *Ips typographicus* molecular data suggest that it did not follow the colonization routes of the spruce that were deduced from the pollen record (Stauffer *et al.* 1999). More studies are needed on this question, to produce comparable molecular phylogeographies of hosts and parasites across their ranges. The answer may be surprising. Southern refugial areas, as in Iberia, Italy, and the Balkans, have supported species through several ice ages, probably due to their varied topography that allows short distance altitudinal displacement to accommodate climate changes (Hewitt 1999). There are reliable fossil pollen data that indicate this in southern Europe and Southeast Asia (Hewitt and Ibrahim 2001). Consequently, in these regions local genomes of the various cohabiting species have probably stayed together much more often. The evidence indicates that these regions contain greater genomic fragmentation and divergence than higher latitudes (Hewitt 2000), and we would expect the sustaining and dependent species genomes to be largely concordant. Once again phylogeographic DNA data are required.

## Global phylogeography

Europe is largely temperate, but these molecular and analytical methods can be applied equally to species in the Boreal, Temperate, Desert, and Tropical regions. Many studies are emerging from North America (Avise 2000), which like Europe is largely Temperate and Boreal, but with Desert and Neotropical components due to its more southerly extension. On present evidence the genetic structure of species that was put in place postglacially, and accumulated over longer time, seems to differ among the major regions of North America (Hewitt 2000).

### Boreal North America

As the ice melted (14 000–8000 years ago), the glaciated parts of North America were colonized from southern refugia, and molecular data show a loss of diversity in northern populations expected of rapid expansion (e.g. Hewitt 1996, 2000; Mila *et al.* 2000). Genetic differences within species indicate that there were a number of refugia, which colonized different northerly regions. The species of the Pacific NW provide a clear example of this, surviving south of the Cordilleran ice sheet and expanding toward Alaska (Brunsfeld *et al.* 2001). Other refugia south of the ice in the center and east of the continent are also evident in the genetic data. Particularly interesting is the molecular evidence for species colonizing the deglaciated north from a Beringian refugium in the NW. These include, for example, fish, beetles, the arctic plant *Dryas integrifolia*, collared lemmings, and brown bears. Beringia contained unglaciated parts of Alaska and the broad land bridge to NE Siberia formed by the lower sea level of the ice age. Not only was this a Neoarctic refuge, but it also acted as an intercontinental invasion route, as seen in the phylogenies of several groups (Hewitt 2001).

A number of circumpolar taxa have marked molecular subdivision, which suggests that they have been diverging in refugia over a number of ice ages in the late Pleistocene, for example, guillemots, dunlins, and reindeer. Given the extent of the ice during glacial maxima, these refugia would probably have been distant and disjunct, and the present extensive distribution of the genomes only occurs in brief interglacials like the present. There is considerable interest in Arctic biota and how it evolved, and such genetic studies can help greatly to satisfy this, as these few studies show. Unfortunately, less work like this has been done on Boreal Eurasian species such as lemmings, and more is badly needed. While this region was very cold in the ice age, it was much less glaciated. Consequently the survival and colonization dynamics were likely very different from species in temperate or deglaciated regions. Their genetic diversity and genome structure should be rather different also (Hewitt 2001).

### Southeastern USA

The southeast USA is another region for which there is very good phylogeographic information; indeed the first mtDNA studies were in these parts (Avise 1994) with marine, freshwater and coastal species studied in particular. Species genomes are subdivided and divergent, while their northern expansions are much less so. This indicates that the SE sector provided refugia through the Pleistocene climatic oscillations. With its moderate temperature, varied topography and extended coastline, many species could survive by going up and down mountains, rivers, and shores within the region, and retain much diversity (Hewitt 2000). It is farther south than Mediterranean Europe, and would seem as a consequence to be perhaps genetically richer. Charles Remington identified one of his major suture zones between closely related taxa across this region, and the molecular phylogeographies of species confirm and extend this. For example, the distribution of disjunct molecular lineages in species of fish and turtles reflecting east and west drainages are broadly coincident (Avise 2000). Such concordance among species phylogeographies clearly demonstrates that the components of the biota have a shared biogeographical history. It was produced by common refugia and rapid postglacial climatic amelioration, which allowed a general advance and contact. As in Europe, molecular data are greatly increasing our understanding of how diversity was created and maintained.

### Western North America

The mountains and deserts of the North American West and Southwest have provided a dissected and diverse environment through the many Pleistocene changes in climate. Much of the region was generally even drier in the last ice age, and the complex shifts in habitat are difficult to discern. Molecular studies in several species show extensive genome subdivision and divergence (Brunsfeld *et al.* 2001; Hewitt and Ibrahim 2001), which might be expected from such a checkered history. But because of this the region offers an exciting challenge to genetic investigation, which should be scientifically rewarding. Recent DNA work on the montane grasshopper *Melanoplus* of the "sky islands" of the Rocky Mountains, which were affected by the Cordilleran ice sheet to the north, shows low sequence divergence among species. The grasshopper would have undergone many altitude and range shifts, and cogent simulation analysis indicates that speciation was rapid (Knowles 2001).

### Savannah

Worldwide, little phylogeography exists for drier regions and it is badly needed. The savannah and similar grasslands expanded during the ice age with its drier climate, and contracted in wetter times. Such climatic changes are reflected in the DNA phylogeographies of African bovids, like the hartebeest, topi, and wildebeest. For example, the gene tree **topology** for the wildebeest suggests a colonization of East Africa from South Africa in the last ice age, and its fossil record indicates that it had previously occupied East Africa, but went extinct. Colonization from South Africa to East Africa is also indicated by such DNA approaches for the kuda, and possibly the impala, and they also reveal previously unrecognized genetic distinction of endangered populations (Nersting and Arctander 2001). Once again the addition of molecular phylogeography is providing new and clear explanations of how species evolved and present diversity was created.

### Tropics

The Tropics contain the greatest diversity of species, and what few molecular studies there are indicate considerable genetic diversity within species. There are intraspecific phylogenies from tropical parts of Africa, South America, Central America, Australia, and SE Asia, which in general have deep DNA lineage divergence surviving from the Pliocene or before. Many of the same processes seem to operate as in southern temperate regions, but perhaps over a longer time and with smaller range displacement. In Central America the phylogeographies of birds and fishes have diverged in the Pliocene or Miocene, with several overlaid colonizations. Salamander genomes occupy relatively small areas and similarly show deep divergence (García-Paris *et al.* 2000; Hewitt 2000). In Amazonia, pioneer work on small mammals has shown a concordant divergence in DNA lineages coincident with the Iquitos Arch, which was formed by Pliocene Andean uplift (da Silva and Patton 1998). This provides strong evidence that such geological events have been important in structuring biodiversity, and argues against previous explanations in this case, such as river barriers. More such studies are needed in this biologically important region to provide substantial evidence on the extent and causes of its biodiversity.

Tropical mountains in South America and Africa contain bird DNA lineages that range from ancient to recently diverged, while the neighboring lowland tropical forests contain lineages that are generally 6 million years old or more. This leads to the suggestion that it is the mountains that act as engines of divergence and speciation, while species of some antiquity survive recent climatic changes in the lowlands (García-Moreno *et al.* 1999; Hewitt 2000). Such comparison will be most interesting for small mammals and particular insect groups as molecular data become available.

It is difficult to combine paleoclimatology and phylogeny in the Tropics as has been done in Temperate lands; the pollen record and other paleobiological indicators in Amazonia and elsewhere are so sparse. However, recent work in the tropical rainforest of NE Australia demonstrates the value of such combined study. Here, DNA phylogeographies of some frogs, reptiles, and birds show concordant divergence from the Pliocene that is coincident with a particular repeated ice age break in the forest. The dissection of the wet forest during these dry cold periods would have reduced population size, and this is signaled by reduced genetic diversity within populations and differences amongst them (Schneider *et al.* 1998).

### Oceans and seas

There is so far little phylogeography for organisms from the marine realm, and yet water covers much of the globe. At present, studies indicate little geographic genetic structure within oceanic species, as for example the butterfly fish in the tropical Pacific, and pelagic fishes in general have low geographic differentiation. This might be expected in organisms with such great dispersal potential and reproductive capacity. However, certain regions with varied topography do show considerable genetic variability as well as species diversity. Such a picture is emerging for the seas of Indonesia, Sunda Shelf, and Wallacea, which have many islands and are also the junction of the Indian and Pacific Oceans. These shallow seas with islands and deep trenches may act rather like tropical mountains as refuges and engines of speciation for marine organisms (Hewitt 2000). Nowhere does the genetic basis of diversity need investigation more.

## Conclusions

Molecular techniques, especially those involving DNA, are greatly advancing our knowledge of the present distribution of genetic diversity globally, and our understanding of how it evolved through time. Different sequences can discern divergence events from Pangea to the present. They are proving very useful in understanding biogeography, speciation, and colonization over the last few million years, particularly as applied to the Late Pleistocene and Holocene. There are concomitant advances in analytical methods to extract useful information from these new data. This understanding is greatly enhanced by combination with recent advances in paleoclimatic and paleobotanic studies.

The genetic diversity within species is seen to be highly structured spatially, with a patchwork of genomes divided by hybrid zones. Much of this can be attributed to climatic oscillations associated with ice ages, which have shifted species ranges repeatedly. In temperate regions like Europe and North America there is much more diversity in the south, where it has accumulated in refugia over many ice ages, and much less in the north, where it was lost during postglacial colonization. These northern places have been colonized by species from different southern refugia, and have had little time to become closely coadapted. Such understanding of the distribution of biodiversity carries serious implications for the theory and practice of conservation.

DNA studies emphasize that in certain places there is a concordance of contacts between closely related species, subspecies, and genomes, which can be explained by the general rapid postglacial advance from common refugia. Mountain ranges in warm Temperate and Tropical regions would seem to be important for the survival of lineages through climatic changes, and hence for genome divergence and speciation. More generally the Tropics show deep complex genomic diversity, which has been produced through many climatically induced range changes. Clearly, such findings have great relevance to the debates on the management of biodiversity.

Since modern molecular studies have revealed so much in a few years where they have been applied, it is expected that their use in the Tropics and Oceans will likewise be immensely informative and practically relevant. It is ironic that just when we are learning so much about the structure and evolution of diversity, we are able to destroy it.

PART IV

# Speciation and major evolutionary events

CHAPTER 15

# Allopatric speciation: not so simple after all

Menno Schilthuizen and Bronwen Scott

**Speciation**, the evolutionary event in which one ancestral species splits into two or more daughter species, is a central subject in evolutionary biology. Despite this (or possibly because of this), the subject has been fraught with controversies, conflicts, and misunderstandings. To some extent, this unrest is the result of the difficulties of delineating what is meant by the term "species," because, by definition, the species concept determines which events are considered crucial in speciation.

For most of the past half century, most biologists have adhered to Mayr's (1942) "**biological species concept**," which states that species are groups of organisms that are reproductively isolated from one another (Wu 2001). However, with the advent of sensitive techniques for measuring genetic exchange between populations, it appears that many recognized species maintain their differences while engaging in considerable **gene flow** with related species. For example, **microsatellite** analysis of two European species of oak tree recently showed that in spite of pervasive interspecific hybridization and gene flow, the two **sympatric** species remain morphologically, ecologically, and genetically distinct (Muir *et al.* 2000).

A perusal of the recent literature shows that the biological species concept is currently being replaced by a view in which "cohesion" is crucial. **Species cohesion** may be either the result of the absence of gene flow or the presence of other stabilizing mechanisms, such as differently directed natural selection pressures in different niches (Templeton 1998). Together with this change in perception of what species are, approaches to studying speciation are also changing. Under the biological species concept, speciation was synonymous with "that stage of the evolutionary process at which the once actually or potentially interbreeding array of forms becomes segregated into two or more separate arrays which are physiologically incapable of breeding" (Dobzhansky 1937). The current trend in speciation research, however, recognizes that speciation is driven by evolutionary processes that result in two differently adapted genomes. Such divergence may come about under any amount of ongoing genetic mixing, provided that the differentiating forces (selection and stochastic effects) are strong enough to override it. In many cases, reproductive isolation will eventually evolve, but the crucial steps in the speciation process may have been taken long before.

On the one extreme of this spectrum lies **sympatric speciation**, where new species arise under strong divergent natural selection in the face of high gene flow. Convincing cases of such speciation have been uncovered in last few decades in insects and vertebrates (Orr and Smith 1998; van Alphen and Seehausen 2001). On the other extreme of the gene flow spectrum we find **allopatric speciation**, where full geographic isolation between populations eventually leads to the formation of two new species, due to the effects of random genetic divergence.

Allopatric speciation has traditionally been the least controversial as it has been linked in a one-to-one relation with the biological species concept: If geographic isolation is a prerequisite for speciation, this means that the processes responsible for **allopatric** differentiation would break down under gene flow. So, two species can become sympatric only after reproductive isolation has evolved: In the absence of reproductive isolation, two allopatrically differentiated populations will fuse again upon secondary contact. Therefore, reproductive isolation is the decisive criterion for what constitutes a species, and the evolution of reproductive isolation defines the point where speciation has been completed.

In view of the above-mentioned recent developments in speciation research, however, a reappraisal of allopatric speciation is called for. If full geographic isolation is *not* a prerequisite for speciation, it is legitimate to ask in what respects allopatric speciation is different from sympatric speciation. How certain can we be that gene flow is indeed absent between allopatrically speciating populations? What are the respective roles of **genetic drift**, natural selection, and **sexual selection**, and the interactions between these three? Why would reproductive isolation evolve between populations that never meet each other? In this chapter we will attempt to address these questions. First, in a general sense, and subsequently we will draw attention to a type of case study that appears to be a "classical" example of allopatric speciation, due to the limited possibilities for gene flow: land snails on habitat islands. Finally, we will make some suggestions for aspects that studies of allopatric radiation should pay attention to, if we wish to gain a better understanding of the process.

## Evidence for allopatric speciation

There can be little doubt that speciation commonly occurs in the context of geographic isolation. The great travelling theorists of evolution were already familiar with the fact that dispersal barriers tend to separate related species. Darwin (1859) noticed that each of the Galapagos islands contained endemic but related species of plants and birds, and endemic subspecies of the giant "galápagos," or tortoises. Wallace (1852), upon returning from his travels in South-America, reported to the Zoological Society of London that he had observed that large rivers in the Amazonian basin appear to separate different species of primates (a notion later generalized as the Riverine Barrier Hypothesis); and Mayr (1940) assembled all the data he had collected on bird distribution in New Guinea and the Pacific to show that, more often than not, sea and mountain ranges are boundaries between sister species or subspecies. The review was the basis for his theory of allopatric speciation.

Although the generality of some classical models of allopatric speciation (e.g. the Riverine Barrier Hypothesis) has become more restricted lately (Colwell 2000), the geographic connection continues to turn up in a variety of regions, ecological systems, and organisms. For example, the Isthmus of Panama, the origin of which isolated the Caribbean from the Pacific some 3 million years ago, is now known to be the dividing line between pairs of related species in a wide variety of marine animals (Lessios 1998). Another example comes from the influence of past barriers in the form of land ice during the Pleistocene. Hewitt (2000) reviews the genetic evidence from a wide range of temperate-zone animals and plants that were isolated into separate **refugia** during the last glacial maximum and have expanded since (see also Hewitt, Chapter 14). The available evidence shows that in some cases species or subspecies formation can be linked to this isolation, and in many cases strong genetic diversification was initiated during this event.

## How often are isolated populations really isolated?

At first sight, the inevitability of allopatric speciation is easily grasped: When two populations are totally genetically isolated, even in the absence of natural selection, any mutation and any amount of genetic drift will eventually lead to genetic differentiation between the two populations. After many generations, a sufficient number of genes will have been affected for reproductive isolation and/or widespread ecological, morphological, and behavioral differentiation to have evolved.

However, before accepting that a case of allopatric divergence is due to lack of gene flow, we need to ascertain that gene flow is really zero. In many cases, the impossibility of migration is assumed rather than demonstrated. Sessile organisms on "**ecological islands**" (i.e. not just *real* islands, but all habitat patches surrounded by a dispersal barrier) appear completely isolated from one another. Forest birds on oceanic islands, freshwater fishes in separate stream systems, host-specific internal parasites, calcicolous snails on isolated limestone hills, and cavernicolous arthropods in different cave systems, to name but a few examples, are not expected to engage in gene flow. However, the fact that remote oceanic islands often contain sessile organisms belonging to groups that cannot fly nor swim, is living proof that dispersal is an almost universal biological possibility. A certain amount of gene flow may, thus, be expected between any pair of populations, however remote or isolated they may be.

Nevertheless, if gene flow is sufficiently low, genetic differentiation will still build up due to random genetic drift. Wright has shown that the cut-off point for this to happen is roughly one reproducing migrant exchanged per generation ($Nm = 1$, where $N$ = population size, and $m$ is the migration rate, the fraction of the reproducing population that disperses). More than this amount of gene flow will prevent genetic differences to appear between the two populations.

Thus, it is crucial to have information on migration between allopatric populations before we may be certain that we can infer allopatric speciation. For example, Finston and Peck (1995) found that $Nm$ between populations of *Stomion* beetles in the Galapagos archipelago could be as high as 3.7, despite the fact that these are wingless beetles, sampled from populations on different islands. Such studies show that gene flow may be high even among populations that seem isolated either by their inability to disperse or the insular nature of their habitat, or both.

## What is the role of stochastic effects?

Even in those situations where two populations are genetically sufficiently isolated, it is legitimate to ask whether stochasticity may be held responsible for the differences we see between supposedly allopatrically-speciated species. This question has been addressed over the past 45 years in a number of laboratory experiments, using species of *Drosophila* and houseflies in culture bottles and population cages (reviewed in Rice and Hostert 1993; see also Rundle *et al.* 1998; Meffert *et al.* 1999; Mooers *et al.* 1999). These experiments have looked at the evolution of **pre- or postzygotic isolation** in three different situations of random genetic divergence.

In the first situation, a large population was split into two or more large daughter populations and left to diverge for many generations. These experiments, which mimic the scenario of classic allopatric speciation, always gave negative results (i.e. they produced no evidence of reproductive isolation among daughter populations). In the second kind of experiments, inbred lines were produced from a parent population and reproductive isolation between different inbred lines was assessed after many generations. A few of these experiments, meant to represent situations of extreme genetic drift, gave positive results. The third group of experiments were of the "**founder-flush**" type: singly mated females, drawn from the same, large parent population were foundresses of large ("flush") daughter populations; from these daughter populations, singly mated females were drawn again, etc., resulting in several founder-flush cycles. Again, some of these studies showed reproductive isolation evolving in some of the populations, but in all cases this was of a transient nature and disappeared again after several generations. All in all, the experimental evidence that stochasticity may be responsible for allopatric speciation is considered weak (Rice and Hostert 1993).

## What is the role of natural selection?

Whereas neutral genetic drift, founder-flush and other bottleneck effects may not be very powerful in forcing two isolated populations apart, there is increasing evidence for a significant role for natural selection. A telling example is the experiment by Kilias *et al.* (1980). These researchers studied four large (> 1000 individuals) laboratory populations of *Drosophila melanogaster*. Two of these were kept under dry, dark, and cool conditions, the other two experienced a warm and humid environment and a day/night rhythm. After 5 years, pre-mating sexual isolation of 40 percent had built up between the populations from different environments (some of these populations are still in culture and maintain the same, if not stronger reproductive isolation [Kilias, personal communication]). However, the ones from the same environments never evolved any incompatibility. This result suggests that isolation alone is not important in speciation; **adaptation** is.

Similar data from the field can be interpreted in the same way. The Isthmus of Panama, for example, separated enormous populations of marine organisms into equally enormous Caribbean and Pacific populations. Many of these isolated populations have since diverged to the level of species (Lessios 1998). Given the size of these populations, **neutral evolution** may be excluded. Instead, adaptation to the many biotic and abiotic differences between Caribbean and Pacific (with regard to tide, depth, temperature, current stability, and the distribution of coral reefs) appear to have been the key. Similarly, the fact that many divergent species and subspecies of birds, mammals, and other organisms are found on islands (a fact that has traditionally been explained as the result of **drift** in small populations) may be explained with reference to the fact that island habitats are very different from mainland habitats. Islands are usually drier and windier than mainland, they have low biodiversity and simple ecosystems with fewer top predators. All these factors will exert special selection pressures on island inhabitants.

Finally, strong evidence for the role of natural selection in allopatric speciation has been obtained in studies of the threespine stickleback, *Gasterosteus aculeatus*, complex (Schluter and Nagel 1996). At the end of the Pleistocene, large-bodied marine populations of threespine stickleback that became isolated in postglacial lakes have evolved, in numerous independent cases, into small-bodied daughter-species. Here, parallel allopatric speciation has each time resulted in roughly the same suite of characters, which is strong evidence for adaptation to the ecological requirements from the freshwater environment. Moreover, since sticklebacks mate according to size, the small-bodied form has become reproductively isolated from the ancestral marine species.

### What is the role of sexual selection?

Many differences between allopatric sister species involve traits that play a role in reproduction. Island and mountain endemics in birds usually differ from one another in their male plumages (see, for example, Mayr's (1942) studies of birds of paradise in the islands off the coast of New Guinea), and many allopatric sister species in arthropods differ chiefly in the shape of their male genitalia (Eberhard 1985).

Until quite recently, such differences have often been interpreted as either the result of **founder effects** in small populations or as an accidental byproduct of adaptation to different environments (Mayr 1963). It is now becoming clear, however, that differences in sexual signals can accumulate rapidly in **allopatry**, as the result of sexually antagonistic selection and Fisherian sexual selection, either by themselves or modified by environmental interaction.

Computer simulation shows that sexual traits that are under Fisherian "runaway" sexual selection, continuously evolve in a chaotic and cyclic fashion. In allopatric populations, these cycles are likely to run out of phase quickly, resulting in different sets of sexual signals in each locality (Iwasa and Pomiankowski 1995). A similar divergence is likely to happen under an alternative, "chase-away" model of sexual selection (Holland and Rice 1998), where males and females coevolve antagonistically because of conflicting sexual interests (Arnqvist *et al.* 2000). At the same time, such cycles may be modified by interaction with the environment. For example, birds will be able to develop more striking colors on an island that has no visually searching predators (Zuk and Kolluru 1998), and more extreme development of sexually selected traits may be expected in places with high population density, where more encounters between sexes take place (e.g. Gage 1995). So, it is indeed quite likely that sexual traits strongly diverge in allopatric populations, as a result of sexual selection, modified by environmental interaction.

## Case studies

In the following paragraphs, we will give a few examples of cases of speciation that at first sight seem to conform with the classical model (i.e. fully isolated populations diverging stochastically). The organisms chosen are short-range endemic land snails on small habitat islands. Land snails are organisms that have almost no powers for active dispersal, which, especially for the habitat-specialists described below, should easily result in strict isolation of populations. Second, the differences between allopatric species (mostly shell and genitalia shape) have usually not been interpreted in an adaptive way or have been considered explicitly non-adaptive (Gould 1984; Gittenberger 1991; but see Eberhard 1985; Goodfriend 1986), and are, thus, potentially the results of random evolution. Although for many of the cases below relevant data are still lacking, we present them here to stimulate thought on alternative interpretations for the speciation processes that may have given rise to them, and hopefully to encourage further study into these systems.

### Ningbing ranges

The limestone hills of the Ningbing and Jeremiah Ranges lie on the eastern shores of the Ord River delta in tropical Western Australia. Squat outcrops rising up from savannah and scrub dotted with baobabs (*Adansonia*), they are "home to what is perhaps the greatest concentration of short range endemic species found anywhere in the world" (Solem 1988, p. 59). Twenty-six species of land snails in four genera are restricted to this area, all occupying very short, largely allopatric ranges. They belong to one family, the Camaenidae, a group that is unsurpassed among Australian snails for short-range endemism. The species differ mainly in shell shape and the genitalia.

The Ningbing Ranges are composed of three major blocks: northern (NNR), which is bisected by the narrow Utting Gap; central (CNR); and southern (SNR). The blocks are separated by alluvial plain and intermittent water courses. Two clusters of outcrops—the Gorge and the Pillars—lie between the CNR and SNR. The Jeremiah Hills are a scattered collection of small outcrops lying to the SE of the SNR, separated from them by 14 km of alluvial flats.

Almost all camaenid species here are allopatric. Some are restricted to tight clusters of outcrops, others extend across several such clusters. In a few cases, most notably in the genus *Ningbingia*, species are partially sympatric, with the edges of ranges overlapping. However, several congeneric pairs have parapatric distributions, demonstrating an abrupt transition from one to the other on the same "island." *Ningbingia res* and *N. australis* occupy the same island in the NNR, but are never sympatric. Similarly, *Cristilabrum primum* and *C. grossum* occur on an outcrop at the southern edge of the SNR, but their ranges do not overlap. In both cases, the transition from one species to the other takes place over a few meters.

### Central ranges—Red Heart

In the center of Australia, rows of sandstone hills run east–west across the desert, dissected by sporadically flowing waterways to form narrow gorges and gaps. Thirty-nine camaenid snail species are known from the central ranges around Alice Springs, out of a total of 83 from central Australia (Scott 1997). Most of these species are rock-dwellers.

By far the most diverse centralian genus is *Semotrachia*, a group of small snails with flattened shells and constricted apertures that adhere tightly to rock faces during estivation. They differ in shell shape and anatomy of the genitalia. Of the 25 described species, 19 species are found in the central ranges. Almost all species from the genus have restricted distributions, with several (e.g. *S. jessieana*, *S. emiliana*) known only from single gorges or gaps.

Allopatry is the common pattern in *Semotrachia* (Fig. 15.1). Only four instances of **sympatry** are known, two of which are overlaps between the widely distributed *S. setigera* and short-range endemics. Most species appear to occupy discrete areas, either gorges or cliff faces, along the ranges. All species occur on sandstone or quartzite, where they live in rock piles and the leaf litter that accumulates beneath the desert figs *Ficus platypoda*. The single exception is *Semotrachia euzyga*, which lives beneath granite boulders on hills around the Alice Springs Telegraph Station.

### Karst in Borneo

Limestone in Borneo comes in the form of spectacular outcrops, often of the "tower karst" type.

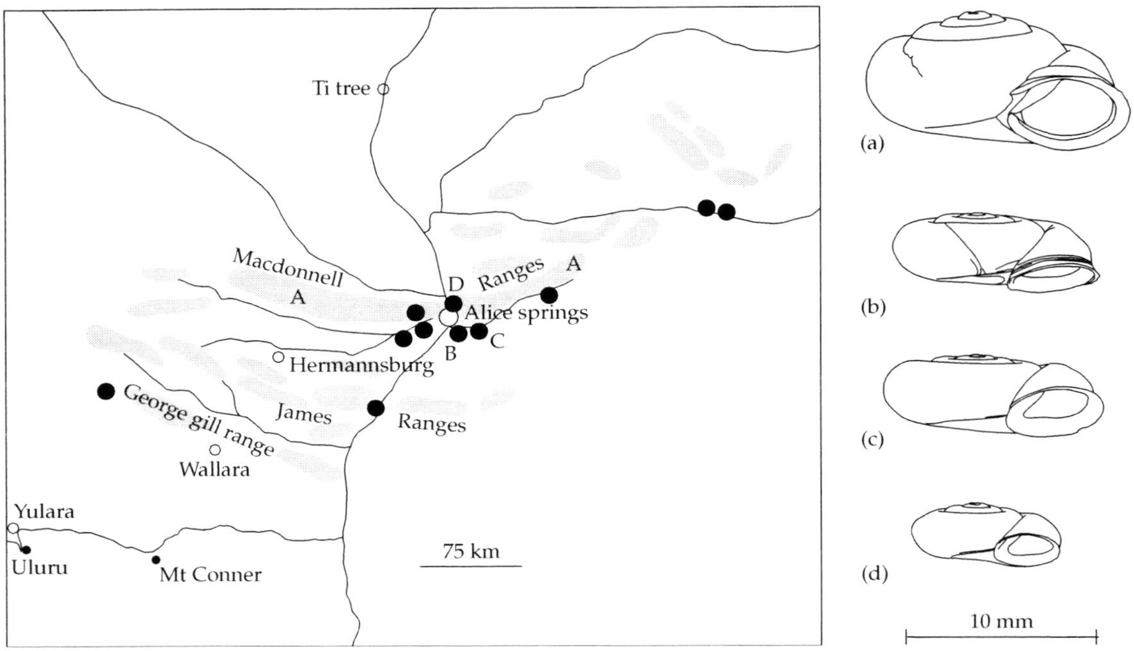


**Figure 15.1** Distributions of selected species of the endemic genus *Semotrachia* in central Australia. *Sermotrachia setigera* (a) is widespread in the MacDonnell Ranges. Solid circles represent the distribution of eleven species of short range endemics. (b) *S. jessieana*, Jessie Gap; (c) *S. emiliana*, Emily Gap; (d) *S. euzyga*, Alice Springs telegraph station. Shells redrawn from Solem (1993). Map based on Thompson (1991).

These hills are usually small (on average less than 1 km in diameter), isolated and widely scattered, with acidic, noncalcareous soils in between. Although some clusters of hills exist, belonging to the same limestone deposit, individual limestone hills are usually tens of kilometers apart and lenticular in origin, which means that they are outcrops of small isolated limestone sediments and have never been connected. The Malaysian states of Sabah and Sarawak alone, together occupying one-third of the island, have an estimated 300 hills, together covering less than 0.5 percent of the total land surface.

Limestone hills harbor a rich land snail fauna, both in terms of population densities and species diversities. This is generally believed to be the result of snails' requirements for high calcium concentration and low acidity, which are met in the alkaline, calcium-rich karst environment. The areas in between limestone hills, which usually are acidic and calcium-poor, are regarded as poor in snails, both in terms of abundance and diversity.

Certain groups, in particular Diplommatinidae, appear to be composed entirely of calcicolous species in Borneo. In fact, until recently, virtually no records away from limestone were available for these families. Hence, for these groups, limestone hills would be habitat islands, beyond which they cannot disperse. Consequently, endemism is high here. For example, out of the 158 known species of Bornean diplommatinids, 56 occur only on a single hill (Vermeulen, personal communication). Diversity in shell shape (anatomy has not been studied in many Bornean species yet) is great. This includes the presence and absence of radial ribs on the whorls, large flanges on radial ribs and on the aperture, and various orientations of the aperture. Figure 15.2 shows some of the more curious morphologies encountered in the Bornean representatives of this family.

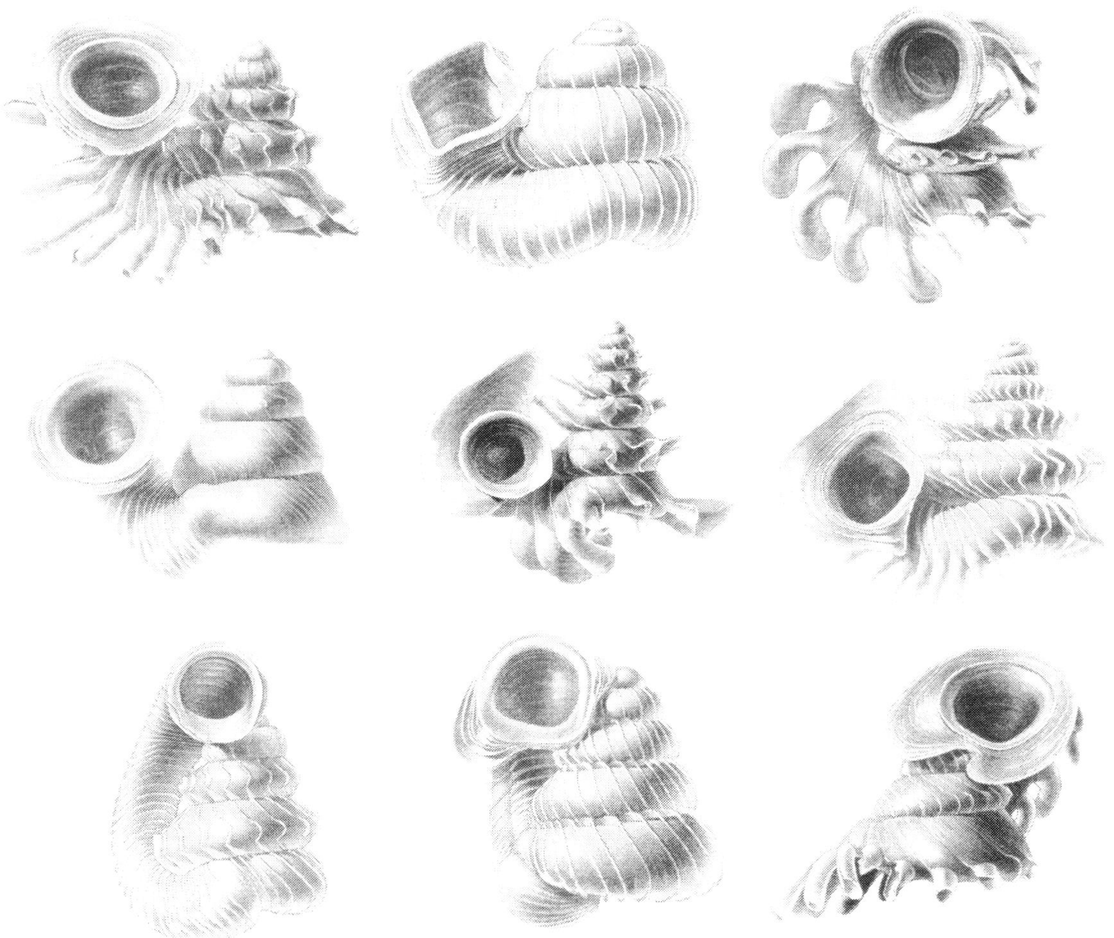

**Figure 15.2** The habitus of nine selected species of Bornean limestone diplommatinids, showing the wide range of shell shapes found in just one subgenus (*Plectostoma*). Drawings by J. J. Vermeulen.

## Discussion and conclusion

### Case studies

The three case studies presented above conform to the pattern expected under traditional allopatric speciation: full isolation on "islands," no gene flow, and random divergence as a result. The snail studies all share the same geographic setting: They are rock-dwelling species, living on rocky outcrops that are isolated from each other by inhospitable habitats. Consequently, they have diverged into large numbers of short-range endemics, differing in shell and genital morphology. However, a closer look at the cases reveals that things are not as simple as they seem.

To begin with, at first glance, the combination of a high diversity, allopatry, and a fragmented habitat, suggests causality. However, on close inspection, many of the Australian camaenids are actually not restricted to a single outcrop but either occupy only part of an outcrop or are distributed over several, which suggests that their distribution is dictated by other factors than just the patchiness of their habitat, and that dispersal between outcrops across the inhospitable alluvial plains must have taken place. In Borneo, something similar applies. A molecular study on an "obligate" rock-dwelling vertiginid from the Malay peninsula shows that populations on isolated outcrops conform to an isolation-by-distance population structure, which suggests that, in fact, populations must exist in between outcrops as well (Schilthuizen *et al.* 1999). Recent collections from nonrocky, noncalcareous habitats in Borneo support this: Almost twenty species of diplommatinid, which until now were considered to live exclusively on limestone outcrops, have already been found (at low densities) away from limestone (Schilthuizen, 2004).

All this suggests that if geographic barriers play a role in speciation in these groups, these may not be just the barriers that are obvious to the human eye. In the case of the Bornean diplommatinids, different species on different hills may even be interconnected by low-density populations, making the speciational setting parapatric, rather than allopatric. Second, sexual selection could have had a strong influence on the morphological disparity in Australian camaenids and Bornean Diplommatinidae. In the former group, most diversity is in the shape of the male genitalia. Even though pulmonate land snails are hermaphrodites, and hence are often considered to respond poorly to sexual selection (Darwin 1859), it has been shown that, as long as sperm remains an easily produced type of gamete, and sperm competition is high, sexual selection can strongly affect the reproductive systems of these animals as well (Eberhard 1985). In the Bornean diplommatinids, shell shape is highly divergent. In theory, these may be the result of **random drift** and founder effects after an initial colonization by a small number of **propagules**. However, if it is true that populations extend at low densities beyond the limits of the limestone outcrops, such colonizations by founders which then experience population explosions in isolation may not be a realistic scenario. Rather, selection may be invoked. Given that different species on the same limestone hill do not converge on the same shell shape and that no correlation with the environment is obvious, natural selection may not be the driving force. Instead, sexual selection on shell shape might be an unorthodox, yet plausible, proposition. Population densities on limestone hills can be extremely high, with hundreds of individuals on a single square meter. This greatly increases the number of potentially sexual encounters, which creates suitable conditions for sexual selection. Also, copulation in land snails is known to involve tactile contact between the foot of the one partner and the shell of the other. Under these circumstances, a role for the shell as a sexual signal is not fully improbable.

## Reevaluating allopatric speciation

We hope that, by highlighting the land snail case studies above, we have made it clear that, even in situations that at first sight appear to conform to an allopatric scenario, complications arise upon closer study. Gene flow between "islands" may well exist and should be measured either directly or indirectly. If found to be high, natural and/or sexual selection may be responsible for their morphological divergence rather than neutral evolution. Natural selection may be studied by transplantation experiments, or by studies of correlations between shell morphology and environmental parameters. A role for sexual selection may be made plausible by direct experimentation (mating success after manipulation of shell ornamentation, for example) or indirectly, by comparing the development of possibly sexually selected traits in species with different mating systems.

We advocate such a research program in all studies of allopatric radiation. Zero gene flow should not be assumed a priori, but should be tested by indirect methods with neutral molecular markers or by direct capture–mark–recapture studies. We predict that in many cases, substantial gene flow will be found between "allopatrically" diverged populations, which necessitates investigations into the selection pressures that have caused their divergence. It may well be that eventually the distinction between sympatric and allopatric speciation will become blurred. Rather, restriction of gene flow and strong selection (be it sexual or natural) may well be the common ingredients for all speciation processes.

We feel that inquisitive studies of sessile organisms in fragmented habitats, be they land snails on rocky outcrops, cave-dwelling arthropods or flightless aquatic insects in different watersheds, will help in reaching a more unified body of speciation theory.

We gratefully acknowledge Victoria University, Melbourne, for funding a visit by MS to the School of Life Sciences and Technology, which made the writing of this chapter possible. We also wish to thank two anonymous reviewers for their comments, which greatly improved the coherence of the text.

## Further reading

Howard, DJ and Berlocher, SH (1998). *Endless forms. Species and speciation*. Oxford University Press, Oxford.

Magurran, AE and May, RM (1999). *Evolution of biological diversity*. Oxford University Press, Oxford.

Schilthuizen, M (2001). *Frogs, flies, and dandelions. The making of species*. Oxford University Press, Oxford.

Schluter, D (2000). *The ecology of adaptive radiation*. Oxford University Press, Oxford.

Solem, A and Bruggen, AC van (1984). *World-wide snails. Biogeographical studies on non-marine Mollusca*. Brill/Backhuys, Leiden, The Netherlands.

CHAPTER 16

# Introgression and hybrid speciation via transposition

**Antonio Fontdevila**

Since the advent of the **biological species** concept (Dobzhansky 1937), defined as a set of interbreeding populations reproductively isolated from other similar sets, interspecific hybrids have been stigmatized as low **fitness** genotypes due to the supposedly highly effective reproductive isolation mechanisms inherent to true species. Consequently, hybrids were viewed as lineages devoid of evolutionary significance. Ironically, some early naturalists, including Linnaeus and later several neo-Darwinian evolutionists such as Anderson, Stebbins, and Grant, considered hybridization as a source of new species (Arnold 1997). Recently, thanks mostly to the use of genome-wide molecular markers, large numbers of unsuspected cases of interspecies crossing are being uncovered, some of them leading to well characterized species of hybrid origin (reviewed in Arnold 1997). In addition, some instances of natural **reticulate evolution**, where divergent species share genes due to interspecific crossing, have been experimentally reproduced, showing that genome reorganization after hybridization is rapid and allows for endogenous selection to overcome fertility barriers (Rieseberg *et al.* 1996). The purpose of this chapter is to analyze the possible causes of this reorganization and to propose a plausible scenario of hybrid evolution. My aim is not only to dispel the notion that hybrids are evolutionary dead ends, but also to summarize the evidence that hybrids are, in fact, a positive source of evolutionary innovations.

Mobile DNA sequences, collectively designated as **transposable elements** (TEs) or **transposons** due to their ability to disperse (transpose) copies of themselves throughout the genome, are putative agents of hybrid genome reorganization. There are many reasons why transposons can be characterized in this way. First, TEs comprise a large fraction of the eukaryotic genome. For example, TEs make up more than 40 percent of the genome of many plant species, maize being one of the most extreme cases with amounts ranging from 50 to 80 percent (Kumar and Bennetzen 1999). Their contribution to genome size is also significant in animals. In *Drosophila*, TEs make up 10–15 percent of DNA and in humans the Line 1 (*L1*) **retrotransposon** (retrotransposons are transposons that transpose via an RNA intermediate) amounts to 15 percent of our genome. Often, TEs are responsible for the mobilization (and amplification) of other DNA sequences that also contribute to enlarge genome size. As an example, *L1* retrotransposons are responsible for 25 percent of the human genome (Moran *et al.* 1999).

Second, the contribution of TEs to genome shaping is far more important than their mere repetitive ubiquity suggests (McDonald 1995). TE insertions are responsible for many gene mutations, ranging from 15 to 20 percent in mammals to more than 80 percent in *Drosophila*. Moreover, these insertion mutations often generate novel patterns of gene expression that contribute significantly to genome evolution. Many novel regulatory patterns are tissue specific and respond to signals associated with TE insertions. In support of the evolutionary role of TEs in genome organization, many regulatory TE insertions have been documented in *Drosophila*, *Antirrhinum*, maize, sea urchin, and humans (Kidwell and Lisch 1997), some of them of quite ancient origin (Britten 1996). Last, but not least, the importance of TEs in the shaping of genomes is documented by their role in generating large genome rearrangements. This is accomplished by recombination between **homologous** TE sequences located in different chromosomal sites. Products of this recombination are **duplications, deletions**, and **inversions**, as has been documented in artificial and natural conditions in yeast and *Drosophila* (see later).

The above paragraphs summarize the pervasive evidence that TE spreading in the genome is not only the explanation for the redundant, excess DNA, often referred to as "**parasitic**" or "**junk**" **DNA**, but also, and most relevant to evolution, a main causative agent responsible for profound and continuous genome building and repatterning. Most interesting, the abundant "footprints" that are left in the genome by TEs reveal that many TE insertions that are initially neutral for the host genome may evolve to new host functions at later evolutionary stages in a process known as **cooption** (see Kidwell and Lisch 2001, for a review). Interestingly, searching for the agents responsible for these footprints has revealed that TE dynamics is not steady through evolutionary time, often showing an alternation of periods of high **transposition** bursts with periods of low, background transposition, as documented in studies of the evolution of maize (SanMiguel *et al.* 1998). This chapter is devoted to describing how TEs may play a decisive role in hybrid evolution, focusing on the bursts of transposition that occur during hybridization, especially in relation to their putative implication in rapid genome reorganizations that may lead, eventually, to speciation.

## Mutation rate increases in interspecific hybrids: some previous case studies

Hybridization between highly differentiated populations followed by $F_2$ and successive generations increases variability due to segregation and recombination among genes that differed in the parental populations. However, pioneer investigators, such as Sturtevant (1939), noticed that such crosses could also induce mutations at a higher rate than expected, contributing to the observed enhanced variability. Sturtevant crossed two so-called "races" of *Drosophila pseudoobscura* and obtained, in backcrosses, mutation frequencies of about 9 percent of lethals and 0.5 percent of sex-linked visible mutants, two figures much higher than normal spontaneous mutation rates. In Sturtevant's own words, "there is, . . . , a persistent feeling that perhaps interracial crossing also induces the production of new mutations" (1939, p. 308). Later, these races were recognized as two different species (*D. pseudoobscura* and *D. persimilis*) partially isolated by male hybrid sterility. So, what Sturtevant actually witnessed was an increase of mutation rates in backcross interspecific hybrids. As far as I know, this is the best early documented case of mutation increase in hybrids, but similar experiments with other interspecific *Drosophila* hybrids have produced similar, though less clear-cut results. More recently, cases of increases in rates of chromosomal rearrangements in hybrids have been reported in *Nicotiana* species, *Caledia captiva* subspecies, and *Chironomus thummi* subspecies (see references in Labrador and Fontdevila 1994). Most relevant to the hypothesis being considered here is the relationship between production of new rearrangements and TE transposition observed in hybrids between two subspecies of *C. thummi* (Schmidt 1984). Evidence of increased rates of transposition in hybrids has been reported for the mariner element in *D. simulans*/*D. mauritiana* hybrids, but other experiments using *Drosophila* found no increases in transposition rates or the increases were limited to somatic cells (see references in Labrador and Fontdevila 1994).

Evidence of enhanced transposition in the latter experiments was obtained largely by indirect methods, based on observations of morphological reverse mutations due to TE excision. Direct analyses of transposition in interspecific hybrids by chromosomal ***in situ* hybridization** had not yet been systematically conducted. Perhaps the first well-documented case of increased transposition rate in species hybrids by direct observation of new insertions was reported by Labrador and Fontdevila (1994) in *D. koepferae*/*D. buzzatii* hybrids. Following is a brief summary of these experiments.

## A hybrid instability story in *Drosophila*

*Drosophila buzzatii* and *D. koepferae* are two sibling species coexisting in the arid zones of Bolivia and

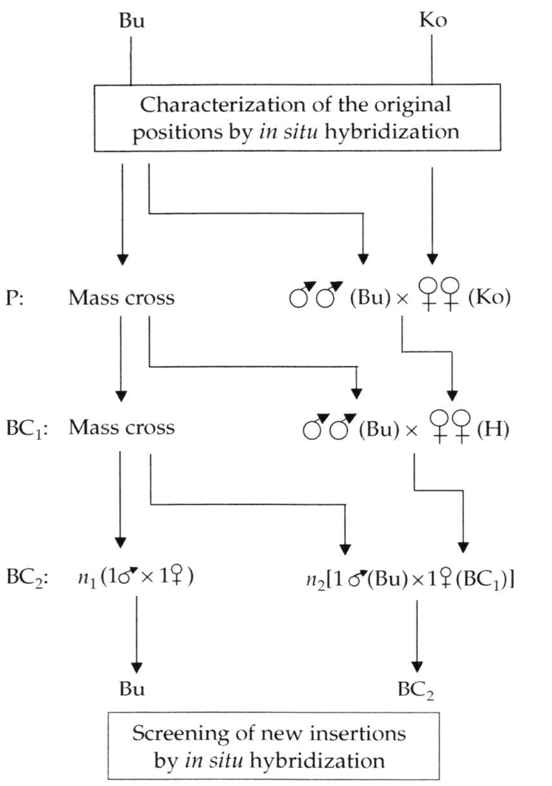


**Figure 16.1** Flow chart of crosses performed to detect *Osvaldo* transposition. One *D. buzzatii* line (Bu) and one *D. koepferae* line (Ko) are characterized for their *Osvaldo* occupation sites by polytene chromosome *in situ* hybridization. Thereafter, mass crossing between Bu males and Ko females (P generation) is performed, obtaining hybrid fertile females (H) and sterile males. The first backcross generation ($BC_1$) is obtained by mass crossing hybrid females (H) with Bu males. A second backcrossing ($BC_2$) is performed by ($n_2$) individual pair matings between one $BC_1$ female and one Bu male. Bu line is maintained by mass crossing in P and $BC_1$ generations and by ($n_1$) single pair matings in $BC_2$ generation. Samples of $BC_2$ offspring larvae per single-pair crosses in Bu line and backcross hybrids ($n_1$ and $n_2$, respectively) were analyzed to detect new insertions by comparison with the original positions. Due to low fertility and viability of $F_1$ (H) and $BC_1$ hybrid females, two backcrossing generations are required to obtain sufficient hybrid mothers to set up a statistically analyzable number of single pair matings (see Labrador *et al.* 1999, for details).

NW Argentina. They are able to hybridize, yielding sterile males and fertile females. Successive backcrosses of hybrid females to *D. buzzatii* males generate a collection of **introgressed hybrids**, that

**Table 16.1** Transposition rates (TR) of *Osvaldo* in *D. buzzatii* (Bu) and in hybrids between *D. buzzatii* and *D. koepferae* (Ko)

| Line/Hybrid | N | LNI | NI | TO | TR = NI/TO |
|---|---|---|---|---|---|
| Bu line | 301 | 15 | 36 | 4224 | $8.5 \times 10^{-3}$ |
| Hybrid with Ko-2.6 line | 163 | 15 | 29 | 1836 | $1.5 \times 10^{-2}$ |
| Hybrid with Ko-SL line | 174 | 39 | 87 | 2230 | $3.9 \times 10^{-2}$ |

N, LNI, and NI stand for the number of larvae examined, the number of larvae with new insertions, and the number of new insertions, respectively. Transposition rate (TR) is defined as the number of transpositions per element per generation. Transposition opportunities (TO) are the grand total of number of times that each element has passed through a chromosomal generation. Results of experiments with two *D. koepferae* lines are shown (adapted from Labrador *et al.* 1999).

is, individuals with a majority of *D. buzzatii* genome that carry genomic portions of *D. koepferae*. An interesting prior observation of Naveira and Fontdevila (1985) was that progenies of some of these introgressed hybrids display a number of new chromosomal rearrangements, indicating that interspecific hybridization induces an episode of genetic instability. Similarity between this burst of gross rearrangements and those observed in previous hybridization experiments (see above), suggested that TE transposition could also be enhanced in this interspecific hybridization, giving rise to periods of genetic instability of potential evolutionary significance (Fontdevila 1988). This prediction was confirmed by detailed studies with *Osvaldo*, a new retrotransposon isolated and characterized from *D. buzzatii* (Pantazidis *et al.* 1999).

Figure 16.1 shows the flow chart of crosses performed by Labrador *et al.* (1999) to quantify transposition rates in species and hybrid lines. The results presented in Table 16.1 show that transposition rates are one order of magnitude higher in hybrids than in parental species. These reports (Labrador and Fontdevila 1994; Labrador *et al.* 1999) are the first quantitatively documented evidence of transposition increase in interspecific hybrids. The strength of these results is based on two experimental approaches. First, insertions are detected by direct cytological methods, namely *in situ* hybridization on chromosome slides using an *Osvaldo* probe, and new insertions are deduced by comparisons between progeny (introgressed lines) and parental individuals (nonhybrid), in which original positions were previously characterized. Second, statistical treatment was performed by paired tests of homogeneity between nonhybrid lines and each of the hybrid lines, showing a highly significant heterogeneity in all cases. Labrador and Fontdevila (1994) showed that *Osvaldo* transposition occurs in bursts, as most of the larvae with new positions contained more than one transposition.

## Is there a general mechanism of transposon mobilization in hybrids?

A similar observation of transposon mobilization and hybrid instability, detected by chromosomal reorganizations, has been reported in the hybrid progeny between *Macropus eugenii* and *Wallabia bicolor*, two kangaroo species (O'Neill *et al.* 1998). These hybrid chromosomal reorganizations include extended **centromeres** resulting from amplification of sequences, and "de novo" rearrangements. Other marsupial hybrids (e.g. *Petrogale assimilis* × *P. inornata*) show similar chromosomal rearrangements. Simultaneously, the **methylation** status of parental and hybrid genomic DNA was compared and an elevated loss of methyl cytosine was found in hybrids. Hybrid unmethylated regions contain DNA sequences with significant **homology** to several genes of **retroviruses**, a class of viruses related to retrotransposons. This finding prompted the authors to conclude that these sequences derived from a novel retroviral element named KERV. Extensive chromosomal analysis showed that many highly repeated copies of KERV are present in the hybrid centromeres, confirming its implication in hybrid genome reorganizations.

Similar results have been obtained in plants, where DNA methylation and retrotransposition have been causally linked. Recently, a series of rice lines introgressed with genomic fractions from wild rice (*Zizania latifolia*) allowed Liu and Wendel (2000) to show the simultaneous presence of novel heritable morphological traits, undermethylation and regulatory changes of retrotransposition. Interestingly, an initial burst of activation of some retrotransposons is correlated with cytosine demethylation occurring within the first few generations following hybridization. These results not only suggest a causal link between hybridization, demethylation, transposition activation, and morphological novelties, but also a rapid host response to genome reorganization followed by an equally fast silencing of TEs.

Although these results are of utmost importance to understand the general mechanism of TE mobilization in hybrids, they are not applicable to some invertebrates, *Drosophila* in particular, because they do not have overall methylation. However, until recently the molecular link between DNA methyl groups and the nucleosome inactive chromatin configuration had not been deciphered (Razin 1998). Recent studies indicate that **deacetylation** may be the primary function in transcriptional repression. In *Drosophila* and yeast, protein factors bind to specific repressor sequences and serve to anchor to the DNA a repression complex responsible for the deacetylation of histones and the following gene repression. Repressor sequences are methylcytosine residues in stable mammalian repression, but they could be other specific DNA sequences that respond to physiological changes in repression. This evidence suggests that deacetylation may be a general mechanism of repression (see Labrador and Corces 1997 for a discussion of regulation of TE expression).

## Hybrid instabilities and reproductive isolation

One of the main consequences of hybridization is a reduction in fertility of the hybrids. This has been shown in interspecific hybrids as well as in crosses between *Drosophila* populations differing in the presence or absence of certain TEs such as *P*, *hobo*, *I*, and others. In fact, hybrid progenies of these population crosses show not only sterility, but also increased mutation (point and chromosomal), recombination, and segregation rates. This set of genetic abnormalities is known as **hybrid dysgenesis**. It has been demonstrated that increased rates of TE transposition are responsible for this syndrome (Berg and Howe 1989). Since its discovery, hybrid dysgenesis has been considered a possible mechanism of speciation for its contribution to reproductive isolation. The increased rate of new chromosomal rearrangements and their posterior differential **fixation** in different populations have also been claimed to contribute to reproductive isolation and chromosomal speciation (Fontdevila 1992). Yet, there are problems with this idea, mainly because TEs can quickly become established in all populations due to incomplete reproductive isolation and spread through hybrids to the whole species. Once a TE is present in all populations, hybrid dysgenesis ceases, as witnessed in the recent worldwide invasion of *D. melanogaster* by the *P* element.

In this chapter I am primarily concerned with genome reorganizations elicited by episodes of instability, mainly in hybridization, that may fuel evolution through speciation. Hybrid dysgenesis provides an initial episode of TE mobilization, but in order to achieve stable genome reorganizations additional population conditions must concur in the evolutionary scenario. Among the most important conditions is the presence of an environment favorable to hybrid **adaptation** through selection and a population structure that favors fixation of new gene reorganizations. It is doubtful that a generalist, cosmopolitan species with high **gene flow** among populations such as *D. melanogaster* will provide the appropriate set of natural conditions conducive for speciation. In what follows, examples with more specialized species, including other *Drosophila* species, are discussed in terms of the importance of TE mobilizations for speciation. Therefore, a brief account of the significance of hybridization to evolution is a useful preface to the ensuing discussion on the dynamics of genomic reorganizations in relation to speciation.

## Hybridization is not an evolutionary dead end

The evolutionary significance of hybrid-induced transposition depends on the ability of hybrids to survive and reproduce. Since interspecific hybrid inviability and/or sterility are taken as the ultimate proof of species integrity by reproductive isolation, hybridization between species has been considered an exception to the biological species concept. Yet, not all biologists have agreed to dismiss "hybridizing morphs" as real species. This is particularly true for botanists, who have repeatedly observed that plant species are able to hybridize in nature without losing their species integrity. Recently, new evidence of natural hybridization among animal species has been reported (see Box 16.1), reinforcing the view that hybridization is a more common event than expected (Arnold 1997). Moreover, detailed measurements of **fitness components** have demonstrated, in many cases, that the overall fitness of hybrids is not inferior to that in their parental species (Table 16.2). Sometimes the decisive fitness component is rather elusive, as in *Carpobrotus* hybrids, where hybrid recruiting superiority is due mainly to the higher resistance of hybrid seeds to the passage through the digestive gut of mammal herbivores that feed on this plant (Vilà *et al.* 2000). This case shows that decreased hybrid fertility does not guarantee that hybrids are less fit than parental species since fitness is the resultant of a set of components, fertility being just one of them, albeit an important one.

### Hybrid speciation: some case studies

The above considerations suggest that hybridization has the potential to fuel further evolutionary episodes that may lead to species formation. Recently, several cases of hybrid speciation have been studied in detail, mainly in plants but also in animals (Arnold 1997). For example, the plant *Iris nelsonni* was initially described as a hybrid species based on chromosomal characteristics and intermediate morphology with respect to the three parental species: *I. fulva*, *I. hexagona*, and *I. brevicaulis*. Later, molecular markers supported the "three way" hybrid origin of *I. nelsonii*. Actually, this hybrid species is the result of a great number of hybrid backcrosses (mainly to *I. fulva*) and shows a higher fitness in new transitional habitats (**ecotones**), as seen by its high viability at the late seed stage of development and other fitness components.

**Box 16.1 Natural hybridization and introgression in *Drosophila***

Although experimental hybridization is common in *Drosophila*, hybrids are rarely detected in nature. Kaneshiro (1990) reviews eight natural cases for which there is genetic information: *aldrichi* × *mulleri*, *montana* × *flavomontana*, *melanogaster* × *simulans*, *metzii* × *pellawae*, *pseudoobscura* × *persimilis*, *setosimentum* × *ochrobasis*, *heteroneura* × *silvestris*, and *malerkotliana* × *bipectinata*. Recently, another case of natural hybridization has been documented in a hybrid zone of the island of Sao Tomé between *D. yakuba* and *D. santomea* (Lachaise *et al*. 2000). Frequencies of hybrid individuals, either true hybrids or resulting from backcross hybridization processes, range from 1 to 2 percent. The hybrid origin is often assessed by mtDNA analysis, but also by segregation of diagnostic phenotypic, chromosomal and/or allozymic characters.

The endemic Hawaiian species, *D. heteroneura* and *D. silvestris*, hybridize in the island of Hawaii, where both species are **sympatric** for most of their range, and may represent a case of high rate of hybridization in *Drosophila*. The extent of hybridization has been detected using distinctive morphological features, such as head shape and abdomen coloration, which readily distinguish the two species. In light of the experimental results, Kaneshiro (1990) concludes that hybridization is frequent and may occur in both directions, its frequency depending on the relative abundance of the two sympatric populations.

When no species-specific genetic (nuclear) or morphological (phenotypic) markers are available, hybrid detection is difficult and mtDNA becomes the most reliable tool to discover introgressive processes. Studies with *D. pseudoobscura* and *D. persimilis*, two sibling species, have shown that both species share more mtDNA **haplotypes** in co-occurring (sympatric) populations than in any two geographically separated (**allopatric**) populations of *D. pseudoobscura* (Powell 1983). In this case, introgression seems to be very frequent and occur in both directions. Experimental studies by Hutter and Rand (1995) with population cages demonstrated that mtDNA is not neutral and fitness differences exist for combinations of nuclear and mtDNA of different origin.

Examples of introgression in *Drosophila* are, probably, not uncommon, but their detection is elusive because of lack of appropriate markers. In general, as more molecular markers become available for phylogenetic studies, more cases of reticulate evolution are revealed (Arnold 1997). The evolutionary history of *D. simulans*, *D. sechellia* and *D. mauritiana* illustrates how introgressive hybridization can explain the phylogenetic discordance between morphology and molecular markers (Solignac and Monerot 1986). This introgression occurred rapidly, most probably promoted by hybrid selective advantage similar to that found in the *pseudoobscura* case. In sum, the examples described here support the idea that even though detection of *Drosophila* hybrids in nature is difficult, in part because many hybridizing species (but not all) are morphologically very similar, introgressive hybridization may have played an important role in *Drosophila* evolution.

Many experiments with *Iris* hybrids have shown the presence of post-pollination barriers to hybrid progeny formation, making the production of $F_1$ hybrids very rare, but not impossible. Some of these rare hybrids show higher fitness in certain environments, allowing them to become established and outcompete the parental species. The extension of hybrid genotypes to new habitats is a common observation not only in plants but also in animals, such as in members of the genus *Geospiza* (Grant and Grant 1996a), making animal hybrid species formation a plausible event.

The genus *Helianthus* provides a thorough case study of hybrid speciation (Rieseberg and Noyes 1998). It comprises a series of self-incompatible annual plants (sunflowers) that live in a variety of soils, ranging from heavy clay (*H. annuus*) to dry–sandy (*H. petiolaris*) soils. These two species are **sympatric** and produce hybrid swarms, with semi-sterile $F_1$ hybrids ($<$ 10 percent pollen viability and

**Table 16.2** Hybrid fitness in some genera known to hybridize in nature. Fitness estimates of the hybrids are relative to both parents

| Genus | Fitness component measured in hybrids | Comparative hybrid fitness |
|---|---|---|
| Plants | | |
| *Quercus* | Fruit maturation | (L–E) |
| *Artemisia* | Developmental stability, herbivory | E |
| *Iris* | Shade tolerance | I (I–H) |
| | Viability of mature seeds | E, L |
| *Eucalyptus* | Reproductive value | I (I–H), L |
| *Carpobrotus* | Recruitment (seeds per plant) | I |
| | Seed germination (after gut passage) | H |
| Animals | | |
| *Hyla* | Developmental stability | E |
| *Sceloporus* | Chromosome segregation in males | E (E–L) |
| *Colaptes* | Clutch and brood size | E |
| *Geospiza* | Survivorship, recruitment, breeding success | H |
| *Allonemobius* | Survivorship | I (L–I) |
| *Mercenaria* | Survivorship | L, E (E–H) |
| *Notropis* | Survivorship | L (L–E) |
| *Bombina* | Viability | L, E |
| *Apis* | Metabolic capacity | L |
| *Gasterosteus* | Foraging efficiency | I |
| *Gambusia* | Development | H, I |

L: lowest fitness; I: intermediate; E: equivalent; H: highest fitness. Most common fitness is given followed by range of fitness values in parentheses; commas separate results from different studies. Adapted from Arnold (1997); data for *Carpobrotus* from Vilà *et al.* (2000).

$<$1 percent seed viability) and $F_2$ individuals showing a wide range of pollen viability (13–97 percent). Repeated hybridization allows for stability in these swarms and the possibility of further evolutionary progress. In fact, *H. anomalus* originated from hybridization between *H. annuus* and *H. petiolaris* and is endemic to xeric habitats within the range of the parental species, providing a good example of hybrid invasiveness of novel habitats. The hybrid origin of *anomalus* is based on multiple studies with molecular markers, including the observation that *anomalus* gene **linkage groups** are interspersed with loci from the two parents in a 50:50 ratio and that *Helianthus* phylogenies show reticulate evolution for ribosomal DNA (rDNA), combining variants from *annuus* and *petiolaris* in *anomalus*, but not for chloroplast DNA (cpDNA) (Rieseberg and Noyes 1998).

These cases of plant hybrid speciation do not rely on asexual or **polyploid** processes following hybridization and, consequently, they are designated as events of **homoploid speciation** to be distinguished from **alloploid speciation**, the formation of new species by chromosomal duplication in interspecific hybrids. In animals, hybrid speciation is related sometimes to asexual processes, but in several cases new hybrid species are **homoploid**, bisexual species (Arnold 1997). Such is the case of *Gila seminuda*, a species belonging to the Cyprinidae (minnow species), a fish family that shows relatively high levels of natural hybridization (11–17 percent). It is well established that *G. seminuda* originated through **introgressive hybridization** between *G. elegans* and *G. robusta*. The entire genus *Gila* is now recognized as evolving via reticulate rather than divergent processes and **introgression** seems to be continuing as evidenced by the extreme mitochondrial (mt) DNA similarity of some *Gila* species. There are other cases of animal hybrid speciation or introgression in *Drosophila* (see Box 16.1), *Sorex*, *Felis*, and *Canis* species. The case of *Canis* is noteworthy because of its putative recent origin. The red wolf (*C. rufus*) originated through introgressive hybridization between the gray wolf (*C. lupus*) and the coyote (*C. latrans*). Extant populations and museum specimens of red wolf contain subsamples of mtDNA and **microsatellites** found in either gray wolf or coyote. The extreme similarity of sequences supports a relatively recent origin of introgression, but does not rule out the ancient hybrid origin.

## Genetic changes after hybridization are rapid

Hybrid homoploid speciation is well established as the origin of many plant and animal species. However, not all hybridization events lead to the origin of new species; many produce either the reinforcement of the species' status or the merging of both parental species. Yet, an alternative outcome is the introgression of genetic material of one species into the other. Introgression is of great importance to the evolution of species, providing new opportunities to incorporate foreign genetic material and opening novel functions. Moreover, detailed analyses of hybrid genomes have shown that reorganizations are rapid after hybridization, leading to a new genomic repatterning that affects the general functions of the organism, namely the recovery of fertility and other fitness components. Interestingly, this rapid repatterning applies not only to homoploids, but also to **alloploids**, suggesting that hybridization must trigger the mechanism of genome reorganization.

### Polyploid repatterning after hybridization

Mapping studies indicate that polyploid species undergo extensive genomic change (Rieseberg and Noyes 1998). This change involves both gross karyotypic rearrangements and DNA sequence repatterning. The genus *Brassica*, descended from an ancestral hexaploid, provides an interesting example of genome reorganization. A minimum of 24 chromosomal rearrangements must be assumed to have taken place since the original **polyploidization** event in order to explain differences in gene order observed among *B. nigra*, *B. rapa*, and *B. oleracea*. Surprisingly, although **polyploidy** is, perhaps, the only universally accepted mechanism of hybrid speciation, little is known about the evolution of the polyploid genome after its formation. Yet, recent experiments with *Brassica* and other organisms are very illustrative in this respect. Following self-fertilization of **synthetic allotetraploids** between pairs of the above *Brassica* species for several generations, extensive genome changes, mostly involving loss and/or gain of parental and novel DNA fragments, were detected by comparing nuclear DNA probes between $F_2$ and $F_5$ progenies. This observation suggests that considerable genetic change accompanies alloploid speciation, but provides no evidence for the mechanisms responsible for these changes. However, studies with the recently formed **allotetraploid** cotton (*Gossypium barbadense*; AD genome) have shown that genomes A and D are similar in gene order in parental diploids but highly divergent in the tetraploid, and that some A-genome DNA dispersed repeats have spread to D-genome chromosomes. Four of the assayed repetitive DNA probes correspond to known transposons and at least 12 hybridize to messenger RNAs (mRNAs) expressed in cotton seedlings, suggesting transposition as the possible mechanism of spread (Zhao *et al.* 1998).

### Homoploid repatterning after hybridization

Rapid genomic reorganization has also been observed in homoploid species, *Helianthus* species being the most thoroughly studied case. The comparison of genetic maps for *H. anomalus* and its parental species, *H. annuus* and *H. petiolaris*, has revealed an extensive repatterning in the hybrid genome, requiring at least three chromosomal breakages, three fusions, and one duplication to explain the differences in gene order between hybrid and parental species. These changes could be responsible for reproductive isolation, adaptation to a novel habitat or increasing hybrid fertility. Riesberg *et al.* (1996) have investigated the dynamics of this new genomic design by mimicking natural hybridization in the laboratory. They were able to obtain three fertile and viable hybrid lineages after five generations of selfing and/or backcrossing, and the genomic composition of these synthetic hybrid lineages was studied using 197 *H. petiolaris* **RAPD markers**. The three lineages, despite their different crossing procedures, showed a high concordance in genomic composition, but, most surprisingly, this genomic composition was statistically concordant with that of the natural hybrid species, *H. anomalus*. This similarity suggests that genomic reorganization in the hybrid is not only rapid but also repeatable, most likely due to endogenous selection for gene blocks that enhance hybrid fertility rather than to ecological (exogenous) selection. The rapid increase in fertility observed in the synthetic lines, from 4 percent to more than 90 percent in only five generations, and the fact that breeding was performed under artificial conditions supports this conclusion. In summary, these results strongly suggest that the genomic organization of hybrid species may be basically fixed in just a few generations after hybridization and stay essentially static thereafter. As far as I know, no such detailed studies have been performed with other homoploid hybrid species, but it is reasonable to think that similar dynamics could be found elsewhere. In particular, the above account of transposon activation followed by rapid repression observed in introgressed rice lines (Liu and Wendel 2000) is in agreement with this hybrid homoploid repatterning process.

## What are the dynamics of genome reorganization?

The above examples illustrate how hybridization can fuel evolutionary processes by genome reorganization. Hybrid species are one example of genomes that show an intensive repatterning and contain a number of associated TE sequences. Several experiments show that this new design is achieved rapidly, and may occur in conjunction with bursts of transposition. This model contrasts, however, with the widespread idea that **synteny** among closely related genomes is almost perfect. Even in cases where a few rearrangements are sufficient to harmonize the maps, these maps are crude representations of genome structure and new reordinations emerge as resolution increases. Molecular markers are increasingly more refined and allow detection of gene reorganizations that were undetectable in the past. In addition, sequence divergence is used to estimate evolutionary times to reconstruct the dynamics of episodes of genome repatterning. This strategy has been successfully applied to the study of the evolutionary history of maize (SanMiguel *et al.* 1998).

Sorghum and maize *Adh1* regions are highly conserved in gene order and orientation, yet three sorghum genes are located elsewhere in the maize genome, suggesting events of transposition. The major difference between both homologous regions is the presence in maize of clusters of retrotransposons interspersed with low copy-number loci that are not present in sorghum. These clusters amount to over 160 kb of maize DNA and are responsible for the size difference between *Adh1* regions in the two species (about 240 kb in maize and 65 in sorghum). Most of these retrotransposons inserted during the last 3 million years in what can be described as a transposition burst. This episode does not coincide with the time of allotetraploid origin of maize 11 million years ago. However, studies like these are essential to understand the history of genome reorganization in many species.

An educated guess using all the available information indicates that genome evolution consists of a series of episodes of reorganization in which transposition plays a significant role. Some of these transposition bursts have been documented in relation to the examples of hybridization described above, but others have been related to different genomic and environmental stress events, of which cell culturing and microbial infections are best documented. Recently, Moran *et al.* (1999) have demonstrated that the *L1* retrotransposon is capable of transposing at a high frequency in human cultured cells. Interestingly, *L1* elements are able to mobilize non-*L1* sequences (**exons** and **promoters**) into existing genes. Since *L1*s make up about 15 percent of the human genome, their mobilization is of utmost importance to genome repatterning.

Transposon mobilization by stress—genomic or environmental—can be interpreted as an adaptation of TEs that ensures that a large number of elements are transmitted to those few hosts that manage to survive. From the perspective of the host, it will be beneficial to increase variability in periods of stress, fuelling further evolution as in the case of hybrid speciation. Nothing mystical exists about a putative selective advantage of those TEs that act in the benefit of the host by cooption for new host functions, as has been profusely documented (see earlier). Rapid genome reorganization by transposition bursts followed by immediate silencing of transposition allows to view host–TE relationships under a broad perspective ranging from extreme parasitism, such as in population invasions by TEs in hybrid dysgenesis, to perfect **mutualism**, related to TE insertion signaling (Kidwell and Lisch 2001).

## The scenario

Several natural scenarios for hybridization are possible, although **hybrid zones** between marginal populations are among the most favored. Hybrid zones are usually associated with ecotones or disturbed areas, marginal populations being a particular case of populations inhabiting an ecologically stressful environment. Under these conditions hybrid fitness relative to parental species is not necessarily reduced (see Table 16.2) and ecotone and disturbed areas provide an array of new open habitats to be occupied successfully by hybrids. Yet, $F_1$ hybrids are rare due to the multiple pre- and postfertilization barriers to the formation of the initial hybrid generation. In experiments with plants, negative hybrid effects are observed on seed production, pollen tube growth, and other fitness components (Arnold 1997). The rarity of $F_1$ hybrids diminishes the probability of their becoming established, but this may be overcome by the multiple opportunities of hybridization provided by contact zones. In fact, most natural hybrid populations consist of a few parental and $F_1$ individuals and an abundance of second-generation $F_2$ hybrids and backcross introgressed individuals. This suggests that $F_1$ hybrids are difficult to form, but also that when a few of them become established, new rounds of hybridization among hybrids and between hybrids and parental individuals are facilitated because barriers to hybridization are less strict in backcrossed and introgressed genotypes. This sets off a runaway process that leads to so-called **hybrid swarms**, commonly observed in plants and also in some animal species. In sum, the establishment of hybrids in ecotones and disturbed areas is facilitated by a two-way process that combines great opportunities for establishment in open newly created habitats with the possibility of repeated crosses and backcrosses among parental and hybrid genotypes.

## The role of selection

Hybrid evolution is dependent not only on adaptation (through exogenous selection) to the environmental array of habitats created in contact zones, but also on endogenous selection to maximize hybrid fitness. Arnold (1997) remarks, however, that this model contradicts the tension zone model championed by Barton and Hewitt (see references in Arnold 1997) in that endogenous purifying selection is not against "all" hybrids but only against certain sterile and/or unviable hybrids, favoring those hybrids that show high levels of fitness. This has been shown in hybrids between *Iris fulva* and *I. brevicaulis* where a high correlation exists between embryo inviability and the number of *I. fulva* genetic markers present in the progeny from *I. brevicaulis*-like maternal plants. This result suggests that endogenous selection is acting against intermediate hybrid individuals, that is, those that contain the highest number of alien genetic elements (Arnold 1997). Another example that reinforces the role of endogenous selection is the experiment performed by Riesberg *et al.* (1996) with *H. anomalus* (see above), in which similar linkage groups of genes are found in several artificial hybrid lines that show enhanced fertility relative to their parental species. This piece of evidence also emphasizes the rapid genome repatterning observed in just a few generations.

## The role of TEs

The scenario proposed here incorporates the mobilization of TEs that follows a hybridization event. As the examples reviewed above document, hybrids show increased rates of TE transposition. This process may also occur in natural hybrid zones, generating an increase in variability that facilitates the operation of endogenous selection to rapidly establish hybrid lineages. Rapidity is crucial to this establishment, since the original variability assisted by recombination may not be sufficient to overcome hybrid breakdown. This is particularly true in speciation by fixation of small chromosomal **underdominant rearrangements** that might mediate reproductive isolation and, eventually, speciation. The probability of fixation of these rare rearrangements is low in large populations under the conventional mutation rates, but greatly enhanced in small hybrid populations, where **genetic drift** acts as a powerful evolutionary force and transposition-mediated mutation and chromosomal change are significantly increased (Fontdevila 1992). Since most of the cases discussed above illustrate processes of rapid genome repatterning in homoploid and

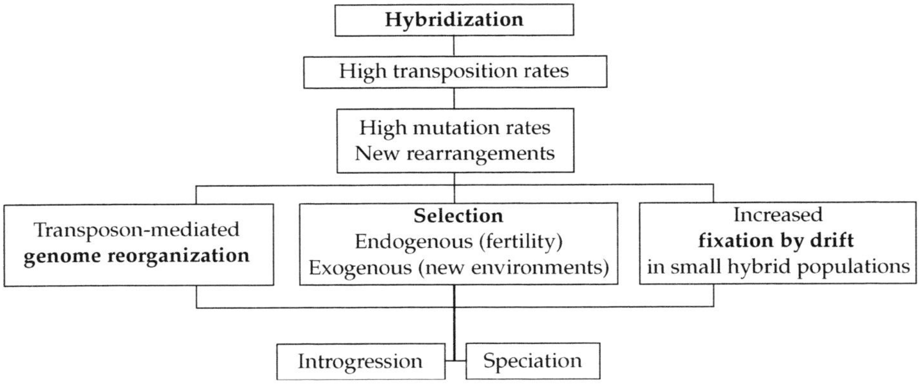


**Figure 16.2** Flow chart of the proposed scenario where hybrid genetic instability may promote introgression or speciation. The first (upper) part of the chart depicts how hybridization induces high transposition rates, which in turn produce new rearrangements and mutations at high rate. The middle part shows the three main genomic changes, namely, (1) the transposon-mediated genome reorganization is subjected to (2) selection for adaptation to new hybrid environments (exogenous or ecological selection) and for increasing levels of fertility and viability (endogenous selection), and (3) fixation by drift in small hybrid marginal populations. The synchronous occurrence of these evolutionary events leads either to introgression or to speciation (lower part).

alloploid species, they are in agreement with the model proposed here.

## Where to look for the scenario?

At face value, the proposed scenario (Fig. 16.2) requires that several episodes occur synchronically. Small marginal populations must hybridize in contact zones where new ecotone habitats are open to hybrids, and, simultaneously, hybrid instability must contribute with new genome repatternings to be selected upon. These are plausible events that occur quite often (see Box 16.2). One class of these events, well documented by ecologists interested in conservation, is the invasion of exotic plants and animals. A possible outcome of these invasions is the formation of hybrids showing a high invasibility that may outcompete the autochthonous species. Another well-known event is the breakdown of natural ecological barriers to hybridization in artificially disrupted habitats. This disturbance, often the result of human activities, promotes hybridization, and the establishment of a hybrid population that may eventually become a new species. Well-documented examples of these two classes of events abound (Abbot 1992). Vilà *et al.* (2000) have summarized the consequences of invasion by hybridization, with particular reference to the case of *Carpobrotus* hybrids in coastal California, which these authors provide as an example of high total fitness. Many hybrid species seem to originate through hybridization events similar to those described under artificial conditions and show, when artificially tested, a guild of new reorganizations in their hybrid genomes. This drastic repatterning is hardly conceivable under the normal rates of recombination and mutation, but seems more plausible if we assume that these rates are enhanced by transposition. Nevertheless, the ultimate proof of the model depends on the simultaneous observation of genome repatterning and TE mobilization bursts in newly formed hybrid species under the ecological setting proposed here. I believe that looking for this natural scenario is a worthwhile endeavor.

## Conclusions

The role of TEs as shapers of genome architecture is being increasingly documented, especially in their contribution to genome repatternings of regulatory importance and long-term evolutionary significance. These ideas were advanced by McClinctock in her early statement: "Since the types of genome restructuring induced by such elements know few limits, their extensive release, followed by stabilization, would give rise to new species or even new

## Box 16.2 The *Caledia* case: is there a link between hybrid instability and natural hybridization?

One example that combines natural observations of dramatic chromosomal reorganizations in clinal and hybrid populations with experimental rapid chromosomal changes of evolutionary value is the grasshopper *Caledia captiva*. This species, cited above as an example of hybrid instability, inhabits the northern and southern seaboards of Australia and comprises several chromosomal taxa due to its extraordinary chromosomal reorganization. Two of these taxa (the Torresian and Moreton subspecies) differ by many pericentric rearrangements and the presence or absence of sets of interstitial and terminal blocks of heterochromatin. The centromeres of Torresian populations are terminal in chromosomes (**acrocentric** and **telocentric**), but Moreton populations show a high variability of centromere localization, exhibiting a latitudinal cline ranging from northern populations with all chromosomes **metacentric**, to southern populations with all chromosomes telocentric, populations in mid-latitudes being intermediate in centromere localization. This karyotype variation in Moretons is not paralleled by the uniform genetic profile found among populations, as described by mtDNA and enzyme variation. This disparity between chromosomal and genetic variation suggested to Shaw (1994) that centromeres are rapidly evolving and chromosome reorganization may play a significant role in speciation. Interestingly, these natural observations can be mimicked, in part, by the chromosomal instability detected in experimental hybrids between both *Caledia* subspecies (Shaw *et al.* 1983). In nature, both subspecies form narrow hybrid zone in which more than 80 percent of individuals are introgressed hybrids. Shaw *et al.* (1983) observed that, although the $F_2$ generation is not viable, experimental backcrossed hybrids showed 50 percent viability, providing ample opportunities for habitat selection. Moreover, around 12 percent of these viables contained novel chromosomal rearrangements. Similar chromosomal rearrangements discovered in the hybrid zone suggest that increased hybrid chromosomal instability may occur in natural conditions as well, providing, again, "de novo" variability of evolutionary value.

In view of the available evidence on the role of TEs as basic components of heterochromatin (O'Neill *et al.* 1998; Dimitri and Junakovic 1999), the observed natural karyotype reorganization of *Caledia* could be explained by some kind of transposon-mediated hybrid instability in the contact zone between the two subspecies, followed by fixation and geographical expansion. The idea that heterochromatin is a genomic wasteland where TEs accumulate because they do not harm genic expression can no longer be sustained. Rather, there is increasing evidence that heterochromatin, centromeric or interstitial, is a preferential target for TEs. Once there, TEs acquire a functional role that may be subject to selection and contribute positively to evolution. They do this in many different ways, from maintaining the integrity of chromosomal ends (telomeres) to influencing the functional and structural roles of heterochromatin. Shaw and his collaborators (Shaw *et al.* 1993) have found a correlation between chromosomal pattern, latitude and developmental time, suggesting that selective adjustment of development to latitude is at the origin of clines in *Caledia*. They also argue in favor of centromere rearrangement as the causative agent of differences in developmental rate. In view of the direct implication of TEs in heterochromatin modification (O'Neill *et al.* 1998), hybrid instabilities in the *Caledia* hybrid zone could be at the base of the observed karyotype variation in centromeres. The idea that centromere repatterning is a consequence of activation of latent centromeres in the Moreton taxon is highly favored by Shaw *et al.* (1993) and is in accordance with the putative role of TEs in shaping the function of centromeric heterochromatin. Moreover, the putative selective value of transposon mediated heterochromatic rearrangement suggested by clinal karyotypic variation is being assessed in other cases as well. Thus, correlation between TE content and climatic differences in an altitudinal cline has been reported in barley (*Hordeum spontaneum*) populations (Kalendar *et al.* 2000), evidencing, once again, the influence of transposon-mediated genome reorganization in clinal adaptation and, eventually, in evolution.

genera" (1980, p. 17). Here, I have focused on the episodic occurrence of bursts of TE transposition in hybridization. These bursts may have a tremendous impact on the genome reorganization that accompanies significant evolutionary events such as introgression and hybrid speciation, and therefore deserve close scrutiny. Although we are still far from understanding what causes these bursts of transposition in hybrids, some putative mechanisms—such as chromatin **acetylation**—have been suggested here. Hybrid speciation and the ensuing genome repatterning are reviewed to emphasize that hybridization is not always a dead end, as the biological species concept might suggest, but a potential source of new hybrid genotypes (hybrid swarms) that may eventually establish themselves in new ecotone habitats as new species, isolated from their parental ancestors. Hybridization is accompanied by rapid genome repatterning, leading to genetic novelties upon which natural selection can act. The rapidity of this genome repatterning is hardly explained by conventional mutation and recombination rates, and suggests an involvement of the transposition bursts that accompany hybridization in rapid hybrid genome reorganizations. The proposed scenario is dependent upon the synchronous occurrence of several events. Contact zones between species allow small marginal populations of both species to merge in hybrid swarms where four simultaneous processes may be responsible for rapid differentiation, namely, (a) hybrid transposon bursts fuel genome reorganization by increasing levels of genome transpositions and point insertions, increasing genotype variability in swarms; (b) exogenous selection operating upon new hybrid genotypes allows the establishment of high fitness hybrid genotypes in novel ecotone and/or disturbed habitats; (c) endogenous selection favors reorganized genomes that show high levels of fertility and viability; and (d) small effective population size of the hybrid zone increases fixation by **drift** of those underdominant rearrangements that show high fitness in homozygous state. There is scattered experimental and natural evidence in support of all these processes, suggesting that the proposed scenario may be a realistic one. However, it will take more work to show that these processes act synchronously, as suggested here. I believe the game is worth the candle.

This work is inspired by many experiments conducted by my research group and endless discussions held with many colleagues and students. Among them, experiments performed by my former students Mariano Labrador, Ignacio Marín, Horacio Naveira, and Alexandros Pantazidis must be acknowledged. Scientific discussions with professors Christian Biémont, Pierre Capy, Esteban Hasson, John McDonald, Joan Modolell, and Mauro Santos have contributed to my enlightenment in this subject. Comments by two anonymous reviewers and the editors are also acknowledged. However, statements and ideas in this paper are my sole responsibility. This work has been possible by funding from the Spanish Ministerio de Ciencia y Tecnología and the Generalitat de Catalunya during the last 15 years.

# CHAPTER 17

# Cooperation and conflict during the unicellular–multicellular and prokaryotic–eukaryotic transitions

**Richard E. Michod and Aurora M. Nedelcu**

Individuals often associate in groups that, under certain conditions, may evolve into higher-level individuals. It is these conditions and this process of individuation of groups that we wish to understand. These groups may involve members of the same species or different species. For example, under certain conditions bacteria associate to form a fruiting body, amoebae associate to form a slug-like slime mold, solitary cells form a colonial group, normally solitary wasps breed cooperatively, birds associate to form a colony, and mammals form societies. Likewise, individuals of different species associate and form symbiotic associations; about 2000 million years ago, such an association evolved into the first mitochondriate eukaryotic cell. The basic problem in an evolutionary transition in individuality is to understand why and how a group of individuals becomes a new kind of individual, possessing heritable variation in **fitness** at a new level of organization.

Certain alliances and associations of individuals are more stable than others, yet not all associations qualify as groups. In groups, interactions occur that affect the fitnesses of both the individuals and the group. Groups are often defined by a group property, usually the group frequency of a phenotype (or some other property reflecting group composition). Groups exist when the fitness of individuals within the group is not a frequency dependent function of the membership of other such groups (Uyenoyama and Feldman 1984). Within a group, member fitness usually is a function of the composition of the group. Initially, group fitness is taken to be the average of the lower-level fitnesses of its members, but, as the evolutionary transition proceeds, group fitness becomes decoupled from the fitness of lower-level components. Witness, for example, colonies of eusocial insects or the cell groups that form organisms; in these cases, some group members have no individual fitness (sterile castes, somatic cells) yet this does not detract from the fitness of the group, indeed it is presumed to enhance it.

The essence of an evolutionary transition in individuality is that the lower-level individuals must as it were "relinquish" their "claim" to fitness, that is to flourish and multiply, in favor of the new higher-level unit. This transfer of fitness from lower to higher-levels occurs through the evolution of cooperation and mediators of conflict that restrict the opportunity for within-group change and enhance the opportunity for between-group change. Until, eventually, the group becomes a new evolutionary individual in the sense of generating heritable variation in fitness (at its level of organization) and being protected from the ravages of within-group change by **adaptations** that restrict the opportunity for defection (Michod 1999). Of course, no individual ever rids itself from the threat of change within, as evidenced by the numerous examples of conflict among different units of selection remaining in evolutionary individuals.

Cooperative interactions are a source of novelty and new functionality for the group. During evolutionary transitions, new higher-level evolutionary

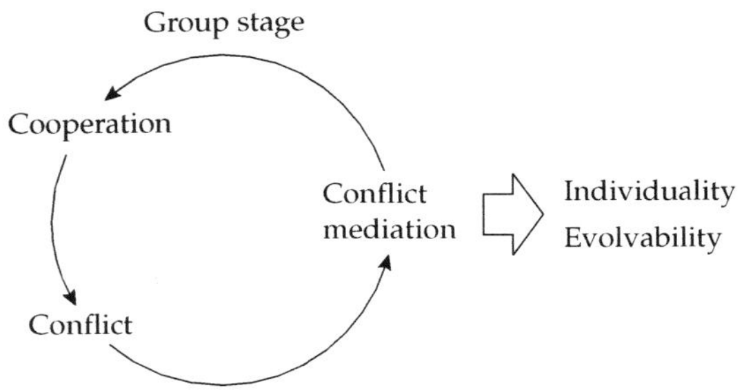


**Figure 17.1** Cooperation and conflict in evolutionary transitions. Stability of the group requires the mediation of conflict resulting from cooperative and conflictual interactions. Conflict may arise directly in response to cooperation as defection spreads within the group. Conflict mediation leads to further increases in cooperation and individuality at the group level. Continued evolvability of the new higher-level unit is fueled by new modes of cooperation and new ways to mediate conflict among component entities leading to new adaptations at the higher level.

units (e.g. multicellular organisms, mitochondriate eukaryotic cells) gain their emergent properties by virtue of the *interactions* among lower-level units (e.g. cells). Cooperation is fundamental to the origin of a new higher-level unit of fitness because cooperation trades fitness from a lower-level (the costs of cooperation) to the higher-level (the benefits for the group) (Michod 1999).

Although eventually lower-level units must cooperate in the formation of a new higher-level unit, initially the fitness interactions within the group may be based on any form of ecological interaction ranging from beneficial interactions, such as **mutualism**, to antagonistic forms such as competition and exploitation (predation, parasitism, pathogenism, slavery) as discussed later (Fig. 17.4; see van Ham *et al.*, Chapter 9). Nevertheless, both mutualism and exploitation involve conflict; exploitation because of its very nature, mutualism because, as do all cooperative types of interaction, it creates the opportunity for defection (Fig. 17.1). Fundamental to the emergence of a new higher-level unit is the mediation of this conflict among lower-level units in favor of the higher-level unit resulting in enhanced cooperation among the lower-level units.

Before the group becomes an individual, cooperation creates conflict and the temptation for defection. This quarrel among units of selection reduces the scope for cooperative interactions and higher-level functions. **Evolvability** (used here to mean the capacity to evolve into more complex forms) of the emerging higher-level unit depends on the invention of new and more intricate forms of cooperation which provide the basis for new adaptations at the higher-level. Conflict mediation leads to enhanced individuality and **heritability** of fitness at the new level. Continued evolvability requires the resolution of this conflict in favor of the higher level so that the continued cooperation so necessary for adaptation is not constantly threatened by conflict within. In the case of multicellular groups, conflict mediation may involve the spread of conflict modifiers producing self-policing, germ line sequestration, or apoptotic responses (see below). In the case of organelle (i.e. mitochondria and chloroplasts) containing eukaryotic cells, conflict mediation may involve the uniparental transmission of organelles.

Until the emergence of the new level is complete (say with the evolution of a structure to "house" the new higher-level unit), interactions among lower-level units are likely to be density and/or frequency-dependent; therefore, there will be problems with rarity, advantages to commonness, and, the constant threat of defection. One of the most basic consequences of frequency-dependent natural selection is that there need not be any benefit for the individuals or the group. In the language of population genetics, the average fitness of the population need not increase under frequency-dependent selection (Wright 1969; Michod 1999). The well-known **Prisoner's Dilemma** game illustrates well the inherent limits of frequency-dependent selection in terms of maintaining the well-being and evolvability of evolutionary units (Michod *et al.* 2003). Natural selection not only fails to maximize the fitness of individuals in the Prisoner's Dilemma game, it minimizes it.

The dilemma of frequency-dependent selection is that while frequency-dependent interactions among members of the group are the basis of higher-level group functions, frequency-dependent selection does not necessarily increase group fitness (Michod 1999). How can frequency-dependent interactions be the basis of higher-level units but not lead to the increase of fitness of those units? This paradox of frequency dependence is the basic problem that must be solved by multilevel selection, both during evolution within a species and during the transition to a new higher-level unit of organization.

## Cooperation

We see the formation of cooperative interactions among lower-level units as the *sine qua non* of evolutionary transitions, even if the groups initially form for exploitative reasons (as may have likely been the case with the origin of the first mitochondriate eukaryotic cell as we discuss below). For this reason we have paid special attention to the evolution of cooperation within groups. As Lewontin (1970) pointed out, a levels-of-selection perspective follows naturally from Darwin's theory of natural selection; however, the role of cooperation in the history of life has been less well appreciated. Thirty years ago, the study of cooperation received far less attention than the other forms of ecological interaction (competition, predation, and parasitism). Scholars generally viewed cooperation to be of limited interest, of special relevance to certain groups of organisms to be sure—the social insects, birds, our own species, and our primate relatives—but not of general significance to life on earth. All that has changed with the study of evolutionary transitions and the appreciation of the importance of population structure in evolution and that selection is usually a multilevel process. What began as the study of animal social behavior some 30 years ago, has now embraced the study of interactions at all biological levels. Instead of being seen as a special characteristic clustered in certain groups of social animals, cooperation is now seen as the primary creative force behind ever greater levels of complexity and organization in all of biology.

As already mentioned, cooperation is special as a form of interaction because it trades fitness from lower- to higher-levels. Cooperation creates new levels of fitness by increasing the fitness of the group, and, if costly at the lower level, by trading fitness from a lower level to a higher level. As cooperation creates new levels of fitness, it creates the opportunity for conflict between levels as deleterious mutants arise and spread. As discussed further, adaptations that restrict the opportunity for conflict between higher and lower levels (what we term **conflict modifiers**) are instrumental in the conversion of the group into a new evolutionary individual. Here, we consider the conversion of two different kinds of cell-groups. In the case of the origin of the first mitochondriate eukaryotic cell (as a symbiotic unit), the cell-group was composed of cells from different species; in the case of the origin of multicellular organisms, the cell-groups are composed of cells belonging to the same species.

Cooperation may be additive (in terms of the cost and benefit as is often assumed in models of altruism) or synergistic. Synergistic forms of cooperation benefit both the cell and the cell-group. In the case of synergistic cooperation, there is no obvious conflict between levels (at least in terms of how cooperation is defined), but if the loss of cooperation harms the higher level more than the lower level, modifiers still evolve that increase the heritability of fitness and evolvability of the group (see fig. 6 of Michod and Roze 2001).

The benefits of cooperative interactions usually depend upon the frequency with which they occur, while the costs of performing a cooperative behavior are usually an inherent property of the behavior itself, not depending on its frequency in the population or group. To the extent that cooperators are frequent in the population, it may pay a particular individual to forgo providing benefits, thereby reaping the benefits bestowed by others while not paying the cost. For these reasons, so long as selection is frequency-dependent, there is always a "temptation" to defect, that is, not help others, and so gain an advantage within the population relative to cooperators.

While it may be easy to agree on the basic role played by cooperation in the diversification of life, cooperation remains a difficult interaction to understand and to model especially when considering the different settings involved in the origin of multicellularity and the origin of first mitochondriate eukaryotic cell. In Tables 17.1 and 17.2, we further discuss the issues introduced previously (Michod and Roze 2001). In Table 17.1, we consider the

**Table 17.1** Issues and contrasts in understanding cooperation

| | Cooperators | | |
|---|---|---|---|
| | **One kind of cooperator** | **More than one kind** | |
| *Level of Interaction* | *Within Species* Behavioral structure. Kin selection models. Cooperate/defect payoff matrix. | *Within species* | Behavioral structure. Models: hypercycle, stochastic corrector |
| | | *Between species* | Behavioral structure. Mutualisms. |
| *Competition* | Is the population viscous so that cooperators compete? Competition reduces advantage of cooperation in viscous populations | Is competitive exclusion possible? Competitive exclusion is usually not possible in the between species case (due to some prior niche displacement), but is possible in within species case (e.g. hypercycle) | |

References: behavioral structure (Michod and Sanderson 1985), viscous populations (Hamilton 1971; Goodnight 1992; Taylor 1992; Wilson *et al.* 1992; Queller 1994; van Baalen and Rand 1998), hypercycle model (Eigen and Schuster 1977, 1978a, 1978b, 1979; Michod 1983, 1999; Frank 1995, 1997), stochastic corrector model (Maynard Smith and Szathmáry 1995; Grey *et al.* 1995).

**Table 17.2** Three contrasts in understanding the benefits of cooperation: additive versus synergistic, exchangeable versus nonexchangeable, and immediate versus delayed

| Benefits | |
|---|---|
| *Additive*<br>Cheating possible | *Synergistic*<br>Cheating *may* not be possible |
| *Exchangeable*<br>Sculling games<br>Sharing<br>Economics of scale | *Nonexchangeable*<br>Rowing games<br>Functional differentiation, reproductive versus nonreproductive specialization |
| *Immediate* | *Delayed*<br>Reciprocation, reciprocal altruism |

Synergistic benefits may be exchangeable or not, immediate or delayed, reproductive or nonreproductive. Likewise with the additive case. References: sculling and rowing games (Maynard Smith and Szathmáry 1995); economics of scale (Queller 1997); reciprocation (Trivers 1971, 1985; Brown *et al.* 1982).

number of different kinds of cooperators, the potential for competition among cooperators and whether the cooperation occurs within or between species. When there is just one kind of cooperator (a single cooperative genotype), the cooperators must belong to the same species; when there are more than one kind of cooperator, the cooperators may belong to the same or different species. The latter situation applies to the origin of the mitochondria-containing eukaryotic cell, while the former situation of a single kind of cooperator is more applicable to the origin of multicellularity. The study of cooperation is often divided by the issue of whether the interactions occur within or between species, because kin selection is possible in the former but not the latter. However, both within and between species cooperation requires spatial and or temporal correlations in the behavior of cooperating individuals. That is to say, there must be **behavioral structure**, that is, structure in the distribution of behaviors (Michod and Sanderson 1985). In the case of within-species interactions, genetic structure may facilitate behavioral structure and this is the basis of kin selection. Because of the need for behavioral structure, competition may also occur among members of cooperative groups and this may reduce the advantages of cooperation and/or lead to the loss of cooperative types. The **hypercycle** is a cooperative group of interacting replicators in which cooperation dynamically stabilizes the densities of the different replicators, thereby resisting competitive exclusion of any of the members (Eigen 1971; Eigen and Schuster 1977, 1978a,b, 1979).

The number of different kinds of cooperators also affects how cooperation is modeled. When there is just a single kind of cooperator, game theoretic payoff matrices are often used to conceptualize the interaction, as in the well studied Prisoner's Dilemma game. The payoff matrix approach can be extended to interactions between two species (Law 1991; Maynard Smith and Szathmáry 1995). When there are multiple members involved in the interaction, different approaches are used such as the hypercycle or stochastic corrector models. The **stochastic corrector** model considers a population of groups of hypercycles containing different numbers of members in a multi-level selection framework (references given in the legend to Table 17.1).

Table 17.2 discusses another major issue in the study of cooperation, the nature of the benefits bestowed by cooperators. A fundamental question is whether cheating (obtaining the benefits of cooperation without paying the costs) is possible. **Synergism** occurs when benefits received from cooperation require the benefactor to participate in the interaction. In other words, it is not possible for an individual to receive the (synergistic) benefits of cooperative acts of others without itself cooperating; defection or cheating is either disadvantageous or not possible. Some of the scenarios for the origin of the eukaryotic cell assume that cooperation is synergistic (López-García and Moreira 1999) as does the explanation for the evolution of cooperation among unrelated ant foundresses (Strassmann and Bernasconi 1999).

Synergism requires nonlinearities in the contribution to fitness of each partner's behavior. If we were to let variables $X$ and $Y$ be the cooperative propensity of each partner, under an additive model of cooperation, fitness of each partner would be a linear function of these propensities. Cheating is possible for linear models, because one individual could have zero propensity to cooperate but still benefit from the cooperative acts of its partner. If we wanted there to be no benefits unless both partners cooperated, we might let each partner gain proportional to the *product* of their cooperative propensities, giving a nonlinear fitness function. If one partner did not cooperate, neither would receive any benefits. Of course, other more realistic functions are possible, our main point is that synergism requires nonlinear models of the fitness effects of the interaction.

A problem with synergism alone as a scenario for the origin of cooperation is that it has difficulty explaining how cooperation gets started in a population of noncooperators. If there is one kind of cooperator, say $C$, interacting with defectors, $D$, we may model the interaction in terms of the familiar payoff matrix

| | $C$ | $D$ |
|---|---|---|
| $C$ | $a$ | $b$ |
| $D$ | $c$ | $d$ |

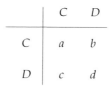

with the understanding that the elements $a, b, c, d$ give the fitness of the strategy on the left when interacting with the strategy on the top. If $a > c$, we say there is synergism (Maynard Smith 1998), cooperation is stable, and cheating is not possible when cooperation is established in the population. However, even in this case ($a > c$), if cooperators pay a cost when their partner is not cooperating, $b < d$, cooperation cannot invade when rare, because most of their interactions are with defectors. One way around this problem is to assume that cooperation is neutral when associated with defection, or $b = d$. Explaining the origin of cooperation is a special virtue of kin selection. Kinship among individuals provides the requisite behavioral structure locally (say, within families), and cooperation can increase (because cooperators tend to be concentrated in certain families), even though cooperators are rare in the global population.

Another important issue in understanding cooperation is whether the benefits contributed by different cooperators are similar or different in kind (Queller 1997). This relates to the issue in Table 17.1 concerning the kinds of cooperation. Sharing food is an example where the cooperating members provide similar benefits that are exchangeable. In contrast, role specialization in the castes of a termite colony, or cell and tissue specialization in a multicellular organism, are both situations where the cooperators provide different kinds of benefits, and, hence, one kind of benefit cannot be exchanged for another. The separation of reproductive functions between germ and soma is another example of non-exchangeable benefits. The distinction made by Maynard Smith and Szathmáry (1995) between rowing and sculling games expresses a similar issue. In **rowing games**, the oarsmen row on different sides of the boat (and so provide different and nonexchangeable functions). In **sculling games**, each oarsmen rows on both sides simultaneously (and so provide similar and exchangeable functions). The distinction is important, because cheating is much more costly in rowing games than in sculling games. In both kinds of games, the cooperators are in the same boat, which is another way of saying that there must be spatial and temporal correlations, that is, behavioral structure.

Synergism may occur between functionally similar (sharing food, sculling games) or dissimilar members (rowing games, interspecies mutualisms). Synergism among functionally similar members must come from the "economics of scale" (Queller 1997). Alliances of similar members must draw their (synergistic) benefits from the numbers of these members, that is, scale. For example, larger things are less likely to be eaten than smaller things (everything else being equal) (Stanley 1973; Shikano *et al.* 1990; Gillott *et al.* 1993; Boraas *et al.* 1998), and this may be one of the advantages to forming groups early during the evolution of multicellularity.

Kin selection operates between genetically similar members. Whether genetically similar individuals are functionally similar depends upon development and differentiation. Indeed, developmental differentiation of cells derived from a single zygote or spore attains the best of both worlds, it achieves the benefits of functional complementation (which will lead to synergism) along with the security of kin selection (security with regard to the spread of selfish mutants). Cooperation is one possible consequence of genetic relatedness among interactants; in addition heightened resource competition and inbreeding may occur. Because genetically related individuals are phenotypically similar, their ecological requirements overlap and competition increases, as in **viscous populations** in which the interactions occur nonrandomly and locally. The benefits of cooperation must overcome these costs of increased competition if cooperation is to spread in viscous populations (Hamilton 1971; Goodnight 1992; Taylor 1992; Wilson *et al.* 1992; Queller 1994; van Baalen and Rand 1998). The theory of sib-competition for the advantage of sex is based on the idea that sexually created variation among offspring helps create dissimilarity and avoid competition. As a result of this variation more offspring survive to reproduce. Depending on the mating system and population structure, genetic relatedness among interactants may also imply genetic relatedness among mates, that is inbreeding (Hamilton 1972; Michod 1979, 1991).

## Origin of multicellular organisms

We illustrate our approach of cooperation, conflict, and conflict mediation, by considering first the origin of multicellular organisms and, in the next section, the origin of the first mitochondriate eukaryotic cell. A multilevel selection approach to evolutionary transitions in individuality begins by partitioning the total change in frequency of phenotypes of lower-level units (and their underlying genes) into within and between-group components.

### Model framework

During the transition from single cells to multicellular organisms, we assume that cells belonging to the same species form groups composed of $N$ cells as in Fig. 17.2 (Michod 1999; Michod and Roze 1999,

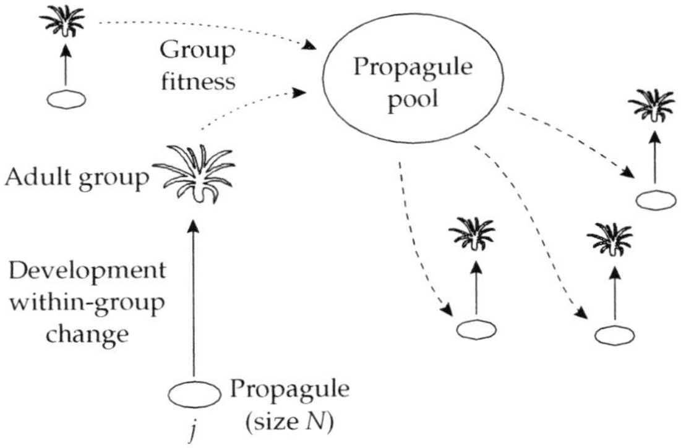


**Figure 17.2** Multilevel framework for the origin of multicellular organisms. The subscript $j$ refers to the number of *cooperating* cells in a propagule; $j = 0, 1, 2, \ldots, N$, where $N$ is the total number of cells in the offspring propagule group, assumed constant for simplicity. The variable $k_j$ refers to the total number of cells at the adult stage of propagules that start out with $j$ cooperating cells. The variable $W_j$ is the fitness of group $j$, defined as the expected number of propagules produced by the group, assumed to depend both on size of the adult group after development and its functionality (or level of cooperation among its component cells) represented by parameter $\beta$ in the model discussed in the text.

2000). The level of kinship among **propagule** cells is determined by the size of the founding group, $N$, and the way in which the group is formed, whether by fragmentation or aggregation and whether sex is involved. During development, cells replicate and die (possibly at different rates depending on cell behavior) to create the adult cell-group. Deleterious mutation may occur during cell division leading to the loss of cooperative cell functions and a decrease in fitness of the adult group. The adult group produces offspring groups of the next generation. We have considered several different modes of reproduction according to how the propagule offspring group is produced: single cell (or spore reproduction), fragmentation, and aggregation (Michod and Roze 2000; Roze and Michod 2001).

Because there are two levels of selection, the cell and the cell-group, there is the opportunity for both *within* and *between-group* selection. Fitness at the cell level involves the rates of cell division and cell death, these in turn depend upon cell behavior. We consider two kinds of cell behavior, cooperation, and defection. Most of our previous work has assumed genetic control of cell behavior, specifically we have assumed that cell behavior is controlled by a single genetic locus with two alleles, $C$ and $D$, for cooperation and defection, respectively. Alternatively, we may assume parental (or spore) control of cell behavior, in this case cell behavior is determined by the genotype of the mother cell. The latter assumption is made in parental manipulation models for the evolution of altruism. As is well known in the theory of kin selection, it is easier for costly forms of cooperation to spread under parental manipulation than under sibling control of the altruistic behavior (Michod 1982).

In the case of genetic control of cell behavior, cell behavior depends upon the cell's genotype, and mutations during development may disrupt cooperative cell functions and harm the group. Uniformly deleterious mutations are assumed to be disadvantageous at both the cell and group levels, while selfish mutations are assumed to be advantageous for the cell and disadvantageous for group. Fitness at cell-group or organism level depends upon the number of propagules produced, which, in turn, depends upon adult size and the level of cooperation among cells. The basic parameters of the model include development time, $t$, the within organism mutation rate per cell division, $\mu$, the effect of mutation on the cell replication rate, $b$ ($b < 1$ or $> 1$ means uniformly deleterious or selfish mutations, respectively), the benefit of cooperation for the group or organism, $\beta$ assumed $> 1$, and the propagule size, $N$. In addition, there is a parameter that tunes the relative effect of group size on fitness. Using this model we have studied the levels of cooperation maintained in populations and the partitioning of fitness among the cell and group levels (Michod 1997, 1999).

Mutation occurs during development and leads to the loss of cooperative group functions (loss of the cooperative benefit $\beta$ at the group level with effect $b$ at the cell level). In our studies of genetic mutations we use a genome wide mutation rate per cell division similar to that in extant microbes of $\mu = 3 \times 10^{-3}$ (Drake 1974, 1991), even though the more relevant rate is that of the primitive single celled ancestors of multicellular organisms. It is likely that the mutation rate has been lowered in modern microbes as a result of the very forces under study in our models. By this, we mean that under most conditions, the model predicts that it is advantageous to lower the mutation rate so as to reduce the scope for selection within-groups and increase the heritability of fitness at the group level. We use this genome wide mutation rate for the single $C/D$ locus, as we imagine this locus to represent all the cooperative functions in the genome. Of course, this is not realistic and we have extended our treatment of mutation using more realistic models based on a random infinite alleles model and the Luria Delbrück distribution (Michod and Roze 2000; Roze and Michod 2001). **Epigenetic mutations** are also likely to be frequent and important in the origin of multicellularity but we have not yet studied them.

## Conflict mediation and programmed cell death

To study how evolution may shape development and the opportunity for selection at the two levels of organization, the cell and cell-group, we assume a second modifier locus that affects the parameters of development and/or selection at the primary cooperate/defect (*C/D*) locus. The evolution of these conflict mediators are the first emergent functions that serve to turn the group into a new higher-level individual. The modifier locus has two alleles (*M* and *m*) and may affect virtually any aspect of the model, such as propagule size, *N* (Michod and Roze 1999, 2000, 2001), and adult size (whether it is determinate or indeterminate). In the case of the evolution of a differentiated germ line (Michod 1996, 1999; Michod and Roze 1997), the development time and or the mutation rate may be lowered in the germ line relative to the soma. In the case of the evolution of self-policing (e.g. the immune system), the modifier affects the parameters of selection at both levels, $b$ and $\beta$, reducing the temptation to defect at a cost to the group (Michod 1996, 1999; Michod and Roze 1997). In the case of the evolution of programmed cell death discussed below, we assume the modifier lowers the replication rate of mutant cells directly to $b - \delta$.

As an illustration of conflict mediation in the case of multicellularity, we consider the evolution of **programmed cell death** (PCD). PCD, sometimes termed **apoptosis**, is an evolutionarily conserved form of cell suicide that enables metazoans to regulate cell numbers and control the spread of cancerous cells that threaten the organism. It is best studied in *Caenorhabditis elegans* and mammals, but similar traits have also been described in unicellular organisms such as slime molds (Ameisen 1996), trypanosomatids (Moreira *et al.* 1996; Barcinski 1998; Welburn *et al.* 1999), and yeast (Madeo *et al.* 1997, 1999; Ligr *et al.* 1998). Presumably, in unicellular organisms, PCD is a form of altruism (Frohlich and Madeo 2000), although there is little direct evidence on this point. We illustrate briefly how it may be viewed as a conflict mediator using our theory. We model the evolution of PCD by using the same two locus modifier methods we have used previously (Michod 1999; Michod and Roze 1999) to study the conditions under which germ line or self-policing modifiers spread and tilt the balance in the units of selection conflict in favor of the cell-group, or organism, thereby enhancing its individuality (Michod 1996, 1999; Michod and Roze 1997, 1999). A PCD modifier lowers the rate of division (or survival) of the mutated cell (parameter *pcd*). We assume this occurs at some cost, $\delta$, to the cell-group, or organism. If there were no costs for the modifier, the modifier would always increase so long as it was introduced in a population in which cooperation was present (the role of cooperation is discussed below).

In Fig. 17.3 we report results for the evolution of PCD modifier alleles, assuming sexual reproduction

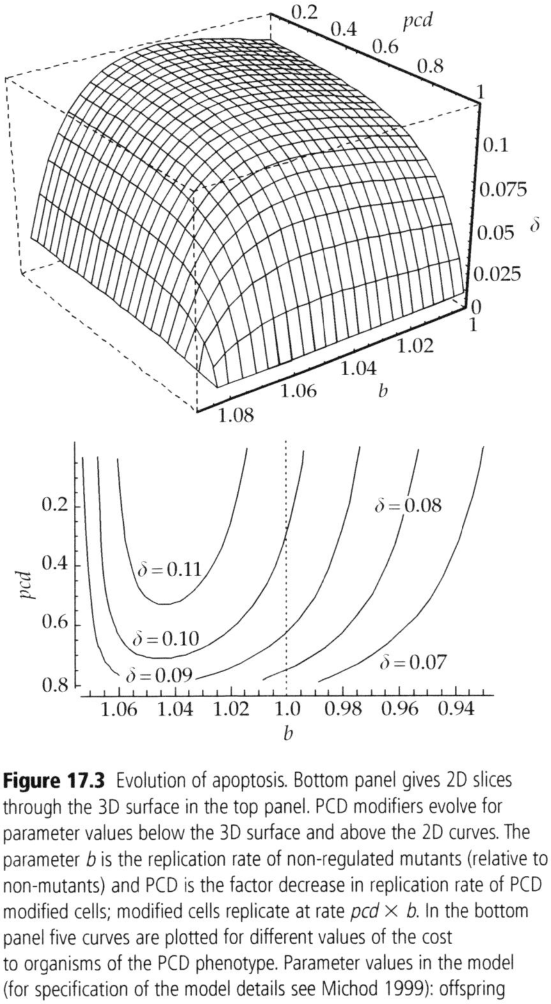


**Figure 17.3** Evolution of apoptosis. Bottom panel gives 2D slices through the 3D surface in the top panel. PCD modifiers evolve for parameter values below the 3D surface and above the 2D curves. The parameter $b$ is the replication rate of non-regulated mutants (relative to non-mutants) and PCD is the factor decrease in replication rate of PCD modified cells; modified cells replicate at rate $pcd \times b$. In the bottom panel five curves are plotted for different values of the cost to organisms of the PCD phenotype. Parameter values in the model (for specification of the model details see Michod 1999): offspring group size $N = 1$, time for development $t = 20$, benefit of cooperation $\beta = 3$, recombination rate between mutated locus and modifier locus $r = 0.2$, survival is incorporated in replication so $s_C = s_D = 1$, and mutation rate $\mu = 0.003$. In the bottom panel selfish mutations lie to the left of the vertical dotted line and uniformly deleterious mutations lie to the right. See text for explanation.

($N = 1$ with sex) and a single class of mutant cells $D$ with fixed effect $b$ (the replication rate of mutant cells without the PCD modifier allele; the replication rate of nonmutant cells is unity). Cells with the modifier allele express the PCD phenotype: mutant cells replicate at rate PCD $\times$ $b$, instead of at rate $b$ in nonmodified cells. A perfect PCD phenotype would mean that all proliferating mutant cells die; in this case we would set $pcd = 0$. Of course, it is unlikely that the first PCD response was perfect, so we consider the entire range of possible values for the PCD phenotype ($0 \leq pcd < 1$). The cost of the PCD phenotype at the organism level is assumed to be $\delta$—the benefit of cooperation is reduced in PCD cells to $\beta - \delta$, instead of $\beta$ in non-PCD cells ($\beta = 3$ in Fig. 17.3).

An interesting feature of the results shown in Fig. 17.3 is that uniformly deleterious mutations (ones that disrupt the functioning of the group and proliferate more slowly than normal cells, $b < 1$), may also select for PCD modifiers, but, to invade, the modifier requires lower costs of the PCD phenotype to the organisms. It is common in the literature on PCD to assume that the risk of selfish mutations has lead to the evolution of the PCD phenotype. However, we see in Fig. 17.3 that both uniformly deleterious and selfish mutations can select for PCD. We have also observed that both kinds of mutations select for the other kinds of modifiers that we have studied, such as germ line and self-policing modifiers.

Why do the curves in Fig. 17.3 fall off rapidly as $b$ increases up towards a value of approximately 1.07? As the proliferation advantage of mutants, $b$, increases, the equilibrium frequency of nonmutant cooperating cells decreases, eventually reaching zero at about 1.07 (when within-group change overpowers between-group selection for cooperation). Without variation at the cell interaction $C/D$ locus the PCD modifier, $M$, is disadvantageous, because when the modifier is introduced the only genotypes are $MD$ and $mD$ (assuming haploidy for explanation purposes; where $D$ is the mutant and $M$ and $m$ are the PCD and non-PCD modifier allele, respectively). Cell-groups initiated by PCD cells ($MD$) end up being smaller than groups initiated by non-PCD cells ($mD$), because of the lower replication rate (or higher death rate) of PCD cells. However, when cooperating cells are maintained in the population before the PCD modifier is introduced, the significant competition is between groups initiated by $CM$ and $Cm$ cells. The cooperating groups carrying PCD modifiers (initiated by $CM$) end up being more functional and having fewer mutant cells in the adult stage and the associated fitness advantage can make up for the cost of PCD, $\delta$ in the regions under the curves shown in the figure). The dependence of the evolution of PCD on the maintenance of cooperation reflects the need for a higher-level unit of selection (the cell-group, or organism). The PCD modifier increases by virtue of tilting the balance in favor of the cell-group, by enhancing its individuality and heritable fitness (Michod 1999).

## Origin of the eukaryotic cell

One of the most significant events in the diversification of biological life is the transition from the prokaryotic to the more complex eukaryotic type of cellular organization. Although the symbiotic monophyletic origin of the eukaryotic cell is now widely accepted, there are many questions yet unanswered. Why did only one particular type of symbiotic association become stable and selected? Initially, what type of partners were involved in this "lucky" association and what was the nature of their interaction? What were the selective pressures that triggered this interaction and its subsequent evolution? Most importantly, how did individuality at the higher level emerge? That is, how did heritability of fitness—the defining characteristic of an evolutionary individual—arise at a new higher level, out of the coevolution of partners who were initially evolutionary individuals in their own right?

Most of the current evolutionary scenarios to explain the origin of the eukaryotic cell considered below (Table 17.3) are based on molecular, cellular, or biochemical data (Cavalier-Smith 1987; Rudel *et al.* 1996; Martin and Müller 1998). Whether eukaryotic features (such as a membrane-surrounded nucleus and a cytoskeleton) evolved before (Cavalier-Smith 1987) or during (Martin and Müller 1998) the acquisition of the mitochondria, is still debatable; nevertheless, this event is recognized as

**Table 17.3** Conflict and conflict mediation in the origin of the eukaryotic cell. nctb: nucleus-cytosol-to-be; mtb: mitochondrion-to-be. References to hypotheses: 1: Martin and Müller (1998); 2: Blackstone (1995); Blackstone and Green (1999); 3: Moreira and López-García (1998); López-García and Moreira (1999); 4: Sagan 1967; Cavalier-Smith (1987); Guerrero (1991); 5: Rudel *et al.* (1996); Kroemer (1997); Frade and Michaelidis (1997).

| Hypothesis | Initial interaction | Conflictual stage | Mediation stage |
|---|---|---|---|
| Hydrogen (1) | Commensalistic: nctb fed on mtb's respiration waste ($H_2$) | nctb limits mtb's ability to import organic substrates required for its growth and reproduction | nctb 'fed' the mtb in exchange for $H_2$; mtb's genes transferred to nctb |
| Units-of-selection (2) | Commensalistic/mutualistic: mtb fed on nctb's excreted carbon; nctb benefits from the re-oxidation of NADH | In aerobic conditions, mtb increased its growth rate and production of oxidants that damaged the nctb | nctb increased its rate of growth, division and recombination |
| Syntrophic (3) | Mutualistic: mtb and nctb fed on each other's waste | Disagreement on efficiency of production of waste | Aerobic metabolism replaced methanogenesis; gene transfers |
| Predatory–prey (4) | Exploitation: nctb fed on living mtb or vice versa | One partner escaped the other partner's digestion | mtb provided energy in exchange for its transmission |
| Pathogen–host (5) | Exploitation: mtb fed on nctb's organic substrates | mtb lysed nctb when nctb's ATP concentration drops | nctb 'stole' ATP from mtb and avoided mtb's lytic mechanisms; gene transfers |

the fundamental step in the prokaryotic–eukaryotic transition, and is the focus of our analysis below. We have approached understanding this major evolutionary transition by translating the different scenarios for the origin of the first mitochondriate eukaryotic cell into the language of multilevel selection theory as a prerequisite to careful population modeling. This approach is intended to help identify the critical factors and thresholds involved in the transition from prokaryotic to mitochondria-containing eukaryotic cells. Our longer-term goal is to evaluate and compare in a common framework the proposed theoretical scenarios for the origin of the first eukaryotic cell.

Our framework is based on understanding the selective and population processes acting during initiation, establishment, and integration of the association so as to understand the emergence of a new unit of evolution with heritable variation in fitness. More specifically, we have considered selective pressures acting on the free-living partners-to-be, which in turn affect the initiation of the association, the initial benefits and costs for the partners versus the free-living relatives, the coevolutionary responses of the partners to each other (and of their association to the environment), the selective forces acting on the group (in relation to the free-living relatives as well as other groups based on different phenotypic associations), and ways of maintaining and integrating the group.

Here, we (i) summarize our multi-level selection approach to investigating the various scenarios regarding the origin of the eukaryotic cell, (ii) pinpoint the key steps in this evolutionary transition and the emergence of individuality at a higher level, and (iii) present the multilevel selection framework that we use in our mathematical modeling (R. E. Michod and A. M. Nedelcu, unpublished). Due to the diversity of scenarios, and especially the multitude of interspecific relations proposed in the current hypotheses, the generic terms "host" and "endosymbiont" as the ancestors of the nucleo-cytosolic compartment and mitochondrion, respectively, can sometime be misleading (i.e. depending on the scenario the "host" is either the "victim", the "predator", or the "prey"). Therefore, we are using the ecologically neutral terms **nucleo-cytosol-to-be** (nctb) and **mitochondria-to-be** (mtb) for the two types of partners, according not to the role they played in the initial interaction but rather to what they are suggested to have evolved into.

There are a variety of verbal scenarios for the origin of the first mitochondriate eukaryotic cell involving almost every form of ecological interaction. Based upon the type of initial interaction between partners, we grouped the current hypotheses in three classes. Commensalistic interactions are invoked in the symbiotic theory (Margulis 1970, 1992), the hydrogen hypothesis (Martin and Müller 1998), and the units-of-evolution hypothesis (Blackstone 1995; Blackstone and Green 1999); mutualistic interactions are suggested in the syntrophic hypothesis (Moreira and López-García 1998); and exploitative interactions are implied in the predator–prey hypotheses (Sagan 1967; Margulis 1981; Cavalier-Smith 1987; Guerrero 1991), as well as pathogen–host hypotheses (Rudel *et al.* 1996; Frade and Michaelidis 1997). This classification reflects only the type of initial interspecific interaction between the nctb and mtb; in fact, some hypotheses involve a succession of commensalistic, exploitative and mutualistic interspecific interactions, for example, the hydrogen and the units-of-evolution hypotheses (Frade and Michaelidis 1997).

To provide a general framework that applies to all the current scenarios, we have identified four stages in the evolution of the association towards the new higher-level unit, the mitochondriate eukaryotic cell. These stages are **initiation** (what type of ecological interactions were present initially), **establishment** (how did these interactions become stabilized in space and over time), **integration** (how did the symbiotic association evolve into an obligate functional unit), and **emergence** of the higher-level individuality (how did the functional unit evolve into an evolutionary individual with fitness heritability).

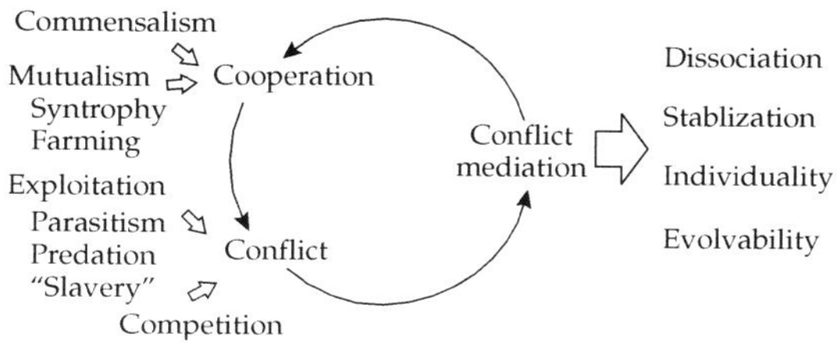


**Figure 17.4** Cooperation and conflict in the origin of the eukaryotic cell. The relation between the various ecological interspecific interactions proposed in various scenarios for the origin of the eukaryotic cell and the stages that characterize the dynamics of behavioral interactions in a group (see Fig. 17.1). Potential outcomes of the conflict mediation stage are also indicated.

We then viewed the scenarios from the point of view of the behaviors associated with the interactions in each stage, namely, cooperation, conflict, and conflict mediation, and mapped them to the evolutionary stages presented above. Interestingly, we found that the partners can enter the cycle in Fig. 17.1 (cooperation, conflict, conflict mediation), through either a cooperative (i.e. mutualism, commensalisms) or conflictual type of interaction (i.e. parasitism, predation, slavery) (Fig. 17.4).

Furthermore, regardless of the initial ecological interaction in each scenario, the association had to face at least one conflictual stage on its way towards integration and higher-level individuality (Table 17.3). This is true even for those scenarios that start out assuming the association was mutualistic or commensalistic to begin with (Margulis 1981; Martin and Müller 1998; López-García and Moreira 1999). Conflict is not associated with a particular stage in the evolution of the association, for it occurs in different stages depending on the scenario analyzed. However, it appears that a conflictual stage is a *sine qua non* condition for the integration of the association and its evolution towards a higher-level individual, and the evolution of the group into a new evolutionary unit depends crucially on the outcome of the conflictual stage. The conflict could result in the dissolution of the association by reverting to free-living state. Alternatively, conflict mediation could stabilize the association by promoting or enhancing cooperation among partners; furthermore, if the cycle is repeated, conflict mediation could later on contribute/result in the integration of the association and the emergence of individuality at the higher level. Last, the way in which the conflict is mediated in each round through the cycle of cooperation, conflict, and conflict mediation, can affect the potential for further evolution (i.e. the evolvability) of the newly emerged evolutionary individual. Below we exemplify our findings with a succinct analysis of three of the scenarios proposed to explain the origin of the first mitochondriate eukaryotic cell.

In the **hydrogen hypothesis** (Martin and Müller 1998), conflict arises during the establishment phase. The selective pressure, that is, the decrease in the level of atmospheric free hydrogen, which initiated a commensalistic interaction between nctbs and mtbs, gained new dimensions as the concentration of free hydrogen continued to drop. To benefit more from the hydrogen released by the mtbs, nctbs surrounded the mtbs to the point where the interaction became detrimental for the latter (by limiting their cell surface, and thus their ability to import the organic substrates required for their growth and reproduction). Consequently, the initially commensalistic interaction changed into an exploitative interaction, and the growth and reproduction of the mtbs were negatively affected. The conflict has likely resulted in the dissolution of many such associations. However, with the decrease in the concentration of free hydrogen, the nctbs became more and more dependent on the mtb-produced hydrogen; therefore, there must have been strong selection for keeping the mtbs alive and functional. Associations in which the nctb found ways to ensure both the survival of the mtb and the mtb's transmission to the nctb offspring were favored. Ways to provide the by now intracellular mtb with organic substrates required (i) the evolution of importers of reduced carbon in the nctb, or (ii) the transfer of the mtb's genes for such importers as well as those for carbohydrate metabolism to the nctb's chromosome. The gene transfer not only resolved the conflict but also resulted in the metabolic and genetic integration of the symbiotic association.

In the **units-of-evolution hypothesis** (Blackstone 1995), conflict arises during the integration phase. The initial association was established through commensalistic interactions; mtbs were mutants with damaged glycolitic mechanisms and, thus, feeding on intermediary metabolites excreted by nctbs. Later, the interaction might have become mutualistic; nctbs benefited from mtbs' oxidation of the reduced cofactor NADH, which allowed the former to produce more ATP and use some of the pyruvate for biosynthesis. The association became and remained stable (but not obligate for both partners) as long as the environmental conditions remained unchanged. However, once the oxygen increased in the environment, the interaction between the nctb and mtb changed dramatically. Due to its aerobic capabilities, the mtb produced a lot more ATP than its host (and, thus, enjoyed a higher rate of growth and reproduction), as well as an increased level of endogenous oxidants for which the nctb did not have the tolerance and the ability to deal with. Consequently, the nctb's fitness decreased and the association became highly unstable. Blackstone (1995) pointed out that a "successful endosymbiosis would depend on successful resolution of units-of-evolution conflicts". By "leaking" ATP, some mtbs contributed to the increase of nctb's growth and division rate, which was in turn beneficial for the mtbs. In addition, some nctbs responded to the oxidative damage inflicted by the mtbs with increased rates of recombination; this has allowed deleterious mutations to be eliminated from the nctb population, and benefited back the mtb by providing novel better-fit genetic **clones** to infect. Because the mtbs became more and more dependent on the nctb, and the nctb benefited from the extra ATP as well as higher rates of growth, division, and recombination, there must have been strong selection for maintaining such associations. In this way, not only did both partners benefit from the resolution of the conflict, but they also became dependent on each other (i.e. the association became an integrated symbiotic unit).

In **pathogen–host scenarios** (Rudel *et al.* 1996; Frade and Michaelidis 1997), conflict arises during the initiation phase. Accidentally engulfed pathogenic mtbs became surrounded by vacuolar membranes produced by the nctb; to ensure their release when the nctb's physiological state deteriorated, the mtbs translocated porin-type membrane channels into the vacuolar membrane and secreted inactive caspase-type proteases into the nctb's cytosol (both porines and caspases are components of the eukaryotic apoptotic machinery). This exploitative association was maintained as long as enough catabolites were present in the nctb and available for the mtbs. When the concentration of catabolites/ATP dropped, the porins opened; this event triggered a series of changes culminating with the death of the nctb and the activation of proteases that eventually digested the nctb. This outcome allowed the mtbs to both leave a weakened/dying host as well as to take advantage of the nutrients released following the caspase activation. This resolution of the conflict was lethal for the nctb, and resulted in the dissolution of the association. The conflict became mediated through the evolution of proteins (such as adenine nucleotide translocators) that the nctb inserted into the mtb's membrane to "steal" ATP from them. As a consequence, the ATP level in the nctb stayed above the threshold that triggered the lytic mechanisms inflicted by the mtbs. Because the mtbs were "leaking" ATP, the association became obligate for such mtbs. Although the direct effects of the new interaction were detrimental for both partners (the nctb looses catabolites to the mtbs, and the mtbs loose ATP to the nctb), they did favor the association (i.e. exchange of catabolites for ATP and vice versa); ultimately the net effect became beneficial for the association relative to other free-living relatives. With the transfer of some genes (including the ones coding for caspases) from the mtb genome to the nctb counterpart, conflict mediation resulted in the metabolic and genetic integration of the two partners into a symbiotic unit.

Table 17.3 succinctly summarizes and compares several of the hypotheses for the origin of the first mitochondriate cell, with respect to the initial interaction, the conflictual stage and its mediation. Our main point is that regardless of the scenario, all these associations between the mtb and the nctb had to pass through a conflictual stage that needed to be mediated, much as we have observed and studied in the unicellular–multicellular transition (Michod 1996, 1997, 1999; Michod and Roze 1997, 1999, 2000). Because conflict mediation is so central to these scenarios, and because the eukaryotic cell is a group of organelles (once independent cells themselves), we plan to extend our multilevel selection methods (Michod 1996), previously used to understand the origin of multicellularity and the origin of life, to study the transition from prokaryotic to complex eukaryotic organizations.

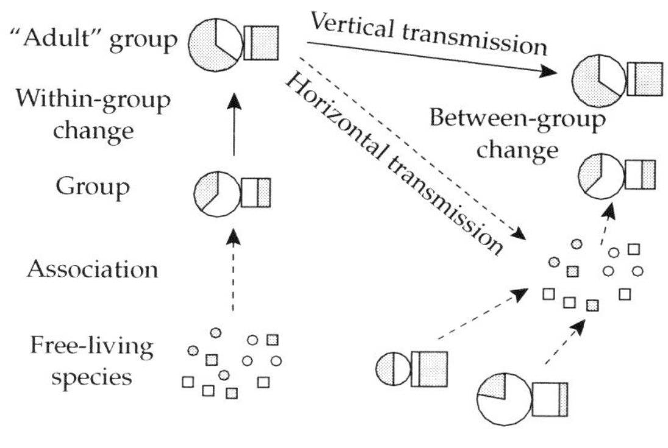


**Figure 17.5** Multilevel selection framework for mixed species groups. Two species are indicated by squares and circles. Within each species there are two phenotypes indicated by open and shaded shapes. A two species group is indicated by touching shapes; the size of each touching shape indicating the relative proportions of each species within the group and the shading of each shape indicating the relative proportion of the two phenotypes within each species in the group.

In Fig. 17.5 we extend the framework of Fig. 17.2 to the origin of the eukaryotic cell. During the transition from bacteria cells to eukaryotic cells, we assume that cells belonging to different species form groups as in Fig. 17.5. Kinship may exist among cells belonging to the same species but not between cells from different species. Initially the interactions between the two species need not be cooperative. The two different species (host and symbiont, predator and prey, two different syntrophic feeders, etc.), are indicated by squares and circles. Within each species there are two phenotypes (cooperate or defect, cultivate or exploit, etc.) indicated by open and shaded regions. A two species group formed by association of members from different species is indicated by the touching shapes; the size of each shape indicating the relative proportions of each species within the group and the shading of each shape indicating the relative proportion of the two phenotypes within each species in the group. After group formation, members may reproduce or survive at different rates leading to within-group change. An additional factor leading to within group change is mutation leading to loss of cooperative cell phenotypes. The total change in frequency of different phenotypes within species and the relative proportions of different species is also affected by the between group change that occurs during horizontal and/or vertical transmission of group properties. Groups may break up and reform as in the case of horizontal transmission, or reproduce as groups in the case of vertical transmission.

## Conclusions

Recall that the basic problem in an evolutionary transition in individuality is to understand why and how a group of individuals becomes a new kind of individual, possessing heritable variation in fitness at a new level of organization. This transfer of fitness from lower to higher levels occurs through the evolution of cooperation and mediators of conflict that restrict the opportunity for within-group change and enhance the opportunity for between-group change. We have illustrated these principles with two major transitions, the origin of multicellular organisms and the origin of the first mitochondriate eukaryotic cell. Although the occurrence of mixed species groups in the case of the eukaryotic cell creates more opportunity for conflict (both ecological interactions and defection), the basic multilevel selection model of cooperation and conflict appears to provide an appropriate conceptual framework for understanding both these evolutionary transitions.

We thank Denis Roze and Cristian Solari for discussion and comments.

CHAPTER 18

# Molecular evidence on the origin of and the phylogenetic relationships among the major groups of vertebrates

**Rafael Zardoya and Axel Meyer**

Vertebrates are a good model to study macroevolutionary patterns and processes because they possess a comparatively well known fossil record (Carroll 1997). Thanks to the detailed investigations of several generations of morphologists and paleontologists over the last two centuries, it has been possible to reconstruct the phylogeny of vertebrates with some degree of confidence (Fig. 18.1). A robust phylogenetic framework of vertebrates is fundamental for comparative studies in this group.

The first major landmark in vertebrate evolution was the origin of jaws from mandibular branchial arches, and dates back to the Cambrian, 540–505 million years ago (mya). Accordingly, vertebrates have been traditionally classified into Agnatha (hagfishes and lampreys) and Gnathostomata (jawed vertebrates) (Fig. 18.1). Among the latter, the major distinction, based on the nature of the skeleton, is between Chondrichthyes (cartilaginous fishes) and Osteichthyes (bony fishes) (Fig. 18.1). Bony fishes are divided into Actinopterygii (ray-finned fishes) and Sarcopterygii (lobe-finned fishes and tetrapods) (Fig. 18.1). The origin of the major lineages within bony fishes as well as the transition to life on land date back to the Devonian, 408–360 mya. The major evolutionary novelty within tetrapods was the origin of the amniote egg, a type of egg with a semipermeable shell, a large amount of yolk, and several embryonic membranes involved in respiration, feeding, and waste disposal. Accordingly, extant tetrapods are classified into Lissamphibia (caecilians, salamanders, and frogs) and Amniota (mammals and reptiles, that is, turtles, squamates, crocodiles, and birds) (Fig. 18.1). The origin of amniotes dates back to the Pennsylvanian, 325–280 mya.

Although the general framework of vertebrate relationships is well supported by morphological and paleontological evidence, and new fossil discoveries continue to refine it, some key branching events in the vertebrate tree remain controversial (see **polytomies**, that is, collapsed nodes in Fig. 18.1). Appearance of vertebrate taxa in the fossil record suggests that the origin of the major lineages of vertebrates occurred as rapid radiation events within a relatively short time frame several hundred million years ago (Carroll 1997). In such cases, it is difficult to recover the exact branching pattern because there was little time during the radiation to accumulate shared derived characters that define lineages, and later much of the phylogenetic information was obliterated during the independent evolution of each lineage. In the particular case of vertebrates, the relative phylogenetic positions of jawless vertebrates, lobe-finned fishes, amphibians, turtles, and monotremes are still under debate.

Recently, the advent of molecular techniques (in particular, the **polymerase chain reaction** or PCR, and automated DNA sequencing) has made it possible to tackle these phylogenetic questions from a different perspective. Two molecular markers, mitochondrial DNA and nuclear rRNA genes, are

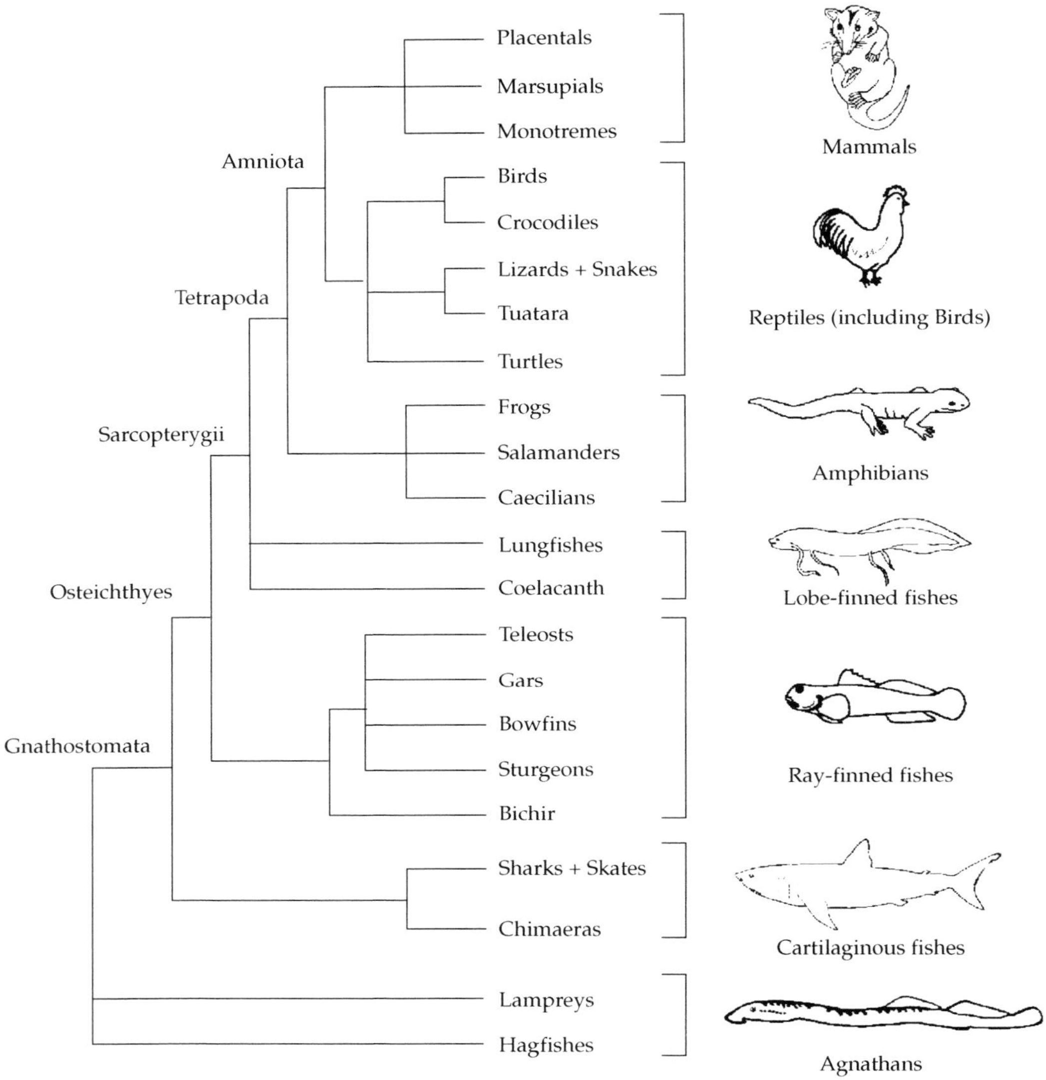


**Figure 18.1** Phylogenetic relationships among the main lineages of vertebrates based on morphological and paleontological evidence (e.g. Carroll 1988; Cloutier and Ahlberg 1997).

the most widely used in phylogenetic studies among distantly related taxa. Phylogenetic analyses of mitochondrial genomes and nuclear rRNA data have largely corroborated the traditional morphology-based phylogeny of vertebrates (Fig. 18.1), and they are contributing to the resolution of some long-standing controversies in vertebrate large-scale systematics. In this chapter, we review the molecular data that have been collected with the explicit goal of resolving competing hypotheses that have been postulated to explain the origin of the main lineages of vertebrates.

## The living sister group of jawed vertebrates

The origin of jaws, a key innovation that allowed gnathostomes to grasp and bite prey, was one of the major events in the history of vertebrates. Among extant vertebrates only the hagfishes (Mixiniformes; 43 species) and the lampreys (Petromyzontiformes; 41 species) remain jawless. The phylogenetic relationships of hagfishes, lampreys, and jawed vertebrates have given rise to one of the longest controversies in vertebrate systematics. There are three problems affecting the recovery of the exact relationships between these taxa (Mallat and Sullivan 1998). First, these three lineages of vertebrates appeared in a time window of less than 40 million years, back in the Cambrian, so it is difficult to find shared derived characters between them. Second, there is a rather poor fossil record to trace this event, and third, both lampreys and hagfishes not only have retained many primitive characters, but also, and because of their apparently fast evolutionary rates, they have accumulated numerous unique characters.

Traditional classifications of vertebrates united hagfishes and lampreys as cyclostomes (Fig. 8.2a). Such grouping was supported by the presence in both taxa of horny teeth, a respiratory velum, and a complex "tongue" apparatus (Delarbre *et al.* 2000). However, more recent morphological analyses found many seemingly shared derived characters between lampreys and jawed vertebrates (Janvier 1981) to the exclusion of hagfishes (Fig. 18.2b). The basal position of hagfishes with respect to lampreys and jawed vertebrates was also supported in a recent analysis of agnathan fossils from China (Shu *et al.* 1999).

Several molecular studies have been conducted to resolve the question of the living sister group of gnathostomes. Phylogenetic analyses of the nuclear 18S and 28S rRNA genes recovered a monophyletic cyclostome (hagfish + lamprey) **clade** with high statistical support (Mallat and Sullivan 1998; Mallat *et al.* 2001; Zardoya and Meyer 2001b) (Fig. 18.2a). Several other nuclear loci also support the cyclostome hypothesis (Kuraku *et al.* 1999) (Fig. 18.2a). In contrast, phylogenetic analyses of mitochondrial protein-coding genes seemed to support lampreys as the closest sister group of jawed vertebrates (Rasmussen *et al.* 1998) (Fig. 18.2b). An important problem in the reconstruction of early vertebrate phylogeny using molecular data is that the hagfish branch is extremely long (Zardoya and Meyer 2001b). This circumstance could spuriously pull the highly divergent hagfish sequence toward the outgroup (the next more divergent sequence in the tree), and may explain the phylogeny recovered by Rassmussen *et al.* (1998). In fact, more recent phylogenetic analyses with additional taxa demonstrated that mitochondrial evidence can support both competing hypotheses on the relative position of jawless vertebrates depending on the choice of method of phylogenetic inference (Delarbre *et al.* 2000).

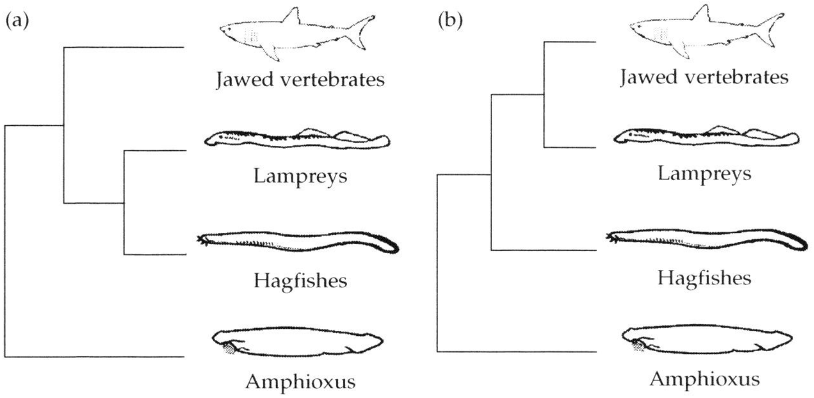


**Figure 18.2** Competing hypotheses on the phylogenetic relationships of extant hagfishes, lampreys, and jawed vertebrates. (a) The Cyclostome hypothesis: hagfishes and lampreys are sister group taxa and equally distant to jawed vertebrates. (b) The Vertebrate hypothesis: lampreys are the living sister group of jawed vertebrates.

## The origin of tetrapods

The origin of land vertebrates occurred as a rapid adaptive radiation back in the Devonian (408–360 mya). The transition from life in water to life on land was a complex evolutionary event that involved successive morphological, physiological, and behavioral changes. It is well established that early tetrapods evolved from lobe-finned fishes (Sarcopterygii), and recent fossil discoveries have shown that panderichthyids are their closest sister group (Cloutier and Ahlberg 1997; Ahlberg and Johanson 1998). The closest relatives of panderichthyids and tetrapods are osteolepiforms (Cloutier and Ahlberg 1997; Ahlberg and Johanson 1998). The other two major groups within sarcopterygians are Dipnomorpha and Actinistia. Dipnomorphs include the extinct porolepiforms, and the air-breathing extant dipnoi (lungfishes). Actinistia or coelacanths were a highly successful group of lobe-finned fishes during the Devonian which now are represented by only two surviving species (*Latimeria chalumnae* and *Latimeria menadoensis*). Although most recent morphological and paleontological evidence support lungfishes as the closest living sister group of tetrapods (Cloutier and Ahlberg 1997; Ahlberg and Johanson 1998), there is still no general agreement regarding which group of sarcopterygians, the Actinistia or the Dipnomorpha, is the one most closely related to the tetrapod lineage. The controversy will continue until new relevant fossils of intermediate forms connecting the three groups are discovered, and agreement among paleontologists about the **homology** of some characters (e.g. the choanae) is achieved (Cloutier and Ahlberg 1997).

Molecular data from lungfishes, the coelacanth, and tetrapods, the only living sarcopterygians, have been collected to bear on the phylogenetic question. There are three competing hypotheses to explain phylogenetic relationships among the living lineages of sarcopterygians: (1) lungfishes as closest living relatives of tetrapods (Fig. 18.3a), (2) the coelacanth as living sister group of tetrapods (Fig. 18.3b), and (3) lungfish and coelacanth equally closely related to tetrapods (Fig. 18.3c).

The first molecular data set that supported lungfishes as the closest living relatives of tetrapods (Fig. 18.3a) was based on two fragments of the mitochondrial 12S rRNA and cytochrome *b* genes (Meyer and Wilson 1990). Further support for this hypothesis was obtained from the phylogenetic analysis of complete 12S and 16S rRNA mitochondrial genes (Hedges *et al.* 1993). However, a reanalysis of this

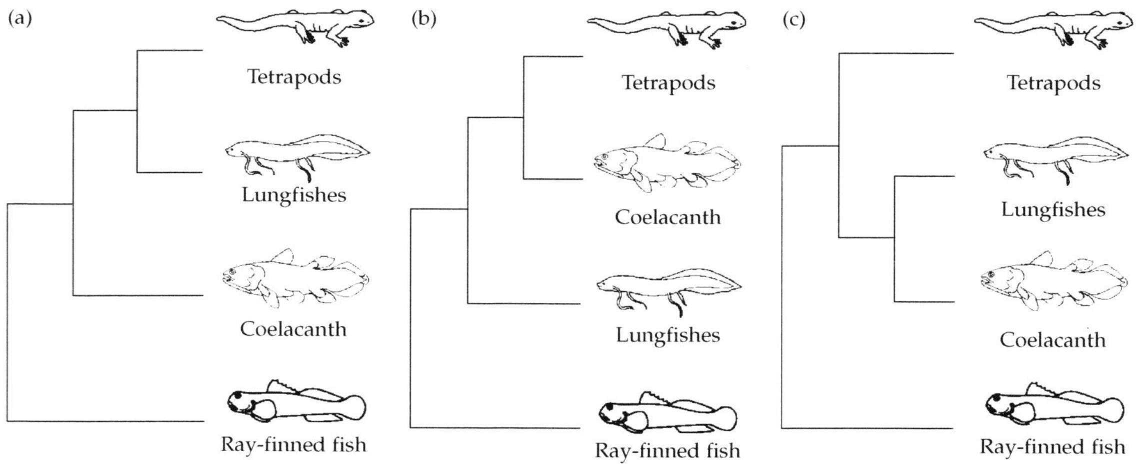


**Figure 18.3** Alternative hypotheses on the phylogenetic relationships of living lobe-finned fishes and tetrapods. (a) Lungfishes as the closest living sister group of tetrapods. (b) Coelacanths as the closest living relatives of tetrapods. (c) Lungfishes and coelacanths equally closely related to tetrapods.

data set with more taxa resulted in an unresolved lungfish + coelacanth + tetrapod trichotomy (Zardoya *et al.* 1998). Phylogenetic analyses of a data set that combined all mitochondrial protein-coding genes recovered lungfishes as sister group of tetrapods (Zardoya *et al.* 1998) (Fig. 18.3a). However, this data set could not statistically reject a lungfish + coelacanth clade (Fig. 18.3c) but could do so for the traditional hypothesis that has been favored by most textbooks for the past 50 years (Fig. 18.3b). Phylogenetic analyses of a data set that combined all mitochondrial tRNA genes supported a close relationship between lungfishes and the coelacanth (Zardoya *et al.* 1998) (Fig. 18.3c). When the mitochondrial protein-coding gene data set was combined with the rest of the mitochondrially encoded (rRNA and tRNA) genes, it also supported lungfishes as the closest living sister group of tetrapods (Zardoya *et al.* 1998). Phylogenetic analyses of nuclear 28S rRNA gene sequences favored a lungfish + coelacanth clade (Zardoya and Meyer 1996). The phylogenetic analyses of the combined mitochondrial and 28S rRNA nuclear data sets were not entirely conclusive. Depending on the method of phylogenetic inference used, both a lungfish + coelacanth or a lungfish + tetrapod clade were supported (Zardoya *et al.* 1998). The hypothesis of the coelacanth as closest living sister group of tetrapods (Fig. 18.3b) received the least support in all phylogenetic analyses of molecular data. Recent phylogenetic analyses of a nuclear gene, the myelin DM20 again supported lungfishes as the sister group of tetrapods (Tohyama *et al.* 2000) (Fig. 18.3a). Moreover, the lungfish + tetrapod clade is supported by a single **deletion** in the amino acid sequence of a nuclear-encoded gene RAG2 that is shared by lungfishes and tetrapods (Venkatesh *et al.* 2001). Overall, most molecular evidence supports lungfishes as the closest living sister group of tetrapods. However, further work on nuclear genes is required to discard definitively a lungfish + coelacanth relationship.

## Phylogenetic relationships of modern amphibians

There is little controversy that modern amphibians (Lissamphibia) are a monophyletic group that likely appeared in the Permian (280–248 mya) (Duellman and Trueb 1994; but see Carroll 1988). However, it is still debated whether the extinct temnospondyls (e.g. Trueb and Cloutier 1991) or the extinct lepospondyls (Laurin and Reisz 1997; Laurin 1998) are their stem group. Moreover, there is no general agreement regarding the phylogenetic relationships among the three living orders of amphibians, that is, Gymnophiona (caecilians), Caudata (salamanders), and Anura (frogs). Most morphological and paleontological studies support that salamanders are the living sister group of frogs (and form the clade Batrachia) and that caecilians are more distantly related to both (Trueb and Cloutier 1991; Duellman and Trueb 1994) (Fig. 18.4a). However, other studies seem to suggest that salamanders can be the sister group of caecilians to the exclusion of frogs (Carroll 1988) (Fig. 18.4b). The latter hypothesis finds support in Laurin's (1998) analysis, although the author suggests that this may be a spurious result. Because all three lineages of extant amphibians acquired their distinctive body plans early on their evolutionary history, there are few reliable shared derived characters among them. Moreover, a rather poor Permian–Triassic fossil record complicates the search for putative Lissamphibian relatives (Carroll 1988).

Early phylogenetic analyses of this question using nuclear and mitochondrial rRNA data suggested that caecilians are the closest living relatives of salamanders to the exclusion of frogs (Feller and Hedges 1998) (Fig. 18.4b). However, phylogenetic analyses of the complete mitochondrial genomes of a salamander (*Mertensiella luschani*), a caecilian (*Typhlonectes natans*), and a frog (*Xenopus laevis*) supported with high statistical support the Batrachia hypothesis, that is, a sister group relationship of salamanders and frogs (Zardoya and Meyer 2001a) (Fig. 18.4a), in agreement with most morphological evidence rather than with earlier molecular studies. The current overall support of the Batrachia hypothesis both from morphology and molecules provides a phylogenetic framework that will be helpful in many comparative studies of living amphibians. Unfortunately, molecular data cannot provide information on the question of the closest relative of Lissamphibia which requires the

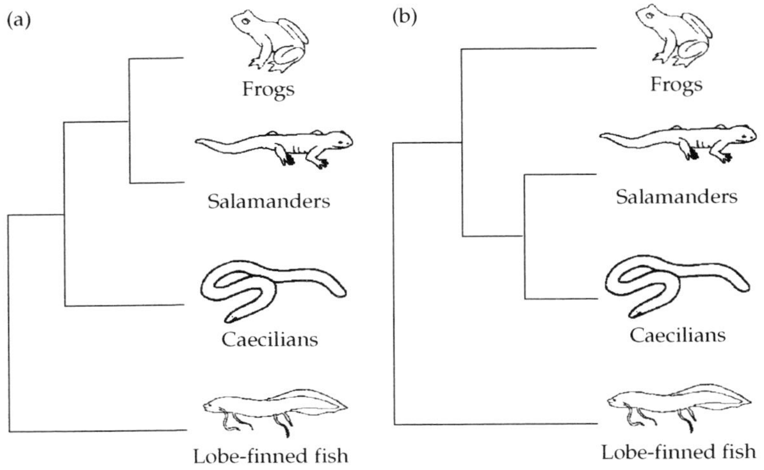


**Figure 18.4** Phylogenetic relationships of the three living groups of amphibians. (a) The Batrachia hypothesis: frogs as the closest living sister group of salamanders. (b) Caecilians as the closest extant relatives of salamanders.

discovery of key Permian–Triassic fossils, and their rigorous phylogenetic analysis.

## The phylogenetic position of turtles and amniote relationships

Living amniotes are classified, based on the presence and type of temporal fenestration of the skull, in anapsids (those that show a completely roofed skull), diapsids (those that have two fenestrae in the temporal region of the skull), and synapsids (those that present a single lower temporal hole in their skulls). Turtles are considered to be the only living representatives of anapsids; the tuatara, lizards, snakes, crocodiles, and birds are diapsids; and mammals are synapsids. The classical view of amniote phylogeny supported by morphological and fossil data considers synapsids as the most basal lineage, and places diapsids in a derived position relative to anapsids (Laurin and Reisz 1995; Lee 1997) (Fig. 18.5a).

However, turtles exhibit such a unique morphology that they only share a few characters with any other group of amniotes. As a result, it is difficult to determine the exact phylogenetic position of turtles within the amniota. Recent analyses reveal that support for the anapsid affinities of turtles is rather weak (deBraga and Rieppel 1997). Alternatively, turtles might be the closest living relatives of the Lepidosauria (tuatara, lizards, and snakes) (Fig. 18.5b) (deBraga and Rieppel 1997), or the sister group of Archosauria (crocodiles and birds) (Fig. 18.5c) (Hennig 1983).

If turtles are anapsid reptiles, their closest relatives may be procolophonids (Laurin and Reisz 1995) or pareiasaurs (Gregory 1946; Lee 1997). If turtles are placed as advanced diapsid reptiles, they may be closely related to extinct Sauropterygia (marine plesiosaur and pliosaur reptiles), and Lepidosauria would be their closest living relatives (deBraga and Rieppel 1997) (Fig. 18.5b). However, there is also morphological evidence that places plesiosaurs and pliosaurs basal to the Archosauria (Merck 1997). Hence, it is possible that turtles could be closely related to the Archosauria (Hennig 1983) (Fig. 18.5c) rather than to the Lepidosauria. In both cases, if turtles are diapsids, then the anapsid condition of the turtle skull was developed secondarily.

Although early analyses of mitochondrial complete 12S and 16S rRNA gene data sets supported the traditional anapsid position of turtles, that is, outside diapsids (e.g. Hedges 1994) (Fig. 18.5a), recent reanalyses of the same genes with additional taxa (including representatives of the two major lineages of turtles, Pleurodira and Cryptodira) recover a turtle + Archosauria clade with moderately high statistical support (Zardoya and Meyer 1998) (Fig. 18.5c). However, alternative hypotheses, that is, turtles as

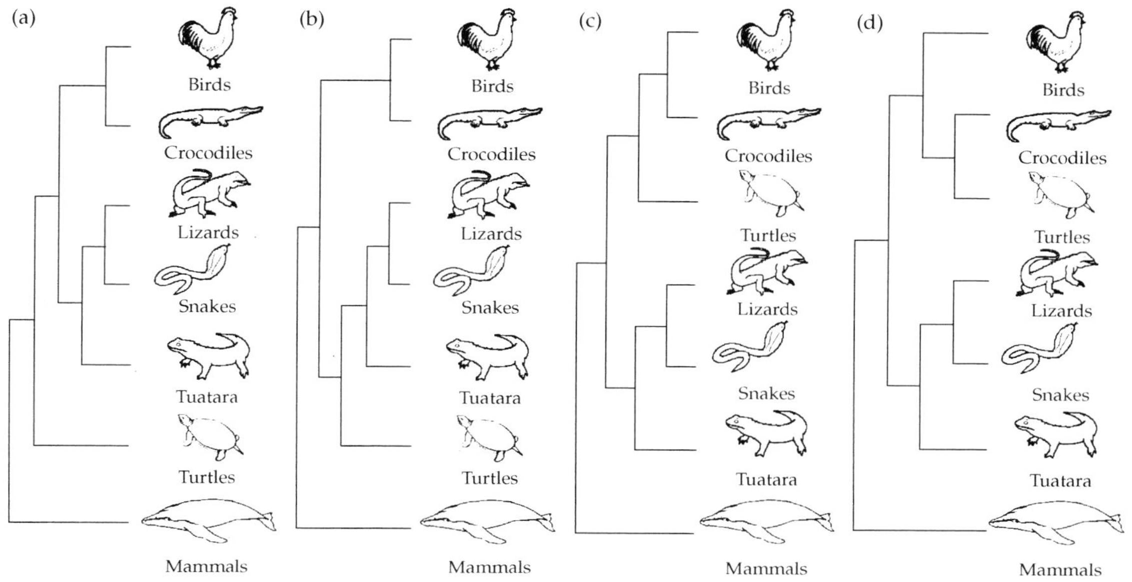


**Figure 18.5** The phylogenetic position of turtles within living amniotes. (a) Turtles as the only living representatives of anapsid reptiles, and basal to diapsid reptiles, that is, Lepidosauria (tuatara, snakes, and lizards) and Archosauria (crocodiles and birds). (b) Turtles have diapsid affinities, and are the sister group of Lepidosauria. (c) Turtles have diapsid affinities, and are the sister group of Archosauria. (d) Turtles have archosaurian affinities, and are the sister group of crocodiles.

anapsids (Fig. 18.5a) or turtles as sister group of lepidosaurs (Fig. 18.5b), could not be statistically rejected based on this data set (Zardoya and Meyer 1998). Recent phylogenetic analyses of relatively large mitochondrial and nuclear sequence data sets further support the diapsid affinities of turtles, and only differ on their relative position with respect to Lepidosauria and Archosauria. Molecular evidence based on complete mitochondrial protein-coding genes confirmed the archosaurian affinities of turtles, and statistically rejected alternative hypotheses (Kumazawa and Nishida 1999; Janke *et al.* 2001) (Fig. 18.5c). Phylogenetic analyses of a data set including complete mitochondrial protein-coding, rRNA, and tRNA genes also strongly supported the phylogenetic position of turtles as sister group of archosaurs (Zardoya and Meyer 2001b) (Fig. 18.5c). In agreement with mitochondrial evidence, nuclear pancreatic polypeptide data support archosaurs as the living sister group of turtles (Platz and Conlon 1997).

Phylogenetic analyses of eleven nuclear proteins, as well as the nuclear 18S and 28S rRNA genes, supported crocodiles as the closest living relatives of turtles (Hedges and Poling 1999) even to the exclusion of birds (Fig. 18.5d). Furthermore, a phylogenetic analysis that combined mitochondrial and nuclear data also recovered a crocodile + turtle grouping (Cao *et al.* 2000). However, morphological and paleontological evidences clearly support the monophyly of archosaurs. Interestingly, both crocodiles and turtles show significantly long branches which might introduce biases in the phylogenetic analyses. Hence, a sister group relationship of crocodiles and turtles needs to be treated as tentative, and further molecular work lies ahead.

## The origin of placental mammals

Mammals are tetrapods well suited for life on land. In particular, the acquisition of a placenta, a membrane to nourish the fetus, was a major breakthrough in mammalian evolution that may partly explain their success and rapid radiation after the mass extinction of dinosaurs in the uppermost Cretaceous. The classical view of mammalian phylogeny supported by morphological and fossil data

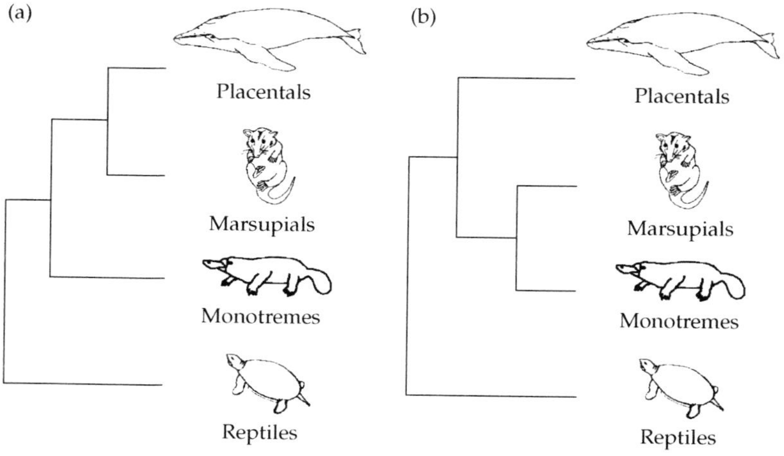


**Figure 18.6** Alternative hypotheses explaining the phylogenetic relationships of monotremes, marsupials, and placental mammals. (a) The Theria hypothesis: marsupials are the closest living sister group of placentals. (b) The Marsupionta hypothesis: the marsupials are closely related to monotremes, and both groups are equally distant to placentals.

places monotremes (the platypus and echidnas) as the most basal lineage of mammals, with the marsupials as closest relatives of eutherians (placental mammals) (Carroll 1988) (Fig. 18.6a). In this regard, there are many morphological features that have been interpreted as shared derived characters between marsupials and placentals. However, some workers advocate a close relationship of monotremes and marsupials (the Marsupionta hypothesis) based on similar tooth-replacement patterns (Gregory 1947; Kühne 1973), to the exclusion of placentals (Fig. 18.6b). The poor fossil record for monotremes (Carroll 1988) complicates the analysis of their phylogenetic relationships to marsupials and placentals, and leaves open the question on the closest relative of Eutheria.

The complete nucleotide sequences of the mitochondrial genomes of the platypus and the opossum were determined to clarify the debate based on morphological data (Janke *et al.* 1996). Phylogenetic analyses of a data set that combined the inferred amino acid sequences of the mitochondrial protein-coding genes favored with high statistical support that monotremes and marsupials are sister groups (Janke *et al.* 1996) (Fig. 18.6b). Further phylogenetic analyses with additional tetrapod taxa (including the wallaroo, *Macropus robustus*) seemed to confirm the mitochondrial support for the Marsupionta hypothesis (Janke *et al.* 2001). However, some workers have noted that the support of the mitochondrial protein data for the Marsupionta hypothesis varies considerably depending on the outgroup and phylogenetic methods used (Wadell *et al.* 1999). Phylogenetic analyses of DNA-hybridization data on several amniotes also supported the monotreme + marsupial clade (Kirsch and Mayer 1998). However, it has been suggested that monotremes show relatively high GC contents, and that this bias might be shared by marsupials, but not by placentals (Kirsch and Mayer 1998). If confirmed, a base-compositional bias rather than a true phylogenetic signal could be responsible for the monotreme + marsupial grouping in the DNA-hybridization analyses. Recently, a large nuclear gene, the mannose 6-phosphate/insulin-like growth factor II receptor, was sequenced from representatives of all three mammalian groups to clarify the controversy (Killian *et al.* 2001). Phylogenetic analyses of this nuclear gene sequence favored with statistical support that marsupials are the sister group of eutherians to the exclusion of monotremes (Fig. 18.6a). Hence, new molecular data seems to corroborate morphological evidence. Future molecular studies (including, for example, more nuclear gene sequence data), and the confrontation of molecular and morphological phylogenies will certainly improve our understanding of the origin of placental mammals.

## Summary

Vertebrates offer the opportunity to study long-term evolutionary patterns and processes because their phylogeny is comparatively well known. Most major events that have occurred throughout the evolution of vertebrates are well documented in the fossil record, and phylogenetic analyses of such paleontological data have made it possible to reconstruct rather resolved trees that explain vertebrate phylogenetic relationships. However, some nodes in the vertebrate tree, at the origin of the main lineages, remain controversial. The origin of the main lineages of vertebrates was accompanied by key morphological innovations and rapid radiation events. Gaps in the fossil record associated to some of these events, difficulties in the interpretation of what has been preserved, and the existence of few morphological shared derived characters between the putative living sister groups and the radiated groups hamper the inference of the exact phylogenetic relationships between the taxa involved in the origin of the major groups of vertebrates.

Recent advances in molecular techniques as well as more powerful phylogenetic algorithms and faster computers have made it possible to infer phylogenetic relationships using sequence data. Two molecular markers, mitochondrial DNA and nuclear rRNA genes, have been widely applied to phylogenetic inference of vertebrate relationships. Besides the corroboration of the traditional morphology-based phylogeny, new molecular data have been particularly helpful in discerning among alternative hypotheses to explain the origin of the major lineages of vertebrates. Examples are the recent molecular evidence that supports a sister group relationship of hagfishes and lampreys, that groups lungfishes with tetrapods to the exclusion of coelacanths, that favors the Batrachia hypothesis (salamanders as sister group of frogs), that places turtles as diapsid reptiles, and that suggests marsupials as the closest relatives of placental mammals to the exclusion of monotremes.

In most cases, molecular data corroborate morphological evidence, but in some cases molecular and morphological signals conflict. Ultimately, comparisons of conflicting signals should enable evolutionary biologists to detect anomalies that result in misinterpreting one of the two types of data. Understanding the sources of signal conflict will definitively improve phylogenetic inference and may contribute to settling open debates on vertebrate systematics.

We thank Michel Laurin and an anonymous reviewer for providing helpful comments on the manuscript. This work received partial financial support from grants from the Ministerio de Ciencia y Tecnología to RZ (REN2001-1514/GLO), and the Lion Foundation, the Deutsche Forschungsgemeinschaft, the University of Konstanz, and the US National Science Foundation to AM.

CHAPTER 19

# Mass extinctions and evolutionary radiations

**Douglas H. Erwin**

Mass extinctions are a critical driving force for evolutionary change. Whether the patterns of extinction are random or selective, they prune the tree of life and the subsequent evolutionary rebounds provide an opportunity for previously minor **clades** to become dominant, for dramatic reorganizations of ecosystems, and for the appearance of significant evolutionary innovations. Not all mass extinctions trigger such reorganizations, however. Some have relatively little long-term influence on the course of evolution. Yet, however much we may regret the current, human-driven loss of biodiversity, the fossil record clearly demonstrates the evolutionary impact of mass extinctions. What happens during a mass extinction? Do mass extinctions share a common cause? Are the processes and patterns of extinction similar between different events or is each episode unique? And what of the aftermath of mass extinctions? How does the biota recover, and how does both the extinction and recovery influence the evolutionary significance of mass extinctions?

Following the Alvarez *et al.* (1980) hypothesis that the end-Cretaceous (K–T) extinction of (nonavian) dinosaurs and other organisms was triggered by the impact of an extraterrestrial object, there was an enormous increase in scientific interest in mass extinctions, not only by paleontologists, but also by geochemists, geologists, and (too many) physicists. The resulting field studies, laboratory analyses, theoretical studies, and critical syntheses have produced a far better understanding of most of the five great mass extinctions than existed 20 years ago, even if some important issues remain unresolved. Studies of smaller biotic crises have added important details, and compilations of global histories of marine fossils have suggested the possibility of a periodic pattern to mass extinctions over the past 250 million years. Perhaps most significantly, the demonstration of an impact associated with the K–T mass extinction has convinced all but a few skeptics that such events may be catastrophic, and may be triggered by extraterrestrial influences. Of course, it comes as no surprise to students of human nature that the disciples have far outstripped the prophets and many recent impact hypotheses lack the plausible mechanisms and evidence advanced by Alvarez and colleagues in 1980. The advent of new tools and scientific approaches that allow paleontologists and others interested in extinctions to approach mass extinctions more rigorously have been the other great advance of the past two decades. Consequently, we are better able to frame and test hypotheses, and, one hopes, less likely to be beguiled by speculation.

Rather than use a historical survey of each event (for which see Hallam and Wignall 1997; Erwin 2001a), this chapter employs a comparative approach to issues common to the five great mass extinctions, including the rates, patterns, and causes of these events, and the process of biotic recovery which follow.

## Mass extinctions

### What are mass extinctions?

Not all losses of biodiversity are mass extinctions. Mass extinctions involve a significant loss of generic or family diversity (commonly more than 25 percent) across many different groups in a relatively brief interval of time. Five events punctuate

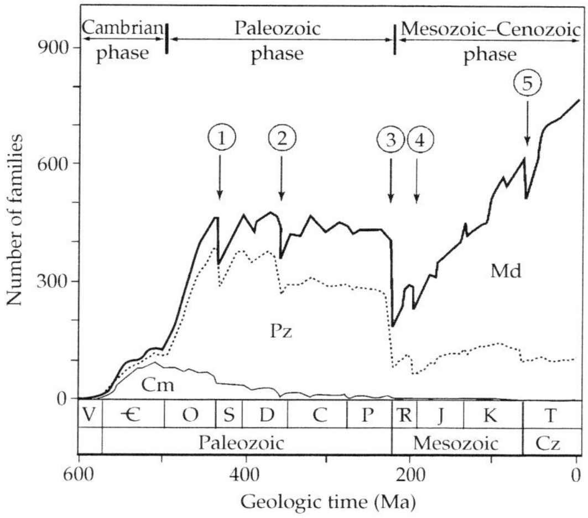


**Figure 19.1** Diversity of marine families through the Phanerozoic (the past 550 million years). The thick black line shows the total number of durably skeletonized marine families found in the fossil record in each of 83 geologic stages. The circles correspond to the five major mass extinctions: 1, Late Ordovician; 2, Late Devonian; 3, Late Permian; 4, End Triassic; 5, Cretaceous-Tertiary. The lower solid line shows family diversity from the Cambrian evolutionary fauna (Cm); the dotted line shows the family diversity of the Paleozoic Evolutionary Fauna (Pz); the remainder of the graph is assignable to the Modern Evolutionary fauna (Md). V, Vendian; C, Cambrian; O, Ordovician; S, Silurian; D, Devonian; C, Carboniferous; P, Permian; Tr, Triassic; J, Jurassic; K, Cretaceous; T, Tertiary; Cz, Cenozoic. After Sepkoski (1984).

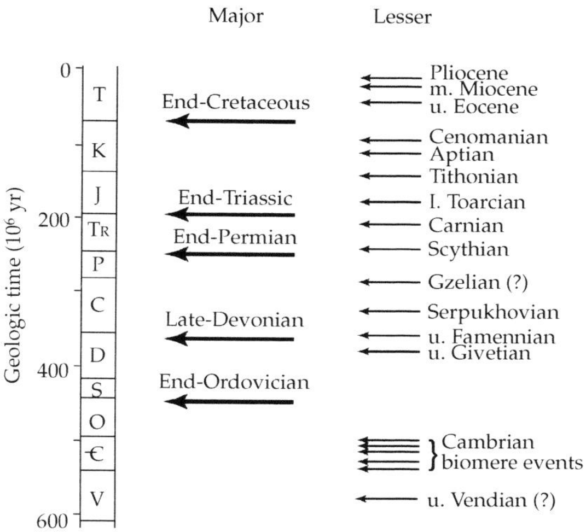


**Figure 19.2** Major and lesser extinctions during the Phanerozoic. Most of the lesser biotic crises shown here are not discussed in the text. Some of these putative lesser biotic crises only appear on global compilations and are likely to be preservational or statistical artifacts. See the legend to Fig. 19.1 for abbreviations.

the history of life by eliminating many fossil groups and are widely accepted as mass extinctions (see Figs 19.1 and 19.2): the end-Ordovician (439 million years ago: mya), Late Devonian (*c.* 370 mya), end-Permian (251 mya), end-Triassic (199 mya) and end-Cretaceous (65 mya). With the exception of the end-Triassic mass extinction, which remains a bit obscure, the others were all apparent to geologists over a century ago, although relatively little attention was actually given to these events. The human-induced loss of many large terrestrial mammals and endemic island birds over the past 50 000 years also does not qualify as one of the five great mass extinctions because the total number of taxa lost is relatively small (although hardly insignificant) and they were largely concentrated among terrestrial mammals and birds.

Are mass extinctions truly different from other extinctions? Jablonski (1986) has argued that the patterns of survival are distinct between mass extinctions and extinctions during the intervals between mass extinctions (**background extinctions**). His data for Cretaceous molluscs, and studies by others on a variety of different groups at other mass extinctions (summarized in Jablonski 1989) suggest that during background times survivorship is enhanced by local abundance, broad geographic distribution of species, and numerous species within a clade. But during at least some mass extinctions, broad geographic range of the clade, independent of the geographic range of the component species appears to have been the critical variable (Jablonski and Raup 1995; Jablonski 2001). In contrast Raup has argued that ranking extinctions by intensity to produce a "kill curve" shows no dramatic break between events that are classically defined as mass extinctions and smaller biotic crises (Raup 1991), suggesting the distinction may be artificial. Rather, there is a gradation of biodiversity crises from small, local events up to global catastrophes.

Missing from any of these discussions has been consideration of how the structure of the sedimentary record may influence paleontologists' perception of mass extinctions and other biotic crises. Although aware of small-scale biases in the fossil record, paleontologists have generally argued that the patterns of large-scale diversity that reveal mass extinctions are real, and not produced by any sort of bias. Smith *et al.* (2001) have recently challenged this assumption. Their combination of a detailed phylogenetic analysis of Cretaceous echinoids with documentation of the preservational patterns of the fossils shows that the apparent magnitude of the Cenomanian–Turonian extinction (a smaller mid-Cretaceous biotic crisis) was considerably magnified by taphonomic bias. Smith *et al.* (2001) raise the possibility that preservational biases may affect much of the extinction record. In general, the issue is not whether the largest mass extinctions actually occurred, they clearly did, but whether the estimated magnitudes of these events have been inflated, and whether many of the smaller events may be artifacts. Although they do not address this issue directly, the apparent continuity of extinction magnitudes documented by Raup's kill curve may simply reflect the variable effects of changes in sea level. This issue of scaling relationships of diversity crises and the influence of biases in the fossil record requires considerable attention in coming years.

## Rates of mass extinctions

The pattern of fossil disappearances during mass extinctions shows a gradual loss of diversity. Until 1980, this reinforced the views of most paleontologists that mass extinctions were gradual events, in some cases stretching over millions of years. Such a view of course favored hypotheses about the causes of mass extinction that emphasized the contribution of gradual forces, including the movement of the continents and climatic change. The Alvarez *et al.* hypothesis changed this in 1980, as evidence rapidly developed that an extraterrestrial impact triggered a sudden climatic shift resulting in a massive loss of biodiversity. Soon after the Alvarez *et al.* proposal, Signor and Lipps (1982) recognized that even a sudden mass extinction would appear gradual in the fossil record because of differences in the likelihood of fossilization between different species. Within a single stratigraphic section of rock, common and easily preserved species would be more likely to be preserved close to the level at which they actually became extinct. In contrast, less common or less easily preserved species are encountered less frequently. Thus, the last appearance of a species may be well before the mass extinction event, even if it actually became extinct during the extinction. Since species differ greatly in their probability of preservation, paleontologists should expect great variation in the frequency with which different species are encountered. In consequence, compiling patterns of species loss before an extinction will show gradual loss, even if the extinction was as catastrophic as the K–T mass extinction.

Subsequent studies have confirmed these results, but have also produced statistical tests that can distinguish between catastrophic and gradual extinctions. The most significant advance is the introduction of tests that take the number of occurrences of a species or genus through a section of rock prior to the extinction and estimate the likely range of the last occurrence of the group in the rock record, to a given level of confidence. To date, these techniques have only been applied to a few events, but they have confirmed that the K–T mass extinction was essentially catastrophic at a section in Spain where most, but not all, species of bivalve and ammonoid disappeared coincident with the signal of the extraterrestrial impact (Fig. 19.3; Marshall and Ward 1996).

The best-preserved marine boundary section across the Permo-Triassic (P–T) boundary is near the village of Meishan up the Yangtze River from Shanghai. Almost two decades of collecting there by teams of Chinese paleontologists has produced detailed records of the precise stratigraphic occurrence of some 333 species (Fig. 19.4). Jin *et al.* (2000) employed this information and the absolute age dates of Bowring *et al.* (1998) to calculate the confidence intervals for the last occurrences of 93 genera. The results demonstrate that the extinction was not stepwise or gradual, but catastrophic, coinciding with an abrupt shift in carbon isotopes.

Accurate determination of the rapidity of mass extinctions requires not only the use of these statistical methods for fossil occurrences, but also careful

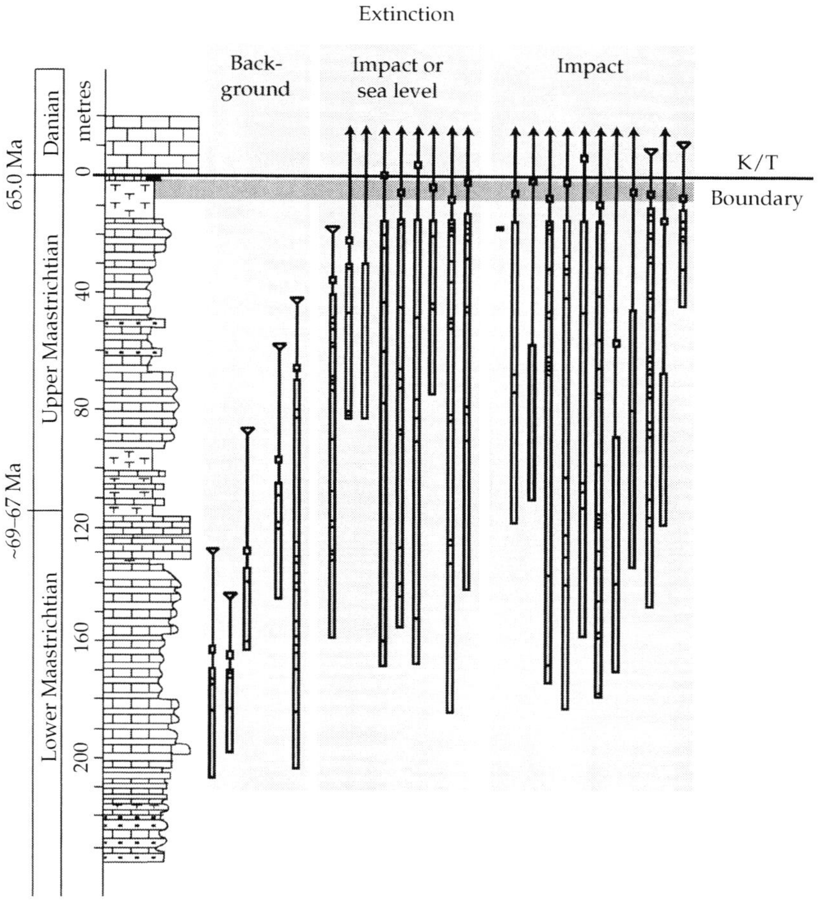


**Figure 19.3** Use of statistical confidence intervals to assess the patterns of disappearance through a Cretaceous–Tertiary boundary interval in Spain. Graph shows the observed temporal ranges (boxes) and occurrences (short horizontal lines) of 26 species of ammonites. Since the observed ranges necessarily underestimate true ranges, confidence intervals provide a means to correct the ranges. The 50 and 95 percent confidence intervals on final occurrences are represented by the small squares and inverted triangles, respectively. Species ranges extending beyond the boundary are indicated by arrows. The species fall into three categories: (1) those that became extinct prior to the boundary, (2) those that may have become extinct due to either impact or a drop in sea-level prior to the mass extinction, and (3) those that became extinct at the K–T boundary. Discussion and further details in Marshall and Ward (1996) (reproduced with permission from Pope *et al.* 1998).

correlation of sections both regionally and globally through biostratigraphy and isotopic shifts, commonly of carbon. Equally important is precise geochronology, but this generally depends on the identification of volcanic ash beds in appropriate sections. Precise dating of such volcanic horizons can yield precise estimates of the duration of an extinction episode. Ideally, multiple ash beds will occur within a single section, which allows the geochronologist to better assess the reliability of the radiometric dates. The Meishan section in south China is such a section, and the combination of the geochronology (Bowring *et al.* 1998), analyses of carbon isotopes, and the statistical analysis of the extinction demonstrates that the main extinction at this locality occurred in much less than

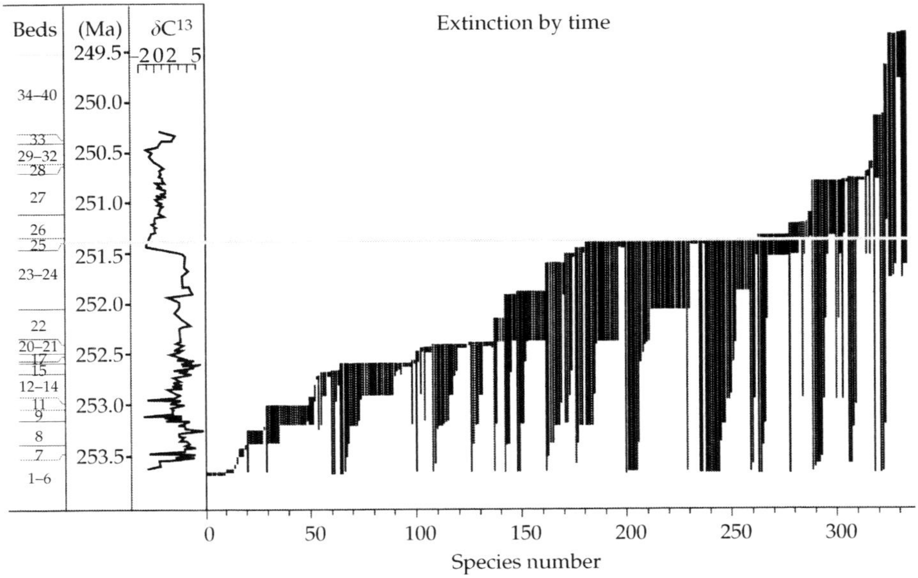


**Figure 19.4** The stratigraphic ranges (indicated by vertical lines) of 333 species through the Permo-Triassic boundary section at Meishan, Zhejiang Province, China. Species ranges are taken from five individual sections and projected onto a single, composite section. Fossil ranges and stratigraphy (left column) are scaled to time using absolute dates from Bowring *et al.* (1998). The carbon isotope record is also shown. Faunal change appears gradual on this plot of first and last occurrences, but statistical treatment of actual occurrences of genera through the section demonstrated that most disappearances above Bed 21 are consistent with a singe extinction event at Bed 25, the Permo-Triassic boundary. Figure after Jin *et al.* (2000), with permission.

500 000 years. The correlations using fossils and carbon isotopes strongly suggest that extinctions elsewhere around the world occurred over a similar timespan.

The rate of P–T mass extinction is probably better resolved than any of the other major mass extinctions. While it is apparent that the K–T impact began instantaneously with the impact of an extraterrestrial object in the Yucatan Peninsula of Mexico, the actual duration of the extinction has been harder to pin down. All the evidence suggests the extinction was catastrophic, and may have happened quicker than geologists can resolve with current methods. It is worth remembering sediment is not preserved at a constant rate through time. A hurricane may destroy 100 years of slow sedimentation and deposit a large volume in only a few days. Consequently geologists can rarely translate rock thickness into time, making calculations of rate dependent upon absolute age dating via geochronology.

For the other great mass extinctions, at the close of the Ordovician, during the late Devonian, and the end-Triassic, the information on duration is less precise. The end-Ordovician event at 439 mya involves two discrete pulses separated by perhaps 0.5–1 million years, with the onset of each pulse fairly rapid. With little geochronology through this interval, the rate of each pulse is not known. Several intervals of heightened extinction occurred during the Late Devonian. McGhee (2001) judges that five distinct extinction horizons are spread over 1–1.5 million years near the boundary between the Frasnian and Famennian stages of the Late Devonian. To date, no statistical tests of these extinction horizons have been performed, and since most of the data is from Europe and North America, the sampling problems described by Smith *et al.* (2001) may well apply to this interval. For the end-Triassic event the age of the extinction is well established at 199 mya, and appears quite rapid as well. Here, the extinction occurs during an interval with considerable high-resolution biostratigraphy and numerous reversals of the magnetic field, so even in the absence of more absolute age dates, the rapidity of the event is reasonably certain.

Ecologists make a useful distinction between press and pulse disturbances, which I extended to mass extinctions (Erwin 1996). **Press disturbances** are those which persist over a sufficient amount of time that adaptive evolution by some or all of the biota to the disturbance is likely. In contrast, a **pulse disturbance** is sudden and rapid; no adaptive response is possible. The K–T impact is the best example of a pulse disturbance. The impact occurred with extinctions following rapidly afterward and then the survivors dealt with the consequences. The end-Ordovician event involved a glaciation followed by deglaciation. As with the Pleistocene events, migration and **adaptation** is likely to have occurred during the onset of glaciation as well as the deglaciation. Due to differences in generation time and other factors that control the ability to respond to selection pressures, some groups may experience a particular event as a press extinction while other groups experience the same event as a pulse extinction. For example, microbes and large vertebrates or vascular plants would have very different responses. Although no one has examined this carefully yet, it is likely that the dynamics of the recovery period will differ between press and pulse events. Press events lead to a change in the composition of the biota during the extinction event itself, which is likely to modify the composition of the survivors and perhaps their ability to respond after the conclusion of the extinction.

## Causes of mass extinctions

The enigma and notoriety of mass extinctions has attracted the attention of many researchers over the past few decades and not a few less well-informed onlookers. Proposed causes of the five great mass extinctions runs from well-founded discussions of glaciations, extraterrestrial impacts, the effects of anoxic oceans and massive continental volcanic eruptions to poorly-grounded speculations involving collisions with antimatter, or the effects of nearby supernova. Regrettably, many hypotheses have been published by workers with only a passing familiarity with the events they purport to explain, producing papers that do little to advance our understanding of these events. The success of the Alvarez *et al.* hypothesis for the K–T extinction has emboldened an impact "mafia" who are convinced that all mass extinctions (and perhaps even many of the minor biotic crises) were caused by impacts. The lack of evidence for such a grand theory of extinctions is of little moment for enthusiasts who conveniently forget that much of the strength of the Alvarez *et al.* hypothesis lay in its testability and the supporting evidence from the spike in the abundance of iridium, a signal of impact, at the K–T boundary.

Constraining the rapidity of a mass extinction is a critical step in evaluating proposed causes. For example, for many years the cause of the P–T event was ascribed to the formation of the great supercontinent of Pangea, which combined most of the continental land mass. The recent improvements in temporal resolution from more detailed biostratigraphy, followed by high-resolution correlations using carbon isotopes and radiometric dating established that Pangea actually formed about 25 million years before the extinction (with no noticeable biotic effect) and that the extinction was far too rapid to be explained by the slow **continental drift**. This has eliminated plate tectonics from serious consideration as a driver of extinction, if not yet from many textbooks. Many other proposed causes of the P–T extinction that are incompatible with the rapid nature of the extinction have also been eliminated.

The K–T, P–T, and end-Triassic mass extinctions all evidently occurred in less than 500 000 years and perhaps in far less time. Such rapidity is consistent with extra-terrestrial impact, but there is, as yet, no convincing evidence of impact at either P–T or the end-Triassic events. That all three events coincide with massive continental flood basalts has also attracted considerable attention. The Siberian flood basalts cover an area about two-thirds the size of the United States with a series of flows that range up to 4000 m thickness. The flows appear to have been generated from at least four distinct centers over 1000 km apart, effectively refuting proposals that an impact may have initiated the volcanism. The Siberian volcanism includes extensive basaltic flows as well as more explosive ash deposits, but how such volcanism would trigger a global mass extinction remains a bit murky. Most suggestions involve either the eruption of large amounts of sulfur, which is converted to sulfuric acid and acid rain, or a combination of global cooling from the large volume of volcanic ash, followed by short-lived global warming driven by the release of large volumes of carbon dioxide and other greenhouse gases. Both models are very difficult to test directly. In particular, the volume of sulfur released from a volcanic eruption is exceedingly difficult to determine even days afterwards. The end-Triassic mass extinction coincides with the eruption of the Central Atlantic Magmatic Province (CAMP). Although CAMP has now been fragmented by the opening of the Atlantic, it appears to have been at least as large as the Siberian volcanics. Impact opponents have long noted the eruption of the Deccan volcanics in India also coincides with the K–T impact. Correlation is not causality however, and it is critical to remember that unlike the evidence for K–T impact, it is unclear how any of these massive continental flood basalts would actually trigger extinction. That the two largest continental flood basalts coincide with the P–T and end-Triassic extinctions is suggestive, but there are numerous oceanic flood basalts that are far larger than these continental eruptions. Yet, these oceanic basalts appear to have had no negative biotic effects (indeed some have suggested they are beneficial, providing important nutrients for marine ecosystems).

Impacts and continental flood basalts are just two of many geological features which have been correlated with multiple mass extinction events. Other paleontologists have invoked correlations with drops in sea level, glaciation, and oceanic anoxia. Many of these apparent correlations have not withstood rigorous study, while others remain problematic. Hallam and Wignall (1997) in their outstanding review of mass extinctions argue strongly, if not always convincingly, for the importance of anoxia in mass extinctions. Several mass extinctions are associated with **black shales** (muds generally deposited under low-oxygen conditions), which they take as evidence of an increase in sea level which spreads low-oxygen waters across the continental shelves, triggering extinction. Black shales and other potential evidence of anoxia appear to be associated with several of the large mass extinctions as well as some smaller biotic crises (such as the Cenomanian–Turonian extinction during the mid-Cretaceous). Unfortunately the evidence for anoxia may also be consistent with other mechanisms. For example, black shales may also reflect high productivity rather than extinction. Closer geological study has shown that the apparent correlation between glacial activity and mass extinctions, and between a drop in sea level and extinctions is less strong than it has appeared. The end-Ordovician mass extinction is, as described above, linked to a glaciation which causes a drop in sea level, then a rise when the glaciers melt. Claims for glaciation during other extinctions have been refuted. Similarly, sea level history is more complex than was understood several decades ago. Indeed what was generally accepted as the best connection between a mass extinction and a drop in sea level at the P–T has reversed, with considerable evidence showing sea level was generally rising at this point (Hallam and Wignall 1997). Although the history of life might be cleaner if a single cause underlay all mass extinctions, there is no reason to believe this is the case.

Patterns of biotic selectivity provide a vital clue to determining the causes of mass extinctions and other biotic crises. Are particular regions or habitats more severely effected? Is survival favored among particular ecological strategies, body sizes, or other adaptations? Although the geographic structure of mass extinctions has received some attention (next section), there is need for more detailed studies of the selectivities of most events. For example, Knoll *et al.* (1996) proposed that the P–T mass extinction preferentially removed species with relatively low metabolic capability and favored those with higher metabolic function. Their preferred cause for this pattern was carbon dioxide poisoning, which could be produced by a number of distinct mechanisms. Further advances in understanding mass extinctions requires establishing critical tests of extinction hypotheses rather than simply telling satisfying stories.

## Geographic structure of mass extinctions

One of the most poorly understood aspects of mass extinctions is how the patterns and processes differ across the globe. For many years paleontologists concentrated on the details of extinctions within local sections or on global compilations of extinction patterns, but paid relatively little attention to differences between regions. Such comparative regional studies require the most data of course, and several have now appeared.

Jablonski and Raup (1995), for example, examined the support for claims that tropical regions suffer greater extinction than other regions. In their detailed comparison of patterns among bivalves during the K–T mass extinction they found no support for these patterns. Reefs, which were excluded from the study, are largely restricted to tropical and subtropical settings and many studies have suggested increased extinction for this habitat. The evidence from Jablonski and Raup, however, suggests this pattern may be restricted to particular habitats rather than to specific biogeographic settings. Similar comparative studies, although without quite the same detailed coverage, have been carried out with other clades, and for other events.

The demise of mammoths, mastodons, the various saber-tooth cats, dire wolves, and the giant sloths along with other elements of Pleistocene land communities is a significant biodiversity crisis that illustrates the crucial role of data on rate and geographic structure in unraveling the causes of a mass extinction. Between 50 000 and about 11 000 years ago many species of the Pleistocene megafauna disappeared. Some paleontologists have argued that climatic changes associated with the Pleistocene glaciations were responsible. Others have noted the close association between the timing of extinction and the arrival of humans in different areas of the earth and suggested that hunting pressure was responsible for the mass extinction. How can we test these two hypotheses? If the climatic hypothesis is correct, we expect a close connection between extinction and evidence of climatic change. The human-induced overkill model suggests that the timing of extinction should be very different across the globe, but will coincide closely with the increases in human population in a region to the point where hunting becomes significant. Precise dating of the extinctions in different regions has now clearly shown that the overkill hypothesis is likely correct. In Australia, recent studies have shown that 55 species of large mammals and birds disappeared about 46 000 years ago, including *Genyornis*, the heaviest bird known (Roberts *et al.* 2001). This occurred only a few thousand years after humans first invaded that continent. Similarly, in North America the extinctions are much later but closely track the migration of humans from Asia and their increase in population size.

## The aftermath of extinction

An over-reliance on equilibrium models has hampered many sciences and none more so than diversity studies. The biotic response following mass extinctions is generally described as recovery, but this assumes that the biota is returning to something similar to the pre-extinction state. For some biotic crises this is true. But the significance of mass extinction lies in their removal of incumbent clades and this by definition prevents a return to a pre-extinction state. Equally important, mass extinctions create opportunities for constructing new ecological relationships. While overall diversity eventually returns, and may even surpass pre-extinction levels, for most mass extinctions there are few other similarities between pre- and post-extinction assemblages. At the level of ecological structure and function there is often little similarity and the concept of recovery is positively misleading. Unfortunately most alternatives to the term recovery are equally problematic, so the term will continue to be used here. The critical questions associated with the biotic recoveries of mass extinction are, what patterns of re-diversification are displayed by different clades, and how are these related to patterns of extinction? what ecological processes drive the recovery? what conditions and processes favor and inhibit the generation of the morphological novelties which often appear during recoveries? why do many recovery intervals show a lag, or survival interval, with no origination of new taxa immediately after the mass extinction?

## Models of biotic recovery

Investigations of post-extinction recovery have been heavily influenced by competition-driven ecological models, particularly the equilibrium models from MacArthur and Wilson's (1967) theory of island biogeography. Sepkoski and others extended this approach to describe the pattern of clade dynamics through the Phanerozoic (543 mya to present), with particular emphasis on the impact of mass extinctions on distinct evolutionary assemblages of clades (Carr and Kitchell 1980; Sepkoski 1984, 1996). In these models there is a logistic recovery phase, and any differences in recovery dynamics between different clades reflects their intrinsic diversity dynamics. Such logistic growth also produces an initial lag phase before the exponential increase in diversity. Absent from such models is any explicit consideration of ecological dynamics other than competition (coupled logistic models incorporate competition between groups with different rates of diversification: Sepkoski 1984, 1996). From this view of recovery flows the most common definition of the recovery interval: from the end of the mass extinction through the phase of exponential growth.

But absent from such models is any more detailed consideration of the rebuilding of actual ecosystems during the recovery process. As the first step towards such models, Solé *et al.* (2002) constructed a model with three trophic levels (primary producers, herbivores, and carnivores). Future models will incorporate omnivory. Even this model exhibited interesting dynamics, including distinct (and testable!) differences in rates of recovery between the different trophic levels, and threshold phenomena where the rate of recovery differed depending on the magnitude of the extinction. More sophisticated models of recovery are urgently needed, as they serve as a vital guide for field studies.

## Patterns of biotic recovery from mass extinction

The specifics of recovery patterns associated with the major mass extinctions have been reviewed elsewhere (Hallam and Wignall 1997; Erwin 1998, 2001b). The general patterns that emerge from such syntheses are (Erwin 2001b), (1) Immediate low-diversity post-extinction assemblages dominated by **eurytopic** species (species with broad environmental tolerances) have been documented for Late Cambrian trilobites, several Silurian groups, Late Devonian corals, a number of earliest Triassic clades, including bivalves, and many early Tertiary groups. Such assemblages have not been found, and may not occur, for the end-Triassic event and several smaller biotic crises. However, studies of extinction dynamics among benthic molluscs for the K–T extinction demonstrate considerable regional heterogeneity (Jablonski 1998). Jablonski's comparison of four different regions showed bursts of opportunistic clades immediately following the extinction in the North American Gulf and Atlantic coastal Plain that were missing from the other three regions. (2) There have been many claims that mass extinctions preferentially remove morphologically ornate species. Saunders *et al.* (1999) did find support for this in their study of Paleozoic ammonoids. Although ammonoid sutural complexity did increase from the Devonian through Triassic, the Late Devonian and P–T mass extinctions reset the trend, so the post-extinction faunas were dominated by morphologically simpler clades. In contrast, Hansen *et al.* (1999) found no evidence that the K–T mass extinction or three other Cenozoic events reset trends in predation-resistant morphologies among molluscs. (3) Reefs are often viewed as particularly susceptible to mass extinctions, but Wood (1999) has observed that the controlling variable may be the presence of carbonate shelves. She argues that the apparent lag in the reappearance of reefs after mass extinctions actually reflects the reestablishment of the carbonate platforms, rather than being an intrinsic feature of reef ecosystems. Moreover, other recent studies suggest considerable regional variation in any "reef gap" in the aftermath of mass extinctions.

The various case studies of specific recoveries remain far fewer than necessary, but they demonstrate a rich variety of patterns, far beyond the simple logistic recovery patterns of early models. This suggests an equally heterogeneous array of processes underlie biotic recoveries. The apparent rapidity of recovery once it begins suggests that positive feedback loops may be significant in the reconstruction of ecological relationships.

How quickly does recovery occur? Views of the rapidity of extinction are highly dependent upon the definition used of recovery, but nonetheless appear to vary widely between recoveries. The **recovery interval** extends from the increase in diversification rate after a biotic crisis until the decline in diversification to pre-extinction levels; the **survival interval** extends from the extinction until diversification rates increase. In few cases is there sufficiently high-resolution geochronology to determine with a high degree of confidence the duration of the recovery interval. As summarized in Erwin (1998), most survival intervals appear to last a few hundred thousand to a million years. The exception is the early Triassic survival interval which appears to extend over several million years for most benthic marine groups as well as terrestrial plants and tetrapods. Recovery intervals commonly last longer than the survival intervals, and the admittedly poorly resolved data suggests a positive relationship between the duration of the survival interval and the duration of the recovery phase. As with mass extinctions, there is a strong need for more highly resolved chemostratigraphic and radiometric dating of recovery intervals and the associated evolutionary patterns.

### The transition from recovery to "background" intervals

Perhaps the most challenging current problem in post-extinction recoveries involves the end of the survival–recovery interval. If the recovery interval is defined on the basis of increased origination rates this transition is marked by a decline to "background" levels of origination (generally coupled to increased extinction rates). Employing a more ecological definition of recovery, this transition could be defined by a return to normally functioning ecosystems. The issue of how one identifies "normally functioning ecosystems" is not simple. Carbon isotope studies can reveal carbon cycling among primary producers and some components of open ocean ecosystems (D'Hondt *et al.* 1998). Simulations of recovery with multiple trophic levels involved in the model suggest that recovery of higher trophic levels may substantially lag primary producers (Solé *et al.* 2002) so isotopic records do not necessarily provide a clean resolutions of this issue. Little attention has been paid to this issue, and thus the end of recovery intervals are fairly fuzzy. This is not surprising, since different clades respond to extinctions in different ways. Ammonoids rediversified within a few hundred thousand years of the P–T event, quickly resuming their normal boom and bust dynamic. But this very quick recovery is not mirrored by other marine or terrestrial clades, as discussed above. Perhaps the most curious aspect of this transition occurs during the lengthy Early Triassic recovery. Among gastropods, bivalves, brachiopods, and several other groups many genera disappeared from the fossil record well before the extinction, only to return, as if from the dead, near the end of the recovery phase. This Lazarus phenomenon is consistent with the hypothesis that positive feedback facilitates the return of some clades.

## Evolutionary impact of mass extinctions

The removal of incumbent taxa and restructuring of ecological relationships has long been recognized as the most significant effect of mass extinctions. As Jablonski (2001) observes, one of the most intriguing issues is the interplay between continuity of long-term evolutionary trends and the disruptions imposed by mass extinctions. None of the mass extinctions have entirely reset the evolutionary clock, and Jablonski (2001) identifies four macroevolutionary patterns associated with mass extinctions:

**(1) Unbroken continuity**, with little change in long-term patterns, including ecological dominance by some groups, escalatory responses between molluscan predators and prey and patterns of offshore expansion and onshore retreat among clades of marine invertebrates.

**(2) Continuity with setbacks**, where mass extinctions perturb long-term trends, but the trends resume. Jablonski identifies the progressive trend toward dominance of cheilostome bryozoans over cyclostomes, the expansion of burrowing bivalves, and the increase in sutural complexity in ammonoids.

**(3) Dead clade walking**, one of the most revealing patterns associated with extinctions and recoveries. A number of clades have been identified that survive the extinction, yet perish during the subsequent recovery. Examples include bellerophontid gastropods and prolecantid ammonoids at the P–T event and speriferoid brachiopods during the end-Triassic. One explanation for this pattern may be that the pattern of recovery altered the ecological structure of communities such that some adaptations were no longer viable. This appears to be the explanation for the demise of the bellerophontid gastropods (likely a polyphyletic assemblage rather than a clade in any case): the growth of mollusc-crushing predators during the Triassic doomed a shell morphology that was easily crushed. The frequency of the dead clade walking phenomenon demonstrates again the complexity of post-extinction processes.

(4) **Unbridled diversification** includes such well-known events as the diversification of placental mammals following the elimination of dinosaurs and other reptiles during the K–T mass extinction. This innovation associated with mass extinctions is perhaps their most significant effect. By removing incumbent clades and allowing the generation of new ecological relationships mass extinctions exert a powerful creative force on the history of life. The demise of the bellerophontid gastropods in the Middle Triassic and the battle for supremacy between large flightless carnivorous birds and placental mammals in the Early Tertiary are both evidence of the contingent nature of biotic recoveries, limiting our ability to predict the course of evolution following a mass extinction.

Studies of the end-Permian mass extinction and biotic recoveries have been funded by the Exobiology Program and the National Astrobiology Institute of NASA and the Walcott fund of the Smithsonian Institution. Studies of biotic recoveries have also received support from the Thaw Charitable Trust through the Santa Fe Institute.

PART V

# Behavior, evolution, and human affairs

CHAPTER 20

# Play: how evolution can explain the most mysterious behavior of all

Gordon M. Burghardt

In a visit to a zoo, one is apt to see gibbons vigorously, but gracefully, swinging through branches or rope "vines" with the greatest of ease. In a nearby enclosure, juvenile chimps chase each other, wrestle, and seem to laugh. In the polar bear exhibit empty beer kegs are tossed around with abandon. Moving on to the bird section, an adult Andean condor repeatedly attacks, grabs, and tosses a small red hard rubber object. All of these actions are enjoyable for the visitors to watch, but why do they even occur? These energetically costly activities are not actions that seem directed at accomplishing important ends such as eating, escaping from enemies, or mating. When considered thoughtfully, their widespread existence in a world where every action, like every animal's form and color, should be shaped and tested through natural or **sexual selection** raises increasingly difficult issues. Perhaps these difficulties underlie why modern textbooks of animal behavior or evolution typically ignore or only briefly discuss play.

The modern study of evolution has certainly made many strides in understanding both the relationships among taxa and the processes underlying them. But whether and, more importantly, how evolution by natural selection can explain the brain, behavior, and mind of animals, particularly the human animal, is still a controversial issue debated since the early days of evolutionary study. Alfred Russel Wallace, co-discoverer with Charles Darwin of natural selection, became increasingly skeptical that material processes could produce the human brain, and eventually became dualistic, even spiritualistic, as his confidence in the power of natural selection to explain behavior waned (Wallace 1901). Today, as at the end of the nineteenth century, there are many exciting attempts to extend Darwinian evolution to the behavior and mental life of humans including the origins of language, emotions, gender differences, cognitive processing, and morality. Still, more than 100 years after Darwin, Wallace's skepticism still casts a shadow over many evolutionary scientists who have as great a difficulty in accepting evolutionary interpretations of human behavior and psychology as do creationists. Thus, Richard Lewontin recoils from the evolutionary analysis of human cognition and mental abilities and has stated "It might be interesting to know how cognition (whatever that is) arose and spread and changed, but we cannot know. Tough luck" (Lewontin 1998, p. 130). Similarly, there are claims that play can never be understood scientifically (Hyland 1984) or evolutionarily because of its "intentional" character (Rosenberg 1990).

Fortunately, the true spirit of science is never to say never; the purview of science is continually growing, for good or ill, as it invades virtually all aspects of modern life; evolutionary science is no exception. Although the study of the evolution of cognition, in spite of Lewontin, is well underway (Bekoff *et al.* 2002), a comparable effort has not been forthcoming in terms of play. A true test of the power of evolutionary analysis might be, then, to apply natural selection and the **comparative method** to understanding this greatly neglected topic. Just as Darwin felt that he needed to be able to explain "instinct" and sterile castes in social insects for his views to have any chance of success

(Darwin 1859), today we also need to take on the tough cases, and not whine about "tough luck" if we truly want to demonstrate the universal application of evolutionary thinking.

Although many fields have claimed to be evolutionary biology's champion in understanding behavior including psychology, genetics, sociobiology, neuroscience, and anthropology, ethology is probably the prime integrative vehicle to accomplish this effectively. In addition to outlining the general agenda of ethology, it is applied here to trying to understand the most evolutionarily enigmatic behavior of all, play.

## Basics of ethology

Ethology is the *naturalistic study of behavior from an evolutionary perspective*. Its origins lie in the post-Darwinian synthesis of comparative biology, comparative psychology, and natural history (Lorenz 1981; Burghardt 1985). Ethology was responsible for reawakening interest in the study of instinctive behavior in animals (Tinbergen 1951; Burghardt 1973). Initially controversial, such behavior patterns are now a feature of most current animal behavior study. Although the labeling of behavior in animals as innate, instinctive, or genetically mediated was hotly debated in the 1960s and 1970s, by the 1990s such concepts again became acceptable in scientific discourse, even in discussions of human behavior. As biology begins to truly understand how genotypes are expressed as phenotypes, how nervous systems evolve and operate, and how behavior serves as the main locus of the interface among animals, their surroundings, and the operation of natural selection, the integrative properties of ethological analysis will have growing impact on many aspects of evolutionary study.

Ethology is based on a careful description of what organisms do in natural contexts. These observations are often critical, particularly for little known species or for events previously unseen in their natural contexts, such as predation, mating, and antipredator defense. However, to form the basis of solid subsequent behavioral research, a more formal behavioral system, often termed an **ethogram**, is necessary. Typically, this is expressed as an objective descriptive catalogue of the behavior patterns of interest, as established in either field or captive environments providing natural settings and stimuli (Lehner 1996). The quality of research using behavioral variables is often limited by the accuracy and objectivity of the behavioral measures used. Non-behavioral scientists adding behavioral measures to their investigations in genetics, neuroscience, and ecology often underestimate the effort and knowledge needed to collect valid ethological data.

Having determined accurate measures of behavioral actions (computers and video/audio recording have been invaluable in this task), the actions can be combined, along with associated environmental stimuli and internal motivational states, into a **behavior system**. These systems involve related activities such as foraging, mating, and parental care. Five specific classes of questions can and should be asked about every type of behavior. Niko Tinbergen, the Nobel Prize winning ethologist formalized four of these *aims* of ethological analysis (Tinbergen 1963). The first aim is to study what factors or mechanisms control behavioral performance. Included here are physiological, sensory, ecological, social, and other processes. The second aim focuses inquiry on the development of the behavior in the life of the individual. The third aim is to study the adaptive function of performing the behavior in terms of enhancing **fitness** at some level (e.g. individual, family, or group). The fourth aim is to inquire into the evolutionary origins and phyletic radiation of a behavior and the processes involved, which may be cultural as well as genetic. Recently, extensive studies on the emotional, cognitive, and other "inner life" processes in animals' lives have led to the addition of a fifth aim, that may be best termed private experience (Burghardt 1997). New methods involving brain imaging, neurotransmitter identification, neuroendocrinology, gene **deletions** and additions, cloning, virtual reality, and computer simulations are making possible scientific breakthroughs in learning how animals perceive, experience, and process events in their lives. The five aims are summarized in the first three columns in Table 20.1.

For ethologists and other students of behavior, none of this is new, although the fifth aim is

**Table 20.1** The five ethological aims as applied to play behavior

| Name | Related terms | Description | Application to play |
|---|---|---|---|
| Control | Causation, mechanism | Internal and external factors underlying behavioral performance | Do animals play more at high temperatures? Is the neocortex of the brain essential for play? |
| Adaptive function | Function, survival value, adaptiveness | Contributions of behavior patterns to individual, group, reproductive, and inclusive fitness | Do animals that play chase more in their youth run faster as adults? Do animals that play fight more have more offspring? |
| Development | Ontogeny | Patterns and processes in behavioral change during individual lifetimes | How does play fighting change in frequency between juvenile and adult stages? |
| Evolution | Phylogeny, genetic, and cultural inheritance | Historical patterns and processes in behavioral change across generations and taxa | Did pretend object play evolve independently in cats and apes? |
| Private experience | Personal world, phenomenal world, subjective experience, heterophenomenology | Patterns and processes in life as experienced | Is all play accompanied by one or a few specific emotions? |

typically the least studied. In the study of play, however, the application of these aims was not well articulated nor stressed until the 1980s (Martin 1984). Once these aims were applied, however, guidance in formulating research questions at different levels was facilitated. Some questions on play derived from the five aims are listed in Table 20.1.

In this chapter, I briefly address a few major questions such as the following: Is playing a distinct behavioral category that can be objectively characterized? Where does playfulness first appear in animal evolution and did it evolve just once or repeatedly? Do all kinds of play share common causal mechanisms deriving from common ancestors? What factors led to play becoming prominent in the lives of so many animals, yet absent in so many others? Is play just an evolutionary sideshow or could it have played, and still play, a critical role in the evolutionary drama. The answers to these questions are neither simple nor obvious and must consider all five ethological aims.

The key element of this ethological approach is that the same behavior must be examined from different perspectives for adequate scientific understanding. In recent decades, sociobiologists and behavioral ecologists emphasized the study of animal behavior from a largely adaptive perspective. This had consequences for the study of play, since the adaptive function of play was not only unclear, but also difficult to empirically demonstrate. This almost sole emphasis on finding an empirically supportable function for play was the major focus of researchers both in animal play and in child development (Power 2000). Furthermore, play seemed to be characterized by subjective factors, such as "having fun" (Spinka *et al.* 2001), which contributed to making play a neglected topic by scientists wanting to be considered rigorous. Since for traditional behaviorists and mechanistic biologists, fun appeared impossible to study in animals, especially "lower ones" such as rodents and birds, the basic premise of studying play as a biological phenomenon was thus highly dubious from the start! We can do better today as emotion becomes a rigorous area of study (e.g. Panksepp 1998), but first we need to know what constitutes play.

## Defining play

The traditional categories of play are locomotor, object (including predatory), and social. These categories, although useful, frequently overlap and are not distinct: tug of war games involve objects and chase games involve locomotion (Fagen 1981; Burghardt 1998b). Thus, in order to compare play across species we must have some way of isolating play behavior using a definition that is not specific to any one kind of play (such as social play). Another important problem is that both the proximate causes and adaptive functions of play may differ for different kinds of play, differ throughout ontogeny, and differ across species. Such complexity suggests that a satisfactory characterization of play must use several criteria, none of which is sufficient alone. In fact, five such criteria seem necessary to characterize play; all must be satisfied in at least one respect (Burghardt 2001).

The first criterion for recognizing play is that the performance of the behavior is not fully functional in the form or context in which it is expressed; that is, it includes elements, or is directed toward stimuli, that do not contribute to current survival. The critical term is "not completely functional," instead of "purposeless," nonadaptive, or having a "delayed benefit."

The second criterion for recognizing play is that the behavior appears to be spontaneous, voluntary, intentional, pleasurable, rewarding, or "done for its own sake." Only one of these often overlapping concepts need apply. Note that this criterion also accommodates any subjective concomitants of play (having fun, enjoyable), but does not make a subjective state essential for *recognizing* play, even if it is a concomitant of much play.

The third criterion for recognizing play is that it differs from the "serious" performance of ethotypic behavior structurally or temporally in at least one respect: incomplete, exaggerated, awkward, precocious, or involves special signals. It also implicitly acknowledges, but does not require, that in many species play is found only in juveniles.

The fourth criterion for recognizing play is that it occurs repeatedly in a similar form, during at least a portion of the animal's ontogeny. Repetition of patterns of movement is found in all play and games in human and nonhuman animals.

The fifth criterion for recognizing play is that it is initiated when the animals are adequately fed, healthy, and free from stress (e.g. predator threat, harsh microclimate, crowding, social instability), or an intense competing behavior system (e.g. feeding, mating). In other words, the animals are in a "relaxed field." This is an essential ingredient of play, as play is one of the first types of behavior to drop out when animals are hungry, threatened, or under inclement environmental conditions.

These criteria seem to cover all types of play and exclude what is generally not considered play, such as obsessive–compulsive and **stereotyped** behavior found in animals in small sterile quarters, and the initial exploratory behavior of animals in strange settings or when confronting unfamiliar objects, species, or conspecifics.

Keeping in mind the nuances underlying each word, a one sentence definition could then read as follows: *Play is repeated, intrinsically rewarding, but incompletely functional, behavior differing from more serious versions structurally, contextually, or ontogenetically, and initiated when the animal is in a relaxed or low stress setting.*

Much play is more complex than the simple gamboling of a colt or the repeated pouncing of a cat at feathers attached to a stick. Social play is often extraordinarily complex and may involve special signals used to solicit or mark off play (Bekoff 1995) and these may be postures, facial expressions, vocalizations, and even odors. Social play may involve role reversals where otherwise larger, older, or more dominant animals will sometimes self-handicap and "lose" the game in order to keep it going. Social play may be used in courtship in adult animals as well as be a prominent feature of the behavior of young dogs, seals, rats, monkeys, deer, and horses. The details of play can vary greatly by species, sex, age, and in different contextual and ecological settings (Power 2000). Such variation has made the study of play so challenging. Complex social play, as described above, has been the focus of most play research. However, it is useful to see such elaborate play as derived from more elementary forms.

## Theories of play

Over the years, many theories of play have been advanced (reviewed in Fagen 1981; Burghardt 1998b, 2003). The most popular view for over a century, formulated by the philosopher Groos, was the **instinct practice theory** that made two claims. The first is that play is so common and elaborate in animals, particularly mammals, that it must be adaptive. Second, since play is primarily found in young animals and is by definition not functional when it is actually occurring, its benefits must be delayed and only apparent in "serious" adult behavior that may necessitate learning for adequate deployment of often complex social, predatory, and defensive abilities. The intuitive appeal of these claims in areas from child development to primate ethology to rodent biopsychology was barely hindered by the virtual complete lack of convincing experimental data supporting either one (Martin and Caro 1985). In many and various guises this future oriented practice view is the most common today in both human and nonhuman animal play research and reflects the view that play must have profound value or it would not exist.

Historically, there were other theories that viewed play as not geared toward practice of adult behavior. Perhaps, as advocated a century ago by the eminent child psychologist, G. S. Hall, play represents ancient instinctive behavior patterns that reappear during ontogeny as a form of **recapitulation** without any necessary long-term functional value. Juvenile play could be developmentally and motivationally transient phenomena as adult behavior patterns become reorganized or developmentally mature from other behavioral systems. Another major view, most forcefully advocated by Herbert Spencer, was that play, being found in behaviorally complex species when well-fed and healthy, is due to having an excess of energy: the **surplus energy theory**. There are, thus, several options: Play is preparation for the future, play is a legacy from the past with at most some lingering immediate function, or play is just a byproduct of metabolic excesses and spare time. Later, I will present a modern integrative and synthetic view termed **surplus resource theory** (SRT) that focuses on the key issue of play's origins, not the often complex versions seen in the most playful species.

**Table 20.2** General categories of proposed benefits of play

| Benefit | Example |
|---|---|
| Enhance motor development | Improve coordination in locomotion |
| Enhance physiological development | Improve cardiovascular system and endurance |
| Enhance perceptual-motor coordination | Improved integration of sensory modalities |
| Practice adult species-typical behavior | Improve prey capture or parental abilities |
| Social-communicative skills | Improve ability to react appropriately to others |
| Establish social roles | Determine dominance–submission status or gender roles |
| Gain information | Learn about what objects/other animals do |
| Enhance neural development | Enhance/consolidate/integrate neural pathways |
| Enhance cognitive abilities | Improve responses to environmental change |
| Stimulate creativity | Source of novel or modified behavioral responses |
| Competence assessment | Allow parents to assess normative development and health |

It should be evident that the theoretical perspective one takes on play will influence the kinds of data one gathers and the experiments one conducts. For example, adherents of the practice view often assume that play is exclusively found in large-brained animals, especially advanced mammals and is a mark, even a cause, of cognitive complexity. Many of these functional interpretations of play are listed in Table 20.2 along with several hypotheses based on more immediate effects such as enhanced cardiovascular or muscle development via exercise (see also Bekoff and Byers 1998; Power 2000). Regardless, the lack of established adaptive functions for play makes this aim not the best one to begin with; the other four ethological aims may provide insights that are more useful.

## The phylogeny of play

A serious conceptual problem is that the search for an adaptive functional basis for play is often conflated with the study of the evolution of play and the origins of play (Burghardt 1984). Remember that a hallmark of ethology is having good descriptions of what animals do. If one has these, then the next useful step is to compare species using a phylogenetic approach. Although there are many descriptions of play in familiar animals, it turns out that for many species data are few, opportunistic, or anecdotal. Nonetheless, using the play criteria developed above and applying them conservatively, but objectively to the data that can be gathered from the literature (Fagen 1981; Power 2000; Burghardt 2003), provides some useful phylogenetic perspectives.

In the first comparison (Fig. 20.1) all three major types of play are mapped on to a modern phylogeny of placental mammals (Liu *et al.* 2001). Play is not equally distributed among the orders but is most prevalent in carnivores, primates, elephants, and a few other groups. This analysis is misleading in several respects, however, as the number of species in these orders varies greatly from one or two to dozens, to hundreds. Thus, while all primates studied show all three major kinds of play, many rodents show very little while others, such as laboratory rats, are very playful (Pellis and Iwaniuk 1999b). Bats are also a very large group of mammals, but little play has been documented in them, object play in fruit bats and perhaps social play in vampire bats being the primary examples known. Note, however, that even with the five play criteria, our lack of knowledge about relatively "alien" species makes it harder for us to recognize play if it does occur. This problem only gets more difficult as

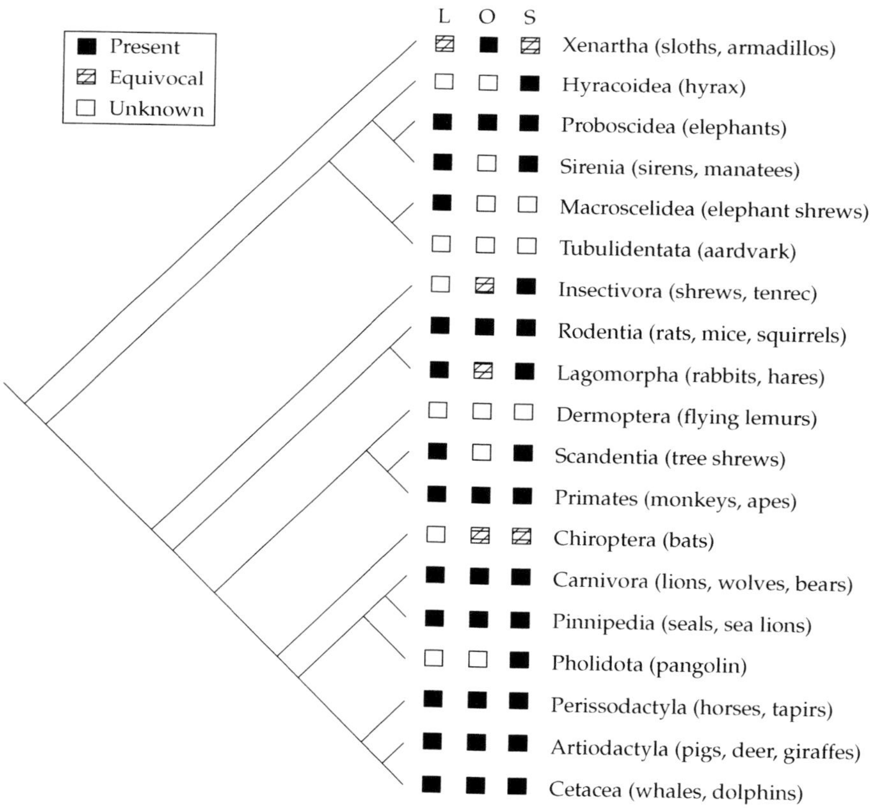


**Figure 20.1** The three major categories of play as found in the existing orders of placental mammals.

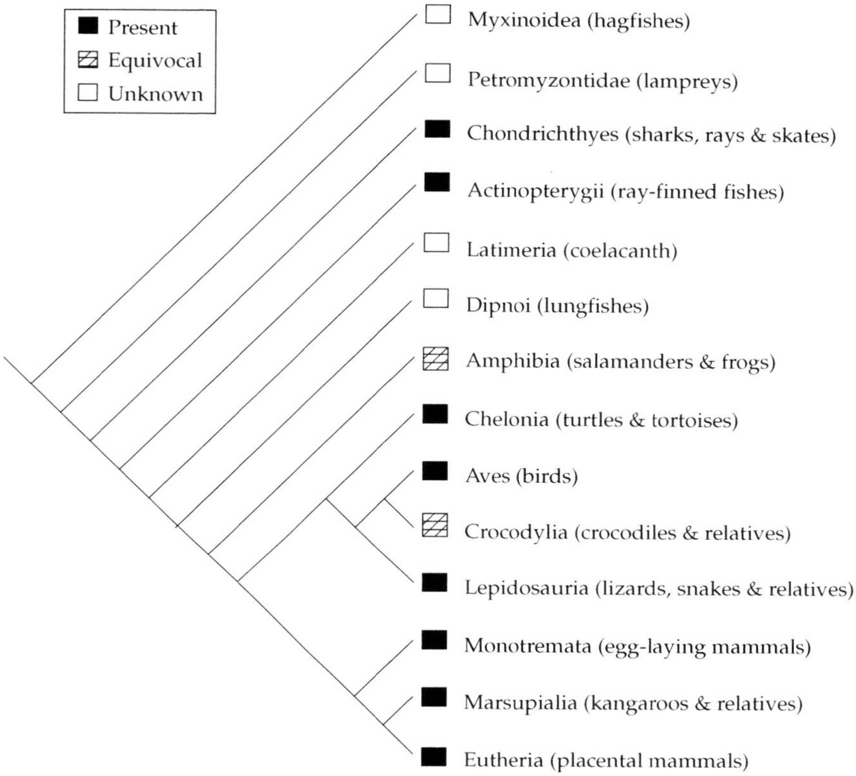


**Figure 20.2** The occurrence of play in the major vertebrate groups.

we consider animals even more phylogenetically remote from us than bats.

Play is not only found in placental mammals (Fig. 20.2). Play has been recorded in many other vertebrates including marsupials, monotremes (the duck-billed platypus), birds, and is now even well documented in some turtles and lizards (Burghardt 1998a). A few turtles and lizards interact with physical objects, including toys, in very similar ways to mammals and birds. Turtles push around balls and swim through hoops and Komodo dragons bang buckets, repeatedly grab and shake shoes, and engage in tug-of-war over metal cans with familiar keepers; all of this without feeding or aggressive motivation. Several disparate groups of fishes show what is arguably locomotor and object play according to the five criteria including manipulating and balancing objects, leaping, and perhaps even chase games. Furthermore, there is increasing evidence for playlike behavior in some invertebrates including insects, crustaceans, and mollusks (e.g. Mather and Anderson 1999). The distribution across phyla again suggests the multiple origins of playlike phenomena (Fig. 20.3).

Taken together, then, the comparative data, even at the level of class and phylum, suggest that play is neither unique to one group of animals nor arisen only once or a few times. It is highly unlikely that play was found in a billion year old common ancestor of today's vertebrates and invertebrates. Thus, playlike behavior most likely arose many times throughout evolutionary time. From what we know to date, however, complex social play is only found in highly derived taxa in major endothermic radiations: placental mammals, marsupials (Watson 1998), and birds (Ortega and Bekoff 1987). Recent

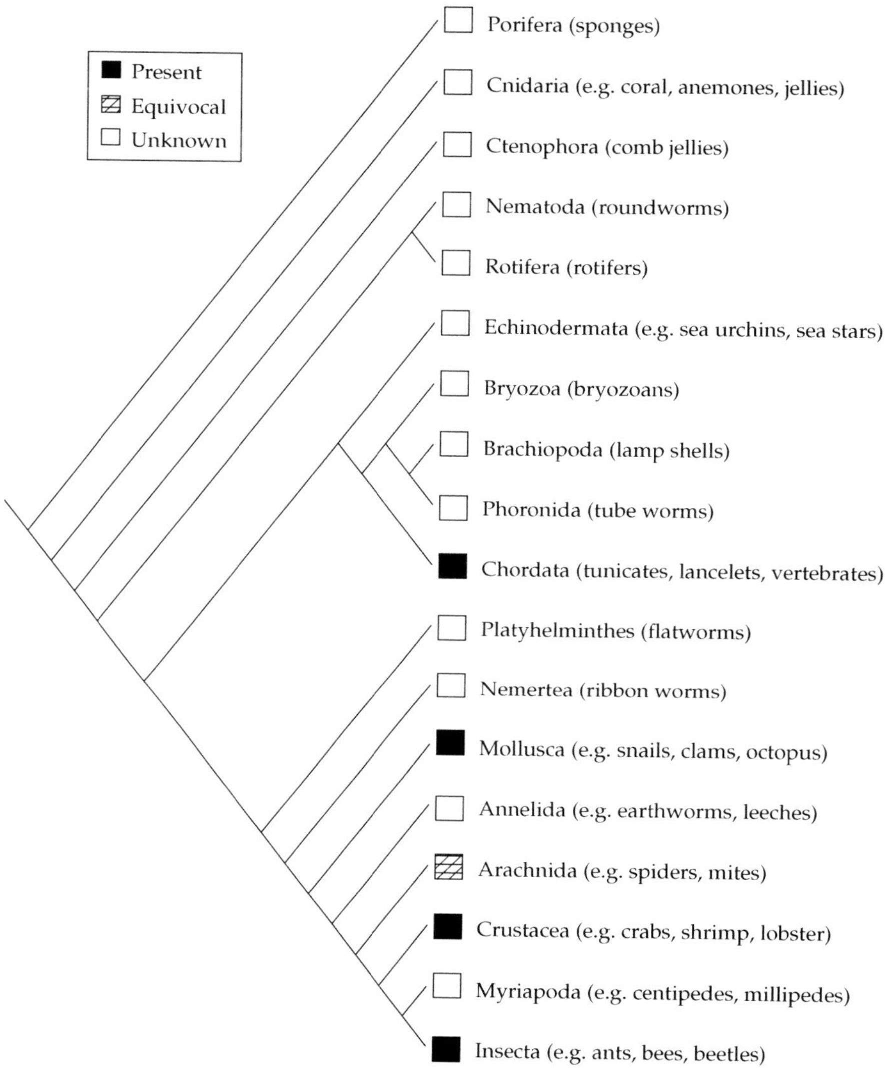


**Figure 20.3** The occurrence of play in the major invertebrate phyla.

phylogenetic analyses at the level of order or lower show marked differences in the extent of different kinds of play within them (Pellis and Iwaniuk 1999a,b). Such data suggest that major features of social play behavior may have been lost and derived within orders and families, suggesting the operation of strong phylogenetic and functional processes.

But how did the myriad forms of play and play-like behavior originate? To answer this question we cannot assume that all behaviors satisfying the play criteria have the same causal (neurobiological) mechanisms or that they are evolutionarily **homologous** and have common ancestral roots. A species' normal behavior, ecology, development, metabolic rate, neural organization, and phylogeny influence the manner in which play is expressed. In fact, it is because play is almost certainly a heterogeneous category that arose repeatedly in the evolution of animals that we need clear criteria to help us sort out the processes and variables influencing playfulness and point the way to experiments and critical phyletic comparisons.

In trying to explain the possible origins of play, it is useful to divide play into three types that outline a broad evolutionary scenario.

**Primary process play**: play behavior as an outcome of factors not related to any direct action of natural selection on the play behavior itself. Such play could be a byproduct of lack of external stimulation, excess metabolic energy, motivational factors, and other sources.

**Secondary process play**: play behavior that, once occurring, has evolved some role, though not necessarily an exclusive or even major one, in the maintenance or normal development of physiological and behavioral capacities.

**Tertiary process play**: play behavior that has gained a major, if not critical, role in modifying and enhancing behavioral abilities and fitness. Such play may also produce variations on species typical behavior that selection could then shape and prune, leading to major behavioral reorganization.

Primary process play is most likely to be found in animals that play rarely or simply. Secondary and tertiary process types of play have been selected to perform some adaptive functions such as those outlined in Table 20.2. These functions are both diverse and often difficult to evaluate. It also needs to be recognized that just because a behavior serves a function currently, it does not mean that it originated as a result of selection for that adaptive function. Which came first in the evolution of birds: feathers for warmth or for flying?

Armed with this comparative information and the conclusions we can draw from it, let us briefly look at some of the other ethological aims through research findings on play that may assist in providing an integrative view of play evolution.

## The control of play

The mechanisms and contexts underlying play include those concerning metabolic rate, behavioral complexity, behavioral capacity (how active can a species be?), body condition and well-being, hormonal state, brain mechanisms, and types and level of environmental stimulation. The nature of social partners, social organization, parental care, and prior experience with the object or play partner are important as well. This brief overview documents that the influences on playfulness are indeed diverse, the pathways among mechanisms complex, and that explaining all types of play is not an easy task. Although only a few phylogenetic analyses are yet published, certain patterns seem evident.

Vigorous play bouts necessitate considerable aerobic metabolic capacity: Baboons and kittens are much more capable of this than sloths and are much more playful. This may be, in turn, related to the nature of the diet: Folivores (leaf eaters) seem less energetic than omnivores and carnivores, and have a lessened capacity for sustained vigorous behavior (McNab 1988). Surviving on a sparse or low energy diet may also require spending more time foraging. There is also an apparent relationship of play with body size: within a lineage small species are generally less playful than moderate to larger species (Burghardt 1988). This, in turn, is related to metabolic rate and thermoregulation: Very small endothermic animals such as shrews and small hummingbirds expend much energy maintaining adequate body temperatures and thus must ingest large quantities of food, which in turn limits the amount of "excess" time for play, if play is not essential (Burghardt 1984). But metabolic rate and size are also related to longevity, so if play has long-term benefits, then they should accrue most to species with longer lifespans, so that any early investment (costs) in play can accrue sufficient interest (benefits) to be selected.

Such considerations, complex as they are, suggest some additional trends that should be found in comparative studies of play. Thus, in animals that do play, play should be found more frequently when adequate food is available and environmental stress is low; research on many taxa support this generalization (Burghardt 1988). Much play appears derived from, or similar to, the species characteristic (instinctive) behavioral repertoire. Animals that are relatively non-specialized in their food habits, especially active predators, extractive foragers (e.g. cebus monkeys), and scavengers (crows, condors) might be expected to show particularly complex object play. Species that are more often prey than predators (deer, mountain goats) typically engage in more locomotor (escape-related) play than object play.

Other contextual factors of a less specific nature may also be involved. Many animals need a certain amount of stimulation for proper perceptual, motor, and neural development. Play may be a means to provide such stimulation at times in an animal's life when this is otherwise not available. As parental care evolved, young of many species spend their early lives in rather boring environments with little stimulation, and this could have provided an important setting factor for engaging in activity "for its own sake." Social play as well as object and locomotor play could all be facilitated under such conditions. Play is also known to be much more prevalent in well-cared for captive animals buffered from needing to provide for their own shelter, defense, or nutrition. Thus, avoiding boredom and increasing sensory experience may be a potent primary process in the origins of play.

Sex differences in social play are commonly found, with males typically the more active and engaging in more play fighting and wrestling. This has been especially noted in polygynous species with intense male–male competition for mates and other resources. The assumption that such fighting is causally related to adult fighting is not, however, proved by such observations. Those predisposed to the practice view often accept such reasoning uncritically, but the recapitulation and surplus energy views can also explain sex differences. Furthermore, there are exceptions and play fighting may actually be more similar to adult sexual, than fighting, behavior (Pellis 1993). Adult social play in primates is found more often in those without a rigid hierarchical social structure (Pellis and Iwaniuk 2000).

Ultimately, however, the main control mechanism for play must be traced to the functioning of the nervous system. The roles of various brain loci and neurochemicals in play can only be briefly touched on here (Panksepp 1998; Siviy 1998; Burghardt 2001). Play, being a protean concept, involves many neural systems and the course of vertebrate brain evolution is still highly controversial (Deacon 1990; Brown 2001; Finlay *et al.* 2001). Furthermore, most research on the role of the brain in play is based on play fighting in rats, a narrow slice of play and species diversity.

Nevertheless, even with a skewed database, the primary message is that many parts of the nervous system are involved in play and that these vary across play types and among aspects of a given type of play. For example, in the diencephalon some hypothalamic and thalamic lesions can reduce play fight initiation or pinning in rats. However, the most promising brain areas to look at for the neural origins of terrestrial vertebrate play are in the telencephalon between the midbrain and neocortex: The striatopallidal complex (basal ganglia), composed of the dorsal and ventral striatum and pallidum and associated structures, and the limbic system, composed of the hippocampus, dentate gyrus, subicular and entorhinal cortices, cingulate gyrus, amygdala, and septum. The basal ganglia seem to be the major locus of motor patterning of much of the major instinctive or "unlearned" behavior in amniote vertebrates. Basal ganglia lesions or dysfunctions interfere with motor patterning, including play fighting in rats, initiating or stopping actions, and in switching from one behavioral sequence to another. Exploratory and appetitive "seeking" behavior are also influenced by the basal ganglia and disruptions in this complex seem to underlie motor difficulties, such as Parkinson's disease and as well as abnormal repetitive stereotyped behavior in animals, obsessive–compulsive disorders, and various addictions in people. The neurotransmitter dopamine and brain receptors for it appear to underlie many of these behavioral dynamics including play. Furthermore, the ventral striatum responds to novel information and changes in neuronal firing in the sensorimotor (caudal) striatum occur during learning and habit formation. Thus, learning is intimately connected with instinctive species typical behavior in the brain.

The limbic system is associated with emotional responses more positive than the rage and fear associated with the striatopallidal complex and may modulate the latter (Panksepp 1998). The limbic system seems involved with addictions and it would be important to find out if play addictions (e.g. gambling) have neural correlates similar to chemical dependencies. Social play seems to depend on the limbic system in that affiliation is required for the physical contact and "bonding" seen among play partners. This affiliation depends on the amygdala and cingulate gyrus in both rodents and primates (Pellis and Pellis 1998). Graybiel (1995) concludes that the basal ganglia and limbic system act as a goal attainment system with the former responsible for the establishment and execution of motor patterns and the latter with the recognition of goals and evaluation of behavioral outcomes. This may be the basis for any linkage of cognitive systems with motor play systems.

In short, the basal ganglia and limbic system are tightly interconnected in mammals, and thus tie the motivation and emotion regions of the brain with both effector systems and more cognitive neocortical (e.g. prefrontal cortex) processing. The basal ganglia, ventral tegmental area, prefrontal cortex, and dopamine systems seem especially involved in reward, anticipation, memory, and goal orientation observed in the often fast paced, contextually sensitive, and anticipatory responses (Pellis and Pellis 1998) of locomotor, object, and social play. A viable hypothesis, then, is that play originated in the initiation and execution of instinctive behavior sequences, in which motor performance was itself rewarding (primary process play), and which through repetition and selection could enhance performance in changing contexts (secondary process play). This could take place in many species, even those with minimal neocortex. The cerebellum in the hindbrain has also been implicated in learning, targeting, and control of rapid movements as well as in play, though apparently not in play motivation (see below).

Finally, what about brain size and play? The comparative survey suggests that playlike phenomena occur in many groups, not just those with large brains. However, the fact that the most complex play (especially object and social) is found in those birds and mammals with large brains suggests that there is a relationship, but the nature of it is difficult to pin down with either total or neocortex relative brain size (reviewed in Burghardt 2001). Any relationship of play with brain size needs to also explain the fact that domesticated species, such as dogs and cats, are often more playful, especially as adults, than their wild counterparts, even though they may have brains about 30 percent smaller in mass, with the neocortex the most proportionately reduced (Kruska 1987). Litter sizes often increase and many behavioral skills shown in wild populations (e.g. wolves) show deterioration in domesticated forms (e.g. dogs).

## The ontogeny of play

In many animals, play is most prominent in early ontogeny, but the sequential details are highly variable among species. Variation in the ontogeny of play can offer insight into both the adaptedness of play and its phylogeny. However, since play can be common in adult animals, especially primates, parrots, ravens, and even among fish and turtles, ontogenetic factors underlying play may themselves be heterogeneous; play occurs for reasons other than for juvenile practice of adult behavior or as stages in normal development.

Unfortunately, most studies of play ontogeny are descriptive and performed to document normal development or to test functional hypotheses with largely negative results (Martin and Caro 1985). Play is generally more prevalent in species with relatively longer periods of parental care, in which there may be more opportunities for play to appear, than for highly precocial species or for those in which little brain development occurs late in ontogeny (Finlay *et al.* 2001).

Some comparative correlational evidence suggests that play may appear during a narrow window in ontogeny as a necessary process to consolidate or even form permanent neural changes in the cerebellum and muscle fiber capacities (Byers and Walker 1995). If confirmed, this would be an example of a derived tertiary type of play. Many other areas of the brain are influenced by experience from play as well, however, and it is unclear why play may not also maintain or enhance other neural and behavioral systems. In cats, ungulates, red pandas, and other species different types of play (locomotor, predatory/object, and social) wax and wane at different ages in early development and it is possible that the appearance of play in ontogeny may be related to consolidating specific neural, perceptual, or motor systems. Whether this relationship is causal, especially in its evolutionary origins, is unclear.

Play often is found when animals begin to add new behavioral components to their repertoire and the drive (motivation) to attempt such behavior can be intense, as in human infants "learning" to crawl, stand, and walk. A derived play process might be that play aids in self-training or self-assessment of behavioral capacities (K. V. Thompson 1998) or aids in the integration of sensory and motor systems due to the unpredictable consequences of interacting with objects, social partners, and even one's own body (Spinka *et al.* 2001). These aspects of play are shared with other behavior systems, of course, but play may backup, facilitate, or refine other developmental systems. In other words, play can involve multiple parallel processing and distributed functions. What is often clear is that animals at certain stages in life are virtually internally driven to play, and doing so often has positive emotional (affective) and self-rewarding consequences. Until we have better physiological measures, however, it is hard to make this argument for species other than the familiar mammals we can often interpret in an uncritically anthropomorphic manner (Burghardt 1997).

In the advanced primates such as apes, play may be an important stage in cognitive development that is itself tied to brain development. Various cognitive and problem solving capacities emerge at different rates in a rather set sequence. This has led to a modern comparative recapitulation model inspired by Piagetian studies on human children and other primates (Parker and McKinney 1999). A key finding is that human children pass through the various stages faster than other primates and attain higher intellectual functioning at relatively earlier life history stages. The onset of complex pretend play is a key stage in their evolutionary scenario; they argue that it is only rarely seen in apes and virtually never in monkeys.

## How play (probably) evolved

In recent years molecular genetics, the focus of much of this volume, has begun to pay attention to the interaction of specific genes with developmental and life-history processes in the expression of organismal traits. Underlying the grand biological diversity in the world is a surprisingly conservative genetic toolkit (Carroll *et al.* 2001). Traits that were thought to have arisen completely independently in animals separated by hundreds of millions of years, such as eyes in flies and mice, are controlled by similar genes: introducing the corresponding mouse gene into a fly induces compound eye tissue, not mouse type eye tissue. Although play has arisen many times in evolution, it may be the result of common environmental contexts that activate a suite of retained genes that, though they may have other functions, could be repeatedly co-opted in the service of play-like traits. Note that I am not claiming that all play is homologous; on the contrary it appears very heterogeneous. It is possible, however, that there are some shared genetic, as well as neurological, traits that facilitate the evolution of play. More broadly, play could evolve whenever physiological (including neural), life history, energetic, ecological, and psychological conditions, in conjunction with a species' behavioral repertoire, reach a threshold level. The task is to tease apart these conditions and rigorously evaluate their contribution to both the origins and elaboration of play, a task barely begun. What is needed first, however, is to isolate those factors that seem to have facilitated the appearance of primary process, or incipient play, and test them across the different radiations in which play has appeared. The aim is to specify with some predictability the taxa in which play appears by presenting a model that integrates much of the comparative material related to the origins of play and generates predictions (as in Burghardt 1988). Here, only a schematic outline for the appearance of play, especially in amniote vertebrates, will be outlined.

SRT is based on the insight that play often appears as superfluous activity very much influenced by the favorable nutritional, health, and general well-being of the player. The basic argument has several components. First, available metabolic energy (both energy stores and the capacity for sustained vigorous activity) is necessary. Second, buffering from serious stress and food shortages is needed to provide the "free" time and lack of conflicting "serious" behavior systems needed for "play." Third, environmental stimulation (objects, companions) may induce behavior out of its normal context while lack of stimulation (boredom) may also do so in the absence of sufficient sensory stimulation. All these relate to the appearance of primary process play. Such play in these contexts could then provide increased or "optimal" levels of arousal that stimulate the animal's brain and physiology including motor development. Fourth, a life style that involves diverse and unpredictable environmental and/or social resources requires more varied and flexible responses and, thus, there are more behavior systems that can be expressed as play and become removed from their original biological function. Thus, generalist species should play more than those with more rigid specialized behavioral repertoires: This seems borne out. Play in all species, then, including children, will be most prevalent when there are excess resources along with appropriate evolved motivational, physiological, and ecological systems. In mammals, however, play may have served as both a cause and consequence of the evolving disparity between periods of juvenile dependence and adult responsibilities; most ectothermic reptiles have highly transient or limited dependence upon parents.

To elaborate: before the evolution of extensive parental care, animals at birth or hatching had to provide for all their own resources and thus had to have a well-honed set of instinctive behavior patterns for survival. Such instincts combined both behavioral responses (feeding, antipredator, social competition) and recognition of biologically significant stimuli (signifying food, enemies, and conspecifics). These abilities, except for reproduction, needed to be deployed at birth or hatching; any learning that took place served to refine abilities that were adequately functional immediately, allowing animals to adapt to the resources and their spatial arrangement in the habitats in which they found themselves. With the advent of parental care (the reasons for this we need not go into here), young animals were buffered from many of the demands for survival as they were fed, defended, kept in secure and warm nests, etc.; they effectively became parasites on their parents. Thus, as with many parasites, natural selection was removed from being as intense as it had been previously on the highly functional and specific motor and perceptual systems such young animals needed for survival. This then led to the deterioration of some neonatal response systems: Motor patterns became less precise and more variable and stimuli triggering feeding, social, and defensive response became both broader in scope and accompanied by lowered stimulus thresholds. However, unlike obligatory parasites, the young animals still would eventually need to perform these behaviors on their own and in serious contexts, so the motivational and behavioral repertoires eventually would be deployed. Furthermore, parents provided large amounts of energy (milk, prey) to their offspring. Such rich diets were needed for the increased aerobic metabolic capacities resulting from endothermy (and larger brains) in birds and mammals. Under benign environmental conditions, however, these diets often provided more resources than were needed for growth and physical development. In addition, animals were often reared in highly protected settings (burrows, nests) in which sensory stimulation was reduced. This set the stage for such animals to engage in behavior, derived from their instinctive repertoire, which simulated the "serious" versions even in the absence of "real" stimuli. Such simulations could also involve the triggering of emotional responses or "virtual emotions" as recently suggested by Sutton-Smith (2003). Such primary process incipient play provided a pool of modified, or even reorganized, developmental, and experiential processes available to endotherms with parental care that could be selected to enhance behavioral and cognitive complexity. With the removal of natural selection from the neonatal instinctive repertoire, the need for practice and skill enhancement may then have become necessary for continued survival of the young by providing continuity for replacing lost, suppressed, or maturationally delayed response systems.

It is in this parental care transition that the need to distinguish between primary and secondary processes in play becomes especially critical. An example of an important primary process derived from SRT would be the role of increased activity in producing "surplus" behavioral "mutants" that could in turn be selected ontogenetically and phylogenetically. Play may have induced novel behavioral variants similar to those produced by mutation, recombination, **duplication**, fragmentation, **translocation**, and **linkage** in genetic systems (Fagen 1974; Sutton-Smith 1999). Such variation can serve as raw material for natural selection to operate upon insofar as the behavioral variation also has some inherited component. Recent molecular genetics methods allowing overexpressed alleles and the elimination of loci (knockouts) allow further genetic metaphors, as well as the promise of identifying some actual genes that could underlie the expression of play, such as dopamine receptors.

A parallel model for testing the evolutionary scenario of SRT is the course of **domestication**; domesticated species, such as dogs, are playful for much of their lives. Many of the processes postulated above as having occurred in juvenile mammals with the onset of parental care and the consequent buffering from the demands of life, are also found in domesticated species. Price (1984) has documented these processes completely independently of the theory outlined here, and the shrinkage in brain size with domestication has been noted already. It would seem that a careful study of play behavior in wild animals undergoing domestication might be a most useful method to see how buffering animals from the harsher aspects of existence may change the amount, type, and frequency of play as well as cognitive capacities. Similarly, studies of feral animals undergoing the reverse process would also be useful. The increase in play in well-fed captive wild animals has already been noted and provides support for the proposed incipient stages.

An apparent secondary process derived from a detailed consideration of primary processes is based on the recent claim by Byers and Walker (1995) of a correlation between the onset of vigorous motor play with the age at which permanent long-term changes occur in the muscular and cerebellar systems of several species of domesticated animals (rat, mouse, cat). If play is essential or even useful in establishing these permanent physiological systems, then a derived process has been established. If the play behavior is a mere accompaniment to this developmental process with no causal role, then it is but another primary process. On the other hand, this secondary process could be at least a partial response to domestication.

The study of the role of play in development should initially focus on the primary processes leading to behavior satisfying the five criteria for play. Increased endurance, functional endothermy, parental care, and lack of sufficient external stimulation facilitate incipient playlike behavior. As the trends favoring this incipient play expanded, play acquired secondary functions, including those underlying greater behavioral, social, and cognitive complexity through the evolution of secondary processes. The initial secondary processes were those that maintained the instinctive skills of animals under threat of deterioration due to the lack of natural selection acting to maintain the precision of the predatory, defensive, and social skills necessary for survival in the absence of parental or other protection. Support for this is seen in the far less skillful initial foraging and predatory behavior of many juvenile mammals as compared to turtles, lizards, snakes, and crocodilians. Later, neural and physiological changes resulting from this experience-based learning and plasticity opened up new possibilities for cognitive and emotional complexity in many mammals and some birds as compared to nonavian reptiles. Could play, so ridiculed or ignored by most biologists as unimportant, have played an essential role in cognitive evolution?

According to SRT, play could have originated via initially nonadaptive primary processes, but later incorporated secondary processes that allow behavior and psychological attributes to shape new behavior and capacities, perhaps through a process of positive feedback loops. A hallmark of mammalian behavioral evolution is the rapid diversification of behavior and increased forebrain size in a relatively short (geologically) time span as compared to many ectothermic vertebrates and invertebrates, whose core behavior patterns and abilities may have changed little over millennia in spite of often rapid microevolutionary **adaptation**. Processes involved in play may have been major engines in the rapid cascade of evolutionary change that led to increased cognitive complexity. An extension of the SRT approach that tackles the shift from behavioral to "mental" play (as in some apes and human beings) has been proposed elsewhere (Burghardt 2001). The conditions for the appearance of primary play processes may have arisen many times during the evolution of animals, but in endothermic vertebrates, especially mammals, the constellation of facilitating conditions came together most fortuitously for play to have such a breakthrough role.

In addition, play derived from initially highly functional instinctive behavior systems could be incorporated into other more serious endeavors and functions, as in the process of **ritualization** described by the early ethologists that underlie many courtship rituals (Tinbergen 1951). In this way, the once playful behavior may be transformed and "fixed" so that it has shifted to being outside the realm of play. It is possible that after a period of evolutionary reorganization in behavioral ontogeny accentuated by the lengthening of parental care (Burghardt 1988), play facilitated rapid behavioral and cognitive evolution by providing altered phenotypes for natural selection to prune and shape. However, play can also reflect deteriorated or developmentally stalled behavior as found in some domesticated species (Burghardt 1984). Play is paradoxical in many ways, and the path to understanding it is still tangled and enmeshed in an early morning fog slowly rising off Darwin's (1859) tangled bank.

## Summary/Conclusions

Although "play" has not been explained in this chapter, one framework for explaining play through an evolutionary informed analysis has been presented. An integrated ethological analysis is required. The essential starting point is that play, like most behavior, must be viewed as a phenotypic product of evolutionary processes. The evidence outlined here supports the episodic development throughout vertebrate evolution of playlike behavior under suitable ecological, life history, and physiological conditions that provided "surplus resources" in time and energy. Using the five criteria for play, there is compelling evidence that play evolved independently in several lines of invertebrates, fishes, nonavian reptiles, birds, and marsupials, as well as virtually all families of placental mammals (Burghardt 2003). Within these groups, the type (locomotor, object, social) and amount of play can vary dramatically.

It is useful to divide the processes underlying play into primary and derived, with the former being most important for explaining the origins of play from nonplay behavior through SRT. Play is clearly most common in mammals and birds, groups with high metabolic rates, endothermy, parental care, and relatively large brains, although the specific relationships and causal pathways among these variables is not yet well established. Furthermore, the extent of play in nonendothermic vertebrates needs much more study.

Play in terrestrial vertebrates is derived from instinctive behavior patterns whose patterning, motivation, and affective components are controlled by the basal ganglia and limbic system of the telencephalon and structures in the diencephalon, with the cerebellum enhancing detailed timing and behavioral precision. After a period of evolutionary reorganization in behavioral ontogeny, accentuated by the lengthening of parental care (Burghardt 1988), play may have facilitated rapid behavioral change by providing altered phenotypes for natural selection to prune and shape. Nonetheless, play can also reflect deteriorated or developmentally stalled behavior, such as found in some domesticated species such as dogs that retain juvenile behavior indefinitely (Burghardt 1984).

The phenotypic expression of behavior patterns is a complex epigenetic outcome of interactions and feedback occurring at a many levels from gene allele to protein synthesis (gene expression) to behavioral practice to social experience. Selection can operate on all these levels and more, at least indirectly. If so, play may have a subtle, yet profound, role in behavioral ontogeny and phylogeny that we are only beginning to appreciate.

Over the years, I have benefited greatly by the insights, suggestions, writings, and criticisms of many scholars in numerous fields. I am particularly grateful to Enrique Font for his invitation to participate in the Valencia workshop and his reviewers thoughtful comments on my ideas, as well as Matthew Bealor and Randall Small, who aided in preparing the phylogenetic trees. I also thank the National Science Foundation, University of Tennessee, Smithsonian Tropical Research Institute, and the J. S. Guggenheim Foundation for generous support over the years that led to the integration of the many lines of research necessary for attempting a synthetic perspective on evolution, development, and the still somewhat mysterious phenomenon of play.

CHAPTER 21

# The evolutionary psychology of human physical attraction and attractiveness

Randy Thornhill and Steven W. Gangestad

Human physical attractiveness research recently has become a topic surrounded by fanfare, both in the public realm and the scientific circle. Findings about the topic, discovered from hypotheses inspired by the evolutionary approach, have been showcased in numerous television productions worldwide, and in other media channels. The media attention reveals that many people are interested in their preoccupation with looks and the findings are providing them access to its basis, both the physical traits assessed and the evolutionary function of aesthetic judgments. Moreover, the findings promise to place attractiveness research among the major success stories of cognitive science.

Research is ongoing but a considerable body of published findings allow the conclusion that the facial and body features of aesthetic importance as well as the aesthetic judgments of these features evolved because the features were honest indicators of health in the environments of human evolutionary history. Physical attractiveness appears to be a health certification and human ancestors were the individuals in human evolutionary history who conceptualized bodily and facial beauty as health; those individuals historically who deduced physical attractiveness in other terms reproduced less or not at all, and therefore did not become evolutionary ancestors. Simultaneously, this research reveals that the widespread view that human beauty is arbitrary and whimsical is mythological. We discuss evolutionary psychology, the branch of evolutionary biology central to recent discoveries about human attraction and attractiveness. Our focus in this paper, and that of recent research, is on facial attractiveness. We treat also the growing literature on body features.

**Evolutionary psychology** (EP) is the subdiscipline of evolutionary biology that focuses on the evolution of mental features of humans and other animals. Some EP researchers were skeptical of the traditional view of human beauty as, for example, portrayed in Naomi Wolf's best-seller *"The beauty myth"* published in 1992, that human physical beauty is capricious and uncoupled from biological causes including evolution. This uncoupling is thought to arise from arbitrary social or cultural learning that is independent of biological causation. The traditional view has characterized much of the social-science study of physical attractiveness and social commentary about the topic (see Thornhill and Gangestad 1993; Symons 1995). In part, the skepticism from EP stems from the vast evidence across many nonhuman animal species that mate choice is not arbitrary and whimsical but instead is preference of mates that possess functional traits such as health or capability of parental investment. Thus the massive nonhuman literature on beauty judgments of mates reveals functional, not arbitrary, choices in many species that lack culture. Also skepticism of the traditional view of beauty arises necessarily from the fact that humans, like all other species, have had an evolutionary history in which Darwinian selection has brought about most phenotypic (bodily) and genetic change. Therefore, people will almost certainly possess mate choice **adaptations** that are designed for assessment of physical attractiveness.

Selection is not the only cause of evolution, but it is the major cause, and most importantly, selection is the only natural process known that can make adaptations. The human brain, as well as the rest of our bodies, is composed of a multitude of evolved adaptations. Moreover, all behavior and preferences and other mental activity, including learning, arise out of psychological adaptations. Evolved adaptations are bodily mechanisms that are solutions to problems that affected the reproductive success of individuals over the evolutionary history of a species. Adaptations have evolved because they enhanced their bearer's reproductive success, never because of their benefit to the population, society, or species (Thornhill 1997).

Obtaining a mate that will promote one's own ability to reproduce was a problem facing individuals throughout the past generations of human history. Past selection is expected to have created psychological adaptations that are functionally designed to solve this problem in two interrelated ways: (1) evaluate observable facial and bodily traits that varied with components of **mate value** (i.e. traits depicting the various components of health such as age, hormonal status, developmental stability of the body, disease-free skin and genetic quality, as well as traits depicting ability and willingness to invest in a mate and offspring) and (2) cognitively deduce the conception of beauty as traits with highest mate value (Thornhill and Gangestad 1993; Symons 1995). Mate value is what an individual brings to a mating relationship that affects the partner's survival and reproductive success.

Evolutionary psychology emphasizes that selection favors functionally specific adaptations, not general-purpose ones, because only specialized mechanisms can solve the specific problems that are forces of selection. Obtaining a mate that is of high mate value involves many, specific information-processing problems (Symons 1995). Hence, EP anticipates the existence of many, domain-specific, mental adaptations associated with physical attractiveness judgments that function to process cues of evolutionarily historical mate value and that view a beautiful person as one whose physical features promise high reproductive success.

Evolutionary psychology then is the research domain of evolutionary biology that identifies mental adaptations and empirically establishes their functional designs. EP separates mental adaptations from their incidental effects/by-products through evidence of special-purpose function of these adaptations. The demonstration of special-purpose design of a psychological adaptation, or of a nonpsychological adaptation, is simultaneously the demonstration of the type of selection force that made the adaptation (Thornhill 1997).

## Faces and bodies in the environments of evolutionary adaptedness

Evolutionary psychologists conducting beauty research emphasize analysis and interpretation in terms of the **environments of evolutionary adaptedness** (EEA). Each species of organism has an EEA. The human EEA is the subset of evolutionary historical environments over the last several million years that were the actual selective forces that ultimately caused human-specific adaptations (Thornhill 1997), including those involved in signaling and assessing attractiveness in our species.

Many humans live in environments replete with evolutionary novelty (e.g. modern contraception, modern medicine, and middle-aged women who appear nubile because of nulliparity). EEA novelty can disrupt the positive association between reproductive success and a particular adaptation that necessarily existed during the evolutionary historical period in which the adaptation evolved. As an example, in modern society, the relationship between facial attractiveness and male or female reproductive success may not exist, according to a recent study (Kalick *et al.* 1998). This lack of relationship is irrelevant for adaptationist analysis of attractiveness in modern society, because the analysis is of functional design of adaptations, products of effective past selection, rather than of traits' current adaptiveness (current effects on reproductive success). Current adaptiveness is not central because in EEA-novel settings current adaptiveness cannot identify evolved adaptation: Byproducts may be currently adaptive and adaptations may not be (Thornhill 1997).

In EEA-like modern environments, such as seen in some hunter-gatherers, however, facial attractiveness may be currently adaptive. Hence, Kim Hill and Magdalena Hurtado found in their long-term study of the Ache Indians of Paraguay that facially attractive women have more children than facially unattractive women when age effects are controlled (see Thornhill and Gangestad 1999a).

A lack of covariation between attractiveness and reproductive success may arise from the partial uncoupling of what may have been a generally constant relationship in the EEA between attractiveness and health. People consistently rate facially attractive men and women as healthier than their unattractive counterparts implying psychological adaptation functionally designed for these deductions (see e.g. Kalick *et al.* 1998; Rhodes *et al.* 2001). Yet the relationship between attractiveness and actual health gives mixed results in modern society, presumably largely because of modern medicine. For skin health, there is clearly a significant current relationship with skin attractiveness (Fink *et al.* 2001). Also attractive sex-specific **hormone** (androgen- and estrogen-) **markers** (see below) in the face and body depict hormonal health pertaining to fertility and general health and people with an attractive body mass index (weight in kilograms divided by the square of height in meters) are healthier than those with unattractive body mass index (Maisey *et al.* 1999). Moreover, developmental stability, as reflected in facial and body bilateral symmetry (see below), is associated with attractiveness and is widely felt to be a marker of overall developmental health; and there is evidence that more symmetric people have greater physical and mental health than asymmetric people (Thornhill and Møller 1997). Across several studies of facial attractiveness in the United State of America and a variety of health measures (psychological, physiological, health history, and so on) results are equivocal, some supporting the positive relation between attractiveness and health and others not (reviewed in Thornhill and Gangestad 1999a; Rhodes *et al.* 2001).

Cross-cultural research conducted by one of us (SWG) and David Buss indicates that good looks may in general certify health related to resisting parasites, and that people have psychological adaptation that adjusts the priority of having a good looking mate in relation to the abundance of local infectious diseases. The World Health Organization (WHO) tracks infectious disease problems throughout the world. Parasite **prevalence** data from WHO were coupled with questionnaire data from men and women across 29 societies about the importance of good looks in a mate. In each sex, there was a positive correlation across societies between the importance attributed to good looks in mate selection and the prevalence of significant human parasites. Thus, it appears that, where parasites are abundant, both men and women give greater priority to physical attractiveness in mate choice than in cultural environments with fewer parasites (see Thornhill and Gangestad 1999a).

## The handicap principle

The incessant beauty contest of human evolutionary history has resulted in the evolution of signaling adaptations as well as signal-receiving adaptations. EP addresses both types of adaptations: the cognitive workings (information-processing, deduction, and motivation) of psychological adaptations that function in physical attractiveness judgments, as well as the adaptations, psychological and otherwise, that function to create, during an individual's bodily development, the bodily and facial features judged.

One major evolutionary theory of sexual signals is the so-called **handicap principle** proposed by Amotz Zahavi in 1975. It explains the evolution of extravagant and, thus, costly display traits (such as the secondary sexual traits of the adult human face and body that arise at puberty under sex-hormone facilitation) as honest signals of ability to deal with environmental problems. Because such signals are reliable indicators of **fitness** for the EEA, selection favors mate choosers who assess them, rather than assess unreliable traits, and who prefer manifestations of the honest traits that connote highest fitness. A handicap is honest in the sense that only high quality (phenotypically and genetically) individuals can afford it in its full glory. Darwinian selection has crafted species-wide, often sex-specific, signaling adaptations that develop in magnitude across individuals depending upon each individual's relative quality, that is, handicaps are condition-dependent. Handicaps will usually signal both phenotypic and genetic quality of the bearer because both types of quality typically positively co-vary within individuals (see Thornhill and Gangestad 1999a).

## Sex hormone markers

The sex hormone markers of the face and body are hypothesized to be handicaps, specifically condition-dependent, reliable signals of reproductive hormonal health, in two regards. First, the hormones that facilitate the development of the markers divert energy away from growth, repair and maintenance of the body and into sexual behavior and reproduction. This diversion, mediated by both androgens and estrogens, from somatic effort into reproductive effort is a cost to survival of the individual. Second, there is evidence that these hormones compromise the effectiveness of the immune system and, in the case of estrogen, other basic physiological processes by way of toxic breakdown products. Features marking high testosterone in men or estrogen in woman then connote high energy sequestration from the environment and efficient energy utilization and thereby honestly signal the bearer's ability to afford the energetic diversion to reproductive endeavors. In this scenario, such features also reliably signal the ability to cope with immune compromise and the toxic effects of the hormones or their byproducts (Thornhill and Gangestad 1993, 1999a; Thornhill and Grammer 1999).

Men and women's faces develop dimorphically at puberty as a result of a sex-specific ratio of androgens (primarily testosterone) and estrogens. In pubertal males, facilitated by a high testosterone ratio, cheekbones, mandibles, and chin grow laterally, the bones of the eyebrow ridges and central face grow forward, and the lower facial bones lengthen (Symons 1995; Thornhill and Møller 1997). A high estrogen ratio in pubertal females appears to cap the growth of the bony structures of the typical male face just as it caps the growth of long bones, but results in enlargements of the lips and upper cheek area through fat deposition (similar to the estrogen-mediated fat deposition in the thighs, buttocks, and breasts) (Johnston and Franklin 1993; Symons 1995; Thornhill and Grammer 1999).

Sexual dimorphism in certain body features also appears at puberty as a result of the sex-specific hormone ratios. At this time females lay down the reproductive fat in breasts, thighs, and buttocks that will be mobilized during the energetically demanding periods of gestation and lactation (Thornhill and Grammer 1999); males erect the adult male shoulder breadth and musculature, most notably upper body muscles. The waist also is constructed at this time. The waist-to-hip ratio is sexually dimorphic. Women have relatively small waists and large hips compared to men. Devendra Singh's research has shown that low waist-to-hip ratios in women signal high estrogen, fertility, and general health (see Thornhill and Grammer 1999). The adult male shoulder to waist ratio or upper chest to waist ratio also arises developmentally at this time (Maisey *et al.* 1999).

## Hormone markers in facial attractiveness

The sexes differ in the facial and bodily hormone markers preferred in mates. Men appear to consistently find high-estrogen markers most attractive. Women prefer male facial traits reflecting high testosterone effects when at points in ovulatory menstrual cycles of high conception probability but they prefer less masculinized male faces at other points of the cycle.

Michael Cunningham in the mid-1980s first showed the importance of sex-specific facial hormone markers in attractiveness judgments by measuring with calipers features in facial photos of both sexes (reviewed in Cunningham 1995). Later Karl Grammer and one of us (RT) found this same pattern through measurement of facial features of computerized facial images. Victor Johnston and Melissa Franklin (1993) and David Perrett *et al.* (1994) confirmed the importance of estrogen markers for female facial attractiveness using additional methods. In Johnston and Franklin's study, men in the United States of America created their mental picture of a beautiful adult female face using facial feature options (e.g. chins of various sizes) provided as computer images. Johnston has subsequently used a similar procedure employing the World Wide Web to show that men around the globe with access to the web find female faces with high estrogen effects maximally attractive, and more attractive than female composite faces that depict averaged facial features. Perrett *et al.* (1994) used computer techniques to exaggerate the difference between the features of a facial composite of a large sample of women and the features of a composite made from faces of the attractive subset of the same sample. In studies in the United Kingdom and Japan, using both Caucasian and Japanese target faces, they found that highly estrogenized female faces were maximally attractive (also Perrett *et al.* 1998). Doug Jones (1996) and also Cunningham *et al.* (1995) found this same result for women's faces across a number of human societies, including hunter–gatherers with limited Western contact, using caliper measurements. Women fashion models and TV actresses have smaller lower faces than normal women, but this is not the case for male models, and normal women's faces are more attractive when the lower face size is reduced with computer techniques (see review in Thornhill and Gangestad 1999a).

Research on neurobiological correlates of attractiveness judgments confirms the importance of facial hormone markers. Researchers recorded event-related potentials, which measure the electric field generated by neural activity, of men while the men were exposed to a sequence of men and women's faces. The faces used contained computer manipulated hormone markers such that feminized (estrogenized) as well as average facial features occurred in faces of both sexes. Highly estrogenized female facial features resulted in larger electrophysiological scores than average female facial features, but the reverse was found for male faces (i.e. larger scores with average than with feminized face depictions). Also these researchers found that although both males and females reported the same attractiveness preferences for feminized facial features, only the males have a strong emotional response to feminized faces (large P300, a measure of emotional salience) (see Thornhill and Gangestad 1999a for additional discussion).

## Preferences across the menstrual cycle

There has been considerable interest in changes in women's sexual preferences across the menstrual cycle. Evolutionary theory predicts that women possess psychological adaptation(s) that motivates pursuit of sires for their offspring of high genetic quality in the fertile part of the cycle. At infertile cycle points, however, a shift to preference of men who will invest resources is anticipated on theoretical grounds. There is now evidence from multiple studies that women have an olfactory preference for the body scent of symmetrical (i.e. developmentally stable) men only at high conception points of the menstrual cycle (Fig. 21.1). The preference is not seen in the luteal phase or in women who are not ovulating as a result of the contraceptive pill (see review of studies in Thornhill and Gangestad 1999b).

A useful way to scientifically study developmental instability is by degree of **fluctuating asymmetry** (FA). FA is small, random deviations from perfect symmetry in traits that are basically bilaterally symmetric (e.g. ankle and wrist width and length of index finger and ear). FA is caused by perturbations acting during development, both environmental (diseases, food shortages, toxins) and genetic (mutations, homozygosity) insults. FA of the individual is a record of how well the individual has coped with both types of insults during development and the relative prevalence of development-disrupting genes in its genome. Accordingly, FA is a useful measure of individual phenotypic and genetic quality, and the greater the FA, the lower the quality (Gangestad and Thornhill 1999).

FA is related negatively to sexual attractiveness, survival, growth rate and fecundity in a wide range of animal species (Gangestad and Thornhill 1999). In humans specifically, developmental stability measured as FA in nonfacial bodily traits is positively related to mating success variables in men: number of lifetime sex partners, facial attractiveness, number of partners outside pair–bond relationships, number of times chosen as an extra-pair partner by a woman in a relationship, and a mate's copulatory orgasm frequency (which may affect sperm retention). The greater sexual attractiveness of men with relatively

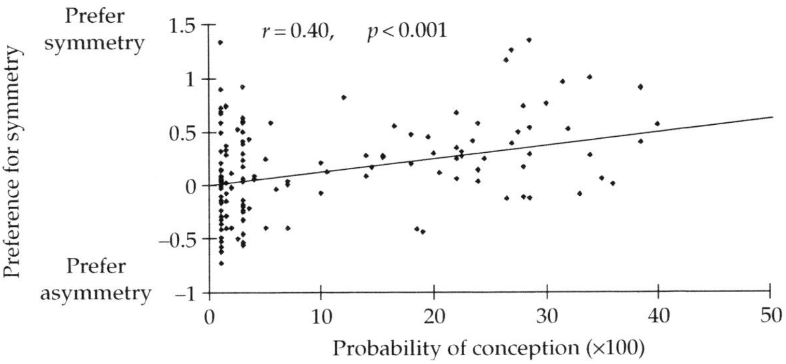


**Figure 21.1** The relationship between non-pill using women's preference for the body scent of symmetrical men and the women's risk of conception across the menstrual cycle. Each point is the regression coefficient of a woman's scent attractiveness ratings of T-shirts worn by men regressed on the men's body asymmetry. Conception risk is based on actual data from a large sample of women. Figured is a summary of data from three separate studies, two discussed in Thornhill and Gangestad (1999b) and one unpublished study; total $N = 141$ women.

low body FA is not the result of the FA itself serving as a cue. Low body FA is associated with enhanced mental competence, social dominance, athleticism and masculinity, and apparently it is these correlates of FA that are the actual cues affecting attractiveness (Gangestad and Thornhill 1997 and below). However, facial FA (see below) is probably an actual cue. In women low body FA seems to predict increased probability of marriage and bodily attractiveness (Thornhill and Møller 1997; Grammer *et al.* 2002). In both sexes, low FA appears to be associated with fertility and above average health, including relatively high IQ and mental health in general (Thornhill and Møller 1997).

Several studies have found that female preference for male hormone markers changes across the menstrual cycle in women who are not using hormone-based contraception. Such women prefer more masculine male facial features at fertile times in the cycle than at infertile cycle times. When infertile, such women prefer less masculinized or even slightly feminized male faces. Women using the contraceptive pill did not show this preference shift (see studies reviewed in Johnston *et al.* 2001; Penton-Voak and Perrett 2001). Also ovulating women's (non-pill users) judgments of male faces across the cycle change to an attractiveness preference for more masculine at high-conception-risk times, but their judgments of male faces that receive attributions of "good dad," "dominant," "intelligent," "masculine," "average," and "androgenous" do not change across the cycle. This result indicates that only women's attractiveness judgments of men and not their other judgments of men change across the cycle, suggesting that fertile women's preference for more masculine faces is not the result of a general change in sensitivity or of enhanced ability to detect facial features in general at high-conception-risk times (Johnston *et al.* 2001). These studies use computer graphics to manipulate the masculinity or femininity of a composite male face by exaggerating or reducing the shape differences between male and female average faces, thereby manipulating the sexually dimorphic features affected by testosterone and estrogen.

Further evidence that women's responses to male faces change across the menstrual cycle was found in research on the P300 response. The magnitude of the P300 response of women in the fertile cycle phase correlated with their rating of male facial attractiveness, but not their ratings of female facial beauty. During the infertile phase, women's responses were undifferentiated and simply covaried with both male and female attractiveness judgments (see Thornhill and Gangestad 1999a).

Female preference for the facial features of men with symmetric bodies also changes across the menstrual cycle (Thornhill and Gangestad, 2003).

University of New Mexico men and women rated facial photos of men from a remote village on the island of Dominica (West Indies) on a scale of physical attractiveness. Nine body features on each of the Dominican men were measured by Gangestad and the features' asymmetries were relativized for feature size and then summed into a composite asymmetry index (see Gangestad and Thornhill 1999 for methods). Such measures are strongly correlated with measures made by a villager who was totally unfamiliar with the hypothesis that symmetric men are more sexually attractive to women than asymmetric men. Each sex of raters found the symmetric men's faces to be more attractive than the faces of asymmetric men. Among women who were not using hormone contraceptives, the strength of their attractiveness attribution for symmetric men is significantly positively correlated with their risk of conception based on menstrual cycle point.

Thus, research indicates that women prefer testosterone-related facial traits in men, as well as facial features and body odor of men that covary with their body symmetry, specifically when conception is most likely. These studies imply that women have psychological adaptations for motivating interest in obtaining a good-genes sire when they are maximally fertile in the menstrual cycle. These studies are only part of the growing evidence that women possess a specialized adaptation for conditional pursuit of copulation with a male other than the regular sexual partner. This adaptation is hypothesized to have evolved in the context of females obtaining genetic benefits that enhance their offspring's health, and hence reproductive success (Thornhill and Gangestad, 2003).

As mentioned, men's facial masculinity is affected by androgen production during development, which, like symmetry, may be a signal of superior condition during development. Masculine features are connected to relatively high developmental health in at least two ways. First, men's facial symmetry and facial masculinization appear to positively covary. Also men's facial masculinity positively covaries with their body symmetry. Hence, facial masculinity and developmental stability tap a common underlying factor. Our working hypothesis is that in the human EEA this factor related to overall condition/general health and heritable fitness and, accordingly, females express preferences for symmetry and masculinity (Gangestad and Thornhill, in press).

Women's preference for men with less masculinized or slightly feminized faces during the infertile period of the menstrual cycle may function to secure certain material benefits from men who are willing to invest as a result of their relatively low phenotypic and genetic quality. Penton-Voak and Perrett (2001) review their research showing that slightly feminized male faces, compared to more masculine male faces, are rated by women as reflecting males who are more cooperative, honest, and good parents. Importantly, they also review findings that these positive personality attributions may correlate with actual behavior of men. Compared to physically unattractive men, attractive men may be better *able* to deliver important resources to females in exchange for mating. However, attractive men are expected on theoretical grounds to be less *willing* to invest as a result of their greater mate value and, therefore, greater choosiness about investment. Attractive men do appear to be actually less investing than unattractive men are (Gangestad and Thornhill 1997; Thornhill and Gangestad 1999a; Penton-Voak and Perrett 2001). Fertile women's preference for symmetrical men and men with highly masculine features indicate that women are willing to trade off between physical attractiveness (and, thus, heritable benefits from mate choice) and material benefits (and willingness to provide those benefits) in mate choice.

Studies have not examined bodily male hormone markers in relation to women's preferences across the menstrual cycle. A high shoulder/chest to hip ratio in men is attractive to both sexes (Maisey *et al.* 1999). But it is reasonable to hypothesize that the preference in women may be especially strong at peak conception risk. Similarly, it would be interesting to examine women's cycle-related attractiveness judgments of male vocal traits and other male behaviors affected by testosterone.

It is also expected on theoretical grounds that women at peak fertility in their cycle will prefer markers of developmental stability in addition to the scent of men's symmetry and men's facial masculinization. Of particular interest in this regard is intelligence and mental functioning in general.

## Covariation of faces and bodies

Thornhill and Grammer (1999) have studied the covariation of facial and body attractiveness in women. They asked in their research, do attractive women's faces tend to occur on bodies that are also attractive and do unattractive faces co-occur with relatively unattractive bodies? They had men rate computer images of the faces, backsides, and front sides (with face and hair blocked from view) of 92 nude young women. The ratings of the three poses were independent in that each rater rated only one pose. The raters, despite considerable cultural diversity among them, generally agreed on the faces, backs, and fronts that were attractive and unattractive. Also the attractiveness of the three features tended to co-occur within individual women. This result implies that women signal their mate value through markers that are significantly consistent in attractiveness across the face, buttocks, thighs, waist, and breasts. A follow-up study found that the attractiveness ratings of the nude women correlate positively with estrogenized facial and body hormone markers and with facial and body symmetry (Grammer *et al.* 2002).

As mentioned, there is evidence that body symmetry positively covaries with men's facial attractiveness. However, more research is needed to examine the relationship between facial and body attractiveness in men.

## Individual differences in attractiveness judgments

We have discussed the value of the evolutionary perspective for elucidating within-woman differences in attractiveness judgments related to point in the menstrual cycle. The evolutionary perspective is potentially just as useful in identifying cognitive and motivational mediators of individual differences in attractiveness judgments in both sexes.

Women pursuing short-term mates seem to value physical attractiveness more than those pursuing long-term mates (Buss 1994). These individual differences may reflect in part alternative-mating adaptations maintained by frequency-dependent selection. In this case, a mating morph is most adaptive when rare and thus increases in frequency until the alternative morph has a rare-frequency advantage. In addition, clearly there is conditionality to the adoption of these alternative-mating behaviors by women. In this case, the alternatives are tactics of one psychological adaptation, and ancestral (EEA) cues bring about a woman's decision to pursue short-term versus long-term mating—for example, the men in her environment are without resources (short-term) or with resources (long-term) that can be invested; also father-absent upbringing, possibly because of its reliable correlation in the EEA with social environments of noninvesting men, seems to cue the short-term tactic in women (see Thornhill and Gangestad 1999a).

Another condition affecting individual differences in attractiveness judgment is personal physical attractiveness. Attractive women place more importance on physical attractiveness than do unattractive women. This is seen in attractive women's relatively greater prioritization of facial symmetry and hormone markers in their judgments of men's faces (Little *et al.* 2001). The researchers that reported this effect offer the reasonable functional hypothesis that unattractive women will be less likely than attractive women to secure non-genetic material benefits from attractive men. Attractive women's greater interest in attractive men allows them to gain the resources and protection that attractive (developmentally healthy, smart, dominant, and physically strong) men can and will disproportionately provide primarily to women of high mate value.

In experimental choice situations, men with greater self-perceived mating success (a component of which is physical attractiveness) are more likely to select mateships with attractive women, especially short-term ones, than are unattractive men. Unattractive men report that they just do not see their desired relationship with an attractive woman as likely to happen. Presumably, men set different standards based on their own looks. But when men's status is rapidly increasing (e.g. nearing completion of medical school), regardless of their looks, they not only report greater choosiness about which women they invest in, but they also seem to show more optimism about obtaining an attractive mate(s) (see review of studies in Buss 1994; Thornhill and Gangestad 1999a).

## Facial symmetry

Another major area of research is the role of facial developmental stability (symmetry) in facial attractiveness. This research employs various approaches and has yielded mixed results, but overall it appears that facial asymmetry, even in small amounts, reduces facial attractiveness in both sexes. Here, we mention only a few salient studies (for critical review of these and other studies see Thornhill and Gangestad 1999a; Penton-Voak and Perrett 2001).

One approach introduced by Karl Grammer and one of us (RT) examines the influence of naturally varying facial asymmetry on facial attractiveness judgments of facial photographs. In one small study Grammer and Thornhill controlled metrically measured facial averageness and found symmetry predicted attractiveness in both sexes (see below for discussion of research on facial averageness). In other studies without this control, facial symmetry shows variable relationships in terms of sign with facial attractiveness ratings and the summary effect is very small, if not zero. The approach of using natural facial asymmetry may be a weak test of facial symmetry's role in facial attractiveness, because covariation of other traits may obscure symmetry effects. As mentioned above, there is some evidence that facial symmetry correlates with sex-specific attractive manifestations of facial hormone markers (e.g. men with symmetrical faces have larger jaws; Gangestad and Thornhill 1997, 2003). Also, facial asymmetry shows a relationship with age, and age in certain investigations may serve as a confound along with facial hormone markers.

Some later studies manipulated facial symmetry and included the control of facial expression (Penton-Voak and Perrett 2001). One study used pairs of face images comprising an original face and a more symmetrical version of the original. The symmetric version was calculated by averaging more than 200 corresponding facial locations on the two sides of the face, and then remapping the original face to create the symmetric depiction. Raters made attractiveness comparisons of each pair of faces, and symmetric faces were preferred, including by raters who were not consciously aware that symmetry was involved in the study (the majority of the raters). An additional experiment of the study presented all faces in a haphazard order to raters, whereas in the original experimental design, face stimuli were presented in pairs, potentially increasing raters' awareness of the symmetry manipulations. Raters in all the experiments gave the same results. This study indicates that perfectly symmetric faces are perceived to be more attractive than normal levels of facial asymmetry. It is conceivable, however, that raters were responding to the high degrees of symmetry because of a preference for novelty, rather than symmetry *per se*.

Gillian Rhodes and her colleagues were especially interested in whether people can detect subtle differences in facial asymmetry and whether these differences influence facial attractiveness ratings. Rhodes *et al.* (1998) used four versions of each computerized face: a normal face, a perfectly symmetrical face (made by combining a normal face with its mirror image), a high-symmetry face (made by *reducing* the differences between the perfectly symmetric face and the normal face by 50 percent), and a low-symmetry face (made by *increasing* this difference by 50 percent). The high-symmetry face, then, had more symmetry than normal faces, but lacked perfect symmetry; the low-symmetry face had more asymmetry than normal faces. Raters assessed faces for attractiveness and appeal as a long-term mate. Symmetry was important in both sexes and in both contexts, and ratings correlated with degree of symmetry, indicating that any preference for novelty was not involved.

## Facial averageness

Research revealing the attractiveness of nonaverage manifestations of facial secondary sexual traits discussed above (e.g. larger than average brow ridges and chin in men and smaller than average mandible in women) does not necessarily contradict the hypothesis that averageness in certain facial traits will positively relate to facial attractiveness. Nor is the averageness hypothesis counter to the view that physical attractiveness is a health certification. Variation in facial averageness, just like degree of symmetry or hormone marker size, may honestly signal phenotypic and genetic quality for dealing with the EEA. Facial averageness may be attractive, because averageness is associated with above-average performance of facial features in tasks such as chewing and breathing. Said differently, natural selection on facial features is stabilizing (favors the mean) and sexual selection in the form of mate choice reinforces this (see Langlois and Roggman 1990). Thornhill and Gangestad (1993) suggested an alternative hypothesis, that a preference for average traits in some facial features (not the secondary sex traits) may have evolved, because in continuously distributed, heritable traits, averageness is associated with genetic heterozygosity, which may connote a fit mate; one that is outbred or one with high genetic diversity (useful in defense against parasites). Facial attractiveness is continuously distributed and facial features comprising it appear heritable (Thornhill and Møller 1997; Gangestad and Thornhill 1999). Also, symmetry and heterozygosity may often be positively related, which would result in averageness and symmetry also being positively correlated. A recent study provided evidence that facial averageness is associated with health in children and young adults (Rhodes *et al.* 2001).

Initial evidence that averageness matters in attractiveness was provided by Langlois and Roggman (1990). They found that average faces in each sex, created by compositing, are more attractive than many of the individual faces from which composites were made. The attractiveness of average faces was found also by Grammer and Thornhill, who used composites and calculated facial averageness, and by Jones (1996), who used calculated averageness. Critics, however, appropriately have pointed to the importance of non-average features (e.g. the hormone markers) and the possibility of confounding factors of increased facial symmetry or smoothness and reduction of facial blemishes with the computer-generated composites (see Thornhill and Gangestad 1999a).

However, facial averageness itself appears to be a cue of facial attractiveness. Rhodes and Tremewan (1996) showed this using line drawing composites that eliminate various methodological problems with computerized composites and a computerized caricature generator to vary overall facial averageness. Caricatures (extremes) are not attractive and attractiveness ratings increased as a result of anti-caricaturing (averaging) a face. Rhodes *et al.* (1999) showed experimentally, using computerized facial images, that averageness and symmetry made independent, positive contributions to attractiveness, and that manipulated facial averageness co-varied with attractiveness even in perfectly symmetric faces. These studies seriously question the confounding role of novelty in the form of high or perfect averageness as a mate choice cue. This is revealed in that the extremes of caricatures are novel cues, but averageness is preferred.

Thus, still unidentified facial features are attractive when they are near the population average. The averageness effect does not include the facial secondary sex traits or women's eyes and noses. Women's eyes are more attractive when large and their noses are more attractive when small. Nose and eyes are not secondary sex traits (they do not change dramatically or arise *de novo* at puberty; see Thornhill and Gangestad 1999a).

Donald Symons (1995) has proposed that people possess a psychological adaptation that erects through learning a mental composite of the faces of each sex in the social environment and these average-face templates serve as reference and comparison in deductions about facial attractiveness. Such a mechanism could give rise to viewing certain facial features as attractive when average in size and facial secondary sexual traits attractive when larger than average.

## Attractiveness of healthy skin

There is more to attractive skin than youthful appearance. Various skin diseases discolor and erode skin, leading to asymmetry in skin color and texture, and create acne and other skin lesions, warts, cysts and tumors. Also hirsutism in women is associated with relatively low estrogen and high testosterone. People are consciously aware that skin textural smoothness and uniform color, skin health, and skin attractiveness go together. The cosmetic makeup industry exists as a result of this association. Fink *et al.* (2001) hypothesized that skin features were honest signals of immune system quality and general health in the human EEA and that past selection favored individuals with a preference for smooth-textured, uniformly colored skin. They manipulated skin texture in facial pictures of young women and then had them rated for attractiveness by men. Skin homogeneity positively correlates with facial beauty ratings. Homogeneous skin is uniform in color and smooth texture, and shows a relative absence of hair growth. On theoretical grounds men's facial attractiveness should also positively correlate with skin features associated with skin health.

Bare skin patches often are salient sexual signals in birds and mammals. The visibility of skin in humans as a result of relative hairlessness may be the basis of an entire body covering that functions to honestly signal immunocompetence.

## Species- and sex-typicality of attractiveness

Despite individual differences in attractiveness judgments and even within individual variation shown by women across the menstrual cycle, attractiveness judgments show significant agreement in general. In both sexes and sexual orientations and across a diversity of ethnic groups and ages ranging from infants to the elderly, facial attractiveness assessments are more similar than different, with correlations between photograph raters typically around 0.3 and higher (Cunningham *et al.* 1995; Jones 1996). Consistency in facial ratings is seen also within and between human groups with little or no contact with Western standards of beauty (Jones 1996). Furthermore, males and females of diverse ages, races, and cultures show similarity in their judgments of bodily attractiveness (Magro 1999; Grammer *et al.* 2002). The human-typical aesthetic evaluation of faces and nonfacial body features is evidence of species-typical psychological adaptations that evaluate and make cognitive inferences based on similar or identical cues.

Sexually differentiated judgments are thought widely to include the greater importance that men, compared to women, place on physical attractiveness cues, including facial attractiveness, in mating and romance (Buss 1994). It is well established that, in contrast, women generally place far more weight than do men on cues of a potential mate's status, dominance and resource holdings—a result of selection having favored females in human evolutionary history who preferred good providers and protectors as mates (Buss 1994). Although these sex differences are reliable generally for the human species, in considering the magnitude of any sex difference in the importance of good looks in mate selection, it is critical to consider the variation among women in conception risk and pursuit of short-term relationships.

For photographic ratings, the variance in attractiveness of the opposite sex is frequently sex differentiated and women's ratings are more variable. A proximate hypothesis for this pattern is that women's ratings more than men's reflect personal circumstances (menstrual cycle point, pursuit of short- versus long-term relationship) and variable willingness to trade off between physical attractiveness (and, thus, heritable benefits from mate choice) versus material benefits in mate choice (Thornhill and Gangestad 1999a).

Youth is a salient aspect of attractiveness in both sexes, but the preference in men is stronger and for younger mates because fertility is much more tightly linked to age, specifically to young adulthood, in women than in men (Buss 1994; Symons 1995; Jones 1996). Although women find younger men more attractive than older men, they do not find pubertal or adolescent men very attractive. Men, however, find pubertal females about as attractive as females in their early 20s. The adult female preference appears to be a male in his mid-to-late 20s or 30s; the adult male's is for a mid-to-late teen to early-20s woman (Thornhill and Gangestad 1999a).

In both sexes, age after reaching adulthood is positively correlated with the detrimental effects of senescence, and hence with both reduced health and lower performance. Young men have maximum strength and athletic ability, and thus presumably were better EEA hunters and protectors than old men. Phenotypic correlates of young adulthood in both sexes, then, are honest signals of mate value. If this involves the handicap mechanism *per se*, high-quality individuals will physiologically maintain youthful appearance and performance longer than low-quality ones under the same regime of environmental insults (such as weather, toxins, and disease in both sexes and parity in women), because that maintenance is less costly to the high-quality individual. Consistent with this is the role of estrogen, a candidate handicapping hormone, in maintaining youthful appearance and behavior in women (Thornhill and Møller 1997). The hypothesis that age-related facial cues are involved in honest signaling is the antithesis of Jones's (1996) hypothesis that attractive women's faces deceptively signal youth.

## Conclusions

Although social psychologists had demonstrated by the 1980s the important role of looks in many areas of people's everyday lives, specifically biases in favor of attractive others, they did not entertain the question of why the biases exist, and thus could not explore the answer's implications for further understanding attraction and attractiveness. Consideration of this evolutionary question and its empirical implications has resulted in an explosion of research on human beauty over the last decade. Evolutionary psychologists have examined a number of hypotheses subsumed under the general notion that the evolutionary function of physical attractiveness judgments is discrimination of an individual's phenotypic and genetic condition and, broadly speaking, health status, and that various features of bodily and facial attractiveness have evolved to signal health status. This paper reveals that these research endeavors have been quite fruitful.

Charles Darwin hypothesized that human sexual attraction evolved because it led to obtaining a mate that yields offspring with a mating advantage solely as a result of their physical attractiveness. Sexual attractiveness to Darwin was explained by mate choice for attractiveness *per se*, not for traits that are attractive because they signal **fitness components** other than mating success such as health, survival, and genetic quality. Darwin's hypothesis (and R. A. Fisher's later similar one) is difficult to sustain in light of the accumulated evidence that signaling and perceiving looks revolve around health. Likewise, the hypotheses in the literature that human facial preferences: (1) are incidental effects of sensory biases arising from psychological adaptation for general object perception or another task, or (2) arise from adaptations that function to limit mating to conspecifics (but not high mate-value ones) and thereby prevent maladaptive matings with heterospecific hominids and other apes, or function to secure a mate that is unambiguously of the opposite sex, appear unable to account for this same evidence (see Thornhill and Gangestad 1993, 1999a).

There is probably more prejudice and differential treatment in the context of physical attractiveness than in the contexts of the sex and ethnicity of individuals combined. Many humans are no more aware consciously of why they have their prejudices in favoring attractive others than of why they have an elegantly designed heart or foot. Functional design, whether in the mind or in other body parts, is the product of evolution by Darwin's demon, highest reproductive success of individuals with traits that best solve ecological and social problems. Scientific knowledge of why we are designed to feel and behave the way we do is available solely through education in modern evolutionary biology. Increased awareness and knowledge of one's beauty biases may assist people who desire to behave more democratically and thus contrary to the criteria that determined which individual would be ancestors throughout human evolutionary history. Also widespread knowledge of why people are biased may produce a society in which there is less tolerance of prejudice. It should be obvious to readers that there is nothing in the facts or hypotheses in the evolutionary psychology of beauty that can serve to justify prejudice based on looks.

R. Thornhill is grateful to the organizers, especially A. Moya and E. Font, of the workshop on evolution held at the University of Valencia for their invitation to lecture and warm hospitality.

## Further reading

Andersson, M (1994). *Sexual selection*. Princeton University Press, Princeton.

Buss, DM (1999). *Evolutionary psychology: the new science of the mind*. Allyn and Bacon, New York.

Etcoff, NL (1999). *Survival of the prettiest: the science of beauty*. Doubleday, New York.

Møller, AP and Swaddle, JP (1997). *Asymmetry, developmental stability and evolution*. Oxford University Press, Oxford.

Thornhill, R (1998). Darwinian aesthetics. In C Crawford and D Krebs, eds *Handbook of evolutionary psychology: ideas, issues and applications*, pp. 243–272. Lawrence Erlbaum Associates, Mahwah, NJ.

Zahavi, A and Zahavi, A (1997). *The handicap principle: a missing piece of Darwin's puzzle*. Oxford University Press, Oxford.

CHAPTER 22

# Genome views on human evolution

**Jaume Bertranpetit and Francesc Calafell**

It has long been acknowledged that understanding the biological nature of humans does not require a special scientific framework. Nonetheless, it is fascinating how often this sentence is repeated, suggesting that it still needs to be restated against mythological explanations of human nature. The number of human traits that are included in biological studies has been growing with time. From the inclusion of humans in the animal kingdom to recent advances in the study of the molecular bases of consciousness, challenges to our own understanding seem an endless endeavor; the combined efforts from a wide range of disciplines are needed to understand the biological nature of humans.

The scientific disciplines concerned with the study of human evolution fall into two groups: historical (both biological and social) and comparative (again, both biological and social). The main disciplines that adopt a historical view of human evolution are paleoanthropology, archaeology, and historical linguistics. Comparative studies of present-day humans include molecular biology and genetics, physiology, and anatomy (usually approached from biological anthropology), and studies of cultural traits that may be used to assess similarities and differences among human populations.

With such a variety of approaches converging on the study of human evolution it is essential to formulate a coherent set of questions, focus on them, and use the information provided by each discipline accordingly. An unorganized confluence of illustrated interdisciplinary issues does not assure a high level of understanding. Thus, asking the proper questions is a key issue in advancing our knowledge of human evolution.

What does it mean to study human evolution from a molecular perspective? In fact, there are many ways in which molecular studies can throw new light into our origins and evolutionary history:

**(1)** Adopting a molecular perspective may mean to describe and understand the tempo and mode of diversification of humans in relation to other closely related species. An initial question is the time depth of separation of the human lineage; other relevant questions include whether this process is special in any sense, such as in its speed or in the total amount of change as seen from a comparative perspective.

**(2)** It may also mean to recognize the specific functional changes that define our own species. These changes can be identified either by newly acquired functions (or by their loss) or by an accelerated evolutionary pace of some biological trait in relation to the overall tempo of evolutionary change in our lineage, most of which may be nonfunctional or unrelated to the abilities that we associate to humanity. This implies the need to acknowledge that the singular traits that make us human and that seem so clear at the morphological and functional levels have a genetic basis that needs to be unraveled.

**(3)** Once the singularity was produced and divergence started, the next question may be the evolutionary relationship of the initial humans (or humans-to-be) with modern humans. Or, in other words, inquiring about the origin of contemporary human populations. Tracing back the ancestry of current diversity may be a suitable approach for unraveling the place and time of origin of our ancestors. Moreover, the pattern of genetic variation may inform us of demographic events close to the time of human origins and may provide answers to long-standing questions, such as do our origins trace back to a large or small population?

A hypothetical origin from a single couple has been suggested both from mythology and chromosomal studies, but now we know this makes no sense. Thus, human origins and the ancestry of contemporary humans are two quite different questions that comparative studies approach differently, through divergence between humans and other species the former and through divergence among humans the latter.

**(4)** Within the short time scale where human evolution and human history merge, can we trace back the processes that gave rise to contemporary human populations? Which is the finest geographic or temporal detail in human history that a molecular approach can resolve? Here, questions may go from broad patterns (such as the settlement of entire continents) to very local and specific ranges (such as the demographic impact of a cultural transition or of settlement in specific areas). This is a vast area of human population genetics that will not be dealt with here.

Even at a finer level of detail, the functional significance of differences among populations and among individuals (i.e. the biological bases of individuality) can be subjects for a comparative approach. Our ignorance of the genetic makeup of individual variation is vast, and most of what we know lies in the realm of pathology, far from the average, small differences among any given pair of humans. There is at least one claim to the understanding of the genetic bases of "novelty-seeking," a psychological trait defined from test scores (Benjamin *et al.* 1996), although genetic variation explains at most 5 percent of the psychological variation in this trait.

We do know about one evolutionary mechanism that promotes differences among individuals: Selection triggered by pathogens has favored heterozygotes for the major HLA (human leucocyte antigen) loci, resulting in the highest inter-individual diversity known for any nuclear genome region (Przeworski *et al.* 2000). But this is likely to be more the exception than the norm for genetic differences. In fact, an evolutionary perspective into the biology of individuality is far from being feasible nowadays.

## The human genome and its dynamics

Evolution does not act on an ethereal substance, but on genetic material with its own biochemical, structural, and functional constraints which may vary across the genome. Thus, it is essential to understand the dynamics of genomic change before leaping to conclusions regarding the processes and products of evolutionary forces.

According to traditional population genetics, genetic differences arise by **mutation, recombination**, and **gene conversion**, and are modulated by **selection** and **drift**. That is, evolution is fueled by mechanisms that are embedded in the genome (such as the possibility of chemically substituting one base for another, or of bringing together chromosome segments that were previously in different strands), but the product of those processes is then exposed to interactions with the wider genetic and general environment, which can just ignore it (if the resulting variant is selectively neutral), propel it in the path to **fixation** (if the resulting variant increases the Darwinian **fitness** of its bearers), or send it to oblivion. The latter possibility may sieve out instantly many mutations that arise but that happen to be incompatible with life.

Once through this first filter imposed by selection, genetic variants are at the mercy of the vagaries imposed by population factors such as **genetic drift**, population structure, or **gene flow**. The stochastic nature of these factors leads to fluctuations in allele frequency with time, which are wider in smaller populations.

Changes in effective population size determine the overall shape of the **coalescent tree**, that is, the tree showing the lines of descent of a sample of a gene or genome region taken from a number of individuals in a population. Coalescent trees are deeper in larger populations: it takes more time to reach the common ancestor of the sample of genes. But a population expansion changes radically the shape of the coalescent tree: travelling back from the present to the past, the population size suddenly shrinks and a flurry of coalescent events occur. Thus, most branches stand alone for most of the depth of the tree and then coalesce in a short time span, as rays bursting from a star (Fig. 22.1b).

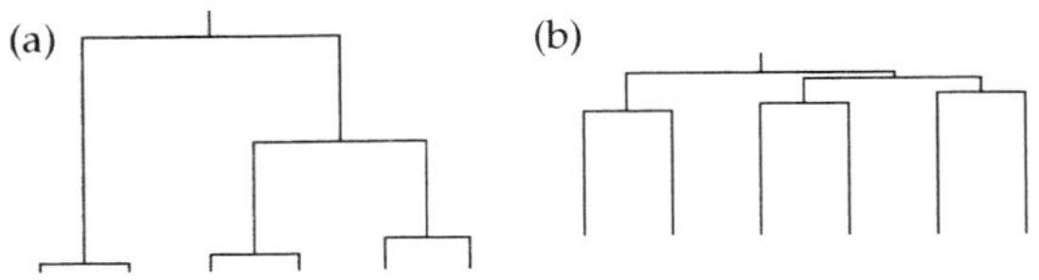


**Figure 22.1** Coalescent trees for six gene copies simulated with (a) a constant population size, and (b) with an expanding population size.

Such differences in tree shape are crucial for understanding the way genomic variation accretes. In a starlike tree, most lineages have been isolated from each other for most of their existence, and genetic variation has accumulated independently in each branch. Thus, most mutations would lead to alleles borne by one or a few individuals. Trees with more regular, elongated shapes, such as those established in constant populations, contain often deep, long, basal splits (Fig. 22.1a). Mutations in those deep branches will be shared by all of their descendants, that is, by a large fraction of the gene sample.

We have just described how changes in population size forge the process of accumulation of genetic variation and result in different patterns. The reverse inference is one of the tenets of evolutionary genetics: we may be able to recognize population processes by the distinctive patterns they leave in extant genetic variation. Such statistical inference has been framed in terms of statistical testing, in which the distribution of some genetic parameter is known or inferred under a null hypothesis; such parameters can be measured in an actual population, and the null hypothesis (often the **neutral model** of evolution; see Ohta, Chapter 1) be then accepted or rejected.

Population processes affect the genome as a whole; however, their effects are often intertwined with those of genome-context specific mechanisms, such as those resulting from the interaction of selection and **linkage**. This combination of factors will leave footprints that are often indistinguishable from those left by population processes alone. For instance, the effects of positive selection mimic those of a population expansion. When a selectively advantageous allele arises, selection will increase its frequency to the point of fixation, wiping out all other alleles in the process. The dynamics of the expansion of a selectively advantageous allele are equivalent to those of the demographic expansion of a population and its consequences for the patterns of extant genetic variation are very similar (see Aguadé *et al.*, Chapter 2). Most, if not all, tests designed to detect departures from the neutral model (such as Tajima's $D$), will yield significant results both under population expansion and under positive selection. In fact, very often a given test result is interpreted in selective or demographic terms depending on the background of the researcher. In human population genetics, the usual interpretation favors population history, as if humans were absolutely impervious to natural selection. The same test result would usually be interpreted by *Drosophila* population geneticists as the inescapable legacy of natural selection. However, this *culture gap* is being quickly bridged, and the effects of natural selection on the human genome are increasingly being recognized.

The effects of selection may go beyond the gene that is under its effects. Eukaryotic genomes are organized in linear chromosomes, which constitute **linkage blocks**, broken only by recombination. Thus, if an allele in a gene is being selected for, not only this allele increases in frequency, but a whole genome segment around it. The length of this segment is an inverse function of the recombination rate: A low recombination rate implies that a longer genome segment will be inherited together and will undergo selection as a block. Thus, positive selection at a gene will decrease genetic variability at that gene and at a larger region centered on that gene. This has resulted in a genome-wide trend: Genetic diversity in humans is positively correlated with recombination rate (Nachman 2001). That is, the amount of genetic variability observed in any given region is dependent on the genomic context; this observation is often ignored, and the genetic patterns tend to be interpreted in terms of genome-wide processes, such as population history, disregarding local genomic effects.

An example of genome-driven evolution was found in the GBA **pseudogene** (psGBA; Martínez-Arias *et al.* 2001). One can imagine that pseudogenes, which are by definition noncoding and thus free of the constraints imposed by natural selection, should accumulate genetic variation in larger amounts than functional genes do. However, a study of a sequence of 5.6 kb in 100 human chromosomes showed that psGBA had a genetic variation among the lowest observed so far in an **autosomal** locus, lower even than that of many functional genes. psGBA lies in a genome region with one-fourth the average genome recombination rate; thus, its low variation was interpreted as the result of selective pressure not on itself, but on one of the many neighboring genes. This is an example of how the combined action of selection and linkage can shape genetic variation at other loci irrespective of their coding or functional status.

The same property makes it difficult, once the action of selection on genetic variability has been detected, to pinpoint exactly what has been selected for. But then it should be remembered that selection acts on function, not on sequence. One of the most revealing signs of positive selection is the ratio of **nonsynonymous** to **synonymous substitutions** (i.e. the $K_a/K_s$ ratio). Given the composition of a sequence and the genetic code, it is easy to model how many synonymous and nonsynonymous substitutions are expected under a purely random mutation process. If a gene is compared in two or more closely related species and a significant bias in favor of nonsynonymous substitutions is found, this constitutes evidence for selection having shaped the process of evolutionary change. In particular, a significantly high $K_a/K_s$ ratio is one of the signs that should be sought for in any attempt to pinpoint which genes are responsible for the relevant functional changes in the emergence of humans.

## The phylogenetic position of humans

The position of humans in the natural world has been one of the longer lasting questions in evolutionary biology. As noted by Darwin, this is a problem for comparative biology rather than for paleontology. It is a comparative issue in the sense that human evolution has to be reconstructed in a framework that takes into account the species that are more similar to humans. The great apes, our putative relatives, are included in a family called Pongidae and differentiated from our own family, the Hominidae. The Pongidae include four species and three genera: orang-utans (*Pongo pygmaeus*), gorillas (*Gorilla gorilla*), chimpanzees (*Pan troglodytes*), and bonobos (*Pan paniscus*).

If taxonomy reflected phylogeny, Pongidae and Hominidae should be sister **clades**, which is not the case. As we will see, the closest evolutionary relationship is to be found between humans and the common clade of chimpanzee and bonobo. It is interesting to note how the perceived relatedness among these species has changed over time, due to the subjective nature of observations. Humans have sometimes been seen as having nothing to do with other species whereas at other times the commonalities have been stressed to the point of suggesting that there were no differences (Fig. 22.2). As early as 1863, Huxley recognized the problem and, while stressing that differences between apes and humans were significant, he suggested that a balanced approach was needed. However, his main contribution was that he claimed that differences should actually be sought in order to investigate whether they are large or small.

This balance was not reached until molecular analyses entered the study of human evolution and comparisons between species were undertaken at levels other than morphology. One of the main lessons of evolutionary psychology is that it is extremely difficult to be objective in recognizing differences in outward appearance (mainly in facial traits) whose perception has been so deeply shaped by natural selection (see Thornhill and Gangestad, Chapter 21). As we are so extremely good at recognizing minimal differential traits among humans, especially in faces, it may be difficult to ask ourselves to be objective in recognizing and weighting differences that include individuals of other species that are so similar to ourselves.

Moreover, morphology can change at various evolutionary speeds, obscuring the expected correlation between morphological similarity and evolutionary events. Thus, the molecular approach, after a fierce initial rejection, has been a primary source in establishing the tempo and mode of primate

**Figure 22.2** The subjective distinction among species can be clearly seen by comparing these two engravings from the nineteenth century, where the similarities or differences between human and great apes are greatly exaggerated.

evolution, and in resolving the **topology** of human and ape evolution.

An initial insight came from the discovery that blood groups were also present in some primates. Even if all apes have responses to some ABO antigens that are similar to those in humans, we now know that the similarity provides information on the evolution of blood group genes (quite complex in itself) rather than on the evolution of the species concerned. This is made evident by the fact that the same immunological specificities have arisen more than once, as in the case of the O allele which is polyphyletic even in humans.

The strength of the immunological reaction between species was first assayed in a classical paper by Sarich and Wilson (1967) which provided the first evidence of the strong similarity between African apes and humans. For the first time, immunological reactions were interpreted in terms of divergence time and even if the results were dependent on the immunological recognition of specific proteins (and, thus, of **epitopes** on which selection has been acting), this work opened the field both to the use of other molecules and to more precise and easy to interpret techniques.

Comparative analyses of proteins opened widely the field, with pioneer work by M. Goodman, who showed the power of sequence data in reconstructing the phylogeny of several genes in the globin family in vertebrates. A major challenge in protein studies (and later in analyses of gene sequences) is understanding the extent to which the evolutionary history of the genes, rather than that of the species, is reconstructed. The final product of these studies is a tree that is the result of selective factors, and may tell us little on tempo and mode of species evolution. In fact, when the similarity or even identity in globin sequence between apes and humans was highlighted in an effort to understand human evolution, the reply from some paleontologists was: what is the point of describing similarities when the differences are so clear in the living species and all through the fossil record? The information derived from the study of single molecules should be separated (or in some cases integrated, as discussed below) from the data that may explain the general trend of genome evolution and that may be used to understand species evolution.

A major step forward came with DNA hybridization studies, in which a total measure of genome differentiation could be obtained for pairwise comparisons of species through the differences in melting temperature of interspecies **heteroduplexes**. These studies gave similar results to subsequent analyses of DNA sequences based on multiple loci (reviewed in Ruvolo 1997). Soon, it became evident that humans and *Pan* are more closely related to each other than any of them is to the gorilla. In fact, whether the gorilla lineage split first and, if that was the case, how long the branch connecting *Gorilla* to the *Pan-Homo* clade was were matters of much debate; the joint analysis of 14 independent DNA data sets (Ruvolo 1997) clearly rejects a tricotomy between the three lineages.

Recently, more data have been produced in order to reconstruct the phylogeny of humans and apes. In a study by Chen and Li (2001), care was taken to choose intergenic nonrepetitive DNA segments, which yielded low sequence divergence estimates between humans and apes. The results are fully congruent with previous sequence and hybridization data, but with higher statistical power and in complete agreement with the **molecular clock hypothesis**: the divergence between humans and *Pan* has been established around 6 million years ago (mya; 1.24 percent sequence divergence), with an earlier separation of the gorilla 7–8 mya (1.62 percent sequence divergence between gorilla and the branch leading to humans and chimpanzees). Thus the two branches seem to have been separated by around 2 million years. Orang-utans diverged much earlier, between 12 and 16 mya (divergence to the other three species around 3.0 percent) and mtDNA and globin gene studies in chimpanzees and bonobos tend to indicate a relative recent divergence, around 2.5 mya, but with an estimate based on *X* chromosome sequence of just around 1 million years (Kaessmann *et al.* 1999). Besides the framework that this tree gives to other disciplines studying evolution (mainly paleontology) it is valuable because it provides a phylogenetic context for the study of genomic evolution and molecular evolutionary phenomena (Ruvolo 1997).

## Intraspecific variation in humans and apes

Although a large number of studies, starting in the 1920s, have produced allele frequencies for the so-called **classical polymorphisms** (i.e. expressed products such as blood groups, proteins, and the HLA system), good estimates of the overall amount of genetic variation in humans have only become available in the last decade. For a long time most researchers focused on describing variation at a single locus or for small set of markers in specific populations. Only when large amounts of data were pooled was it possible to gain insights on human populations rather than on the genetic forces shaping variation at a single locus (see Cavalli-Sforza *et al.* 1994 and references therein). Moreover, this research has for a long time suffered from a population bias: Most of the studies were conducted with populations of European origin, as can be seen in maps annexed to Cavalli-Sforza *et al.* (1994).

DNA studies have been fundamental for recognizing the extent of genetic variation in humans. Initially undertaken at anonymous **restriction fragment length polymorphisms** (RFLPs), genetic analyses moved through **minisatellites**, into **microsatellites** (essential nowadays in forensic genetics), and lately into nucleotide variation. Two strategies are currently in use: typing of previously known **single nucleotide polymorphisms** (SNPs), and sequencing of a given genome region for a set of individuals. The latter provides a good estimate of the existing variation, most in the form of SNPs but also as insertions/deletions (**indels**) or other reorganizations.

The simplest (though highly informative) parameter used to measure human genetic variation is the **average nucleotide diversity** $\pi$, which is defined as the average relative number of nucleotide differences per site between two sequences; it is equivalent to the gene diversity or heterozygosity at the nucleotide level. Its inverse ($1/\pi$) gives the expected length of sequence that will contain one variant.

Any estimate of $\pi$ that aspires to capture diversity within the human species as a whole must contain African samples, since those populations are known to harbor most human diversity, as discussed below. Several such estimates exist, and sequencing efforts (including the Human Genome Project) have enabled us to recognize global levels of genetic variation for specific genomic regions or for whole regions, chromosomes, or the entire genome. As a simple rule of thumb, human variation is in the order of one per thousand, as seen in two large surveys:

**(1)** The Human Genome Project produced highly interesting variation data, with more than 1.4 million SNPs published in the original report. The overall result is $\pi = 7.51 \times 10^{-4}$ (or 0.7 per thousand), meaning that, on average, one SNP can be found in every 1330 bp surveyed in two human chromosomes drawn from the panel gathered by the US National Institutes of Health (The International SNP Map Working Group 2001).

**(2)** Many research groups are producing studies of human sequence variation for specific genome regions, throwing light into the extent of nucleotide variation. Although nucleotide diversity values depend on the specific genome region surveyed, and also on sample size and sampling procedures, the average value is around 0.8 per thousand ($8.1 \times 10^{-4}$). It is worth noting that diversity values vary widely depending on the genome region analyzed, and that those values are not easily related to the putative effects of selection. As already mentioned, one of the lowest values ($\pi = 0.0004$) has been described for a pseudogene (Martínez-Arias *et al.* 2001); this is an example of how simple correlations between function of the gene product and amount of variation that would be explained by purifying selection cannot be guessed beforehand.

The values obtained in different studies are quite similar and give a general picture of human genome variation; but the point is to understand the magnitude of the value in relation to the genomic and evolutionary factors that shaped it. Thus, a comparative perspective is needed.

Some nucleotide diversity data are available for the primate species closest to humans. For a few loci, such as mtDNA (Gagneux *et al.* 1999) and a noncoding region in the *X* chromosome (Kaessmann *et al.* 2001), quite complete data are available. These studies are based on large samples and thus provide a fairly good estimate of overall genetic variation, a difficult task for primate species with fragmented habitats and many different breeding groups of still uncertain systematic status.

Results for mtDNA are straightforward and underscore the comparatively low genetic diversity of humans (Gagneux *et al.* 1999). Chimpanzees show much higher diversity than humans, with diversity values that vary with the subspecies (or even population) considered: 2.8 times higher in the eastern chimpanzee than in humans, 5.8 in the central, 8.3 in the western, and, surprisingly, 15.8 in the Nigerian chimpanzee, which has recently been accorded subspecies status (*Pan troglodytes vellerosus*). Thus, values are not only higher than in humans but also very heterogeneous depending on chimpanzee groups. The bonobo shows a nucleotide diversity value in the range of chimpanzees (3.4 times higher than in humans) whereas for the gorilla (including individuals from the two subspecies, that are considered very homogeneous in morphology, behavior, and ecology) the value is extremely high: 30 times more variation than in humans. Sampling in this survey was extremely comprehensive, which enabled the authors to capture most of the genetic variability in great apes, but mtDNA may show genome specific features as compared to nuclear DNA.

A survey of the *X* chromosome yielded similar, though not so extreme, trends: nucleotide diversity is 3.5 times higher in chimpanzees than in humans, 4.6 times in gorillas, and 7.3 times in orang-utans (Kaessmann *et al.* 2001). Even if differences with mtDNA data are not easily understood as sampling was very different in the two studies, the message is clear: when comparing human and apes, the amount of variation is far lower in humans than in any of the other species (or even subspecies).

Beyond mere description, it is interesting to interpret these differences. The usual explanation for this pattern is the short time of diversification of modern humans. However, we shall see that this is an indirect, rather than a direct consequence of the difference in nucleotide diversities among primate species. Under a neutral, constant-population model, nucleotide diversity is a function of effective population size: The larger a population has been historically, the more diversity it can harbor. This equation is robust to fluctuations in population size; in that case, nucleotide diversity is proportional to the harmonic mean of population size through time. Thus, it could be argued that the small diversity in humans is simply the result of a small population size, which could have been produced either through one or several bottlenecks or strong population subdivision, lowering the total effective population size.

We know, however, that estimates of the time to the most recent common ancestor (TMRCA) in a coalescent model are a function of effective population size. Thus, both views can be reconciled: the genetic homogeneity of humans is the result of a recent bottleneck. Nonetheless, other studies reject the existence of a bottleneck in human history: several loci show an enormous amount of variation whose coalescence is very old, to such an extent that a gene genealogy of alleles for humans and apes shows a complete intermixing of branches for the various species. This picture is clear for several HLA loci, indicating that some human alleles are more closely related to those of apes than to those of other humans (Ayala 1995a). However, it is generally acknowledged that HLA loci are under **balancing selection**, which would tend to preserve larger than expected amounts of diversity even through narrow population bottlenecks.

The apparent discrepancy between the small long-term effective population size of humans as inferred from extant genetic diversity and the perceived total population of a species like ours that ranged over much of the globe may be reconciled by considering that many human lineages have become extinct, taking with them their genetic diversity. This happened with the Neanderthals: The few mtDNA sequences that have been obtained from them fall clearly beyond the range of extant human variation (Krings *et al.* 2000). And, in the Neolithic revolution, many hunter–gatherer populations perished, while living African hunter–gatherers are among the most internally diverse human populations (Calafell *et al.* 1998).

It should be borne in mind that the relationship between $\pi$ and effective population size was derived under the assumption of a constant population size. In the event of a population expansion, as explained above, the accumulation of genetic variation is a function of the time since expansion rather than of effective population size. Historically, mtDNA was the first genome segment to be extensively sequenced in humans, and its analysis gave unequivocal signs of population expansion, from which an expansion time was estimated (Rogers and Harpending 1992). However, as discussed in the following, this may be an oddity of mtDNA rather than a norm across the genome.

As a summary of this section, Fig. 22.3 presents data on divergence and polymorphism in a noncoding sequence at Xq13.3. The branches of the tree are proportional to the relative amount of divergence among species (i.e. the average nucleotide difference between two chromosomes sampled in different species). Given the molecular clock and after estimating the mutation rate, the scale of divergence could be converted into a time scale. At the tips of the tree, we have represented, using the

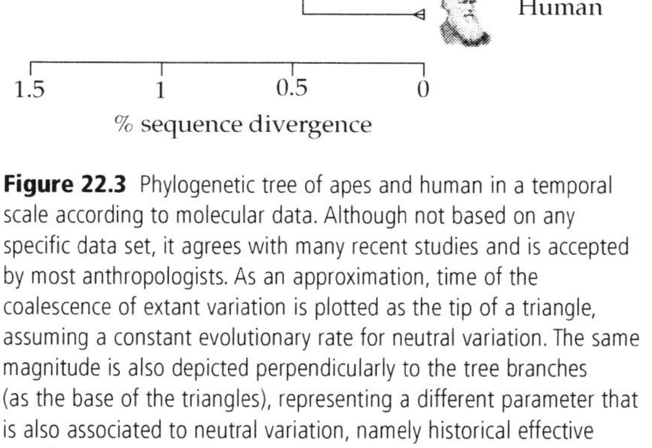

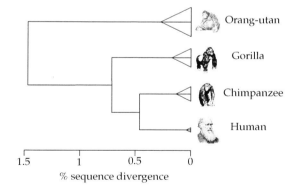


**Figure 22.3** Phylogenetic tree of apes and human in a temporal scale according to molecular data. Although not based on any specific data set, it agrees with many recent studies and is accepted by most anthropologists. As an approximation, time of the coalescence of extant variation is plotted as the tip of a triangle, assuming a constant evolutionary rate for neutral variation. The same magnitude is also depicted perpendicularly to the tree branches (as the base of the triangles), representing a different parameter that is also associated to neutral variation, namely historical effective population size. Data from noncoding sequence at Xq13.3 (reworked from Kaessman *et al.* 2001).

same scale, polymorphism within each species (i.e. the average difference between pairs of chromosomes within each species). The latter magnitude is represented both as breadth (illustrating the close relationship between polymorphism and effective population size) and as depth (in the same evolutionary scale as divergence, representing the TMRCA for each species). This is a graphical example of how small polymorphism is in humans compared to other primates, which may be interpreted as the result of a recent bottleneck.

## What makes a human human? An evolutionary perspective

Most of the analyses mentioned so far rely on putatively neutral variation from which the general pace of evolution can be reconstructed. But, what genetic differences account for the observable specific human traits? Until recently, the evolutionary study of the genome has focused on showing how similar the human and ape genomes are and has paid little attention to detecting and understanding the differences. In fact, to understand a species from its genomic information is an endeavor far beyond the possibilities of contemporary biology; we do not possess any artificial life software application that would read a genome sequence and recognize, from the information contained in it, which life form it codes for, specially in the case of complex genomes. Nonetheless, we may wonder what is the extent of coding and relevant differences between genomes of humans and apes that may account for the emergence of specific human traits.

We know little about this issue, but studies of gene expression, posttranscriptional editing of proteins, and functional relevance of protein sequence variation are likely to provide interesting insights in the near future. An example of these studies is the work of Enard *et al.* (2002a) comparing patterns of gene expression in human, chimpanzee, orangutan, and macaque tissues, including brain tissue.

Further insights could be gained from adopting an evolutionary perspective. One could, for example, attempt to estimate the overall amount of variation expected in the coding part of the genome. Taking as a mean value for autosomes a divergence of 1.2 percent between humans and chimpanzees, there is heterogeneity among genome regions. The lowest value, as expected, is to be found in coding regions: 1.11 percent for synonymous and 0.80 percent for nonsynonymous substitutions. Differences between the two are not extremely large, indicating that the general action of mutation was only slightly modulated by selection. Neutral substitutions are expected to be one-fourth synonymous and three-fourths nonsynonymous, which would yield an expected 0.88 percent weighted average for coding regions.

If we take 1300 bp as the mean length of the coding part of a gene, we should expect to find, on average, 11.4 substitutions. If substitutions were neutral, 8.5 of them would be nonsynonymous, that is, amino acid substitutions. For the whole genome, this means (depending on the number of genes) 0.8 percent of changes reflected in the protein sequence. For 35 000 genes, this represents around 300 000 differences between humans and chimpanzees. Thus, many differences exist, but the problem is to know which of them are relevant for specific (and interesting) traits; a challenge for which we do not have an answer and, which, undoubtedly, needs an enormous development of bioinformatic tools both to infer function from structure and to predict changes in function due to small amino acid sequence changes.

A second approach using the evolutionary perspective may be to identify specific genes that are active or inactive in humans as opposed to apes. There are some cases of gene inactivation in the human lineage but not in African great ape lineages (Gagneux and Varki 2000; Hacia 2001). A good example is the inactivation in humans of the CMP-sialic hydroxilase, the enzyme that transforms Neu5Ac (*N*-acylneuraminic acid), a form of sialic acid, into Neu5Gc (*N*-glycoylneuraminic acid); this reaction modifies cell surface glycolipids. The lack of Neu5Gc has immunological importance, may be relevant in cell–cell interactions, and could explain differences in susceptibility to certain bacteria and viruses. It has also been hypothesized that lack of Neu5Gc may alter glycoprotein function in the human brain and potentially affect brain development, but this is quite a speculative conclusion. It may be extremely difficult to recognize the phenotypic importance of differences between humans and apes, as either a human-specific character could have been fixed through strong selection (and thus, be relevant) or just randomly fixed under drift on a neutral (or nearly neutral) character (and, thus, less likely to be interesting).

Besides the presence or absence of a function, relevant quantitative differences may also be found. For example, Enard *et al.* (2002b) found two nonsynonymous substitutions in humans compared to chimpanzees in the *FOXP2* gene, which may be related to language abilities. However, the functional relevance of those changes remains unknown.

A third approach is to look for direct footprints of selection in our species, as selection may have driven phenotypic changes and specific **adaptations**. The ratio between nonsynonymous and synonymous substitutions may detect such a selective pressure (see above) and is used to understand species-specific DNA sequences of functional relevance. In general values of $K_a/K_s$ above 1 deserve attention.

For example, one may hypothesize that serotonin may have some specific function in humans; thus the comparative analysis of genes involved in the serotonin functional pathway may be helpful. Of six receptors analyzed, all but one (receptor HTR1A) yield very low $K_a/K_s$ values (unpublished results by A. M. Andrés and N. Saitou). Thus, HTR1A is worth being further analyzed for possible differential functions in chimpanzees and humans.

Among the many gene studies (Chen and Li 2001), protamine genes have very high $K_a/K_s$ values, a likely indicator of rapid evolution of male reproductive proteins. Nonetheless, even in this case the interpretation is not straightforward (see references in Hacia 2001).

Not every case of positive selection is going to be detected through this type of analysis, as positive selection for a single point mutation having functional importance will not draw attraction by a high $K_a$ value; only the footprint of a general loss of diversity will persist. Although plenty of possibilities exist in evolutionary genetics to provide insights on what makes a human human (or, in other words, how a genome creates a species), most work is still to be done.

## A genetic approach to the origin of modern humans

The origin in time and space of our species, *Homo sapiens sapiens*, has been debated for long. The interpretation of fossil evidence has led to the formulation of two hypotheses on the origin of anatomically modern humans: the multiregional and the substitution hypotheses. The multiregional model posits that modern humans evolved roughly at the same time from local archaic lineages all over the Old World. Selectively advantageous traits would have spread by gene flow, keeping morphological change in synchrony over this vast territory. According to the multiregional hypothesis, the last common ancestors of all humans lived more than 1 mya. In contrast, the replacement or "Out of Africa" model envisages a much more recent origin (*c.* 150 000 years ago) in Eastern Africa, from where humans emerged and colonized the rest of the Old World replacing local archaic populations such as Neanderthals in Europe and *Homo erectus* in Asia. There is a raging, often bitter debate between paleoanthropologists adhering to either hypothesis; the fragmentary nature of fossil evidence has probably prevented a reasonable closure of this controversy.

Genetics has been called into the fray since the two models predict different demographic histories for humankind, which in turn may have led to completely different genetic landscapes. Thus, a reverse inference process can be attempted: from the extant population genetic patterns, the demographic history of humans is reconstructed, at least to the point of declaring its compatibility with one of the competing hypotheses.

One of the most recurrent findings in human population genetics is the observation of greater variability within African populations as compared to non-Africans. African populations tend to be more diverse by any measure: They contain more alleles, expected heterozygosity and nucleotide diversity are higher, and they have more population-specific alleles. This has been observed for mtDNA sequences (Vigilant *et al.* 1991), microsatellites (Pérez-Lezaun *et al.* 1997; Calafell *et al.* 1998), minisatellites (Armour *et al.* 1996) and their flanking regions (Alonso and Armour 2001), and many autosomal sequences (reviewed in Przeworski *et al.* 2000). Moreover, rare variants in autosomal sequences and extreme alleles in microsatellite distributions are more frequent in African populations.

The higher levels of genetic diversity within African populations can be explained by a history of population subdivision with a larger effective population size in Africa than elsewhere. This piece of information would still be claimed to be compatible with both hypotheses on the origins of modern humans. However, there is further information on the **phylogeographic** structure of genetic variation that appears to tip the scales in favor of one of the two hypotheses. In a number of loci, the deepest split in their allele phylogeny separates allele subsets that are exclusively African from others that are found in Africans and in other humans. This has been reported for mtDNA (Vigilant *et al.* 1991), the *Y* chromosome (Underhill *et al.* 2000), autosomal loci such as psGBA (Martínez-Arias *et al.* 2001), and the flanking region of minisatellite MS205 (Alonso and Armour 2001). Thus, genetic variation in non-African populations is a phylogenetic subset of that found in Africans, which hints that non-African humans are themselves a recent splinter of Africans, in line with the replacement hypotheses of the origin of modern humans.

The subsetting process of human variation in non-African populations was a random event, which may have affected intensely some loci while leaving others roughly untouched. Examples include sequences at genes CD4 and DM, where the sampling of genetic variation that occurred when the first anatomically modern humans left Africa was rather stringent, leading to an increased **linkage disequilibrium** in non-Africans. A counter-example is found in the CFTR gene, where linkage disequilibrium is similar in Africans and non-Africans. The relative proportions of both types of loci would provide a measure of the narrowness in a population bottleneck that may have preceded the "Out of Africa" expansion (Mateu *et al.* 2001).

Przeworski *et al.* (2000) recomputed Tajima's *D* (see earlier) in a number of human sequence population datasets and found that, as a rule, Tajima's *D* was closer to that expected under constant population size in non-Africans that in Africans. A phase of population reduction (i.e. a bottleneck, as discussed above) prior to the colonization of the rest of the Old World is compatible with such a result. However, there are some conspicuous exceptions to that rule: mtDNA (Excoffier and Schneider 1999), the *Y* chromosome (Underhill *et al.* 2000), and the Xq13.3 region (Kaessmann *et al.* 1999). It should be noted that all three genome regions share low or null recombination rates, which makes them prone, as discussed above, to the effects of **selective sweeps**.

Genomic effects seem also to confuse the estimate of time to the most recent common ancestor in different genomic regions. Conflicting dates are obtained, with very recent values for the *Y* chromosome (down to 50 000 years ago; discussed in Bertranpetit 2000), and for mtDNA (150 000 years ago; Ingman *et al.* 2000). However, dates based on autosomes tend to be much older, up to 800 000 years ago (Harding *et al.* 1997). The larger effective population size of any autosomal region (roughly four times that of mtDNA or the *Y* chromosome) generates deeper coalescence trees for any sample of genes; thus, the reconstruction of autosomal phylogenies opens a window that extends further into the past as compared to mtDNA or the *Y* chromosome.

## Conclusion

A genome view on human evolution is a rich approach as it may throw light from very different perspectives. One such perspective is the comparative study of the evolution of the species, undertaken either through differences between humans and closely related species (which can resolve the evolutionary tree) or through differences among present-day humans (which can resolve the origin of contemporary human populations). As discussed earlier, the phylogenetic position of our species does not provide an insight on the ancestry of modern humans. In fact, the time depth of current variation is much shallower than the genealogy of the species due to demography, particularly effective population size. It is clear from the available genetic information that a limited effective population size has driven human molecular evolution. Even if a reduced population size through one or several successive bottlenecks could explain the present findings, a strong population subdivision seems more plausible, as it could explain the maintenance of a large variety of alleles for some loci. When this population reduction took place and how narrow it was remain to be explained, and our demographic past is one of the main issues to be understood through molecular data.

In a different perspective, a genome view on human evolution may help us understand the evolution of specific genes in our lineage that at the very end could explain the emergence of human specific traits and, thus, the genetic basis of our specificities. Even if there is nothing special in our evolutionary past, it may be of interest to recognize the biological bases of our uniqueness as a species; in this endeavor, molecular evolutionary analyses may be highly relevant in the future, even if they have not offered many relevant results yet. A quantum leap will undoubtedly take place when the genomes of other primates (including chimpanzees) are known.

It has been claimed that modern molecular biology has taken a highly descriptive, rather than explicative path. Description is necessary: Inventories of individuals and species for the ecologist, DNA sequences and protein lists for the molecular biologist. We would like to argue that molecular description provides us with the unexpected possibility of undertaking comparative studies in an evolutionary framework. It is an illuminating perspective for the future understanding of life and for the understanding of our own species.

We thank David de Lorenzo (University of Munich), Andrés Moya (Universitat de València), and David Comas (Universitat Pompeu Fabra) for helpful theoretical discussion. Aida M. Andrés and Marta Soldevila (Universitat Pompeu Fabra) read an earlier version of this manuscript and made helpful suggestions.

## CHAPTER 23

# Could there be a Darwinian account of human creativity?

**Daniel C. Dennett**

Weaverbirds create intricate nests; sculptors and other artists and artisans also create intricate, ingenious constructions out of similar materials. The products may look similar, and outwardly the creative processes that create those processes may look similar, but there are surely large and important differences between them. What are they, and how important are they? The weaverbird nestmaking is "instinctual," and "controlled by the genes" some would say, but we know that this is a crude approximation of a more interesting truth, involving an intricate interplay between genetic variation, long-term developmental and environmental interaction and short-term environmental variation—in opportunities and materials accessible at the time of nest building. And on the side of the human creator, a similarly complex story must be told. Genes play some role surely (think of the likelihood of heritable differences in musical aptitude, for instance), but so do both long-term and short-term environmental interactions. The myth of the artist "blessed" by a spark of "divine genius" is even cruder and more distorted than the myth of the birdnest as a simple product of a gene—as if it were a protein.

Our thinking about human creativity is pulled out of shape somewhat by a famous contrast introduced to the world by Darwin. One of his earliest—and most outraged—critics summed it up vividly:

> In the theory with which we have to deal, Absolute Ignorance is the artificer; so that we may enunciate as the fundamental principle of the whole system, that, IN ORDER TO MAKE A PERFECT AND BEAUTIFUL MACHINE, IT IS NOT REQUISITE TO KNOW HOW TO MAKE IT. This proposition will be found, on careful examination, to express, in condensed form, the essential purport of the Theory, and to express in a few words all Mr. Darwin's meaning; who, by a strange inversion of reasoning, seems to think Absolute Ignorance fully qualified to take the place of Absolute Wisdom in all the achievements of creative skill. (MacKenzie 1868)

Darwin's "strange inversion of reasoning" promises—or threatens—to dissolve the Cartesian *res cogitans* as the wellspring of creativity, and then where will we be? Nowhere, it seems. It seems that if creativity gets "reduced" to "mere mechanism" *we* will be shown not to exist at all. Or, we will exist, but we will not be thinkers, we will not manifest genuine "Wisdom in all the achievements of creative skill." Whenever we zoom in on the act of creation, it seems we lose sight of it, the intelligence or genius replaced at the last instant by stupid machinery, an echo of Darwin's shocking substitution of Absolute Ignorance for Absolute Wisdom in the creation of the biosphere. Many people dislike Darwinism in their guts, and of all the ill-lit, murky reasons for antipathy to Darwinism, this one has always struck me as the deepest, but only in the sense of being the most entrenched, least accessible to rational criticism. There are thoughtful people who scoff at Creationism, dismiss dualism out of hand, pledge allegiance to academic humanism—and then get quite squirrelly when it is suggested that a Darwinian theory of creative intelligence might be in the cards, and might demonstrate that all the works of human genius can be understood in the end to be products of a cascade of generate-and-test procedures that are, at bottom, algorithmic, mindless. Absolute Ignorance? Artificial Intelligence? Fie on anybody who would thus put "A" and "I" together!

Besides, would not a Darwinian theory of human creativity be covertly self-contradictory? The Darwinian mechanism of natural selection is famously mindless, purposeless, lacking all foresight and intention—the blind watchmaker (Dawkins 1986). If natural selection is "the opposite" of God, a strange inversion of the traditional vision of creativity, then it must be "the opposite" of us, too, since God is made in our image! Human creative endeavors are obviously both foresighted and purposeful, so, then, they are not Darwinian processes. What could be more obvious?

But there is a tension, is there not? A key part of Darwin's great revolution is that we are part of it. Human beings are just one species among many, fully biological, and hence capable of no miracles, restricted to the same sorts of processes and methods as the other species. Our creative processes are surely natural (not supernatural!), so in that bland sense they are as biological as the creative processes of the weaverbird and the beaver.

William Poundstone (1985) puts the inescapable challenge succinctly in terms of "the old fantasy of a monkey typing *Hamlet* by accident." He calculates that the chances of this happening are "1 in 50 multiplied by itself 150 000 times."

> In view of this, it may seem remarkable that anything as complex as a text of *Hamlet* exists. The observation that *Hamlet* was written by Shakespeare and not some random agency only transfers the problem. Shakespeare, like everything else in the world, must have arisen (ultimately) from a homogeneous early universe. Any way you look at it *Hamlet* is a product of that primeval chaos. (Poundstone 1985, p. 23)

## Credit assignment for creativity

Where does all that Design come from? What processes could conceivably yield such improbable "achievements of creative skill?" What Darwin saw is that Design is always both *valuable* and *costly*. It does not fall like manna from heaven, but must be accumulated the hard way, by time-consuming, energy-consuming processes of mindless search through "primeval chaos," automatically preserving happy accidents when they occur. This broadband process of Research and Development is breathtakingly inefficient, but—this is Darwin's great insight—*if* the costly fruits of R and D can be thriftily conserved, copied, stolen, and reused, they can be *accumulated* over time to yield "the achievements of creative skill." "This principle of preservation, I have called, for the sake of brevity, Natural Selection." (Darwin 1859, p. 127)

There is no requirement in Darwin's vision that these R and D processes run everywhere and always at the same tempo, with the same (in-)efficiency. If we think of design work as lifting in Design Space (an extremely natural and oft-used metaphor, exploited in models of hill-climbing and peaks in adaptive landscapes, to name the most obvious and popular applications), then we can see that the gradualistic, frequently backsliding, maximally inefficient basic search process can on important occasions yield new conditions that speed up the process, permitting faster, more effective local lifting (Maynard Smith and Szathmáry 1995). Call any such product of earlier R and D a **crane**, and distinguish it from what Darwinism says does not happen: **skyhooks** (Dennett 1995). Skyhooks, like manna from heaven, would be miracles, and if we posit a skyhook anywhere in our "explanation" of creativity, we have in fact conceded defeat.

What, then, is a mind? The Darwinian answer is straightforward. A mind is a crane, made of cranes, a mechanism of not quite unimaginable complexity that can clamber through Design Space at a giddy—but not miraculously giddy—pace, thanks to all the earlier R and D, from all sources, that it exploits. What is the anti-Darwinian answer? It is perfectly expressed by one of the twentieth century's great creative geniuses (though, like MacKenzie, he probably did not mean by his words what I intend to mean by them):

> *Je ne cherche pas; je trouve.* (Pablo Picasso)

Picasso purports to be a genius indeed, someone who does not need to engage in the menial work of trial and error, generate-and-test, R and D; he claims to be able to *leap* to the summits of the peaks—the excellent designs—in the vast reaches of Design Space without having to guide his trajectory (he searches not) by sidelong testing at any way stations. As an inspired bit of bragging, this is *non pareil*, but I do not believe it for a minute. And anyone who has strolled through an exhibit of Picasso drawings (as I recently did in Valencia, while attending the conference that led to this volume) looking at literally dozens of variations on a single theme, all signed—and sold—by the artist, will appreciate that whatever Picasso may have meant by his *bon mot*, he could not truly claim that he did not engage in a time-consuming, energy-consuming exploration of neighborhoods in Design Space. At best he could claim that his own searches were so advanced, so efficient, that it did not seem—to himself—to be design *work* at all. But then what did he have within him that made him such a great designer? A skyhook, or a superb collection of cranes? (I have been unable to discover the source of Picasso's claim, which is nicely balanced by a better known remark by a more down-to-earth creative genius, Thomas Edison (1932): "Genius is one percent inspiration and ninety-nine percent perspiration.")

We can now characterize a mutual suspicion between Darwinians and anti-Darwinians that distorts the empirical investigation of creativity. Darwinians suspect their opponents of hankering after a skyhook, a miraculous gift of genius whose powers have no decomposition into mechanical operations, however complex and informed by earlier processes of R and D. Anti-Darwinians suspect their opponents of hankering after an account of creative processes that so diminishes the Finder, the Author, the Creator, that it disappears, at best a mere temporary locus of mindless differential replication. We can make a little progress, I think, by building on Poundstone's example of the creation of the creator of *Hamlet*. Consider, then, a little thought experiment.

Suppose Dr Frankenstein designs and constructs a monster, Spakesheare, that thereupon sits up and writes out a play, *Spamlet*. Who is the author of *Spamlet*? First, let us take note of what I claim to be irrelevant in this thought experiment. I have not said whether Spakesheare is a robot, constructed out of metal and silicon chips, or, like the original Frankenstein's monster, constructed out of human tissues—or cells, or proteins, or amino acids, or carbon atoms. As long as the design work and the construction were carried out by Dr Frankenstein, it makes no difference to the example what the materials are. It might well turn out that the only way to build a robot small enough and fast enough and energy-efficient enough to sit on a stool and type out a play is to construct it from artificial cells filled with beautifully crafted motor proteins and other carbon-based nanorobots. That is an interesting technical and scientific question, but not of concern here. For exactly the same reason, if Spakesheare is a metal-and-silicon robot, it may be allowed to be larger than a galaxy, if that is what it takes to get the requisite complication into its program—and we will just have to repeal the speed limit for light for the sake of our thought experiment. Since these technical constraints are commonly declared to be off-limits in these thought experiments, so be it. If Dr Frankenstein chooses to make his AI robot out of proteins and the like, that's his business. If his robot is cross-fertile with normal human beings, and hence capable of creating what is arguably a new species by giving birth to a child, that is fascinating, but what we will be concerned with is Spakesheare's purported brainchild, *Spamlet*. Back to our question: Who is the author of *Spamlet*?

In order to get a grip on this question, we have to look inside and see what happens in Spakesheare. At one extreme, we find inside a file (if Spakesheare is a robot with a computer memory) or a basically memorized version of *Spamlet*, all loaded and ready to run. In such an extreme case, Dr Frankenstein is surely the author of *Spamlet* (unless we find there is a Ms Shelley who is the author of Dr Frankenstein!), using his intermediate creation, Spakesheare, as a mere storage-and-delivery device, a particularly fancy word processor. All the R and D work was done earlier, and copied to Spakesheare by one means or another. Now look at the other extreme, in which Dr Frankenstein leaves most of the work to Spakesheare. The most realistic scenario would surely be that Spakesheare has been equipped by Dr. Frankenstein with a virtual past, a lifetime stock of pseudo-memories of experiences on which to draw while responding to its Frankenstein-installed obsessive desire to write a play. Among those pseudo-memories, we may suppose, are many evenings at the theater, or reading books, but also some unrequited loves, some shocking close calls, some shameful betrayals and the like. Now what happens? Perhaps some scrap of a "human interest" story on the network news will be the catalyst that spurs Spakesheare into a frenzy of generate-and-test, ransacking its memory for useful tidbits and themes, transforming—transposing, morphing—what it finds, jiggling the pieces into temporary, hopeful structures that compete for completion, most of them dismantled by the corrosive processes of criticism that, nevertheless, expose useful bits now and then, and so forth, and all of this multi-leveled search would be somewhat guided by multilevel, internally generated evaluations, including evaluation of the evaluation . . . of the evaluation functions as a response to evaluation of . . . the products of the ongoing searches.

Now if the amazing Dr Frankenstein had actually anticipated all this activity down to its finest grain at the most turbulent and chaotic level, and had hand-designed Spakesheare's virtual past, and all its search machinery, to yield just this product, *Spamlet*, then Dr Frankenstein would be, once again, the author of *Spamlet*, but also, in a word, God. Such Vast (not literally infinite, but Very much more than Astronomical—Dennett 1995, p. 109) foreknowledge would be simply miraculous. Restoring a smidgen of realism to our fantasy, we can consider a rather less extreme position and assume that Dr Frankenstein was unable to foresee all this in detail, but rather delegated to Spakesheare most of the hard work of completing the trajectory in Design Space to one literary work or another, something to be determined by later R and D occurring within Spakesheare itself.

## Real artificial creators

We have now arrived in the neighborhood of reality itself, for we already have actual examples of impressive artificial authors that vastly outstrip the foresight of their own creators. Nobody has yet created an artificial playwright worth serious attention, but an artificial chess player—IBM's Deep Blue—and an artificial composer—David Cope's EMI—have both achieved results that are, in some respects, equal to the best that human creative genius can muster.

Who beat Garry Kasparov, the reigning World Chess Champion? Not Murray Campbell or any of his IBM team. Deep Blue beat Kasparov. Deep Blue designs better chess games than any of them can design. None of them can author a winning game against Kasparov. Deep Blue can. Yes, but. Yes, but. I am sure many of you are tempted to insist at this point that when Deep Blue beats Kasparov at chess, its brute force search methods are entirely unlike the exploratory processes that Kasparov uses when he conjures up his chess moves. But that is simply not so—or at least it is not so in the only way that could make a difference to the context of this debate about the universality of the Darwinian perspective on creativity. Kasparov's brain is made of organic materials, and has an architecture importantly unlike that of Deep Blue, but it is still, so far as we know, a massively parallel search engine which has built up, over time, an outstanding array of heuristic pruning techniques that keep it from wasting time on unlikely branches. There is no doubt that the investment in R and D has a different profile in the two cases; Kasparov has methods of extracting good design principles from past games, so that he can recognize, and know enough to ignore, huge portions of the game space that Deep Blue must still patiently canvass *seriatim*. Kasparov's "insight" dramatically changes the shape of the search he engages in, but it does not constitute "an entirely different" means of creation. Whenever Deep Blue's exhaustive searches close off a type of avenue that it has some means of recognizing (a difficult, but not impossible task), it can reuse that R and D whenever it is appropriate, just as Kasparov does. Much of this analytical work has been done for Deep Blue by its designers, and given as an innate endowment, but Kasparov has likewise benefitted from hundreds of thousands of person-years of chess exploration transmitted to him by players, coaches, and books. It is interesting in this regard to contemplate the suggestion recently made by Bobby Fischer, who proposes to restore the game of chess to its intended rational purity by requiring that the major pieces be randomly placed in the back row at the start of each game (random, but mirror image for black and white). This would instantly render the mountain of memorized openings almost entirely obsolete, for humans and machines alike, since only rarely would any of this lore come into play. One would be thrown back onto a reliance on fundamental principles; one would have to do more of the hard design work in real time—with the clock running. It is far from clear whether this change in rules would benefit human beings more than computers. It all depends on which type of chess player is relying most heavily on what is, in effect, rote memory—reliance with minimal comprehension on the R and D of earlier explorers.

The fact is that the search space for chess is too big for even Deep Blue to explore exhaustively in real time, so like Kasparov, it prunes its search trees by taking calculated risks, and like Kasparov, it often gets these risks pre-calculated. Both presumably do massive amounts of "brute force" computation on their very different architectures. After all, what do neurons know about chess? Any work they do must be brute force work of one sort or another.

It may seem that I am begging the question in favor of a computational, AI approach by describing the work done by Kasparov's brain in this way, but the work has to be done somehow, and no other way of getting the work done has ever been articulated. It will not do to say that Kasparov uses "insight" or "intuition" since that just means that Kasparov himself has no privileged access, no insight, into how the good results come to him. So, since nobody knows how Kasparov's brain does it—least of all Kasparov—there is not yet any evidence at all to support the claim that Kasparov's means are "entirely unlike" the means exploited by Deep Blue. One should remember this when tempted to insist that "of course" Kasparov's methods are hugely different. What on earth could provoke one to go out on a limb like that? Wishful thinking? Fear?

But that's just chess, you say, not art. Chess is trivial compared to art (now that the world champion chess player is a computer). This is where David Cope's EMI comes into play (Cope 2001; Dennett 2001b). Cope set out to create a mere efficiency-enhancer, a composer's aid to help him over the blockades of composition any creator confronts, a high-tech extension of the traditional search vehicles (the piano, staff paper, the tape recorder, etc.). As EMI grew in competence, it promoted itself into a whole composer, incorporating more and more of the generate-and-test process. When EMI is fed music by Bach, it responds by generating musical compositions in the style of Bach. When given Mozart, or Schubert, or Puccini, or Scott Joplin, it readily analyzes their styles and composes new music in their styles, better pastiches than Cope himself—or almost any human composer—can compose. When fed music by two composers, it can promptly compose pieces that eerily unite their styles, and when fed, all at once (with no clearing of the palate, you might say) all these styles at once, it proceeds to write music based on the totality of its musical experience. The compositions that result can then also be fed back into it, over and over, along with whatever other music comes along in MIDI format, and the result is EMI's own "personal" musical style, a style that candidly reveals its debts to the masters, while being an unquestionably idiosyncratic integration of all this "experience." EMI can now compose not just two-part inventions and art songs but whole symphonies—and has composed over a thousand, when last I heard. They are good enough to fool experts (composers and professors of music) and I can personally attest to the fact that an EMI–Puccini aria brought a lump to my throat—but then, I am on a hair trigger when it comes to Puccini, and this was a good enough imitation to fool me. David Cope can no more claim to be the composer of EMI's symphonies and motets and art songs than Murray Campbell can claim to have beaten Kasparov in chess.

To a Darwinian, this new element in the cascade of cranes is simply the latest in a long history, and we should recognize that the boundary between authors and their artifacts should be just as penetrable as all the other boundaries in the cascade. When Richard Dawkins (1982) notes that the beaver's dam is as much a part of the beaver phenotype—its extended phenotype—as its teeth and its fur, he sets the stage for the further observation that the boundaries of a human author are exactly as amenable to extension. In fact, of course, we've known this for centuries, and have carpentered various semi-stable conventions for dealing with the products of Rubens, of Rubens' studio, of Rubens' various students. Wherever there can be a helping hand, we can raise the question of just who is helping whom, what is creator and what is creation. How should we deal with such questions? To the extent that anti-Darwinians simply want us to preserve some tradition of authorship, to have some rules of thumb for determining who or what shall receive the honor (or blame) that attends authorship, their desires can be acknowledged and met, one way or another (which does not necessarily mean we should meet them). To the extent that this is not enough for the anti-Darwinians, to the extent that they want to hold out for authors as an objective, metaphysically grounded, "natural kind," they are looking for a skyhook.

## Does the author disappear?

There is a persistent problem of imagination management in the debates surrounding this issue: people on both sides have a tendency to underestimate the resources of Darwinism, imagining simplistic alternatives that do not exhaust the space of possibilities. Darwinians are notoriously quick to find (or invent) differences in genetic **fitness** to go with every difference they observe, for instance. Meanwhile, anti-Darwinians, noting the huge distance between a beehive and the *St. Matthew Passion* as created objects, are apt to suppose that anybody who proposes to explain both creative processes with a single set of principles must be guilty of one reductionist fantasy or another: "Bach had a gene for writing baroque counterpoint just like the bees' gene for forming wax hexagons" or "Bach was just a mindless trial-and-error mutator and selector of the musical memes that already flourished in his cultural environment." Both of these alternatives are nonsense, of course, but pointing out their flaws does nothing to support the idea that ("therefore") there must be irreducibly non-Darwinian principles at work in any account of Bach's creativity. In place of this dimly imagined chasm with "Darwinian phenomena" on one side and "non-Darwinian phenomena" on the other side, we need to learn to see the space between bee and Bach as populated with all manner of mixed cases, differing from their nearest neighbors in barely perceptible ways, replacing the chasm with a traversable gradient of non-minds, protominds, hemi–demi–semi minds, magpie minds, copycat minds, aping minds, clever-pastiche minds, "pathfinding" minds, "ground-breaking" minds, and eventually, genius minds. And the individual minds, of each caliber, will themselves be composed of different sorts of parts, including, surely, some special-purpose "modules" adapted to various new tricks and tasks, as well as a cascade of higher-order reflection devices, capable of generating ever more rarefied and delimited searches through pre-selected regions of the Vast space of possible designs.

It is important to recognize that genius is itself a product of natural selection and involves generate-and-test procedures all the way down. Once you have such a product, it is often no longer particularly perspicuous to view it solely as a cascade of generate-and-test processes. It often makes good sense to leap ahead on a narrative course, thinking of the agent as a self, with a variety of projects, goals, presuppositions, hopes, . . . In short, it often makes good sense to adopt the **intentional stance** (Dennett 1971, 1987) towards the whole complex product of evolutionary processes. This effectively brackets the largely unknown and unknowable mechanical microprocesses as well as the history that set them up, and puts them out of focus while highlighting the patterns of rational activity that those mechanical microprocesses track so closely. This tactic makes especially good sense to the creator himself or herself, who must learn not to be oppressed by the revelation that on close inspection, even on close introspection, a genius dissolves into a pack rat, which dissolves in turn into a collection of trial-and-error processes over which nobody has ultimate control.

Does this realization amount to a loss—an elimination—of selfhood, of genius, of creativity?

Those who are closest to the issue—the artistic and scientific geniuses who have reflected on it—often confront this discovery with equanimity. Mozart (in an oft-quoted but possibly spurious passage; see Dennett 1995, pp. 346–7) is reputed to have said of his best musical ideas: "Whence and how do they come? I don't know and I have nothing to do with it." The painter Philip Guston is equally unperturbed by this evaporation of visible self when the creative juices start flowing:

> When I first come into the studio to work, there is this noisy crowd which follows me there; it includes all of the important painters in history, all of my contemporaries, all the art critics, etc. As I become involved in the work, one by one, they all leave. If I'm lucky, every one of them will disappear. If I'm really lucky, I will too.

Resistance to extending Darwinian thinking into human creativity and human culture is not restricted to closet Creationists and anti-scientific humanists. Two highly visible Darwinian spokespersons—Stephen Jay Gould and Steven Pinker—who agree on precious little else, find common ground in their doubts about this:

> I am convinced that comparisons between biological evolution and human cultural or technological change have done vastly more harm than good—and examples abound of this most common of intellectual traps . . . Biological evolution is powered by natural selection, cultural evolution by a different set of principles that I understand but dimly. (Gould 1991, p. 63)

> To say that cultural evolution is Lamarckian is to confess that one has no idea how it works. The striking features of cultural products, namely their ingenuity, beauty, and truth (analogous to organisms' complex adaptive design), come from the mental computations that "direct"—that is, invent—the "mutations," and that "acquire"—that is, understand—the "characteristics." (Pinker 1997, p. 209)

Pinker has imputed the wrong parallel; it is not Lamarck's model, but Darwin's models of unconscious and methodical (artificial) selection (as special cases of natural selection) that accommodate the phenomena he draws to our attention in this passage (Dennett 2001a). And it is ironic that Pinker overlooks this, since the cultural phenomena he himself has highlighted as examples of evolution-designed systems, linguistic phenomena, are almost certainly not the products of foresightful, ingenious, deliberate human invention. Some designed features of human languages are no doubt genetically transmitted, but many others—such as changes in pronunciation, for instance—are surely culturally transmitted, and hence products of cultural, not genetic, evolution.

## Conclusion

The cranes of human culture did not just open up Design Space; they opened up perspectives on Design Space that permitted "directed" mutation, foresighted mutation, reflective mutation, both in cultural and, most recently, genetic innovation. This nesting of different processes of natural selection now has a new member: genetic engineering. How does it differ from the methodical selection of Darwin's day? It is less dependent on the pre-existing variation in the gene pool, and proceeds more directly to new candidate genomes, with less overt trial and error. Darwin (1859, p. 38) had noted that in his day,

> Man can hardly select, or only with much difficulty, any deviation of structure excepting such as is externally visible; and indeed he rarely cares for what is internal.

But today's genetic engineers have carried their insight into the molecular innards of the organisms they are trying to create. There is ever more accurate foresight, but even here, if we look closely at the practices in the laboratory, we will find a large measure of exploratory trial and error in their search of the best combinations of genes. (In fact, biochemists and molecular biologists are finding that artificial evolutionary processes are more efficient R and D procedures than their foresightful hand-work efforts by orders of magnitude. In other words, they are finding that the *breeding* of domesticated microorganisms and polymers is the best way to conduct their creative searches.)

Are the products of genetic engineering "Darwinian" products? They are produced not by blind or random trial-and-error variation, but by highly intelligent, guided, foresightful processes. Nevertheless, these processes are themselves the products of earlier design work accomplished by Darwinian R and D, and if we look closely at the microprocesses that compose their current, local search, we will still find plenty of instances of random (undesigned, chaotic) generation of candidates for further scrutiny.

It may seem, however, that we have now passed the Pickwickian limits of Darwinian orthodoxy. Does a Darwinian gloss actually supplement or adjust the traditional intellectualist ways of thinking? I think it does, because without the steady pressure of the Darwinian "strange inversion of reasoning," it is all too tempting to revert to the old essentialist, Cartesian perspectives. For instance, there is always the temptation, often succumbed to, to establish "principled" boundaries, or to erect a polar contrast between insightful and blind processes of search, as we saw in the unsupportable assertion that Kasparov's methods are fundamentally unlike Deep Blue's. If Deep Blue's methods are ultimately "blind and mechanical," then so, ultimately, are Kasparov's—his neurons are as blind and mechanical as any circuit board. The foresighted, purposeful breeding of domesticated plants and animals is obviously not a damning counterexample to Darwin's theory of natural selection as a foresightless, purposeless process, because his theory shows (as we are beginning to learn) how such foresight and purpose could itself evolve by blind natural selection. Kasparov's creative genius (or Bach's or Shakespeare's) is for the same reason no counterexample to the Darwinian theory of creativity, but rather one of the most recent blooms on the tree of life that we still need to account for in Darwinian terms.

Portions of this paper are drawn from Dennett (2003). I have been unable to locate the source of Philip Guston's quote, but I have found much the same remark attributed to the composer, John Cage, a close friend and contemporary of Guston's, who (is said to have) said this about painting: "When you are working, everybody is in your studio—the past, your friends, the art world, and above all, your own ideas—all are there. But as you continue painting, they start leaving, one by one, and you are left completely alone. Then, if you are lucky, even you leave." Like all other creators, Guston and I like to reuse what we find, adding a few touches from time to time.

## Further reading

Campbell, DT (1960). Blind variation and selective retention in creative thought as in other knowledge processes. *Psychological Review*, 67, 380–400.

Mithen, SJ (ed.). (1988). *Creativity in human evolution and prehistory*. Routledge, London.

Mithen, SJ (1996). *The prehistory of the mind: the cognitive origins of art, religion, and science*. Cambridge University Press, Cambridge.

Toulmin, S (1972). *Human understanding. Vol. 1: The collective use and development of concepts*. Clarendon Press, Oxford.

# Glossary

**Acetylation**. A chemical process that incorporates acetyl moieties to the free amino group of lysine. Histone **deacetylation** is associated with the remodeling of the chromatin to a state that prevents gene transcription.

**Acquired Immune Deficiency Syndrome** (AIDS). Chronic, fatal disease of humans due to progressive loss of immune function and the acquisition of opportunistic infections. Caused by infection with the **human immunodeficiency virus** (HIV).

**Acrocentric**. A chromosome having the **centromere** located near one end (cf. **metacentric**).

**Adaptation**. The process of evolutionary modification that results, in a species, in improved survival or reproduction. Any trait that enables or enhances survival or reproduction of an organism.

**AFLPs** (amplified fragment length polymorphisms). A diagnostic fingerprinting technique based on selective **polymerase chain reaction** amplification of restriction fragments from a total digest of genomic DNA. This technique detects the presence or absence of restriction fragments and not length differences (cf. **restriction fragment length polymorphism**).

**Allopatric speciation**. Speciation that occurs as a result of two or more populations diverging while spatially separated for at least part of the speciation process (cf. **sympatric speciation**).

**Allopatry** (**allopatric**). Occupying different, non-adjacent geographic locations (cf. **sympatry**).

**Alloploid speciation**. The formation of new species by chromosomal duplication in interspecific hybrids (cf. **homoploid speciation**).

**Alloploid**. Hybrid arising by **alloploid speciation**, thus having a larger number of chromosome sets than either of the parental species (cf. **homoploid**).

**Allotetraploid**. An individual that is diploid for two genomes, each from a different species.

**Alternative splicing**. A mechanism that generates different proteins from a single gene by splicing together different nonconsecutive **exons** during the transcription process.

**Amictic**. See **apomixis**.

**Amphoteric**. Females able to produce eggs both through meiosis and mitosis.

**Anhydrobiosis**. A form of dormancy involving loss of internal water, dehydrating the animal.

**Annotated**. See **annotation**.

**Annotation**. The process of appending structural, functional and other pertinent information (e.g. comments, literature references) to a genome sequence. A sequence that contains all of these elements is said to be **annotated**.

**Apomixis** (**apomictic**). Form of reproduction not involving meiosis or fusion of a male and female haploid gamete in organisms reproducing by eggs.

**Autogenic succession**. The replacement of organisms in a community driven by internal factors, that is, by the activity of the organisms themselves.

**Autosomal**. A gene or DNA sequence located on a chromosome other than a sex chromosome.

**Background selection**. Negative or purifying selection against deleterious mutations that reduces the amount of genetic variability at linked neutral sites.

**Balanced polymorphism**. Polymorphism in which genetically distinct forms are more or less permanent components of the population, maintained by selection in favor of diversity as in the case of selection for heterozygotes.

**Balancing selection.** Selection that maintains a **balanced polymorphism** in a population.

**Basic reproductive rate** ($R_0$). The average number of infections caused by a single index case in an entirely susceptible population. Epidemic infections require $R_0 > 1$.

**Behavior system**. The sensory, neural, and motor components underlying a class of behavior patterns such as feeding, courtship, parental care, and predator defense/avoidance.

**Biological species**. A set of interbreeding populations reproductively isolated from other similar sets.

**CD4+ T-helper cells**. Immune cells bearing the CD4 receptor that are a vital part of the immune system's defense against viral infections.

**CD8+ cytotoxic T-lymphocyte** (CTL). Key component of the cellular immune response against viral infections. Vital in controlling HIV **viral load**.

**Centromere**. The constricted region of a chromosome that includes the site of attachment to the mitotic or meiotic spindle.

**Circulating recombinant forms** (CRFs). Intersubtype recombinant strains of HIV-1. At very high frequencies in sub-Saharan Africa.

***cis*-regulatory sequence**. A discrete region of DNA that affects the rate of transcription of a gene located nearby on the same chromosome.

**Clade**. A monophyletic group of taxa sharing a closer common ancestor with each other than with members of any other such group.

**Classical polymorphism**. Any polymorphism, such as blood groups and allozymes, which is detected in expression products (i.e. proteins) rather than on DNA itself.

**Clear-water phase**. Predictable, short period of very clear water in otherwise phytoplankton rich temperate lakes, caused by intensive grazing of herbivorous cladocerans.

**Clonal reproduction**. Reproduction not involving meiosis or fusion of a male and female haploid gamete, thus resulting in a lineage of genetically identical individuals (i.e. a **clone**). See **apomixis**.

**Clone** (**clonal**). A genetically identical lineage or assemblage of individuals that are products of vegetative growth or **parthenogenesis** from a single zygote or cell.

**Co-infection**. The (near) simultaneous infection of an individual with multiple pathogen strains.

**Coadaptation**. The term given to the action of selection in producing new adaptive combinations of traits in a lineage. See **adaptation**.

**Coalescent theory**. A theory that predicts what gene genealogies are expected to look like if populations have different demographic and evolutionary histories—that is, how genealogies are affected by changes in population size and structure, selection, migration, etc.

**Coalescent tree**. Tree describing the genealogical relationships between copies of a gene in a population, from the present to their most recent common ancestor.

**Comparative method**. A method for testing evolutionary hypotheses based on comparisons among species of known evolutionary relationships.

**Conservation genetics**. The use of genetic markers and DNA sequence information together with population genetic and ecological theory for designing and implementing conservation strategies.

**Constrained mutation**. A mutation that is subject to selection.

**Continental drift**. The theory that continental landmasses have drifted apart over the course of geological time.

**Continuing horizontal transfer**. The hypothesis that organisms are currently receiving and transferring genes horizontally, that is, from or to other species (cf. **massive early transfer**).

**Continuous growth model**. In population dynamics, a model describing population growth due to constant reproduction, as opposed to reproduction at discrete time intervals (cf. **discrete growth model**).

**Cooption**. An adaptive evolutionary process that endows a trait (e.g. structure, behavior pattern) with a new function different from its original one.

**Copepodite**. A juvenile developmental stage following the **nauplius** stage of copepods.

**Crane**. A subprocess or feature of a design process that can be demonstrated to permit the local speeding up of the basic, slow process of natural selection, and that can be demonstrated to be itself the predictable (or retrospectively explicable) product of the basic process (cf. **skyhook**).

**Cryptic species**. Closely related and frequently **sympatric** species which are morphologically indistinguishable but which are reproductively isolated.

**Cyclical parthenogenesis**. Reproduction by a series of **parthenogenetic** generations interspersed with sexually reproducing generations.

**Deacetylation**. See **acetylation**.

**Deletion**. A genetic rearrangement that involves the loss of part of a chromosome.

**Demographic sweep**. Population expansion following a bottleneck, so that all extant individuals are derived from a few founders.

**Diapause**. A form of dormancy or metabolic arrest occurring at a specific developmental stage. A diapausing individual resumes development after receiving an appropriate stimulus.

**Diapausing egg**. See **diapause**.

**Directional selection**. Selection causing a directional shift in gene frequencies of a trait and leading to increased adaptation.

**Discrete growth model**. In population dynamics, a model describing population growth due to reproduction at specific time intervals. This type of model is a good choice to describe populations where births occur in pulses (e.g. every spring) (cf. **continuous growth model**).

**Domestication**. A genetic process in which organisms are selected, often by accident, to have traits that make them more docile, more fecund, more adaptable to human environments, and alter in many other ways their physiology, morphology, and behavior.

**Dormancy**. A state of relative metabolic arrest, such as estivation, cryptobiosis, **diapause**, hibernation. Also, a state in which viable seeds, spores or buds fail to germinate under conditions favorable for germination and vegetative growth.

**Drift**. See **genetic drift**.

**Duplication**. A genetic rearrangement that involves the doubling or repetition of part of a gene or chromosome.

**Ecotone**. The transitional zone between adjacent communities, often characterized by one or several dominant species.

**Environments of evolutionary adaptedness** (EEA). The period or periods of its evolutionary past to which each species' biology is adapted. Each species has its own EEA. For humans, this period usually corresponds to the upper Paleolithic (40 000 to 10 000 years ago) or to the entire Pleistocene (1.8 million years ago to 10 000 years ago).

**Ephippium** (pl. **ephippia**). **Resting egg** capsules produced by cladocerans. The ephippium is formed as the dorsal part of the carapace covering the brood chamber thickens. The ephippium is shed as the female molts. In most species, each ephippium can carry two resting eggs.

**Epigenetic mutation**. In developmental or physiological genetics this term is often used to describe a mutation whose phenotype is unaffected by another mutation (cf. **epistasis**).

**Epilimnion**. Upper, usually warmer layer of a stratified lake that can be mixed by the wind (cf. **hypolimnion**).

**Epistasis** (**epistatic**). The interaction of nonallelic genes in which one gene (epistatic gene) masks the expression of another at a different locus.

**Epitope**. Part of a protein that is recognized by an antigen.

**Ethogram**. A descriptive catalog of behavior patterns shown by a species along with key terms to identify each pattern.

**Eurytopic**. Tolerant of a wide range of habitats; physiologically tolerant.

**Eutrophic lake**. A lake rich in inorganic nutrients with high productivity and algal biomass, low transparency and oxygen depletion in the **hypolimnion** during summer.

**Evolutionarily stable strategy**. A strategy that, if adopted by all members of a population, prevents the invasion of the population by a mutant strategy under natural selection.

**Evolvability**. The ability to respond to long-term exposure to unfavorable environments through adaptive changes, usually by evolving into more complex forms.

**Exon**. The part of the eukaryotic gene that is represented in the spliced mRNA and in the resulting amino acid sequence of the protein product (cf. **intron**).

**Fitness**. The contribution to the next generation of a genotype, relative to that of other genotypes, reflecting its probability of survival and its reproductive output.

**Fitness component**. Each functional part of the life cycle that contributes to organismal **fitness**, such as fertility, viability, developmental time, mating propensity, etc.

**Fixation**. Increase in frequency of an allele up to the point that all other alleles are lost.

**Fixation index** ($F_{ST}$). An index used to measure the extent of population subdivision, the reduction of heterozygosity by **genetic drift**, or the amount of **gene flow** between subpopulations (e.g. high levels of gene flow will result in low values of $F_{ST}$).

**Flagellates**. Small, single-celled photosynthetic or nonphotosynthetic organisms characterized by whip-like flagella that propel them through the water.

**Fluctuating asymmetry**. Small, random deviations from perfect symmetry in traits that are bilaterally symmetrical.

**Footprinting**. A method for determining the DNA sequence to which a particular DNA-binding protein binds.

**Founder effect**. Changes in gene frequencies that take place when a small sample of a larger population establishes itself as a newly isolated entity and its gene pool carries only a fraction of the genetic diversity present in the parental population.

**Founder event**. The founding of a new population by a small sample of individuals drawn from a larger population. Founder events may result in a loss of genetic variation and abrupt changes in allele frequencies.

**Gametic imprinting**. Genes in gametes marked by DNA methylation in the germ line of parents and which are not expressed in the early development of the offspring.

**Gene conversion**. A recombination process in which two sequences interact in such a way that one is converted by the other. It is a non-reciprocal process, because one sequence is changed whereas the other is not.

**Gene flow**. The exchange of genes between different populations of the same species produced by migrants, and commonly resulting in simultaneous changes in gene frequencies at many loci in the recipient gene pool.

**Genet**. The organism developed from a zygote. Used in modular organisms and members of a **clone** to define a genetic individual (cf. **ramet**).

**Genetic drift**. Random changes in allele frequencies in a population due to random sampling of genes.

**Genetic load**. The relative difference between the actual mean **fitness** of a population and the mean fitness that would exist if the fittest genotype presently in the population were to become fixed.

**Genomics**. The systematic study of all or a large portion of an organism's genome.

**Handicap principle**. The hypothesis that some signals, particularly the sexually selected displays and physical ornaments of males in many species, are handicaps for the animal carrying them (i.e. they are costly to produce and maintain), and thus are honest indicators of **fitness** (i.e. they are advertisements for good genes).

**Haplodiploidy**. Sex determining mechanism by which males develop from unfertilized eggs, therefore males are haploid and produced by **parthenogenesis** and females are produced by the fusion of reduced eggs and sperm.

**Haplotype**. A unique combination of closely linked genetic markers present in a chromosome.

**Haplotype clade**. A monophyletic group of taxa sharing a closer common ancestral **haploytpe** with one another than with members of any other **clade**.

**Heritability**. The proportion of variability in a particular trait that can be attributed to genetic differences among individuals. The proportion of the total phenotypic variance due to genetic variance. This is a statistical description that applies to a specific population and may change if the environment is altered.

**Heteroduplex**. A DNA generated by base pairing between complementary single strands derived from the different parental duplex molecules.

**Hitchhiking**. In a genetic context, hitchhiking refers to the idea that the frequency of specific alleles (e.g. of **neutral markers**) can be influenced by natural selection acting on linked loci. In the case of **clonal** organisms, the whole genome acts as one linkage group, so that hitchhiking can be very strong.

**Homologous**. Two or more traits found in different taxa that are derived from the same or equivalent trait in their most recent common ancestor, but do not necessarily retain similarity of structure or function. Two traits that are genealogically linked within a single line of descent.

**Homologue**. See **homologous**.

**Homology**. Possession by two or more species of a trait derived, with or without modification, from their common ancestor.

**Homoplasy**. Structural resemblance due to parallelism or convergent evolution rather than to common ancestry (cf. **homology**).

**Homoploid**. Hybrid arising by **homoploid speciation**, thus having the same ploidy (i.e. number of chromosome sets) as the parental species (cf. **alloploid**).

**Homoploid speciation**. Hybrid speciation without an increase in ploidy (i.e. number of chromosome sets) (cf. **alloploid speciation**).

**Human immunodeficiency virus** (HIV). Human retrovirus which causes AIDS. Comes in two forms (type 1 and type 2) that have different epidemiological profiles.

**Hybrid dysgenesis**. Inability of certain strains of *D. melanogaster* to interbreed because the hybrids are sterile as a result of not sharing the same set of **transposable elements**.

**Hybrid zone**. The zone of overlap between adjacent populations, subspecies, or species in which interbreeding occurs.

**Hypolimnion**. Volume of deep, cool water that does not mix with the surface water during summer temperature stratification of a lake (cf. **epilimnion**).

**Indel**. Polymorphism in which the alleles differ by the gain (insertion) or loss (deletion) of a DNA fragment.

***In situ* hybridization**. A molecular technique that allows binding a visually detectable probe to specific nucleic acid sequences in morphologically preserved chromosomes, cells or tissue sections.

**Intentional stance**. The strategy of interpreting a complex entity or phenomenon as if it were a rational agent, by attributing beliefs and desires to it and thereby predicting how it will act, given those beliefs and desires.

**Introgression**. A process of incorporating by recombination genomic portions of one species into another following hybridization.

**Introgressive hybridization**. See **introgression**.

**Intron** (**intronic**). A segment of DNA that is transcribed, but removed from within the transcript by splicing together the sequences **(exons)** on either side of it. Its DNA sequence is not represented in the spliced mRNA or in the amino acid sequence of the resulting protein.

**Inversion**. A genetic rearrangement in which part of a chromosome is reversed, so that the genes within that part are in inverse order.

**Isolation by distance**. The isolation of two populations so that they are prevented from interbreeding due to the distance between them.

**Iteroparous**. See **iteropary**.

**Iteropary**. Reproductive strategy involving multiple cycles of reproduction. Contrasts with semelpary, in which there is only one reproductive bout, followed by senescence and mortality.

**Junk DNA**. See **parasitic DNA**.

**Kairomone**. An interspecific chemical signal that benefits the receiver but not the emitter.

**Leptokurtic dispersal**. Pattern of dispersal in which the distances moved by dispersers follow a leptokurtic distribution (higher central peak and larger tails than a normal distribution).

**Linkage**. In genetics, the greater association in inheritance of two or more nonallelic genes than is to be expected from independent assortment. Linkage is usually due to genes residing in physical proximity on the same chromosome.

**Linkage block**. Any segment of a chromosome thought to be in tight **linkage**.

**Linkage disequilibrium**. Nonrandom association of alleles at different loci.

**Linkage group**. See **linkage block**.

**Linked selection**. See **hitchhiking**.

**Maladaptation**. Lack of **adaptation**.

**Massive early transfer**. According to this hypothesis, during the early stages of prokaryotic evolution there was a mass transfer of genes, before the diversification into the current prokaryotic groups (cf. **continuing horizontal transfer**).

**Maternal effects**. A nonlasting influence of the genotype or phenotype of the mother upon the phenotype of the immediate offspring.

**Metacentric**. A chromosome having the **centromere** located at or near the middle (cf. **acrocentric, telocentric**).

**Methylation**. A chemical process that adds methyl groups mostly to DNA cytosines. In eukaryotes the majority of CG sequences are methylated. Methylation of genetic control regions renders genes inactive by preventing transcription whereas undermethylation and demethylation activate gene expression.

**Microcomplement fixation**. An immunological technique used to estimate genetic variation by comparing the strength of the antigen-antibody reaction in proteins from different taxa.

**Microsatellite**. Tandem repeated sequence of a 2–6 basepair unit, repeated 5–50 times, usually polymorphic in number of copies.

**Mictic**. See **mixis**.

**Minisatellite**. Tandem repeated sequence of a 5-64 basepair unit, repeated hundreds of times, usually polymorphic in sequence and number of copies.

**Mixis**. Form of reproduction involving meiosis and fusion of a male and female haploid gamete. In rotifers, mictic females are females that produce sexual eggs.

**Molecular clock hypothesis**. The hypothesis that nucleotide substitutions occur at a sufficiently regular rate to permit the dating of phylogenetic dichotomies; it assumes a direct relationship between the extent of molecular divergence and the time of ancestral separation of the two branches.

**Motif**. A short sequence with important functional or structural roles present in a protein or nucleic acid.

**M-tropic HIV strain**. Strain of HIV-1 that preferentially replicates in macrophage cells.

**Müller's ratchet**. Gradual accumulation of deleterious mutations in a finite asexual population, leading to loss of **fitness**.

**Multi-gene family**. A set of genes descended by duplication and variation from some ancestral gene and retaining the same function. Such genes may be clustered together on the same chromosome or dispersed on different chromosomes.

**Mutational deterministic hypothesis**. The hypothesis that sex is advantageous because it eliminates deleterious mutations more efficiently than asexual reproduction. This hypothesis does not assume finite population size (in contrast to **Müller's ratchet** hypothesis). It requires the combined effects of different mutations on **fitness** to be negatively synergistic.

**Mutational load**. The genetic instability sustained by a population due to the accumulation of deleterious genes generated by recurrent mutation.

**Mutational meltdown**. Extreme decline in the average **fitness** of a population, eventually leading to its extinction, due to the accumulation of deleterious mutations. See **Müller's ratchet**.

**Mutualism** (**mutualistic**). A mutually beneficial relationship between two organisms.

**Nauplius** (pl. **nauplii**). The larval stage in the development of copepodes.

**Nearly neutral theory**. A theory of molecular evolution that incorporates mutations with very small selective coefficients.

**Negative pleiotropy**. Negative correlation between different **fitness components** affected by the same gene.

**Neutral evolution**. See **neutral theory**.

**Neutral marker**. Genetic marker that is not subject to natural selection because it is selectively neutral.

**Neutral model**. Any population genetics model in which selection is not considered.

**Neutral theory**. A theory according to which the majority of nucleotide mutations that segregate within populations in the course of evolution are neutral. According to this theory, nucleotide substitutions would be the result of the random fixation of neutral mutations, rather than the result of positive Darwinian selection.

**Nonsynonymous substitution**. Nucleotide change in a protein-coding region that results in an amino acid change in the encoded protein (cf. **synonymous substitution**).

**Nucleotide diversity**. In any given genome position, probability that two random chromosomes carry different nucleotides. It is usually averaged over a genome sequence.

**Oligotrophic lake**. A nutrient-poor lake with low algal biomass, clear water, and no oxygen depletion.

**Orthologous**. **Homologous** traits in different taxa determined by a common ancestral gene present in their most recent common ancestor.

**Orthologue**. See **orthologous**.

**Overdominant gene**. The phenomenon of heterozygotes having a more extreme phenotype than either homozygote. Overdominance generally refers to the situation in which heterozygous individuals are more fit that homozygous individuals.

**Panmixis** (**panmictic**). Random mating; a population in which genetic exchange occurs at random.

**Paralogous**. **Homologous** traits that arose by gene duplication and evolved in parallel within a single line of descent.

**Paralogue**. See **paralogous**.

**Parasitic DNA**. Redundant, excess DNA in the eukaryotic genome, whose function is commonly unknown and often increases its amount by replication in a way similar to viruses.

**Parthenogenesis** (**parthenogenetic**). Form of asexual reproduction in which the egg develops into a new individual without fertilization.

**PCR**. See **polymerase chain reaction**.

**PCR-based hypervariable markers**. Highly polymorphic molecular markers whose variants can be detected using the **polymerase chain reaction**.

**Pheromone**. A chemical released to the environment that is used for communication between members of the same species (cf. **kairomone**).

**Phylogenetic footprinting**. A method for the discovery of regulatory elements in a set of **orthologous** regulatory regions from two or more species. It does so by identifying the best conserved **motifs** in those orthologous regions.

**Phylogenetic nonindependence**. The problem arising when species comparisons do not take into account that species, due to phylogenetic history, are not statistically independent entities, thus violating the assumptions of standard statistical tests.

**Phylogenetically independent contrasts**. A method of comparative analysis that attempts to circumvent the problem of **phylogenetic nonindependence**. For any group with a known phylogeny, character values are subtracted from one another for each terminal species pair and for each ancestral node. Species themselves are not statistically independent, but the differences between them (i.e. contrasts) are.

**Phylogeography** (**phylogeographic**). A scientific discipline concerned with the principles and processes governing the geographic distribution of genealogical lineages, especially those within and among closely related species. Using **neutral markers**, the history of colonization and spread of a species over its area of occurrence can be reconstructed and related to historic, geologic, or climatologic phenomena.

**Phytoplankton**. Community of small, freely drifting photosynthetic organisms (algae, cyanobacteria).

**Polygenic**. The effects of a large number of different genes, each of which has a slight influence on the phenotype.

**Polymerase chain reaction** (PCR). A technique in which cycles of denaturation, annealing with primers, and extension with DNA polymerase, are used to amplify the number of copies of a target DNA sequence by more than one million times.

**Polyphenism**. Multiple morphological or physiological forms in otherwise genetically identical individuals, members of a **clone**. For example, the sexual/asexual females in rotifers, and the winged/wingless aphids.

**Polyploid** (**polyploidy**). An individual having more than two sets of chromosomes. See **polyploidization**.

**Polyploidization**. Process of becoming **polyploid**.

**Polytomy**. A node in a phylogenetic tree that has a degree greater than three (i.e. it has one ancestor and more than two immediate descendants).

**Postzygotic isolation**. A mechanism preventing interbreeding between two or more populations that is effective after zygote formation (cf. **prezygotic isolation**).

**Preadaptation**. The phenomenon whereby traits in ancestors, whether selected for a certain function or selectively neutral, are fortuitously suited for transformation into new **adaptations** in descendants. See **adaptation.**

**Presexual form**. A specific form in the life cycle of aphids; females that give rise to the sexual females.

**Prevalence**. The proportion of individuals in a population that have a disease or parasite at a given time (usually expressed as a percentage).

**Primary infection**. Very early stage of HIV infection, comprising the first 6-12 weeks of infection. During this period, the infected individual lacks a strong immune response to viral infection so **viral load** is high.

**Primate lentiviruses**. The group of retroviruses that contains both the human (HIV) and simian (SIV) immunodeficiency viruses.

**Prisoner's Dilemma**. A two-player game theory model in which the **fitness** payoffs are set such that mutual cooperation between the players generates a lower return than defection, which occurs when one individual accepts assistance from the other but does not return the favor.

**Programmed cell death**. An evolutionarily conserved form of cell suicide that enables metazoans to regulate cell numbers and control the spread of cancerous cells that threaten the organism.

**Promoter**. The region of a gene near the start site of transcription to which the general transcriptional machinery binds.

**Propagule**. A term used for a structure, like a seed or spore, from which a new individual may arise. Originally used for plants, but recently more widely used. Given this definition, a propagule is also the unit of dispersal.

**Pseudogene**. Copy of a gene inactivated by mutation.

**Quantitative trait locus**. A genetic locus or chromosomal region that contributes to variability in a complex quantitative trait (e.g. body weight), as identified by statistical analysis.

**Ramet**. A module formed by vegetative growth or **parthenogenesis** which is capable of physiological independence (cf. **genet**).

**Random drift**. See **genetic drift**.

**RAPD marker** (random amplified polymorphic DNA). Polymorphic DNA sequence produced by **polymerase chain reaction** amplification using short oligonucleotides of random sequence.

**Reaction norm**. The range of phenotypic variation that a particular genotype is able to produce in various environments.

**Recapitulation**. An early theory in evolutionary biology that claimed that during development an individual organism passes through stages representing adult forms of ancestral species (e.g. gill slits in amniote embryos). A weaker, more recent version holds that this may occur in some cases but is far from the norm.

**Recombination break-point**. The specific site of chromosomal breakage that is associated with a particular chromosomal rearrangement.

**"Red Queen" hypothesis**. The hypothesis that a species evolves as fast as possible to maintain its current level of **adaptation** because its biotic environment is continually deteriorating due to the evolution of other species (e.g. competitors, predators, parasites). In the evolution of sex, the hypothesis that sex provides an advantage for rapid **adaptation** in biotic interactions (e.g. selection by parasites against the most common host genotypes provides an advantage for genetically diverse offspring).

**Refugium** (pl. **refugia**). An area that escaped major climatic changes typical of a region as a whole and acts as a refuge for biota previously more widely distributed; an isolated habitat that retains the environmental conditions that were once widespread.

**Repressor**. The product of a regulatory gene that blocks the function of another gene.

**Resting egg** (syn. **dormant egg**, **diapausing egg**). An egg that does not develop immediately, but either only develops after a certain time has elapsed (refractory phase) or upon receiving an additional stimulus. Although the term "egg" is often used, in many organisms young embryos (e.g. blastula stage) are involved.

**Resting egg bank**. An accumulation of **resting eggs** in sediments or soils. The resting egg bank can contribute to the active population through hatching of part of the eggs. In plants, resting egg banks are referred to as seed banks.

**Restriction fragment length polymorphism** (RFLP). Genetic variation at the site where a restriction enzyme cuts a piece of DNA, resulting in differences in the lengths of the fragments produced by cleavage with the restriction enzyme.

**Reticulate evolution**. A type of gene evolution in which divergent species' lineages share genes due to interspecific **introgression**. A reticulate gene evolutionary tree yields species groups consisting of distantly related species.

**Retrotransposon**. A **transposable element** that transposes via an RNA intermediate which is transcribed to the original DNA by means of a reverse transcriptase enzyme. Some retrotransposons show direct long terminal repeats (LTRs), like *gypsy* and *Osvaldo*, but others like *L1* and *Alu*, often designated as retroposons, do not.

**Ritualization**. The evolutionary modification of a behavior pattern to serve a communicative function.

**Robustness**. The ability of organisms to minimize the impact of genetic variability in the short term on their phenotype, usually through a buffering mechanism involving regulatory adjustments in other steps of the corresponding path or process, thus achieving a certain degree of homeostasis.

**Selective sweep**. See **hitchhiking**.

**Serial colonization**. Pattern of colonization in which an invader species occupies a series of discontinuous environments (e.g. oceanic islands) in a successive manner.

**Seston**. Sum of all small, suspended organic particles that can be retained on a filter, comprising phytoplankton, nonphotosynthetic flagellates, bacteria and organic debris (detritus).

**Sexual selection**. Selection resulting from the differential abilities of individuals to acquire mates in competition with other individuals of the same sex.

**Silent (nucleotide) sites.** Synonymous sites at protein-coding regions, and noncoding sites.

**Silent (nucleotide) substitution**. See **synonymous substitution**.

**Silent (nucleotide) polymorphism**. Polymorphism at **silent sites**.

**Silent variation**. See **silent polymorphism**.

**SINEs** (short interspersed nuclear elements). A class of non-LTR **retrotransposons** found as short interspersed repeats in mammalian genomes.

**Single nucleotide polymorphism** (SNP). Polymorphism in which two or more different nucleotides can be present at a given genome site.

**Skyhook**. A "mind-first" force or process, an exception to the principle that all design, and apparent design, is ultimately the result of mindless, motiveless mechanicity (cf. **crane**).

**Star genealogy**. A genealogy with very short internal branches compared with external branches. See **star phylogeny**.

**Star phylogeny**. A phylogenetic tree with a complete lack of resolution.

**Stereotyped**. Used in two senses. In one, expressions of species typical behavior patterns such as grooming, courting, flying, nest-building, and fighting may be limited variants and as recognizable as the typical color pattern and markings of a bird, fish, or insect. In the other sense, stereotyped refers to behavior patterns that due to abnormal captive conditions or neurological disorders become almost machine like in their virtually unvarying and highly repetitive nature.

**Subitaneous egg**. An egg that immediately develops into a young individual, without a resting phase.

**Substitution rate**. Number of mutations reaching **fixation** in a population per nucleotide site and per unit time (year or generation).

**Succession**. The gradual and predictable process of progressive community change and replacement, leading towards a stable climax community.

**Super-infection**. Non-simultaneous infection of an individual with multiple pathogen strains. Super-infection means that there is no cross-protective immunity.

**Sympatric speciation**. Speciation that occurs as a result of two or more populations diverging while occupying the same geographical area (cf. **allopatric speciation**).

**Sympatry (sympatric)**. Living in the same geographic location (cf. **allopatry**).

**Synergistic epistasis**. Cooperative action of two or more genes such that the total **fitness** is greater than the sum of the fitnesses of all of them (cf. **epistasis**).

**Synonymous substitution**. Nucleotide change in a protein-coding region that does not result in any amino acid change in the encoded protein (cf. **nonsynonymous substitution**).

**Synteny**. Genes or DNA sequences that are in the same chromosome. From an evolutionary perspective, two **homologous** genes are syntenic if the are located in the same homologous chromosome of two related species.

**Synthetic allotetraploid**. **Allotetraploid** produced artificially by crossing two species and doubling their chromosome number using some kind of chemical treatment in the laboratory.

**Telocentric**. A chromosome having the **centromere** located at one end (cf. **metacentric**).

**Tetrads**. The four products of a meiosis.

**Threshold density of hosts** ($N_T$). Density of susceptible hosts in a population that is necessary for $R_0 > 1$.

**T-tropic HIV strain**. Strain of HIV-1 that preferentially replicates in T-cells.

**Topology**. The branching pattern of a phylogenetic tree.

**Translocation**. A genetic rearrangement in which part of a chromosome is detached by breakage and becomes attached to another part of the same or of a different chromosome.

**Transposable element**. Any kind of DNA sequence capable of inserting itself at a new location in the genome. Transposable elements move by excision and insertion (Class II) or by transcription, reverse transcription, and insertion (Class I; see **retrotransposon**).

**Transposition**. The act of transposing, that is, the movement of any DNA sequence through the genome.

**Transposon**. See **transposable element**.

**Unannotated**. See **annotation**.

**Underdominant rearrangement**. Any chromosomal rearrangement that shows the lowest **fitness** when in heterokaryotypic state.

**Viral load**. The intra-host virus population size.

**Virulence**. The capacity of a pathogen to invade host tissue and reproduce; the degree of pathogenicity.

**Viscous populations**. Populations formed by individuals of relatively slow rate of dispersal, and consequently of slow gene flow.

**Zooplankter**. See **zooplankton**.

**Zooplankton**. Nonphotosynthetic plankton; community of mostly small animals drifting in the open water, consisting primarily of crustaceans, rotifers, and ciliates. Organisms comprising the zooplankton are sometimes referred to as **zooplankters**.

# References


Abbot, RJ (1992). Plant invasions, interspecific hybridization and the evolution of new plant taxa. *Trends in Ecology and Evolution*, **7**, 401–405.

Achaz, G, Netter, P, and Coissac, E (2001). Study of intra-chromosomal duplications among the eukaryotic genomes. *Molecular Biology and Evolution*, **18**, 2280–2288.

Adams, KL, Song, K, Roessler, PG, Nugent, JM, Doyle, JL, Doyle, JJ, and Palmer, JD (1999). Intracellular gene transfer in action: dual transcription and multiple silencings of nuclear and mitochondrial *cox2* genes in legumes. *Proceedings of the National Academy of Sciences USA*, **96**, 13863–13868.

Adams, KL, Daley, DO, Qiu, YL, Whelan, J, and Palmer, JD (2000). Repeated, recent and diverse transfers of a mitochondrial gene to the nucleus in flowering plants. *Nature*, **408**, 354–357.

Adams, KL, Rosenblueth, M, Qiu, YL, and Palmer, JD (2001). Multiple losses and transfers to the nucleus of two mitochondrial succinate dehydrogenase genes during angiosperm evolution. *Genetics*, **158**, 1289–1300.

Aguadé, M (1998). Different forces drive the evolution of the *Acp26Aa* and *Acp26Ab* accessory gland genes in the *Drosophila melanogaster* species complex. *Genetics*, **150**, 1079–1089.

Aguadé, M (1999). Positive selection drives the evolution of the Acp29AB accessory gland protein in *Drosophila*. *Genetics*, **152**, 543–551.

Aguadé, M and Langley, CH (1994). Polymorphism and divergence in regions of low recombination in *Drosophila*. In B Golding, ed. *Non-neutral evolution: theories and molecular data*, pp. 67–76. Chapman and Hall, New York.

Ahlberg, PE and Johanson, Z (1998). Osteolepiforms and the ancestry of tetrapods. *Nature*, **395**, 792–794.

Akashi, H (1995). Inferring weak selection from patterns of poly-morphism and divergence at "silent" sites in *Drosophila* DNA. *Genetics*, **139**, 1067–1076.

Akashi, H (1999). Inferring the fitness effects of DNA mutations from polymorphism and divergence data: statistical power to detect directional selection under stationarity and free recombination. *Genetics*, **151**, 221–238.

Akman, L, Yamashita, A, Watanabe, H, Oshima, K, Shiba, T, Hattori, M, and Aksoy, S (2002). Genome sequence of the endocellular obligate symbiont of tsetse flies, *Wigglesworthia glossinidia*. *Nature Genetics*, **32**, 402–407.

Aksoy, S (1995). *Wigglesworthia* gen. nov. and *Wigglesworthia glossinidia* sp. nov. taxa consisting of the myceotcyte associated, primary endosymbionts of tse-tse flies. *International Journal of Systematic Bacteriology*, **45**, 848–851.

Alley, RB (2000). Ice core evidence of abrupt climatic changes. *Proceedings of the National Academy of Sciences USA*, **97**, 1331–1334.

Alonso, S and Armour, JA (2001). A highly variable segment of human subterminal 16p reveals a history of population growth for modern humans outstide Africa. *Proceedings of the National Academy of Sciences USA*, **98**, 864–869.

Alvarez, LW, Alvarez, W, Asaro, F, and Michel, HV (1980). Extraterrestrial cause for the Cretaceous-Tertiary extinction: experimental results and theoretical interpretation. *Science*, **208**, 1095–1108.

Ameisen, JC (1996). The origin of programmed cell death. *Science*, **272**, 1278–1279.

Amores, A, Force, A, Yan, Y-L, Joly, L, Amemiya, C, Fritz, A, Ho, RK, Langeland, J, Prince, V, Wang, Y-L *et al.* (1998). Zebrafish *hox* clusters and vertebrate genome evolution. *Science*, **282**, 1711–1714.

Amos, W and Balmford, A (2001). When does conservation genetics matter? *Heredity*, **87**, 257–265.

Anders, RF, Coppel, RL, Brown, GV, and Kemp, DJ (1988). Antigens with repeated amino acid sequences from the asexual blood stages of *Plasmodium falciparum*. *Progress in Allergy*, **41**, 148–172.

Anders, RF, McColl, DJ, and Coppel, RL (1993). Molecular variation in *Plasmodium falciparum*: polymorphic antigens of asexual erythrocytic stages. *Acta Tropica*, **53**, 239–253.

Andolfatto, P and Przeworski, M (2000). A genome-wide departure from the standard neutral model in natural populations of *Drosophila*. *Genetics*, **156**, 257–268.

Aparici, E, Carmona, MJ, and Serra, M (1998). Sex allocation in haplodiploid cyclical parthenogens with density-dependent proportion of males. *American Naturalist*, **152**, 652–657.

Aquadro, CF, Begun, DJ, and Kindahl, EC (1994). Selection, recombination, and DNA polymorphism in *Drosophila*. In B Golding, ed. *Non-neutral evolution: theories and molecular data*, pp. 46–56. Chapman and Hall, New York.

Araki, H and Tachida, H (1997). Bottleneck effect on evolutionary rate in the nearly neutral mutation model. *Genetics*, **147**, 907–914.

Arkhipova, I and Meselson, M (2000). Transposable elements in sexual and ancient asexual taxa. *Proceedings of the National Academy of Sciences USA*, **97**, 14473–14477.

Armour, JA, Anttinen, T, May, CA, Vega, EE, Sajantila, A, Kidd, JR, Kidd, KK, Bertranpetit, J, Paabo, S, and Jeffreys, AJ (1996). Minisatellite diversity supports a recent African origin for modern humans. *Nature Genetics*, **13**, 154–160.

Arnold, ML (1997). *Natural hybridization and evolution*. Oxford University Press, New York.

Arnone, MI and Davidson, EH (1997). The hardwiring of development: organization and function of genomic regulatory systems. *Development*, **124**, 1851–1864.

Arnosti, DN, Barolo, S, Levine, M, and Small, S (1996). The eve stripe 2 enhancer employs multiple modes of transcriptional synergy. *Development*, **122**, 205–214.

Arnqvist, G, Edvardsson, M, Friberg, U, and Nilsson, T (2000). Sexual conflict promotes speciation in insects. *Proceedings of the National Academy of Science USA*, **97**, 10460–10464.

Averof, M and NH Patel (1997). Crustacean appendage evolution associated with changes in Hox gene expression. *Nature*, **388**, 682–686.

Averof, M, Dawes, R, and Ferrier, D (1996). Diversification of arthropod Hox genes as a paradigm for the evolution of gene functions. *Seminars in Cell and Developmental Biology*, **7**, 539–551.

Ávila, V and García-Dorado, A (2002). The effects of spontaneous mutation on competitive fitness in *Drosophila melanogaster*. *Journal of Evolutionary Biology*, **15**, 561–566.

Avise, JC (1994). *Molecular markers, natural history and evolution*. Chapman and Hall, New York.

Avise, JC (2000). *Phylogeography: the history and formation of species*. Harvard University Press, Cambridge, MA.

Ayala, FJ (1995a). The myth of Eve: molecular biology and human origins. *Science*, **270**, 1930–1936.

Ayala, FJ (1995b). Adam, Eve, and other ancestors: a story of human origins told by genes. *Pubblicazioni Della Stazione Zoologica di Napoli II*, **17**, 303–313.

Ayala, FJ, Escalante, A, Lal, A, and Rich, S (1998). Evolutionary relationships of human malarias. In IW Sherman, ed. *Malaria: parasite biology, pathogenesis, and protection*, pp. 285–300. American Society of Microbiology Press, Washington, DC.

Ayala, FJ, Escalante, AA, and Rich, SM (1999). Evolution of *Plasmodium* and the recent origin of the world populations of *Plasmodium falciparum*. *Parassitologia*, **41**, 55–68.

Ballard, JWO and Kreitman, M (1994). Unraveling selection in the mitochondrial genome of *Drosophila*. *Genetics*, **138**, 757–772.

Bancroft, I (2001). Duplicate and diverge: the evolution of plant genome microstructure. *Trends in Genetics*, **17**, 89–93.

Bandi, C, Sironi, M, Damiani, G, Magrassi, L, Nalepa, CA, Laudani, U, and Sacchi, L (1995). The establishment of intracellular symbiosis in an ancestor of cockroaches and termites. *Proceedings of the Royal Society of London B*, **259**, 293–299.

Barcinski, MA (1998). Apoptosis in trypanosomatids: evolutionary and phylogenetic considerations. *Genetics and Molecular Biology*, **21**, 21–24.

Barrier, M, Baldwin, BG, Robichaux, RH, and Purugganan, MD (1999). Interspecific hybrid ancestry of a plant adaptive radiation: allopolyploidy of the Hawaiian silversword alliance (Asteraceae) inferred from floral homeotic gene duplications. *Molecular Biology and Evolution*, **16**, 1105–1113.

Barta, JR (1989). Phylogenetic analysis of the class Sporozoea (phylum Apicomplexa Levine, 1970): evidence for the independent evolution of heteroxenous life cycles. *Journal of Parasitology*, **75**, 195–206.

Barta, JR, Jenkins, MC, and Danforth, HD (1991). Evolutionary relationships of avian *Eimeria* species among other Apicomplexan protozoa: monophyly of the apicomplexa is supported. *Molecular Biology and Evolution*, **8**, 345–355.

Bataillon, T, Roumet, P, Poirier, S, and David, J (2000). Measuring genome-wide spontaneous mutation in *T. durum*: implications for long-term selection response and management of genetic resources. In *Quantitative genetics and breeding methods*, pp. 251–256. INRA, Paris.

Baumann, P, Baumann, L, Lai, CY, Rouhbaksh, D, Moran, NA, and Clark, MA (1995). Genetics, physiology, and evolutionary relationships of the genus *Buchnera*: intracellular symbionts of aphids. *Annual Review of Microbiology*, **49**, 55–94.

Baumann, P, Moran, NA, and Baumann, L (2000). Bacteriocyte-associated endosymbionts of insect. In M Dworkin, ed. *The prokaryotes*. Springer-Verlag,

New York (http://link.springer.de/link/service/books/10125/).

Begun, DJ and Aquadro, CF (1992). Levels of naturally occurring DNA polymorphism correlate with recombination rates in *Drosophila*. *Nature*, **356**, 519–520.

Begun, DJ and Whitley, P (2000). Reduced *X*-linked nucleotide polymorphism in *Drosophila simulans*. *Proceedings of the National Academy of Sciences USA*, **97**, 5960–5965.

Begun, DJ, Betancourt, AJ, Langley, CH, and Stephan, W (1999). Is the Fast/Slow allozyme polymorphism at the *Adh* locus of *Drosophila melanogaster* an ancient balanced polymorphism? *Molecular Biology and Evolution*, **16**, 1816–1819.

Begun, DJ, Whitley, P, Todd, BL, Waldrip-Dail, HM, and Clark, AG (2000). Molecular population genetics of male accessory gland proteins in *Drosophila*. *Genetics*, **156**, 1879–1888.

Bekoff, M (1995). Play signals as punctuation: the structure of social play in canids. *Behaviour*, **132**, 419–429.

Bekoff, M and Byers, JA (eds) (1998). *Animal play: evolutionary, comparative, and ecological perspectives*. Cambridge University Press, Cambridge.

Bekoff, M, Allen, C, and Burghardt, GM (eds) (2002). *The cognitive animal: empirical and theoretical perspectives on animal cognition*. MIT Press, Cambridge, MA.

Bell, G (1982). *The masterpiece of nature: the evolution and genetics of sexuality*. University of California Press, Berkeley.

Benjamin, J, Li, L, Patterson, C, Greenberg, BD, Murphy, DL, and Hamer DH (1996). Population and familial association between the D4 dopamine receptor gene and measures of Novelty Seeking. *Nature Genetics*, **12**, 81–84.

Bennett, K (1997). *Evolution and ecology: the pace of life*. Cambridge University Press, Cambridge.

Berg, DE and Howe, MM (1989). *Mobile DNA*. American Society for Microbiology, Washington, DC.

Bergman, C (2001). Evolutionary analyses of transcriptional control sequences. PhD Thesis, University of Chicago.

Berman, BP, Nibu, Y, Pfeiffer, BD, Tomancak, P, Celniker, SE, Levine, M, Rubin, GM, and Eisen, MB (2002). Exploiting transcription factor binding site clustering to identify cis-regulatory modules involved in pattern formation in the *Drosophila* genome. *Proceedings of the National Academy of Sciences USA*, **99**, 757–762.

Bertranpetit, J (2000). Genome, diversity, and origins: the Y chromosome as a storyteller. *Proceedings of the National Academy of Sciences USA*, **97**, 6927–6929.

Birks, HH and Ammann, B (2000). Two terrestrial records of rapid climatic change during the glacial-Holocene transition (14,000–9,000 calendar years B.P.) from Europe. *Proceedings of the National Academy of Sciences USA*, **97**, 1390–1394.

Birky, CW (1967). Studies on the physiology and genetics of the rotifer *Asplanchna*. III. Results of outcrossing, selfing, and selection. *Journal of Experimental Zoology*, **164**, 105–116.

Bjorndal, A, Deng, H, Jansson, M, Fiore, JR, Colognesi, C, Karlsson, A, Albert, J, Scarlatti, G, Littman, DR, and Fenyo, EM (1997). Coreceptor usage of primary human immunodeficiency virus type 1 isolates varies according to biological phenotype. *Journal of Virology*, **71**, 7478–7487.

Blackman, RL (1987). Reproduction, cytogenetics and development. In AK Minks and AP Harrewijn, eds. *Aphids, their biology, natural enemies and control* (*Vol. 2A*), pp. 163–195. Elsevier, Amsterdam.

Blackstone, NW (1995). A units-of-evolution perspective on the endosymbiont theory of the origin of the mitochondrion. *Evolution*, **49**, 785–796.

Blackstone, NW and Green, DR (1999). The evolution of a mechanism of cell suicide. *BioEssays*, **21**, 84–88.

Boersma, M, Spaak, P, and De Meester, L (1998). Predator-mediated plasticity in morphology, life history, and behavior of *Daphnia*: the uncoupling of responses. *American Naturalist*, **152**, 237–248.

Bohonak, AJ (1999). Dispersal, gene flow, and population structure. *Quarterly Review of Biology*, **74**, 21–45.

Boileau, MG, Hebert, PDN, and Schwartz, SS (1992). Non-equilibrium gene frequency divergence: persistent founder effects in natural populations. *Journal of Evolutionary Biology*, **5**, 25–39.

Bonneton, F, Shaw, PJ, Fazakerley, C, Shi, M, and Dover, GA (1997). Comparison of bicoid-dependent regulation of hunchback between *Musca domestica* and *Drosophila melanogaster*. *Mechanisms of Development*, **66**, 143–156.

Boraas, ME, Seale, DB, and Boxhorn, JE (1998). Phagotrophy by a flagellate selects for colonial prey: a possible origin of multicellularity. *Evolutionary Ecology*, **12**, 153–164.

Botstein, D, White, RL, Skolnick, M, and Davis, RW (1980). Construction of a genetic linkage map in man using restriction fragment length polymorphisms. *American Journal of Human Genetics*, **32**, 314–331.

Bowring, SA, Erwin, DH, Jin, YG, Martin, MW, Davidek, KL, and Wang, W (1998). U/Pb zircon geochronology and tempo of the end-Permian mass extinction. *Science*, **280**, 1039–1045.

Bradley, DJ (1999). The last and the next hundred years of malariology. *Parassitologia*, **41**, 11–18.

Bradshaw, AD (1984). The importance of evolutionary ideas in ecology—and vice versa. In B Shorrocks, ed. *Evolutionary ecology*, pp. 1–25. Blackwell, Oxford.

Brady, JP and Richmond, RC (1990). Molecular analysis of evolutionary changes in the expression of *Drosophila* esterases. *Proceedings of the National Academy of Sciences USA*, **87**, 8217–8221.

Brendonck, L, De Meester, L, and Hairston, NG (eds) (1998). *Evolutionary and ecological aspects of crustacean diapause* (*Archiv für Hydrobiologie—Advances in Limnology*, **52**). E. Schweizerbart'sche Verlagsbuchhandlung, Stuttgart.

Brewer, MC (1998). Mating behaviours of *Daphnia pulicaria*, a cyclic parthenogen: comparison with copepods. *Philosophical Transactions of the Royal Society of London B*, **353**, 805–815.

Britten, RJ (1996). DNA sequence insertion and evolutionary variation in gene regulation. *Proceedings of the National Academy of Sciences USA*, **93**, 9374–9377.

Brookfield, JFY (1999). Explanation and prediction and the maintenance of sexual reproduction. *Journal of Evolutionary Biology*, **12**, 1017–1019.

Brown, JS, Sanderson, MJ, and Michod, RE (1982). Evolution of social behavior by reciprocation. *Journal of Theoretical Biology*, **99**, 319–339.

Brown, WM (2001). Natural selection of mammalian brain components. *Trends in Ecology and Evolution*, **16**, 471–473.

Bruce, AE, Oates, AC, Prince, VE, and Ho, RK (2001). Additional *hox* clusters in the zebrafish: divergent expression patterns belie equivalent activities of duplicate *hoxB5* genes. *Evolution & Development*, **3**, 127–144.

Brunsfeld, SJ, Sullivan, J, Soltis, DE, and Soltis, PS (2001). Comparative phylogeography of north-western North America: a synthesis. In J Silvertown and J Antonovics, eds. *Integrating ecology and evolution in a spatial context*. Blackwell Science, Oxford, pp. 319–339.

Buchner, P (1965). *Endosymbiosis of animals with plant microorganisms*. Interscience, New York.

Buetow, KH, Edmonson, M, MacDonald, R, Clifford, R, Yip, P, Kelley, J, Little, DP, Strausberg, R, Koester, H, Cantor, CR, *et al.* (2001). High-throughput development and characterization of a genomewide collection of gene-based single nucleotide polymorphism markers by chip-based matrix-assisted laser desorption/ionization time-of-flight mass spectrometry. *Proceedings of the National Academy of Sciences USA*, **98**, 581–584.

Burghardt, GM (1973). Instinct and innate behavior: toward an ethological psychology. In JA Nevin and GS Reynolds, eds. *The study of behavior: learning, motivation, emotion, and instinct*, pp. 322–400. Scott, Foresman, Glenview, IL.

Burghardt, GM (1984). On the origins of play. In PK Smith, ed. *Play in animals and humans*, pp. 5–41. Basil Blackwell, Oxford.

Burghardt, GM (ed.) (1985). *Foundations of comparative ethology*. Van Nostrand Reinhold, New York.

Burghardt, GM (1988). Precocity, play, and the ectotherm-endotherm transition: superficial adaptation or profound reorganization? In EM Blass, ed. *Handbook of behavioral neurobiology* (*Vol. 9*), pp. 107–148. Plenum, New York.

Burghardt, GM (1997). Amending Tinbergen: a fifth aim for ethology. In RW Mitchell, NS Thompson, and HL Miles, eds. *Anthropomorphism, anecdotes, and animals*, pp. 254–276. SUNY Press, Albany.

Burghardt, GM (1998a). The evolutionary origins of play revisited: lessons from turtles. In M Bekoff and JA Byers, eds. *Animal play: evolutionary, comparative, and ecological perspectives*, pp. 1–26. Cambridge University Press, Cambridge.

Burghardt, GM (1998b). Play. In G Greenberg and M Haraway, eds. *Comparative psychology: a handbook*, pp. 757–767. Garland, New York.

Burghardt, GM (2001). Play: attributes and neural substrates. In EM Blass, ed. *Handbook of behavioral neurobiology* (*Vol. 13: Developmental psychobiology, developmental neurobiology and behavioral ecology: mechanisms and early principles*), pp. 327–366. Plenum, New York.

Burghardt, GM (2003). *The genesis of animal play: testing the limits*. MIT Press, Cambridge, MA.

Burt, A (2000). Sex, recombination, and the efficacy of selection—was Weismann right? *Evolution*, **54**, 337–351.

Buss, DM (1994). *The evolution of desire: strategies of human mating*. Basic Books, New York.

Butlin, RK (1998). What do hybrid zones in general, and the *Chorthippus parallelus* zone in particular, tell us about speciation? In DJ Howard and SH Berlocher, eds. *Endless forms*, pp. 367–378. Oxford University Press, New York.

Byers, JA and Walker, C (1995). Refining the motor training hypothesis for the evolution of play. *American Naturalist*, **146**, 25–40.

Caballero, A, Cusi, A, García, C, and García-Dorado, A (2002). Accumulation of deleterious mutations: additional *Drosophila melanogaster* estimates and a simulation of the effects of selection. *Evolution*, **56**, 1150–1159.

Cáceres, CE (1997). Temporal variation, dormancy and coexistence: a field test for the storage effect. *Proceedings of the National Academy of Sciences USA*, **94**, 9171–9175.

Calafell, F, Shuster, A, Speed, WC, Kidd, JR, and Kidd KK (1998). Short tandem repeat polymorphism evolution in humans. *European Journal of Human Genetics*, **6**, 38–49.

Callaway, DS, Ribeiro, RM, and Nowak, MA (1999). Virus phenotype switching and disease progression in HIV-1 infection. *Proceedings of the Royal Society of London B*, **266**, 2523–2530.

Cao, Y, Sorenson, MD, Kumazawa, Y, Mindell, DP, and Hasegawa, M (2000). Phylogenetic position of turtles among amniotes: evidence from mitochondrial and nuclear genes. *Gene*, **259**, 139–148.

Carius, HJ, Little, T, and Ebert, D (2001). Genetic variation in a host-parasite association: potential for coevolution and frequency-dependent selection. *Evolution*, **55**, 1146–1152.

Carmona, MJ and Snell, TW (1995). Glycoproteins in daphnids—potential signals for mating. *Archiv für Hydrobiologie*, **134**, 273–279.

Carmona, MJ, Gómez, A, and Serra, M (1995). Mictic patterns of the rotifer *Brachionus plicatilis* Müller (1786) in small ponds. *Hydrobiologia*, **313/314**, 365–371.

Carr, TR and Kitchell, JA (1980). Dynamics of taxonomic diversity. *Paleobiology*, **6**, 427–443.

Carroll, RL (1988). *Vertebrate paleontology and evolution*. Freeman, New York.

Carroll, RL (1997). *Patterns and processes of vertebrate evolution*. Cambridge University Press, Cambridge.

Carroll, SB (1994). Developmental regulatory mechanisms in the evolution of insect diversity. *Development*, **Suppl.**, 217–223.

Carroll, SB, Grenier, JK, and Weatherbee, SD (2001). *From DNA to diversity: molecular genetics and the evolution of animal design*. Blackwell, Oxford.

Carvalho, GR (1987). The clonal ecology of *Daphnia magna* (Crustacea, Cladocera). 2. Thermal differentiation among seasonal clones. *Journal of Animal Ecology*, **56**, 469–478.

Carvalho, GR (1994). Genetics of aquatic clonal organisms. In AR Beaumont, ed. *Genetics and evolution of aquatic organisms*, pp. 291–323. Chapman and Hall, London.

Cavalier-Smith, T (1987). Eukaryotes with no mitochondria. *Nature*, **326**, 332–333.

Cavalli-Sforza, LL, Menozzi, P, and Piazza, A (1994). *The history and geography of human genes*. Princeton University Press, Princeton, NJ.

Chakraborty, R and Nei, M (1977). Bottle neck effects on average heterozygosity and genetic distance with the stepwise mutation model. *Evolution*, **31**, 347–356.

Chao, L and Carr, DE (1993). The molecular clock and the relationship between population size and generation time. *Evolution*, **47**, 688–690.

Charlesworth, B, Morgan, MT, and Charlesworth, D (1993). The effect of deleterious mutations on neutral molecular evolution. *Genetics*, **134**, 1289–1303.

Charlesworth, D, Charlesworth, B, and McVean, GAT (2001). Genome sequences and evolutionary biology, a two-way interaction. *Trends in Ecology and Evolution*, **16**, 235–242.

Chavarrías, D, López-Fanjul, C, and García-Dorado, A (2001). The rate of mutation and the homozygous and heterozygous mutational effects for competitive viability: a long-term experiment with *Drosophila melanogaster*. *Genetics*, **158**, 681–693.

Chen, FC and Li, WH (2001). Genomic divergences between humans and other hominoids and the effective population size of the common ancestor of humans and chimpanzees. *American Journal of Human Genetics*, **68**, 444–456.

Chen, Z, Gettie, A, Ho, DD, and Marx, PA (1998). Primary SIVsm isolates use the CCR5 coreceptor from sooty mangabeys naturally infected in west Africa: a comparison of coreceptor usage of primary SIVsm, HIV-2, and SIVmac. *Virology*, **246**, 113–124.

Chiang, EF-L, Pai, C-I, Wyatt, M, Yan, Y-L, Postlethwait, J, and Chung, B-C (2001a). Two *sox9* genes on duplicated zebrafish chromosomes: expression of similar transcription activators in distinct sites. *Developmental Biology*, **231**, 149–163.

Chiang, EF-L, Yan, Y-L, Guiguen, Y, Postlethwait, J, and Chung, B-C (2001b). Two *Cyp19* (P450 aromatase) genes on duplicated zebrafish chromosomes are expressed in ovary or brain. *Molecular Biology and Evolution*, **18**, 542–550.

Chuzhanova, NA, Krawczak, M, Nemytikova, LA, Gusev, VA, and Cooper, DN (2000). Promoter shuffling has occurred during the evolution of the vertebrate growth hormone gene. *Gene*, **254**, 9–18.

Civetta, A and Singh, RS (1998). Sex-related genes, directional sexual selection, and speciation. *Molecular Biology and Evolution*, **15**, 901–909.

Clark, AG (1994). Invasion and maintenance of a gene duplication. *Proceedings of the National Academy of Sciences USA*, **91**, 2950–2954.

Clark, AG and Wang, L (1997). Molecular population genetics of *Drosophila* immune system genes. *Genetics*, **147**, 713–724.

Clark, MS (1999). Comparative genomics: the key to understanding the Human Genome Project. *Bioessays*, **21**, 121–130.

Clausen, J, Keck, DD, and Hiesey, WM (1940). Experimental studies on the nature of species. I. Effect of varied environments on western North American plants. *Carnegie Institution of Washington Publication*, **520**, 1–452.

Claverie, JM and Audic, S (1996). The statistical significance of nucleotide position-weight matrix matches. *Computer Applications in the Biosciences*, **12**, 431–439.

Cloutier, R and Ahlberg, PE (1997). Interrelationships of basal sarcopterygians. In MLJ Stiassny, LR Parenti, and GD Johnson, eds. *Interrelationships of fishes*, pp. 445–479. Academic Press, San Diego.

Coatney, GR (1976). Relapse in malaria—an enigma. *Journal of Parasitology*, **62**, 3–9.

Cockburn, A (1991). *An introduction to evolutionary ecology*. Oxford University Press, Oxford.

Coghlan, A and Wolfe, KH (2002). Fourfold faster rate of genome rearrangement in nematodes than in *Drosophila*. *Genome Research*, **12**, 857–867.

Colbourne, JK and Hebert, PDN (1996). The systematics of North American *Daphnia* (Crustacea: Anomopoda): a molecular phylogenetic approach. *Philosophical Transactions of the Royal Society of London B*, **351**, 349–360.

Colbourne, JK, Hebert, PDN, and Taylor, DJ (1997). Evolutionary origins of phenotypic diversity in *Daphnia*. In TJ Givnish and KJ Systma, eds. *Molecular evolution and adaptive radiation*, pp. 163–188. Cambridge University Press, Cambridge.

Cole, ST, Brosch, R, Parkhill, J, Garnier, T, Churcher, C, Harris, D, Gordon, SV, Eiglmeier, K, Gas, S, Barry, CE, *et al.* (1998). Deciphering the biology of *Mycobacterium tuberculosis* from the complete genome sequence. *Nature*, **393**, 537–544.

Colinvaux, P (1993). *Ecology 2*. Wiley, New York.

Colinvaux, PA, De Oliveira, PE, and Bush, MB (2000). Amazonian and neotropical communities on glacial time-scales: the failure of the aridity and refuge hypotheses. *Quaternary Science Reviews*, **19**, 141–169.

Coluzzi, M (1997). Evoluzione biologica i grandi problemi della biologia. In *Evoluzione biologica i grandi problemi della biologia*, pp. 263–285. Accademia dei Lincei, Rome.

Coluzzi, M (1999). The clay feet of the malaria giant and its African roots: hypotheses and inferences about origin, spread and control of *Plasmodium falciparum*. *Parassitologia*, **41**, 277–283.

Colwell, RK (2000). A barrier runs through it... or maybe just a river. *Proceedings of the National Academy of Science USA*, **97**, 13470–13472.

Conway, DJ, Fanello, C, Lloyd, JM, Al-Joubori, BM, Baloch, AH, Somanath, SD, Roper, C, Oduola, AMJ, Mulder, B, Povoa, MM *et al.* 2000. Origin of *Plasmodium falciparum* malaria is traced by mitochondrial DNA. *Molecular and Biochemical Parasitology*, **111**, 163–171.

Cope, D ed. (2001). *Virtual music: computer synthesis of musical style*. MIT Press, Cambridge, MA.

Corliss, J (1994). An interim utilitarian ("user-friendly") hierarchical classification and characterization of the protists. *Acta Protozoologica*, **33**, 1–51.

Cousyn, C and De Meester, L (1998). The vertical profile of resting egg banks in natural populations of the pond-dwelling cladoceran *Daphnia magna* Straus. *Archiv für Hydrobiologie—Advances in Limnology*, **52**, 127–139.

Cousyn, C, De Meester, L, Colbourne, JK, Brendonck, L, Verschuren, D, and Volckaert, F (2001). Rapid, local adaptation of zooplankton behavior to changes in predation pressure in the absence of neutral genetic changes. *Proceedings of the National Academy of Sciences USA*, **98**, 6256–6260.

Cowman, AF and Lew, AM (1989). Antifolate drug selection results in duplication and rearrangement of chromosome 7 in *Plasmodium chabaudi*. *Molecular and Cellular Biology*, **9**, 5182–5188.

Coyne, JA and Orr, HA (1997). "Patterns of speciation in *Drosophila*" revisited. *Evolution*, **51**, 295–303.

Crampton, JM, Beard, CB, and Louis, C (1997). *The molecular biology of insect disease vectors*. Chapman and Hall, London.

Crease, TJ and Hebert, PDN (1983). A test for the production of sexual pheromones by *Daphnia magna* (Crustacea: Cladocera). *Freshwater Biology*, **13**, 491–496.

Crease, TJ, Stanton, DJ, and Hebert, PDN (1989). Polyphyletic origins of asexuality in *Daphnia pulex* II. Mitochondrial DNA variants. *Evolution*, **43**, 1016–1026.

Crow, JF (1999). The odds of losing at the genetic roulette. *Nature*, **397**, 293–294.

Crow, JF and Kimura, M (1970). *An introduction to population genetics theory*. Harper and Row, New York.

Cuadrado, M, Sacristan, M, and Antequera, F (2001). Species-specific organization of CpG island promoters at mammalian homologous genes. *EMBO Reports*, **2**, 586–592.

Cunningham, MR, Roberts, AR, Wu, CH, Barbee, AP, and Druen, PB (1995). Their ideas of beauty are, on the whole, the same as ours: consistency and variability in the cross-cultural perception of female physical attractiveness. *Journal of Personality and Social Psychology*, **68**, 261–279.

Dale, C, Young, SA, Haydon, DT, and Welburn, S (2001). The insect endosymbiont *Sodalis glossinidius* utilizes a type III secretion system for cell division. *Proceedings of the National Academy of Sciences USA*, **98**, 1883–1888.

Dame, JB, Williams, JL, McCutchan, TF, Weber, JL, Wirtz, RA, Hockmeyer, WT, Maloy, WL, Haynes, JD, Schneider, I, Roberts, D, *et al.* (1984). Structure of the gene encoding the immunodominant surface antigen on the sporozoite of the human malaria parasite *Plasmodium falciparum*. *Science*, **225**, 593–599.

Darwin, C (1859). *On the origin of species by means of natural selection, or the preservation of favoured races in the struggle for life*. Murray, London.

Da Silva, MNF and Patton, JL (1998). Molecular phylogeography and the evolution and conservation of Amazonian mammals. *Molecular Ecology*, **7**, 475–486.

Davies, EK, Peters, AD, and Keightley, PD (1999). High frequency of cryptic deleterious mutations in *Caenorhabditis elegans*. *Science*, **285**, 1748–1751.

Dawkins, R (1982). *The extended phenotype*. Freeman, Oxford.

Dawkins, R (1986). *The blind watchmaker*. Longman, London.

Deacon, TW (1990). Rethinking mammalian brain evolution. *American Zoologist*, **30**, 629–705.

Dean, M, Carrington, M, Winkler, C, Huttley, GA, Smith, MW, Allikmets, R, Goedert, JJ, Buchbinder, SP, Vittinghoff, E, Gomperts, E, *et al*. (1996). Genetic restriction of HIV-1 infection and progression to AIDS by a deletion allele of the CKR5 structural gene. *Science*, **273**, 1856–1862.

deBraga, M and Rieppel, O (1997). Reptile phylogeny and the interrelationships of turtles. *Zoological Journal of the Linnean Society*, **120**, 281–354.

Decaestecker, E, De Meester L, and Ebert, D (2002). In deep trouble: habitat selection constrained by multiple enemies in zooplankton. *Proceedings of the National Academy of Sciences USA*, **99**, 5481–5485.

Declerck, S, Cousyn, C, and De Meester, L (2001). Evidence for local adaptation in neighbouring *Daphnia* populations: a laboratory transplant experiment. *Freshwater Biology*, **46**, 187–198.

De Cock, K (2001). Epidemiology and the emergence of human immunodeficiency virus and acquired immune deficiency syndrome. *Philosophical Transactions of the Royal Society of London B*, **356**, 795–798.

Dedryver, CA, Le Gallic, JF, Gauthier, JP, and Simon, J-C (1998). Life-cycle in the aphid *Sitobion avenae* F. (Homoptera: Aphididae): polymorphism, inheritance of history traits associated to sexual phase and adaptive significance. *Ecological Entomology*, **23**, 123–132.

Dedryver, CA, Hullé, M, Le Gallic, JF, Caillaud, CM, and Simon, J-C (2001). Coexistence in space and time of sexual and asexual populations of the cereal aphid *Sitobion avenae*. *Oecologia*, **128**, 379–388.

Dehal, P, Predki, P, Olsen, AS, Kobayashi, A, Folta, P, Lucas, S, Land, M, Terry, A, Zhou, CLE, Rash, S, *et al*. (2001). Human chromosome 19 and related regions in mouse: conservative and lineage-specific evolution. *Science*, **293**, 104–111.

Delarbre, C, Escriva, H, Gallut, C, Barriel, V, Kourilsky, P, Janvier, P, Laudet, V, and Gachelin, G (2000). The complete nucleotide sequence of the mitochondrial DNA of the agnathan *Lampetra fluviatilis*: bearings on the phylogeny of cyclostomes. *Molecular Biology and Evolution*, **17**, 519–529.

Delmotte, F, Leterme N, Bonhomme J, Rispe C, and Simon, J-C (2001). Multiple routes to asexuality in an aphid species. *Proceedings of the Royal Society of London B*, **268**, 1–9.

Delmotte, F, Sabater, B, Leterme, N, Latorre, A, Sunnucks, P, Rispe C, and Simon, J-C (2003). Phylogenetic evidence for hybrid origins of asexual lineages in an aphid species. *Evolution*, **57**, 1291–1303.

de Martino, S, Yan, Y-L, Jowett, T, Postlethwait, JH, Varga, ZM, Ashworth, A, and Austin, CA (2000). Expression of *sox11* gene duplicates in zebrafish suggests the reciprocal loss of ancestral gene expression patterns in development. *Developmental Dynamics*, **217**, 279–292.

De Meester, L (1992). The phototactic behavior of male and female *Daphnia magna*. *Animal Behaviour*, **43**, 696–698.

De Meester, L (1996). Local genetic differentiation and adaptation in freshwater zooplankton populations: patterns and processes. *Ecoscience*, **3**, 385–399.

De Meester, L and Vanoverbeke, J (1999). An uncoupling of male and sexual egg production leads to reduced inbreeding in the cyclical parthenogen *Daphnia*. *Proceedings of the Royal Society of London B*, **266**, 2471–2477.

De Meester, L, Weider, LJ, and Tollrian, R (1995). Alternative antipredator defences and genetic polymorphism in a pelagic predator–prey system. *Nature*, **378**, 483–485.

De Meester, L, Gómez, A, Okamura, B, and Schwenk, K (2002). The Monopolization Hypothesis and the dispersal-gene flow paradox in aquatic organisms. *Acta Oecologica*, **23**, 121–135.

Deng, HW and Lynch, M (1996). Change in genetic architecture in response to sex. *Genetics*, **143**, 203–212.

Dennett, DC (1971). Intentional systems. *Journal of Philosophy*, **68**, 87–106.

Dennett, DC (1987). *The intentional stance*. MIT Press, Cambridge, MA.

Dennett, DC (1995). *Darwin's dangerous idea*. Simon & Schuster, New York.

Dennett, DC (2001a). The evolution of culture. *The Monist*, **84**, 305–324.

Dennett, DC (2001b). Collision detection, muselot, and scribble: some reflections on creativity. In D Cope, ed. *Virtual music: computer synthesis of musical style*, pp. 283–291. MIT Press, Cambridge, MA.

Dennett, DC (2003). In Darwin's wake, where am I? (Presidential Address, Eastern Division, American Philosophical Association, 29 December 2000). In J Hodge and G Radick, eds. *The Cambridge companion to Darwin*, pp. 357–376. Cambridge University Press, Cambridge.

Denno, RF (1994). Life history variation in planthoppers. In RF Denno and TJ Perfect, eds. *Planthoppers: their ecology and management*, pp. 163–215. Chapman and Hall, New York.

Denno, RF and Peterson, MA (1995). Density-related dispersal and its consequences for population dynamics. In N Cappuccino and PW Price, eds. *Populations dynamics: new approaches and synthesis*, pp. 113–130. Academic Press, New York.

Denno, RF and Peterson, MA (2000). Caught between the devil and the deep blue sea, mobile planthoppers elude natural enemies and deteriorating host plants. *American Entomologist*, **46**, 95–109.

Denno, RF and Roderick, GK (1992). Density-related dispersal in planthoppers: effects of interspecific crowding. *Ecology*, **73**, 1323–1334.

Denno, RF, Olmstead, KL, and McCloud, ES (1989). Reproductive cost of flight capability: a comparison of life history traits in wing dimorphic planthoppers. *Ecological Entomology*, **14**, 31–44.

Denno, RF, Roderick, GK, Olmstead, KL, and Döbel, HG (1991). Density-related migration in planthoppers (Homoptera: Delphacidae): the role of habitat persistence. *American Naturalist*, **138**, 1513–1541.

Denno, RF, Roderick, GK, Peterson, MA, Huberty, AF, Döbel, HG, Eubanks, MD, Losey, JE, and Langellotto, GA (1996). Habitat persistence underlies the intraspecific dispersal strategies of planthoppers. *Ecological Monographs*, **66**, 389–408.

Denno, RF, Hawthorne, DJ, Thorne BL, and Gratton, C (2000). Reduced flight capability in British Virgin Island Populations of a wing-dimorphic insect: role of habitat isolation, persistence, and structure. *Ecological Entomology*, **26**, 25–36.

Denver, DR, Morris, K, Lynch, M, Vassilieva, LL, and Thomas, WK (2000). High direct estimate of the mutation rate in the mitochondrial genome of *Caenorhabditis elegans*. *Science*, **289**, 2342–2344.

Dermitzakis, ET and Clark, AG (2001). Differential selection after duplication in mammalian developmental genes. *Molecular Biology and Evolution*, **18**, 557–562.

Dermitzakis, ET, Bergman, CM, and Clark, AG (2003). Tracing the evolutionary history of *Drosophila* regulatory regions with models that identify transcription factor binding sites. *Molecular Biology and Evolution*, **20**, 703–714.

DeStasio, BT (1989). The seed bank of a freshwater crustacean: copepodology for the plant ecologist. *Ecology*, **70**, 1699–1710.

De Zulueta, J (1994). Malaria and ecosystems: from prehistory to posteradication. *Parassitologia*, **36**, 7–15.

De Zulueta, J, Blazquez, J, and Maruto, JF (1973). Entomological aspects of receptivity to malaria in the region of Navalmoral of Mata. *Revista de Sanidad e Higiene Pública (Madrid)*, **47**, 853–870.

D'Hondt, S, Donaghay, P, Zachos, JC, Luttenberg, D, and Lindinger, M (1998). Organic carbon fluxes and ecological recovery from the Cretaceous–Tertiary mass extinction. *Science*, **282**, 276–279.

Dickinson, WJ (1980). Evolution of patterns of gene expression in Hawaiian picture winged *Drosophila*. *Journal of Molecular Evolution*, **16**, 73–94.

Dickinson, WJ (1991). Evolution of regulatory genes and patterns in *Drosophila*. In MK Hecht, B Wallace, and RJ MacIntyre, eds. *Evolutionary biology (Vol. 25)*, pp. 127–173. Plenum Press, New York.

Dickinson, WJ, Rowan, RG, and Brennan, MD (1984). Regulatory gene evolution: Adaptive differences in expression of alcohol dehydrogenase in *Drosophila melanogaster* and *Drosophila simulans*. *Heredity*, **52**, 215–225.

Dimitri, P and Junakovic, N (1999). Revising the selfish DNA hypothesis: new evidence on accumulation of transposable elements in heterochromatin. *Trends in Genetics*, **15**, 123–124.

Dixon, AFG (1998). *Aphid ecology*. Chapman & Hall, London.

Dobzhansky, T (1937). *Genetics and the origin of species*. Columbia University Press, New York.

Donnelly, P and Tavaré, S (1995). Coalescents and genealogical structure under neutrality. *Annual Review of Genetics*, **29**, 401–421.

Doolittle, WF (1999). Phylogenetic classification and the universal tree. *Science*, **284**, 2124–2128.

Douglas, A (1994). *Symbiotic interactions*. Oxford University Press, Oxford.

Douglas, A (1998). Nutritional interactions in insect-microbial symbioses: aphids and their symbiotic bacteria *Buchnera*. *Annual Review of Entomology*, **43**, 17–37.

Dover, GA (2000). How genomic and developmental dynamics affect evolutionary processes. *Bioessays*, **22**, 1153–1159.

Drake, JW (1974). The role of mutation in bacterial evolution. *Symposium Soc. Gen. Microbiology*, **24**, 41–58.

Drake, JW (1991). A constant rate of spontaneous mutation in DNA-based microbes. *Proceedings of the National Academy of Sciences USA*, **88**, 7160–7164.

Drake, JW and Holland, JJ (1999). Mutation rates among RNA viruses. *Proceedings of the National Academy of Sciences USA*, **96**, 13910–13913.

Drake, JW, Charlesworth, B, Charlesworth, D, and Crow, JF (1998). Rates of spontaneous mutation. *Genetics*, **148**, 1667–1686.

Dubchak, I, Brudno, M Loots, GG, Pachter, L, Mayor, C, Rubin, EM, and Frazer, KA (2000). Active conservation of noncoding sequences revealed by three-way species comparisons. *Genome Research*, **10**, 1304–1306.

Duellman, WE and Trueb, L (1994). *Biology of amphibians*. Johns Hopkins University Press, Baltimore.

Duret, L and Mouchiroud, D (2000). Determinants of substitution rates in mammalian genes: expression pattern affects selection intensity but not mutation rate. *Molecular Biology, and Evolution*, **17**, 68–74.

Eberhard, WG (1985). *Sexual selection and animal genitalia*. Oxford University Press, Oxford.

Ebert, D (1994). Virulence and local adaptation of a horizontally transmitted parasite. *Science*, **265**, 1084–1086.

Ebert, D (1999). The evolution and expression of parasite virulence. In SC Stearns, ed. *Evolution in health and disease*. Oxford University Press, Oxford, pp. 161–172.

Edison, T (1932). Interview in *Life*, ch. 24 (according to the *Oxford Dictionary of Quotations*).

Eigen, M (1971). Self-organization of matter and the evolution of biological macromolecules. *Naturwissenschaften*, **58**, 465–523.

Eigen, M and Schuster, P (1977). The hypercycle, a principle of natural self-organization. Part A: emergence of the hypercycle. *Naturwissenschaften*, **64**, 541–565.

Eigen, M and Schuster, P (1978a). The hypercycle, a principle of natural self-organization. Part B: the abstract hypercycle. *Naturwissenschaften*, **65**, 7–41.

Eigen, M and Schuster, P (1978b). The hypercycle, a principle of natural self-organization. Part C: the realistic hypercycle. *Naturwissenschaften*, **65**, 341–369.

Eigen, M and Schuster, P (1979). *The hypercycle, a principle of natural self-organization*. Springer-Verlag, Berlin.

Eisenberg, D, Marcotte, EM, Xenarios, I, and Yeates, TO (2000). Protein function in the post-genomic era. *Nature*, **405**, 823–826.

Emerson, BC, Oromi, P, and Hewitt, GM (2000). Interpreting colonisation of the *Calathus* (Coleoptera: Carabidae) on the Canary Islands and Madeira through the application of the parametric bootstrap. *Evolution*, **54**, 2081–2090.

Enard, W, Khaitovich, P, Klöse, J, Zöllner, S, Heissig, F, Giavalisco, P, Nieselt-Struwe, K, Muchmore, E, Varki, A, Ravid, R *et al.* (2002a). Intra- and interspecific variation in primate gene expression patterns. *Science*, **296**, 340–343.

Enard, W, Przeworski, M, Fisher, SE, Lai, CS, Wiebe, V, Kitano, T, Monaco, AP, and Pääbo, S (2002b). Molecular evolution of *FOXP2*, a gene involved in speech and language. *Nature*, **418**, 869–872.

Endy, D and Brent, R (2001). Modelling cellular behaviour. *Nature*, **409**, 391–395.

Erwin, DH (1996). Understanding biotic recoveries: extinction, survival and preservation during the end-Permian mass extinction. In D Jablonski, DH Erwin, and J Lipps, eds. *Evolutionary paleobiology*, pp. 398–418. University of Chicago Press, Chicago.

Erwin, DH (1998). The end and the beginning: recoveries from mass extinctions. *Trends in Ecology and Evolution*, **13**, 344–349.

Erwin, DH (2001a). Mass extinctions, notable examples of. In S Levin, ed. *Encyclopedia of biodiversity (Vol. 4)*, pp. 111–122. Academic Press, New York.

Erwin, DH (2001b). Lessons from the past: biotic recoveries from mass extinctions. *Proceedings of the National Academy of Sciences USA*, **98**, 5399–5403.

Escalante, AA and Ayala, FJ (1994). Phylogeny of the malarial genus *Plasmodium*, derived from rRNA gene sequences. *Proceedings of the National Academy of Sciences USA*, **91**, 11373–11377.

Escalante, AA, and Ayala, FJ (1995). Evolutionary origin of *Plasmodium* and other Apicomplexa based on rRNA genes. *Proceedings of the National Academy of Sciences USA*, **92**, 5793–5797.

Escalante, AA, Barrio, E, and Ayala, FJ (1995). Evolutionary origin of human and primate malarias: evidence from the circumsporozoite protein gene. *Molecular Biology and Evolution*, **12**, 616–626.

Escalante, AA, Freeland, DE, Collins, WE, and Lal, AA (1998a). The evolution of primate malaria parasites based on the gene encoding cytochrome *b* from the linear mitochondrial genome. *Proceedings of the National Academy of Sciences USA*, **95**, 8124–8129.

Escalante, AA, Lal, AA, and Ayala, FJ (1998b). Genetic polymorphism and natural selection in the malaria parasite *Plasmodium falciparum*. *Genetics*, **149**, 189–202.

Ewald, PW (1991). Transmission modes and the evolution of virulence, with special reference to cholera, influenza and AIDS. *Human Nature*, **2**, 1–30.

Ewald, PW (1995). The evolution of virulence: a unfiying link between parasitology and ecology. *Journal of Parasitology*, **81**, 659–669.

Ewald, PW (1999). Evolutionary control of HIV and other sexually transmitted diseases. In WR Trevathan, EO Smith, and JJ McKenna, eds. *Evolutionary medicine*, pp. 271–311. Oxford University Press, New York.

Excoffier, L and Schneider, S (1999). Why hunter–gatherer populations do not show signs of pleistocene demographic expansions. *Proceedings of the National Academy of Sciences USA*, **96**, 10597–10602.

Eyre-Walker, A (1999). Evidence of selection on silent site base composition in mammals: potential implications for the evolution of isochores and junk DNA. *Genetics*, **152**, 675–683.

Fagen, R (1981). *Animal play behavior*. Oxford University Press, New York.

Fagen, RM (1974). Selective and evolutionary aspects of animal play. *American Naturalist*, **108**, 850–858.

Fandeur, T, Volney, B, Peneau, C, and de Thoisy, B (2000). Monkeys of the rainforest in French Guiana are natural reservoirs for *P. brasilianum/P. malariae* malaria. *Parasitology*, **120**, 11–21.

Fares, MA, Barrio, E, Sabater-Muñoz, B, and Moya, A (2002a). The evolution of heat-shock protein GroEL from *Buchnera*, the primary endosymbiont of aphids, is governed by positive selection. *Molecular Biology and Evolution*, **19**, 1162–1170.

Fares, MA, Ruiz-González, MX, Moya, A, Elena, SF, and Barrio, E (2002b). GroEL buffers against deleterious mutations. *Nature*, **417**, 398.

Feller, AE and Hedges, SB (1998). Molecular evidence for the early history of living amphibians. *Molecular Phylogenetics and Evolution*, **9**, 509–516.

Fernández, J and López-Fanjul, C (1996). Spontaneous mutational variances and covariances for fitness-related traits in *Drosophila melanogaster*. *Genetics*, **143**, 829–837.

Ferris, SD and Whitt, GS (1977). Duplicate gene expression in diploid and tetraploid loaches (Cypriniformes, Cobitidae). *Biochemical Genetics*, **15**, 1097–1112.

Ferris, SD and Whitt, GS (1979). Evolution of the differential regulation of duplicate genes after polyploidization. *Journal of Molecular Evolution*, **12**, 267–317.

Fickett, JW and Wasserman, WW (2000). Discovery and modeling of transcriptional regulatory regions. *Current Opinion in Biotechnology*, **11**, 19–24.

Fink, B, Grammer, K, and Thornhill, R (2001). Human (*Homo sapiens*) facial attractiveness in relation to skin texture and color. *Journal of Comparative Psychology*, **115**, 92–99.

Finlay, BL, Darlington, RB, and Nicastro, N (2001). Developmental structure in brain evolution. *Behavioral and Brain Sciences*, **24**, 263–308.

Finston, TL and Peck, SB (1995). Population structure and gene flow in *Stomion*: a species swarm of flightless beetles of the Galapagos Islands. *Heredity*, **75**, 390–397.

Fisher, MC and Viney, ME (1998). The population genetic structure of the facultatively sexual parasitic nematode *Strongyloides ratti* in wild rats. *Proceedings of the Royal Society of London B*, **265**, 703–709.

Fisher, RA (1935). The sheltering of lethals. *American Naturalist*, **69**, 446–455.

Fleischer, RC, McIntosh, CE, and Tarr, CL (1998). Evolution on a volcanic conveyor belt: using phylogeographic reconstructions and K-Ar-based ages of the Hawaiian Islands to estimate molecular evolutionary rates. *Molecular Ecology*, **7**, 533–545.

Fleischmann, RD, Adams, MD, White, O, Clayton, RA, Kirkness, EF, Kerlavage, AR, Bult, CJ, Tomb, J-F, Dougherty, BA, Merrick, JM, *et al.* (1995). Whole-genome random sequencing and assembly of *Haemophilus influenzae* Rd. *Science*, **269**, 496–512.

Fontdevila, A (1988). The evolutionary potential of the unstable genome. In G de Jong, ed. *Populations genetics and evolution*, pp. 251–263. Springer-Verlag, Berlin.

Fontdevila, A (1992). Genetic instability and rapid speciation: are they coupled? *Genetica*, **86**, 247–258.

Force, A, Lynch, M, Pickett, B, Amores, A, Yan, Y-L, and Postlethwait, J (1999). Preservation of duplicate genes by complementary, degenerative mutations. *Genetics*, **151**, 1531–1545.

Forró, L (1997). Mating behaviour in *Moina brachiata* (Jurine, 1820) (Crustacea, Anomopoda). *Hydrobiologia*, **360**, 153–159.

Frade, JM and Michaelidis, TM (1997). Origin of eukaryotic programmed cell death: a consequence of aerobic metabolism. *BioEssays*, **19**, 827–832.

Frank, SA (1995). The origin of synergistic symbiosis. *Journal of Theoretical Biology*, **176**, 403–410.

Frank, SA (1997). Models of symbiosis. *American Naturalist*, **150**, S80–S99.

Frech, K, Quandt, K, and Werner, T (1997). Finding protein-binding sites in DNA sequences: the next generation. *Trends in Biochemical Sciences*, **22**, 103–104.

Frohlich, KU and Madeo, F (2000). Apoptosis in yeast—a monocellular organism exhibits altruistic behaviour. *FEBS Letters*, **473**, 6–9.

Fry, JD (2001). Rapid mutational declines of viability in *Drospohila*. *Genetical Research*, **77**, 53–60.

Fry, JD, Keightley, PD, Heinsohn, SL, and Nuzhdin, SV (1999). New estimates of the rates and effects of mildly deleterious mutation in *Drosophila melanogaster*. *Proceedings of the National Academy of Sciences USA*, **96**, 574–579.

Fu, Y-X (1997). Statistical tests of neutrality of mutations against population growth, hitchhiking and background selection models. *Genetics*, **147**, 915–925.

Fu, Y-X (1999). Coalescing into the 21st century: an overview and prospects of coalescent theory. *Theoretical Population Biology*, **56**, 1–10.

Fu, Y-X and Li, W-H (1993). Statistical tests of neutrality of mutations. *Genetics*, **133**, 693–709.

Fukatsu, T and Ishikawa, H (1993). Ocurrence of chaperonin 60 and chaperonin 10 in primary and secondary bacterial symbionts of aphids: implications for the evolution of an endosymbiotic system in aphids. *Journal of Molecular Evolution*, **36**, 568–577.

Funk, DJ, Wernegreen, JJ, and Moran, NA (2001). Intraspecific variation in symbiont genomes: bottlenecks and the Aphid-Buchnera association. *Genetics*, **157**, 477–489.

Gabriel, W and Lynch, M (1992). The selective advantage of reaction norms for environmental tolerance. *Journal of Evolutionary Biology*, **5**, 41–59.

Gage, MJG (1995). Continuous variation in reproductive strategy as an adaptive response to population density in the moth *Plodia interpunctella*. *Proceedings of the Royal Society of London B*, **261**, 25–30.

Gagneux, P and Varki, A (2000). Genetic differences between humans and great apes. *Molecular Phylogenetics and Evolution*, **18**, 2–13.

Gagneux, P, Wills, C, Gerloff, U, Tautz, D, Morin, PA, Boesch, C, Fruth, B, Hohmann, G, Ryder, OA, and Woodroof, DS (1999). Mitochondrial sequences show diverse evolutionary histories of African hominoids. *Proceedings of the National Academy of Sciences USA*, **96**, 5077–5082.

Gallardo, W, Tomita, Y, Hagiwara, A, Soyano, K, and Snell, TW (1997). Effect of some vertebrate and invertebrate hormones on the population growth, mictic female production, and body size of the marine rotifer *Brachionus plicatilis* Müller. *Hydrobiologia*, **358**, 113–120.

Gangestad, SW and Thornhill, R (1997). Human sexual selection and developmental stability. In JA Simpson and DT Kenrick, eds. *Evolutionary social psychology*, pp. 169–195. Lawrence Erlbaum Associates, Mahwah, NJ.

Gangestad, SW and Thornhill, R (1999). Individual differences in developmental precision and fluctuating asymmetry: a model and its implications. *Journal of Evolutionary Biology*, **12**, 402–416.

Gangestad, SW and Thornhill, R (2003). Facial masculinity and fluctuating asymmetry. *Evolution and Human Behavior*, **24**, 231–241.

García-Dorado, A (1997). The rate and effects distribution of viability mutation in *Drosophila*: minimum distance estimation. *Evolution*, **51**, 1130–1139.

García-Dorado, A (2003). Tolerant versus sensitive genomes: the impact of deleterious mutation on fitness and conservation. *Conservation Genetics*, **4**, 311–324.

García-Dorado, A and Caballero, A (2000). On the average coefficient of dominance of deleterious spontaneous mutations. *Genetics*, **155**, 1991–2001.

García-Dorado, A and Caballero, A (2002). The mutational rate of *Drosophila* viability decline: tinkering with old data. *Genetical Research*, **80**, 99–105.

García-Dorado, A and Gallego, A (2003). Comparing analysis methods for mutation accumulation data: a simulation study. *Genetics*, **164**, 807–819.

García-Dorado, A, López-Fanjul, C, and Caballero, A (1999). Properties of spontaneous mutations affecting quantitative traits. *Genetical Research*, **75**, 47–51.

García-Moreno, J, Arctander, P, and Fjeldsa, J (1999). A case of rapid diversification in the Neotropics: phylogenetic relationships among *Cranioleuca* spinetails (Aves, Furnariidae). *Molecular Phylogenetics and Evolution*, **12**, 273–281.

García-Paris, M, Good, DA, Parra-Olea, G, and Wake, DB (2000). Biodiversity of Costa Rican salamanders: implications of high levels of genetic differentiation and phylogeographic structure for species formation. *Proceedings of the National Academy of Sciences USA*, **97**, 1640–1647.

Garnham, PCC (1966). *Malaria parasites and other haemosporidia*. Blackwell Scientific Publications, Oxford.

Gibert, JM, Mouchel-Vielh, E, Queinnec, E, and Deutsch, JS (2000). Barnacle duplicate engrailed genes: divergent expression patterns and evidence for a vestigial abdomen. *Evolution & Development*, **2**, 194–202.

Gil, R, Sabater-Muñoz, B, Latorre, A, Silva, FJ, and Moya, A (2002). Extreme genome reduction in *Buchnera* spp.: towards the minimal genome needed for symbiotic life. *Proceedings of the National Academy of Sciences USA*, **99**, 4454–4459.

Gilbert, JJ (1974). Dormancy in rotifers. *Transactions of the American Microscopical Society*, **93**, 490–513.

Gilbert, JJ (1980). Female polymorphism and sexual reproduction in the rotifer *Asplanchna*: evolution of their relationships and control by dietary tocopherol. *American Naturalist*, **116**, 409–431.

Gilbert, JJ (1993). Rotifera. In KG Adiyodi and RG Adiyodi, eds. *Reproductive biology of invertebrates. Vol. VI, Part A: Asexual propagation and reproductive strategies*, pp. 231–263. Oxford & IBH Publishing Co., New Delhi, India.

Gilbert, JJ (1996). Effect of temperature on the response of planktonic rotifers to a toxic cyanobacterium. *Ecology*, **77**, 1174–1180.

Gilbert, JJ and Schreiber, DK (1998). Asexual diapause induced by food limitation in the rotifer *Synchaeta pectinata*. *Ecology*, **79**, 1371–1381.

Gilbert, SF (1994). *Developmental biology*. Sinauer Associates, Sunderland, MA.

Gillespie, JH (1984). Molecular evolution over the mutational landscape. *Evolution*, **38**, 1116–1129.

Gillespie, JH (1991). *The causes of molecular evolution*. Oxford University Press, New York.

Gillott, M, Holen, D, Ekman, J, Harry, M, and Boraas, ME (1993). Predation-induced *E. coli* filaments: are they multicellular? In G Baily and C Reider, eds. *Proceedings of the 51st annual meeting of the Microscopy Society of America*, p. 420. San Francisco Press, San Francisco, CA.

Gittenberger, E (1991). What about non-adaptive radiation? *Biological Journal of the Linnean Society*, **43**, 263–272.

Gliwicz, ZM (1986). Predation and the evolution of vertical migration in zooplankton. *Nature*, **320**, 746–748.

Goebel, W and Gross, R (2001). Intracellular survival strategies of mutualistic and parasitic prokaryotes. *Trends in Microbiology*, **9**, 267–273.

Gómez, A and Carvalho, GR (2000). Sex, parthenogenesis and the genetic structure of rotifers: microsatellite analysis of contemporary and resting egg bank populations. *Molecular Ecology*, **9**, 203–214.

Gómez, A and Serra, M (1995). Behavioral reproductive isolation among sympatric strains of *Brachionus plicatilis* Müller, 1786: Insights into the status of this taxonomic species. *Hydrobiologia*, **313/314**, 111–119.

Gómez, A, Temprano, M, and Serra, M (1995). Ecological genetics of a cyclical parthenogen in temporary habitats. *Journal of Evolutionary Biology*, **8**, 601–622.

Gómez, A, Carvalho, GR, and Lunt, DH (2000). Phylogeography and regional endemism of a passively dispersing zooplankter: mtDNA variation of rotifer resting egg banks. *Proceedings of the Royal Society of London B*, **267**, 2189–2197.

Gómez, A, Serra, M, Carvalho, GR, and Lunt, DH (2002). Speciation in ancient cryptic species complexes: evidence from the molecular phylogeny of *Brachionus plicatilis* (Rotifera). *Evolution*, **56**, 1431–1444.

González-Crespo, S and Levine, M (1994). Related target enhancers for dorsal and NK-kB signalling pathways. *Science*, **264**, 255–258.

Goodfriend, GA (1986). Variation in land-snail shell form and its causes: a review. *Systematic Zoology*, **35**, 204–223.

Goodnight, KF (1992). The effect of stochastic variation on kin selection in a budding-viscous population. *American Naturalist*, **140**, 1028–1040.

Gould, SJ (1984). Covariance sets and ordered variation in *Cerion* from Aruba, Bonaire and Curaçao: a way of studying nonadaptation. *Systematic Zoology*, **33**, 217–237.

Gould, SJ (1991). *Bully for Brontosaurus*. Norton, New York.

Govindaraju, DR (1988). Relationship between dispersal ability and levels of gene flow in plants. *Oikos*, **52**, 31–35.

Grammer, K, Fink, B, Juette, A, Ronzal, G, and Thornhill, R (2002). Female faces and bodies: N-dimensional feature space and attractiveness. In G Rhodes and LA Zebrowitz, eds. *Facial attractiveness: evolutionary, cognitive, and social perspectives*, pp. 91–125. Greenwood, Westport, CT.

Grant, BR and Grant, PR (1996a). High survival of Darwin's finch hybrids: effects of beak morphology and diets. *Ecology*, **77**, 500–509.

Grant, PR and Grant, BR (1996b). Speciation and hybridization of island birds. *Philosophical Transactions of the Royal Society of London B*, **351**, 765–772.

Grassly, NC, Harvey, PH, and Holmes, EC (1999). Population dynamics of HIV-1 inferred from gene sequences. *Genetics*, **151**, 427–438.

Graybiel, AM (1995). Building action repertoires: memory and learning functions of the basal ganglia. *Current Opinion in Neurobiology*, **5**, 733–741.

Gregory, WK (1946). Pareiasaurs versus placodonts as near ancestors to turtles. *Bulletin of the American Museum of Natural History*, **86**, 275–326.

Gregory, WK (1947). The monotremes and the palimpsest theory. *Bulletin of the American Museum of Natural History*, **88**, 1–52.

Grey, D, Hutson, V, and Szathmáry, E (1995). A re-examination of the stochastic corrector model. *Proceedings of the Royal Society of London B*, **262**, 29–35.

Grimson, MJ, Coates, JC, Reynolds, JP, Shipman, M, Blanton, RL, and Harwood, AJ (2000). Adherens junctions and β-catenin-mediated cell signaling in a non-metazoan organism. *Nature*, **408**, 727–731.

Groisman, EA and Ochman, H (1996). Pathogenicity islands: bacterial evolution in quantum leaps. *Cell*, **87**, 791–794.

Grover, JP (1997). *Resource competition*. Chapman and Hall, New York, USA.

Guerrero, R (1991). Predation as prerequisite to organelle origin: *Daptobacter* as example. In L Margulis and R Fester, eds. *Symbiosis as a source of evolutionary innovation: speciation and morphogenesis*, pp. 106–117. MIT Press, Cambridge, MA.

Guldemond, JA and Dixon, AFG (1994). Specificity and daily cycle of release of sex pheromones in aphids: a case of reinforcement? *Biological Journal of the Linnean Society*, **52**, 287–303.

Gysin, J (1998). Animal models: Primates. In IW Sherman, ed. *Malaria: parasite biology, pathogenesis, and protection*, pp. 419–441. American Society of Microbiology Press, Washington, DC.

Hacia, JG (2001). Genome of the apes. *Trends in Genetics*, **17**, 637–645.

Hagen, JB (1992). *An entangled bank: the origin of ecosystem ecology*. Rutgers University Press, New Brunswick, NJ.

Hagiwara, A, Hino, A, and Hirano, R (1988). Effects of temperature and chlorinity on resting egg formation in the rotifer *Brachionus plicatilis*. *Nippon Suisan Gakkaishi*, **54**, 569–575.

Hahn, BH, Shaw, GM, de Cock KM, and Sharp PM (2000). AIDS as a zoonosis: scientific and public health implications. *Science*, **287**, 607–614.

Hairston, NG (1996). Zooplankton egg banks as biotic reservoirs in changing environments. *Limnology and Oceanography*, **41**, 1087–1092.

Hairston, NG and De Stasio, BT (1988). Rate of evolution slowed by a dormant propagule pool. *Nature*, **336**, 239–242.

Hairston, NG, van Brunt, RA, Kearns, CM, and Engstrom, DR (1995). Age and survivorship of diapausing eggs in a sediment egg bank. *Ecology*, **76**, 1706–1711.

Hairston, NG, Lampert, W, Cáceres, CE, Weider, LJ, Gaedke, U, Fischer, JM, Fox, JA, and Post, DM (1999). Rapid evolution revealed by dormant eggs. *Nature*, **401**, 446.

Hairston, NG, Holtmeier, CL, Lampert, W, Weider, LJ, Post, DM, Fischer, JM, Cáceres, CE, Fox, JA, and Gaedke, U (2001). Natural selection for grazer resistance to toxic cyanobacteria: evolution of phenotypic plasticity? *Evolution*, **55**, 2203–2214.

Haldane, JBS (1933). The part played by recurrent mutation in evolution. *American Naturalist*, **67**, 5–9.

Haldane, JBS (1937). The effect of variation on fitness. *American Naturalist*, **71**, 337–349.

Hales, DF, Tomiuk J, Wöhrmann K, and Sunnucks P (1997). Evolutionary and genetic aspects of aphid biology: a review. *European Journal of Entomology*, **94**, 1–55.

Hallam, A and Wignall, PB (1997). *Mass extinctions and their aftermath*. Oxford University Press, Oxford.

Halushka, MK, Fan, JB, Bentley, K, Hsie, L, Shen, N, Weder, A, Cooper, R, Lipschutz, R, and Chakravarti, A (1999). Patterns of single-nucleotide polymorphisms in candidate genes for blood-pressure homeostasis. *Nature Genetics*, **22**, 239–247.

Hamilton, WD (1971). Selection of selfish and altruistic behavior in some extreme models. In JF Eisenberg and WS Dillon, eds. *Man and beast: comparative social behavior*, pp. 59–91. Smithsonian Press, Washington DC.

Hamilton, WD (1972). Altruism and related phenomena mainly in social insects. *Annual Review of Ecology and Systematics*, **3**, 193–232.

Hanazato, T (1991). Effects of repeated application of carboryl on zooplankton communities in experimental ponds with or without the predator *Chaoborus*. *Environmental Pollution*, **74**, 309–324.

Hancock, JM (1999). Microsatellites and other simple sequences: genomic context and mutational mechanisms. In DB Goldstein and C Schlötterer, eds. *Microsatellites, evolution and applications*, pp. 1–9. Oxford University Press, Oxford.

Hansen, TA, Kelley, PH, Melland, VD, and Graham, SE (1999). Effect of climate-related mass extinctions on escalation in molluscs. *Geology*, **27**, 1139–1142.

Hansen, TF, Carter, AJ, and Chiu, C-H (2000). Gene conversion may aid adaptive peak shifts. *Journal of Theoretical Biology*, **207**, 495–511.

Harada, H, Oyaizu, H, and Ishikawa, H (1996). A consideration about the origin of aphid intracellular symbiont in connection with gut bacterial flora. *Journal of General and Applied Microbiology*, **42**, 17–26.

Harding, RM, Fullerton, SM, Griffiths, RC, Bond, J, Cox, MJ, Schneider, JA, Moulin, DS, and Clegg JB (1997). Archaic African and Asian lineages in the genetic ancestry of modern humans. *American Journal of Human Genetics*, **60**, 772–789.

Hastings, KEM (1996). Strong evolutionary conservation of broadly expressed protein isoforms in the troponin I gene family and other vertebrate gene families. *Journal of Molecular Evolution*, **42**, 631–640.

Hawthorne, DJ and Via, S (2001). Genetic linkage of ecological specialization and reproductive isolation in pea aphids. *Nature*, **412**, 904–907.

Hebert, PDN (1987). Genotypic characteristics of cyclic parthenogens and their asexual derivatives. In SC Stearns, ed. *The evolution of sex and its consequences*, pp. 175–195. Birkhauser Verlag, Basle.

Hebert, PDN and Finston, TL (1996). Genetic differentiation in *Daphnia obtusa*: a continental perspective. *Freshwater Biology*, **35**, 311–321.

Hebert, PDN and Wilson, CC (1994). Provincialism in plankton: endemism and allopatric fragmentation in Australian *Daphnia*. *Evolution*, **48**, 1333–1349.

Hebert, PDN, Finston, TL, and Foottit, R (1991). Patterns of genetic diversity in the sumac gall aphid, *Melaphis rhois*. *Genome*, **34**, 757–762.

Heddi, A, Charles, H, Khatchadourian, C, Bonnot, G, and Nardon, P (1998). Molecular characterization of the principal symbiotic bacteria of the weevil *Sitophilus oryzae*: a peculiar G + C content of an endocytobiotic DNA. *Journal of Molecular Evolution*, **47**, 52–61.

Hedges, SB (1994). Molecular evidence for the origin of birds. *Proceedings of the National Academy of Sciences USA*, **91**, 2621–2624.

Hedges, SB and Poling, LL (1999). A molecular phylogeny of reptiles. *Science*, **283**, 998–1001.

Hedges, SB, Hass, CA, and Maxson, LR (1993). Relations of fish and tetrapods. *Nature*, **363**, 501–502.

Hedrick, PW (2001). Conservation genetics: where are we now? *Trends in Ecology and Evolution*, **16**, 629–636.

Hennig, W (1983). Testudines. In W Hennig ed. *Stammesgeschichte der Chordaten*, pp. 132–139. P. Parey, Hamburg.

Henter, H and Via, S (1995). The potential for coevolution in a host-parasitoid system. I. Genetic variation within an aphid population in susceptibility to a parasitic wasp. *Evolution*, **49**, 427–438.

Hentschel, U and Hacker, J (2001). Pathogenicity islands: the tip of the iceberg. *Microbes and Infection*, **3**, 545–548.

Hewitt, GM (1993). Postglacial distribution and species substructure: lessons from pollen, insects and hybrid zones. In DR Lees and D Edwards, eds. *Evolutionary patterns and processes*, pp. 97–123. Academic Press, London.

Hewitt, GM (1996). Some genetic consequences of ice ages, and their role in divergence and speciation. *Biological Journal of the Linnean Society*, **58**, 247–276.

Hewitt, GM (1999). Post-glacial recolonization of European biota. *Biological Journal of the Linnean Society*, **68**, 87–112.

Hewitt, GM (2000). The genetic legacy of the Quaternary ice ages. *Nature*, **405**, 907–913.

Hewitt, GM (2001). Speciation, hybrid zones and phylogeography—or seeing genes in space and time. *Molecular Ecology*, **10**, 537–549.

Hewitt, GM and Ibrahim, KM (2001). Inferring glacial refugia and historical migrations with molecular phylogenies. In J Silvertown and J Antonovics, eds. *Integrating ecology and evolution in a spatial context*, pp. 271–294. Blackwell Science, Oxford.

Holland, B and Rice, WR (1998). Chase-away sexual selection: antagonistic seduction versus resistance. *Evolution*, **52**, 1–7.

Holmes, EC (2001). On the origin and evolution of the human immunodeficiency virus (HIV). *Biological Reviews*, **76**, 239–254.

Houle, D, Hoffmaster, DK, Assimacopoulos, S, and Charlesworth, B (1992). The genomic mutation rate for fitness in *Drosophila*. *Nature*, **359**, 58–60.

Houle, D, Hoffmaster, DK, Assimacopoulos, S, and Charlesworth, B (1994). Correction: the genomic mutation rate for fitness in *Drosophila*. *Nature*, **371**, 358.

Houle, D, Morikawa, B, and Lynch, M (1996). Comparing mutational variabilities. *Genetics*, **143**, 1467–1483.

Houle, D, Hughes, KA, Assimacopoulos, S, and Charlesworth, B (1997). The effects of spontaneous mutations on quantitative traits. II. Dominance of mutations with effects on life-history traits. *Genetical Research*, **70**, 27–34.

Huang, Y, Paxton, WA, Wolinsky, SM, Neumann, AU, Zhang L, He, T, Kang, S, Ceradini, D, Jin, Z, Yazdanbakhsh, K, *et al.* (1996). The role of a mutant CCR5 allele in HIV-1 transmission and disease progression. *Nature Medicine*, **2**, 1240–1247.

Hudson, RR (1990). Gene genealogies and the coalescent process. *Oxford Surveys in Evolutionary Biology*, **7**, 1–44.

Hudson, RR, Kreitman, M, and Aguadé, M (1987). A test of neutral molecular evolution based on nucleotide data. *Genetics*, **116**, 153–159.

Hudson, RR, Bailey, K, Skarecky, D, Kwiatowski, J, and Ayala, FJ (1994). Evidence for positive selection in the *Superoxide Dismutase* (*Sod*) region of *Drosophila melanogaster*. *Genetics*, **136**, 1329–1340.

Huelsenbeck, JP, Larget, B, and Swofford, D (2000). A compound Poisson process for relaxing the molecular clock. *Genetics*, **154**, 1879–1892.

Huey, RB, Partridge, L, and Fowler, K (1991). Thermal sensitivity of *Drosophila* melanogaster responds rapidly to laboratory natural selection. *Evolution*, **45**, 751–756.

Huff, CG (1938). Studies on the evolution of some disease-producing organisms. *Quarterly Review of Biology*, **13**, 196–206.

Hughes, AL (1993). Coevolution of immunogenic proteins of *Plasmodium falciparum* and the host's immune system. In N Takahata and AG Clark, eds. *Mechanisms of molecular evolution*, pp. 109–127. Sinauer Associates, Sunderland, MA.

Hughes, AL (1994). The evolution of functionally novel proteins after gene duplication. *Proceedings of the Royal Society of London B*, **256**, 119–124.

Hughes, AL and Nei, M (1988). Pattern of nucleotide substitution at major histocompatibility complex class I loci reveals overdominant selection. *Nature*, **335**, 167–170.

Hughes, AL and Verra, F (2001). Very large long-term effective population size in the virulent human malaria parasite *Plasmodium falciparum*. *Proceedings of the Royal Society of London B*, **268**, 1855–1860.

Hughes, MK and Hughes, AL (1993). Evolution of duplicate genes in a tetraploid animal, *Xenopus laevis*. *Molecular Biology and Evolution*, **10**, 1360–1369.

Hughes, MK and Hughes, AL (1995). Natural selection on *Plasmodium* surface proteins. *Molecular and Biochemical Parasitology*, **71**, 99–113.

Hughes, RN (1989). *A functional biology of clonal organisms*. Chapman and Hall, London.

Hutter, CM and Rand, DM (1995). Competition between mitochondrial haplotypes in distinct nuclear genetic environments: *Drosophila pseudoobscura* vs *D. persimilis*. *Genetics*, **140**, 537–548.

Hyland, DA (1984). *The question of play*. University Press of America, Lanham, MD.

Ibrahim, KM, Nichols, RA, and Hewitt, GM (1996). Spatial patterns of genetic variation generated by different forms of dispersal during range expansion. *Heredity*, **77**, 282–291.

Ingman, M, Kaessmann, H, Pääbo, S, and Gyllensten, U (2000). Mitochondrial genome variation and the origin of modern humans. *Nature*, **408**, 708–713.

Innes DJ and Dunbrack, RL (1993). Sex allocation variation in *Daphnia pulex*. *Journal of Evolutionary Biology*, **6**, 559–575.

Innes, DJ and Hebert, PDN (1988). The origin and genetic-basis of obligate parthenogenesis in *Daphnia pulex*. *Evolution*, **42**, 1024–1035.

International Human Genome Sequencing Consortium (2001). Initial sequencing and analysis of the human genome. *Nature*, **409**, 860–921.

Iwasa, Y and Pomiankowski, A (1995). Continual change in mate preferences. *Nature*, **377**, 420–422.

Jablonski, D (1986). Background and mass extinction: the alternation of macroevolutionary regimes. *Science*, **231**, 129–133.

Jablonski, D (1989). The biology of mass extinction: a paleontological view. *Philosophical Transactions of the Royal Society of London B*, **325**, 357–368.

Jablonski, D (1998). Geographic variation in the molluscan recovery from the End-Cretaceous extinction. *Science*, **279**, 1327–1330.

Jablonski, D (2001). Lessons from the past: evolutionary impacts of mass extinctions. *Proceedings of the National Academy of Sciences USA*, **98**, 5393–5398.

Jablonski, D and Raup, DM (1995). Selectivity of End-Cretaceous marine bivalve extinctions. *Science*, **268**, 389–391.

Jaccoud, D, Peng, K, Feinstein, D, and Kilian, A (2001). Diversity Arrays: a solid state technology for sequence information independent genotyping. *Nucleic Acids Research*, **29**, e25.

Jacob, F and Monod, J (1961). Genetic regulatory mechanisms in the synthesis of proteins. *Journal of Molecular Evolution*, **3**, 318–356.

Jain, R, Rivera, MC, and Lake, JA (1999). Horizontal gene transfer among genomes: the complexity hypothesis. *Proceedings of the National Academy of Sciences USA*, **96**, 3801–3806.

Janke, A, Gemmell, NJ, Feldmaier-Fuchs, G, von Haeseler, A, and Pääbo, S (1996). The mitochondrial genome of a monotreme-the platypus (*Ornithorhynchus anatinus*). *Journal of Molecular Evolution*, **42**, 153–159.

Janke, A, Erpenbeck, D, Nilsson, M, and Arnason, U (2001). The mitochondrial genomes of the iguana (*Iguana iguana*) and the caiman (*Caiman crocodylus*): implications for amniote phylogeny. *Proceedings of the Royal Society of London B*, **268**, 623–631.

Janvier, P (1981). The phylogeny of the Craniata with particular reference to the significance of fossil "agnathans". *Journal of Vertebrate Paleontology*, **1**, 121–159.

Jenkins, DL, Ortori, CA, and Brookfield, JF (1995). A test for adaptive change in DNA sequences controlling transcription. *Proceedings of the Royal Society of London B*, **261**, 203–207.

Jin, YG, Wang, Y, Wang, W, Shang, QH, Cao, CQ, and Erwin, DH (2000). Pattern of marine mass extinction near the Permian–Triassic boundary in South China. *Science*, **289**, 432–436.

Johnston, VS and Franklin, M (1993). Is beauty in the eye of the beholder? *Ethology and Sociobiology*, **14**, 183–199.

Johnston, VS, Hagel, R, Franklin, M, Fink, B, and Grammer, K (2001). Male facial attractiveness: evidence for hormone-mediated adaptive design. *Evolution and Human Behavior*, **22**, 251–267.

Jones, D (1996). *Physical attractiveness and the theory of sexual selection: results from five populations*. Museum of Anthropology, University of Michigan, Ann Arbor, MI.

Juan, C, Emerson, BC, Oromi, P, and Hewitt, GM (2000). Colonization and diversification: towards a phylogeographic synthesis for the Canary Islands. *Trends in Ecology and Evolution*, **15**, 104–109.

Kaessmann, H, Wiebe, V, and Pääbo, S (1999). Extensive nuclear DNA sequence diversity among chimpanzees. *Science*, **286**, 1159–1162.

Kaessmann, H, Wiebe, V, Weiss, G, and Pääbo, S (2001). Great ape DNA sequences reveal a reduced diversity and an expansion in humans. *Nature Genetics*, **27**, 155–156.

Kalendar, R, Tanskanen, J, Immonen, S, Nevo, E, and Schulman, AH (2000). Genome evolution of wild barley (*Hordeum spontaneum*) by *BARE*-1 retrotransposon dynamics in response to sharp microclimatic divergence. *Proceedings of the National Academy of Sciences USA*, **97**, 6603–6607.

Kalick, SM, Zebrowitz, LA, Langlois, JH, and Johnson, RM (1998). Does human facial attractiveness honestly advertise health? Longitudinal data on an evolutionary question. *Psychological Science*, **9**, 8–13.

Kaneshiro, KY (1990). Natural hybridization in *Drosophila*, with special reference to species from Hawaii. *Canadian Journal of Zoology*, **68**, 1800–1805.

Kanki, PJ, Travers, KU, M'boup, S, Hsieh, CC, Marlink, RG, Guèye-Ndiaye, A, Siby, T, Thior, I, Hernández-Avila, M, Sankalé, J-L, *et al.* (1994). Slower heterosexual spread of HIV-2 than HIV-1. *Lancet*, **343**, 943–946.

Kanki, PJ, Hamel, DJ, Sankalé, J-L, Hsieh, C-C, Thior, I, Barin F, Woodcock, SA, Guèye-Ndiaye, A, Zhang, E,

Montano, M, *et al.* (1999). Human immunodeficiency virus type 1 subtypes differ in disease progression. *Journal of Infectious Diseases*, **179**, 68–73.

Kaplan, NL, Hudson, RR, and Langley, CH (1989). The "hitchhiking effect" revisited. *Genetics*, **123**, 887–899.

Karlin, S, Campbell, AM, and Mrazek, J (1998). Comparative DNA analysis across diverse genomes. *Annual Review of Genetics*, **32**, 185–225.

Kauffman, SA (1993). *The origins of order*. Oxford University Press, Oxford.

Kawamoto, F, Win, TT, Mizuno, S, Lin, K, Kyaw, O, Tantular, IS, Mason, DP, Kimura, M, and Wongsrichanalai, C (2002). Unusual *Plasmodium* malariae-like parasites in southeast Asia. *Journal of Parasitology*, **88**, 350–357.

Kawashima, T, Amano, N, Koike, H, Makino, S-i., Higuchi, S, Kawashima-Ohya, Y, Watanabe, K, Yamazaki, M, Kanehori, K, Kawamoto, T, *et al.* (2000). Archaeal adaptation to higher temperatures revealed by genomic sequences of *Thermoplasma volcanicum*. *Proceedings of the National Academy of Sciences USA*, **97**, 14257–14262.

Keightley, PD (1996). Nature of deleterious mutation load in *Drosophila*. *Genetics*, **144**, 1993–1999.

Keightley, PD and Bataillon, TM (2000). Multigeneration maximum-likelihood analysis applied to mutation-accumulation experiments in *Caenorhabditis elegans*. *Genetics*, **154**, 1193–1201.

Keightley, PD and Caballero, A (1997). Genomic mutation rate for lifetime reproductive output and lifespan in *Caenorhabditis elegans*. *Proceedings of the National Academy of Sciences USA*, **94**, 3823–3827.

Keightley, PD and Eyre-Walker, A (1999). Terumi Mukai and the riddle of deleterious mutation rates. *Genetics*, **153**, 515–523.

Keightley, PD and Eyre-Walker, A (2000). Deleterious mutations and the evolution of sex. *Science*, **290**, 331–333.

Keightley, PD, Davies, EK, Peters, AD, and Shaw, R (2000). Properties of ethilmethane sulfonate-induced mutations affecting life-history traits in *Caenorhabditis elegans* and inferences about bivariate distributions of mutation effects. *Genetics*, **156**, 143–154.

Kidwell, MG and Lisch, D (1997). Transposable elements as sources of variation in animals and plants. *Proceedings of the National Academy of Sciences USA*, **94**, 7704–7711.

Kidwell, MG and Lisch, D (2001). Perspective: transposable elements, parasitic DNA, and genome evolution. *Evolution*, **55**, 1–24.

Kilias, G, Alahiotis, SN, and Pelecanos, N (1980). A multifactorial genetic investigation of speciation theory using *Drosophila melanogaster*. *Evolution*, **34**, 730–737.

Killian, JK, Buckley, TR, Stewart, N, Munday, BL, and Jirtle, RL (2001). Marsupials and Eutherians reunited: genetic evidence for the Theria hypothesis of mammalian evolution. *Mammalian Genome*, **12**, 513–517.

Kim, J (2001). Macro-evolution of the hairy enhancer in *Drosophila* species. *Journal of Experimental Zoology*, **291**, 175–185.

Kimura, M (1962). On the probability of fixation of mutant genes in a population. *Genetics*, **47**, 713–719.

Kimura, M (1968). Evolutionary rate at the molecular level. *Nature*, **217**, 624–626.

Kimura, M (1983). *The neutral theory of molecular evolution*. Cambridge University Press, Cambridge, UK.

Kimura, M and Ohta, T (1969). The average number of generations until fixation of a mutant gene in a finite population. *Genetics*, **61**, 763–771.

King, CE (1980). The genetic structure of zooplankton populations. In WC Kerfoot, ed. *Evolution and ecology of zooplankton communities*, pp. 315–328. University Press of New England, Hanover (NH).

King, CE and Schonfeld, J (2001). The approach to equilibrium of multilocus genotype diversity under clonal selection and cyclical parthenogenesis. *Hydrobiologia*, **446/447**, 323–331.

King, JL and Jukes, TH (1969). Non-Darwinian evolution: random fixation of selectively neutral mutations. *Science*, **164**, 788–798.

Kirsch, JAW and Mayer, GC (1998). The platypus is not a rodent: DNA hybridization, amniote phylogeny and the palimpsest theory. *Philosophical Transactions of the Royal Society of London B*, **353**, 1221–1237.

Knoll, AH, Bamback, RK, Canfield, DE, and Grotzinger, JP (1996). Comparative earth history and Late Permian mass extinction. *Science*, **273**, 452–457.

Knowles, LL (2001). Did the Pleistocene glaciations promote divergence? Tests of explicit refugial models in montane grasshoppers. *Molecular Ecology*, **10**, 691–701.

Koehn, RK and Hilbish, JJ (1987). The adaptive importance of genetic variation. *American Scientist*, **75**, 134–141.

Komaki, K and Ishikawa, H (1999). Intracellular bacterial symbionts of aphids possess many genomic copies per bacterium. *Journal of Molecular Evolution*, **48**, 717–722.

Kondrashov, A (1988). Deleterious mutations and the evoluation of sexual reproduction. *Nature*, **336**, 435–440.

Kondrashov, A (1995). Contamination of the genome by very slightly deleterious mutations: why have we not died 100 times over? *Journal of Theoretical Biology*, **175**, 583–594.

Kondrashov, AS (1999). Comparative genomics and evolutionary biology. *Current Opinion in Genetics and Development*, **9**, 624–629.

Koonin, EV (2000). How many genes can make a cell: the Minimal-Gene-Set concept. *Annual Review of Genomics and Human Genetics*, **1**, 99–116.

Koonin, EV (2001). Computational genomics. *Current Biology*, **10**, R155–R158.

Korber, B, Muldoon, M, Theiler, J, Gao, F, Gupta, R, Lapedes, A, Hahn, BH, Wolinksy, S, and Bhattacharya, T (2000). Timing the ancestor of the HIV-1 pandemic strains. *Science*, **288**, 1789–1796.

Korswagen, HC, Herman, MA, and Clevers, HC (2000). Distinct $\beta$-catenins mediate adhesion and signalling functions in *C. elegans*. *Nature*, **406**, 527–532.

Kostrikis, LG, Huang, Y, Moore, JP, Wolinsky, SM, Zhang, LQ, Guo, Y, Deutsch, L, Phair, J, Neumann, AU, and Ho, DD (1998). A chemokine receptor CCR2 allele delays HIV-1 disease progression and is associated with a CCR5 promoter mutation. *Nature Medicine*, **4**, 350–353.

Kreitman, M (1983). Nucleotide polymorphism at the *alcohol dehydrogenase* locus of *Drosophila melanogaster*. *Nature*, **304**, 412–417.

Kreitman, M and Hudson, RR (1991). Inferring the evolutionary histories of the *Adh* and *Adh-dup* loci in *Drosophila melanogaster* from patterns of polymorphism and divergence. *Genetics*, **127**, 565–582.

Krings, M, Capelli, C, Tschentscher, F, Geisert, H, Meyer, S, von Haeseler, A, Grossschmidt, K, Possnert, G, Paunovic, M, and Pääbo, S (2000). A view of Neandertal genetic diversity. *Nature Genetics*, **26**, 144–146.

Kroemer, G (1997). Mitochondrial implication in apoptosis: towards an endosymbiont hypothesis of apoptosis evolution. *Cell Death and Differentiation*, **4**, 443–456.

Kruska, D (1987). Mammalian domestication and its effect on brain structure and behavior. In HJ Jerison and I Jerison, eds. *Intelligence and evolutionary biology* (*Vol. G17, Nato ASI Series*), pp. 211–250. Springer-Verlag, Berlin.

Ku, HM, Vision, T, Liu, J, and Tanksley, SD (2000). Comparing sequenced segments of the tomato and *Arabidopsis* genomes: large-scale duplication followed by selective gene loss creates a network of synteny. *Proceedings of the National Academy of Sciences USA*, **7**, 9121–9126.

Kühne, WG (1973). The systematic position of monotremes reconsidered (Mammalia). *Zeitschrift für Morphologie der Tiere*, **75**, 59–64.

Kumar, A and Bennetzen, JL (1999). Plant retrotransposons. *Annual Review of Genetics*, **33**, 479–532.

Kumazawa, Y and Nishida, M (1999). Complete mitochondrial DNA sequences of the green turtle and blue-tailed mole skink: statistical evidence for archosaurian affinity of turtles. *Molecular Biology and Evolution*, **16**, 784–792.

Kuraku, S, Hoshiyama, D, Katoh, K, Suga, H, and Miyata, T (1999). Monophyly of lampreys and hagfishes supported by nuclear DNA-coded genes. *Journal of Molecular Evolution*, **49**, 729–735.

Kusumi, J, Tsumura, Y, Yoshimura, H, and Tachida, H (2002). Molecular evolution of nuclear genes in Cupressacea, a group of conifer. *Molecular Biology and Evolution*, **19**, 736–747.

Labrador, M and Corces, V (1997). Transposable element-host interactions: regulation of insertion and excision. *Annual Review of Genetics*, **31**, 381–404.

Labrador, M and Fontdevila, A (1994). High transposition rates of *Osvaldo*, a new *Drosophila buzzatii* retrotransposon. *Molecular and General Genetics*, **245**, 661–674.

Labrador, M, Farré, M, Utzet, F, and Fontdevila, A (1999). Interspecific hybridization increases transposition rates of *Osvaldo*. *Molecular Biology and Evolution*, **16**, 931–937.

Lachaise, D, Harry, M, Solignac, M, Lemeunier, F, Bénassi, V, and Cariou, ML (2000). Evolutionary novelties in islands: *Drosophila santomea*, a new melanogaster sister species from Sao Tomé. *Proceedings of the Royal Society of London B*, **267**, 1487–1495.

Lampert, W (1987). Predictability in lake ecosystems: the role of biotic interactions. In ED Schulze and H Zwölfer, eds. *Ecological Series 61*, pp. 333–346. Springer-Verlag, Berlin.

Lampert, W (1993). Ultimate causes of diel vertical migration of zooplankton: new evidence for the predator avoidance hypothesis. *Archiv für Hydrobiologie—Advances in Limnology*, **39**, 79–88.

Lampert, W and Brendelberger, H (1996). Strategies of phenotypic low-food adaptation in *Daphnia*: filter screens, mesh sizes, and appendage beat rates. *Limnology and Oceanography*, **41**, 216–223.

Lampert, W and Sommer, U (1997). *Limnoecology*. Oxford University Press, Oxford.

Lampert, W and Trubetskova, I (1996). Juvenile growth rate as a measure of fitness in *Daphnia*. *Functional Ecology*, **10**, 631–635.

Lampert, W, McCauley, E, and Manly, BFJ (2003). Trade-offs in the vertical distribution of zooplankton: ideal free distribution with costs? *Proceedings of the Royal Society of London B*, **270**, 765–773.

Lan, R and Reeves, PR (2000). Intraspecies variation in bacterial genomes: the need for a species genome concept. *Trends in Microbiology*, **8**, 396–401.

Langellotto, GA, Denno, RF, and Ott, JR (2000). A trade-off between flight capability and reproduction in males of a wing-dimorphic insect. *Ecology*, **81**, 865–875.

Langley, CH and Fitch, WM (1974). An examination of the constancy of the rate of molecular evolution. *Journal of Molecular Evolution*, **3**, 161–177.

Langlois, JH and Roggman, LA (1990). Attractive faces are only average. *Psychological Science*, **1**, 115–121.

Laurin, M (1998). The importance of global parsimony and historical bias in understanding tetrapod evolution. Part I. Systematics, middle ear evolution and jaw suspension. *Annales des Sciences Naturelles*, **1**, 1–42.

Laurin, M and Reisz, RR (1995). A reevaluation of early amniote phylogeny. *Zoological Journal of the Linnean Society*, **113**, 165–223.

Laurin, M and Reisz, R (1997). A new perspective on tetrapod phylogeny. In SS Sumida and KL Martin, eds. *Amniote origins*, pp. 9–59. Academic Press, New York.

Law, R (1991). The symbiotic phenotype: origins and evolution. In L Margulis and R Fester, eds. *Symbiosis as a source of evolutionary innovation*, pp. 57–71. MIT Press, Cambridge, MA.

Lee, M-H, Shroff, R, Cooper, SJB, and Hope, R (1999). Evolution and molecular characterization of a $\beta$-globin gene from the Australian echidna *Tachyglossus aculeatus* (Monotremata). *Molecular Phylogenetics and Evolution*, **12**, 205–214.

Lee, MSY (1997). Pareiasaur phylogeny and the origin of turtles. *Zoological Journal of the Linnean Society*, **120**, 197–280.

Lehner, PN (1996). *Handbook of ethological methods*. 2nd Edition. Cambridge University Press, Cambridge.

Lenski, RE and Bennett, AF (1993). Evolutionary response of *Escherichia coli* to thermal stress. *American Naturalist*, **142**, S47–S64.

Lessios, HA (1998). The first stage of speciation as seen in organisms separated by the Isthmus of Panama. In DJ Howard and SJ Berlocher, eds. *Endless forms. Species and speciation*, pp. 186–201. Oxford University Press, Oxford.

Leung, JY, McKenzie, FE, Uglialoro, AM, Flores-Villanueva, PO, Sorkin, BC, Yunis, EJ, Hartl, DL, and Goldfeld, AE (2000). Identification of phylogenetic footprints in primate tumor necrosis factor-alpha promoters. *Proceedings of the National Academy of Sciences USA*, **97**, 6614–6618.

Levin, BR and Bull, JJ (1994). Short-sighted evolution and the virulence of pathogenic micro-organisms. *Trends in Microbiology*, **2**, 76–81.

Levin, BR, Bull, JJ, and Stewart, FM (2001). Epidemiology, evolution, and the future of the HIV/AIDS pandemic. *Emerging Infectious Diseases*, **7**, 505–511.

Levine, N (1988). *The protozoan phylum apicomplexa*. CRC Press, Boca Raton, FL.

Levinson, G and Gutman, GA (1987). Slipped-strand mispairing: a major mechanism for DNA sequence evolution. *Molecular Biology and Evolution*, **4**, 203–221.

Lewontin, RC (1970). The units of selection. *Annual Review of Ecology and Systematics*, **1**, 1–18.

Lewontin, R (1998). The evolution of cognition: questions we will never answer. In D Scarborough and S Sternberg, eds. *Invitation to cognitive science (Vol. 4: Methods, models, and conceptual issues)*.pp. 110–132. MIT Press, Cambridge, MA.

Li, W-H (1985). Accelerated evolution following gene duplication and its implication for the neutralist-selectionist controversy. In T Ohta and K Aoki, eds. *Population genetics and molecular evolution*, pp. 333–352. Springer-Verlag, Berlin.

Li, W-H (1997). *Molecular evolution*. Sinauer Associates, Sunderland, MA.

Li, W-H and Sadler, LA (1991). Low nucleotide diversity in man. *Genetics*, **129**, 513–523.

Ligr, M, Madeo, F, Frohlich, E, Hilt, W, Frohlich, KU, and Wolf, DH (1998). Mammalian Bax triggers apoptotic changes in yeast. *FEBS Letters*, **438**, 61–65.

Lin, Y and Waldman, AS (2001). Capture of DNA sequences at double-strand breaks in mammalian chromosomes. *Genetics*, **158**, 1665–1674.

Lister, JA, Close, J, and Raible, DW (2001). Duplicate *mitf* genes in zebrafish: complementary expression and conservation of melanogenic potential. *Developmental Biology*, **237**, 333–344.

Little, AC, Burt, DM, Penton-Voak, IS, and Perrett, DI (2001). Self-perceived attractiveness influences human female preferences for sexual dimorphism and symmetry in male faces. *Proceedings of the Royal Society of London B*, **268**, 39–44.

Liu, B and Wendel, JF (2000). Retrotransposon activation followed by rapid repression in introgressed rice plants. *Genome*, **43**, 874–880.

Liu, FGR, Miyamoto, MM, Freire, NP, Ong, PQ, Tennant, MR, Young, TS, and Gugel, KF (2001). Molecular and morphological supertrees for eutherian (placental) mammals. *Science*, **291**, 1786–1789.

Liu, T, Wu, J, and He, F (2000). Evolution of *cis*-acting elements in 5′ flanking regions of vertebrate actin genes. *Journal of Molecular Evolution*, **50**, 22–30.

Lively, CM (1999). Migration, virulence, and the geographic mosaic of adaptation by parasites. *American Naturalist*, **153**, S34–S47.

Loeb, LA, Essigmann, JM, Kazazi, F, Zhang, J, Rose, KD, and Mullins, JI (1999). Lethal mutagenesis of HIV with mutagenic nucleoside analogs. *Proceedings of the National Academy of Sciences USA*, **96**, 1492–1497.

Long, M and Langley, CH (1993). Natural selection and the origin of jingwei, a chimeric processed functional gene in *Drosophila*. *Science*, **260**, 91–95.

Loomis, WF and Kuspa, A (1992). Spontaneous generation of enhancers by point mutations. *Trends in Genetics*, **8**, 229.

Loose, CJ (1993). *Daphnia* diel vertical migration behavior: response to vertebrate predator abundance. *Archiv für Hydrobiologie—Advances in Limnology*, **39**, 29–36.

López-Antuñano, F and Schumunis, F (1993). Plasmodia of humans. In J Kreier, ed. *Parasitic protozoa*, pp. 135–265. Academic Press, New York.

López-García, P and Moreira, D (1999). Metabolic symbiosis at the origin of eukaryotes. *Trends in Biochemical Sciences*, **24**, 88–93.

Lorenz, KZ (1981). *The foundations of ethology*. Springer-Verlag, New York.

Ludwig, MZ and Kreitman, M (1995). Evolutionary dynamics of the enhancer region of even-skipped in *Drosophila. Molecular Biology and Evolution*, **12**, 1002–1011.

Ludwig, MZ, Patel, N, and Kreitman, M (1998). Functional analysis of eve stripe 2 enhancer evolution in *Drosophila*: rules governing conservation and change. *Development*, **125**, 949–958.

Ludwig, MZ, Bergman, C, Patel, N, and Kreitman, M (2000). Evidence for stabilizing selection in a eukaryotic *cis*-regulatory element. *Nature*, **403**, 564–567.

Lynch, M (1980). The evolution of cladoceran life histories. *Quarterly Review of Biology*, **55**, 23–42.

Lynch, M (1984a). Destabilizing hybridisation, general-purpose genotypes and geographic parthenogenesis. *Quarterly Review of Biology*, **59**, 257–290.

Lynch, M (1984b). The genetic structure of a cyclical parthenogen. *Evolution*, **38**, 186–203.

Lynch, M (2002). Chromosomal repatterning by gene duplication. *Science*, **297**, 945–947.

Lynch, M and Conery, JS (2000). The evolutionary fate and consequences of duplicate genes. *Science*, **290**, 1151–1154.

Lynch, M and Conery, JS (2001). Response. *Science*, **293**, 1551a.

Lynch, M and Conery, JS (2003). The evolutionary demography of duplicate genes. *Journal of Structural and Functional Genomics*, **3**, 35–44.

Lynch, M and Deng, H (1994). Genetic slippage in response to sex. *American Naturalist*, **144**, 242–261.

Lynch, M and Force, A (2000a). The probability of duplicate-gene preservation by subfunctionalization. *Genetics*, **154**, 459–473.

Lynch, M and Force, A (2000b). Gene duplication and the origin of interspecific genomic incompatibility. *American Naturalist*, **156**, 590–605.

Lynch, M and Gabriel, W (1983). Phenotypic evolution and parthenogenesis. *American Naturalist*, **122**, 745–764.

Lynch, M and Walsh, B (1997). *Genetics and analysis of quantitative traits*. Sinauer, Sunderland, MA.

Lynch, M, Burger, R, Butcher, D, and Gabriel, W (1993). The mutational meltdown in asexual populations. *Journal of Heredity*, **84**, 339–344.

Lynch, M, Conery, J, and Bürger, R (1995). Mutation accumulation and the extinction of small populations. *American Naturalist*, **146**, 489–518.

Lynch, M, Latta, L, Hicks, J, and Giorgianni, M (1998). Mutation, selection and the maintenance of life-history variation in a natural population. *Evolution*, **52**, 727–733.

Lynch, M, Blanchard, J, Houle, D, Kibota, T, Schulz, S, Vassilieva, L, and Willis, J (1999). Perspective: spontaneous deleterious mutation. *Evolution*, **53**, 645–663.

Lynch, M, O'Hely, M, Walsh, B, and Force, A (2001). The probability of fixation of a newly arisen gene duplicate. *Genetics*, **159**, 1789–1804.

MacArthur, RH and Wilson, EO (1967). *The theory of island biogeography*. Princeton University Press, Princeton.

MacIntyre, RJ (1982). Regulatory genes and adaptation: past, present and future. In MK Hecht, B Wallace, and RJ MacIntyre, eds. *Evolutionary biology (Vol. 15)*. pp. 247–285. Plenum Press, New York.

MacKenzie, RB (1868). *The Darwinian theory of the transmutation of species examined* (published anonymously 'By a Graduate of the University of Cambridge'). Nisbet & Co., London.

Madeo, F, Frohlich, E, and Frohlich, KU (1997). A yeast mutant showing diagnostic markers of early and late apoptosis. *Journal of Cell Biology*, **139**, 729–734.

Madeo, F, Frohlich, E, Ligr, M, Grey, M, Sigrist, SJ, Wolf, DH, and Frohlich, KU (1999). Oxygen stress: a regulator of apoptosis in yeast. *Journal of Cell Biology*, **145**, 757–767.

Magro, AM (1999). Evolutionary-derived anatomical characteristics and universal attractiveness. *Perceptual and Motor Skills*, **88**, 147–166.

Maier, D, Preiss, A, and Powell, JR (1990). Regulation of the segmentation gene fushi tarazu has been functionally conserved in *Drosophila. EMBO Journal*, **9**, 3957–3966.

Maisey, DS, Vale, ELE, Cornelissen, PL, and Tovee, MJ (1999). Characteristics of male attractiveness for women. *Lancet*, **353**, 1500.

Mallat, J and Sullivan, J (1998). 28S and 18S rDNA sequences support the monophyly of lampreys and hagfishes. *Molecular Biology and Evolution*, **15**, 1706–1718.

Mallat, J, Sullivan, J, and Winchell, CJ (2001). The relationship of lampreys to hagfishes: a spectral analysis of ribosomal DNA sequences. In PE Ahlberg, ed. *Major events in early vertebrate evolution*, pp. 106–118. Taylor and Francis, London.

Manwell, R (1955). Some evolutionary possibilities in the history of the malaria parasites. *Indian Journal of Malariology*, **9**, 247–253.

Marcus, NH, Lutz, R, Burnett, W, and Cable, P (1994). Age, viability and vertical distribution of zooplankton resting eggs from an anoxic basin: evidence of an egg bank. *Limnology and Oceanography*, **39**, 154–158.

Margulis, L (1970). *Origin of eukaryotic cells*. Yale University Press, New Haven.

Margulis, L (1992). *Symbiosis in cell evolution*. 2nd Edition. W.H. Freeman, San Francisco.

Margulis, L, McKhann, H, and Olendzenski, L (1993). *Illustrated guide of protoctista*. Jones and Bartlett, Boston.

Mark Welch, D and Meselson, M (2000). Evidence for the evolution of Bdelloid rotifers without sexual reproduction or genetic exchange. *Science*, **288**, 1211–1215.

Mark Welch, DB and Meselson, MS (2001). Rates of nucleotide substitution in sexual and ancient asexual rotifers. *Proceedings of the National Academy of Sciences USA*, **98**, 6720–6724.

Marlink, R, Kanki, P, Thior, I, Travers, K, Eisen, G, Siby, T, Traore, I, Hsieh, C-C, Dia, MC, Gueye, E-H, *et al.* (1994). Reduced rate of disease development after HIV-2 infection as compared to HIV-1. *Science*, **265**, 1587–1590.

Marshall, CR and Ward, PD (1996). Sudden and gradual molluscan extinctions in the latest Cretaceous of western European Tethys. *Science*, **274**, 1360–1363.

Marshall, CR, Raff, EC, and Raff, RA (1994). Dollo's law and the death and resurrection of genes. *Proceedings of the National Academy of Sciences USA*, **91**, 12283–12287.

Martin, P (1984). The (four) whys and wherefores of play in cats: a review of functional, evolutionary, developmental, and causal issues. In PK Smith, ed. *Play in animals and humans*, pp. 71–94. Basil Blackwell, Oxford.

Martin, P and Caro, TM (1985). On the function of play and its role in behavioral development. *Advances in the Study of Behavior*, **15**, 59–103.

Martin, W and Müller, M (1998). The hydrogen hypothesis for the first eukaryote. *Nature*, **392**, 37–41.

Martin, W, Stoebe, B, Goremykin, V, Hansmann, S, Hasegawa, M, and Kowallik, KV (1998). Gene transfer to the nucleus and the evolution of chloroplasts. *Nature*, **393**, 162–165.

Martínez-Arias, R, Calafell, F, Mateu, E, Comas, D, Andrés, A, and Bertranpetit, J (2001). Sequence variability of a human pseudogene. *Genome Research*, **11**, 1071–1085.

Maside, X, Assimacopoulos, S, and Charlesworth, B (2000). Rates of movement of transposable elements on the second chromosome of *Drosophila melanogaster*. *Genetical Research*, **75**, 275–284.

Mason, SJ, Miller, LH, Shiroishi, T, Dvorak, JA, and McGinniss, MH (1977). The Duffy blood group determinants: their role in the susceptibility of human and animal erythrocytes to *Plasmodium knowlesi* malaria. *British Journal of Haematology*, **36**, 327–335.

Mateu, E, Calafell, F, Lao, O, Bonné-Tamir, B, Kidd, JR, Pakstis, A, Kidd, KK, and Bertranpetit, J (2001). Worldwide genetic analysis of the CFTR region. *American Journal of Human Genetics*, **68**, 103–117.

Mather, JA and Anderson, RC (1999). Exploration, play, and habituation in octopuses (*Octopus dofleini*). *Journal of Comparative Psychology*, **113**, 333–338.

Maynard Smith, J (1998). *Evolutionary genetics*. Oxford University Press, Oxford.

Maynard Smith, J and Szathmáry, E (1995). *The major transitions in evolution*. W.H. Freeman, San Francisco.

Mayr, E (1940). Speciation phenomena in birds. *American Naturalist*, **74**, 249–278.

Mayr, E (1942). *Systematics and the origin of species*. Columbia University Press, New York.

Mayr, E (1963). *Animal species and evolution*. Harvard University Press, Cambridge, MA.

McClintock, B (1980). Modified gene expressions induced by transposable elements. In WA Scott, R Werner, DR Joseph, and J Schultz, eds. *Proceedings of the Miami Winter Symposium (Vol. 17: Mobilization and reassembly of genetic information)*, pp. 11–19. Academic Press, New York.

McCutchan, TF and Waters, AP (1990). Mutations with multiple independent origins in surface antigens mark the targets of biological selective pressure. *Immunology Letters*, **25**, 23–26.

McDonald, JF (1995). Transposable elements: possible catalysts of organismic evolution. *Trends in Ecology and Evolution*, **10**, 123–126.

McDonald, JH and Kreitman, M (1991). Adaptive protein evolution at the *Adh* locus in *Drosophila*. *Nature*, **351**, 652–654.

McFall-Ngai, MJ (1999). Consequences of evolving with bacterial symbionts: lessons from the squid-*Vibrio* associations. *Annual Review of Ecology and Systematics*, **30**, 235–256.

McGhee, GR (2001). Late Devonian extinction. In DEG. Briggs and PR Crowther, eds. *Palaeobiology II*, pp. 223–236. Blackwell, Oxford.

McLysaght, A, Hokamp, K, and Wolfe, KH (2002). Extensive genomic duplication during early chordate evolution. *Nature Genetics*, **31**, 200–204.

McNab, BK (1988). Complications inherent in scaling the basal rate of metabolism in mammals. *Quarterly Review of Biology*, **63**, 25–54.

Meffert, LM, Regan, JL, and Brown, BW (1999). Convergent evolution of the mating behaviour of founder flush populations of the housefly. *Journal of Evolutionary Biology*, **12**, 859–868.

Merck, J (1997). A phylogenetic analysis of the Euryapsid reptiles. *Journal of Vertebrate Paleontology Supplement*, **17**, 65A.

Merriman, JL and Kirk, KL (2000). Temporal patterns of resource limitation in natural populations of rotifers. *Ecology*, **81**, 141–149.

Merritt, TJ and Quattro, JM (2001). Evidence for a period of directional selection following gene duplication in a neurally expressed locus of triosephosphate isomerase. *Genetics*, **159**, 689–697.

Meyer, A and Wilson, AC (1990). Origin of tetrapods inferred from their mitochondrial DNA affiliation to lungfish. *Journal of Molecular Evolution*, **31**, 359–364.

Michod, RE (1979). Genetical aspects of kin selection: effects of inbreeding. *Journal of Theoretical Biology*, **81**, 223–233.

Michod, RE (1982). The theory of kin selection. *Annual Review of Ecology and Systematics*, **13**, 23–55.

Michod, RE (1983). Population biology of the first replicators: on the origin of the genotype, phenotype and organism. *American Zoologist*, **23**, 5–14.

Michod, RE (1991). Inbreeding and the evolution of social behavior. In N Wilmsen Thornhill and WM Shields, eds. *The natural history of inbreeding and outbreeding: theoretical and empirical perspectives*, pp. 74–96. University of Chicago Press, Chicago.

Michod, RE (1996). Cooperation and conflict in the evolution of individuality. II. Conflict mediation. *Proceedings of the Royal Society of London B*, **263**, 813–822.

Michod, RE (1997). Cooperation and conflict in the evolution of individuality. I. Multi-level selection of the organism. *American Naturalist*, **149**, 607–645.

Michod, RE (1999). *Darwinian dynamics, evolutionary transitions in fitness and individuality*. Princeton University Press, Princeton, NJ.

Michod, RE and Roze, D (1997). Transitions in individuality. *Proceedings of the Royal Society of London B*, **264**, 853–857.

Michod, RE and Roze, D (1999). Cooperation and conflict in the evolution of individuality. III. Transitions in the unit of fitness. In CL Nehaniv, ed. *Mathematical and computational biology: computational morphogenesis, hierarchical complexity, and digital evolution (Vol. 26)*, pp. 47–92. American Mathematical Society, Providence, Rhode Island.

Michod, RE and Roze, D (2000). Some aspects of reproductive mode and the origin of multicellularity. *Selection*, **1**, 97–109.

Michod, RE and Roze, D (2001). Cooperation and conflict in the evolution of multicellularity. *Heredity*, **81**, 1–7.

Michod, RE and Sanderson, MJ (1985). Behavioural structure and the evolution of social behaviour. In JJ Greenwood and M Slatkin, eds. *Evolution—Essays in honour of John Maynard Smith*, pp. 95–104. Cambridge University Press, Cambridge,

Michod, RE, Nedelcu, AM, and Roze, D (2003). Cooperation and conflict in the evolution of individuality IV. Conflict mediation and evolvability in *Volvox carteri*. *BioSystems*, **69**, 95–114.

Mila, B, Girman, DJ, Kimura, M, and Smith, TB (2000). Genetic evidence for the effect of a postglacial population expansion on the phylogeography of a North American songbird. *Proceedings of the Royal Society of London B*, **267**, 1033–1040.

Millen, RS, Olmstead, RG, Adams, KL, Palmer, JD, Lao, NT, Heggie, L, Kavanagh, TA, Hibberd, JM, Gray, JC, Morden, CW, *et al.* (2001). Many parallel losses of *infA* from chloroplast DNA during angiosperm evolution with multiple independent transfers to the nucleus. *Plant Cell*, **13**, 645–658.

Miller, LH, Mason, SJ, Clyde, DF, and McGinniss, MH (1976). The resistance factor to *Plasmodium vivax* in blacks. The Duffy-blood-group genotype, FyFy. *New England Journal of Medicine*, **295**, 302–304.

Miller, LH, Roberts, T, Shahabuddin, M, and McCutchan, TF (1993). Analysis of sequence diversity in the *Plasmodium falciparum* merozoite surface protein-1 (MSP-1). *Molecular and Biochemical Parasitology*, **59**, 1–14.

Miller, RL, Ikram, S, Armelagos, GJ, Walker, R, Harer, WB, Shiff, CJ, Baggett, D, Carrigan, M, and Maret, SM (1994). Diagnosis of *Plasmodium falciparum* infections in mummies using the rapid manual ParaSight-F test. *Transactions of the Royal Society of Tropical Medicine and Hygiene*, **88**, 31–32.

Miller, W (2001). Comparison of genomic DNA sequences: solved and unsolved problems. *Bioinformatics*, **17**, 391–397.

Mitchell, SE and Lampert, W (2000). Temperature adaptation in a geographically widespread zooplankter, *Daphnia magna*. *Journal of Evolutionary Biology*, **13**, 371–382.

Mitsialis, SA and Kafatos, FC (1985). Regulatory elements controlling chorion gene expression are conserved between flies and moths. *Nature*, **317**, 453–456.

Mooers, AØ, Rundle, HD, and Whitlock, MC (1999). The effects of selection and bottlenecks on male mating success in peripheral isolates. *American Naturalist*, **153**, 437–444.

Moore, MV, Holt, CL, and Stemberger, RS (1996). Consequences of elevated temperatures for zooplankton assemblages in temperate lakes. *Archiv für Hydrobiologie*, **135**, 289–319.

Moran, JV, DeBerardinis, RJ, and Kazazian, HH (1999). Exon shuffling by L1 retrotransposition. *Science*, **283**, 1530–1534.

Moran, NA (1992). The evolution of aphid life cycles. *Annual Review of Entomology*, **37**, 321–348.

Moran, NA (1996). Accelerated evolution and Muller's ratchet in endosymbiotic bacteria. *Proceedings of the National Academy of Sciences USA*, **93**, 2873–2878.

Moran, NA and Baumann, P (1994). Phylogenetics of cytoplasmically inherited microorganisms of arthropods. *Trends in Ecology and Evolution*, **9**, 15–20.

Moran, NA and Wernegreen, JJ (2000). Lifestyle evolution in symbiotic bacteria: insights from genomics. *Trends in Ecology and Evolution*, **15**, 321–326.

Moreira, D and López-García, P (1998). Symbiosis between methanogenic archaea and delta-proteobacteria as the origin of eukaryotes: the syntrophic hypothesis. *Journal of Molecular Evolution*, **47**, 517–530.

Moreira, MEC, DelPortillo, HA, Milder, RV, Balanco, JMF, and Barcinski, MA (1996). Heat shock induction of apoptosis in promastigotes of the unicellular organism *Leishmania*. *Journal of Cellular Physiology*, **167**, 305–313.

Morgan, KK, Hicks, J, Spitze, K, Latta, L, Pfrender, ME, Weaver, CS, Ottone, M, and Lynch, M (2001). Patterns of genetic architecture for life-history traits and molecular markers in a subdivided species. *Evolution*, **55**, 1753–1761.

Morrison, DA and Ellis, JT (1997). Effects of nucleotide sequence alignment on phylogeny estimation: a case study of 18S rDNAs of apicomplexa. *Molecular Biology and Evolution*, **14**, 428–441.

Moses, K, Heberlein, U, and Ashburner, M (1990). The Adh gene promoters of *Drosophila melanogaster* and *Drosophila orena* are functionally conserved and share features of sequence structure and nuclease-protected sites. *Molecular and Cellular Biology*, **10**, 539–548.

Muenchow, G (1978). A note on the timing of sex in asexual/sexual organisms. *American Naturalist*, **112**, 774–779.

Muir, G, Fleming, CC, and Schlötterer, C (2000). Species status of hybridizing oaks. *Nature*, **405**, 1016.

Mukai, T (1964). The genetic structure of natural populations of *Drosophila melanogaster*. I. Spontaneous mutation rate of polygenes controlling viability. *Genetics*, **50**, 1–19.

Mukai, T (1969). The genetic structure of natural populations of *Drosophila melanogaster*. VII. Synergistic interaction of spontaneous mutant polygenes controlling viability. *Genetics*, **61**, 749–761.

Mukai, T and Yamazaki, T (1968). The genetic structure of natural populations of *Drosophila melanogaster*. V. Coupling-repulsion effects of spontaneous mutant polygenes controlling viability. *Genetics*, **59**, 513–535.

Mukai, T, Chigusa, SI, Mettler, LE, and Crow, JF (1972). Mutation rate and dominance of genes affecting viability in *Drosophila melanogaster*. *Genetics*, **72**, 333–355.

Muller, HJ (1950). Our load of mutations. *American Journal of Human Genetics*, **2**, 111–176.

Murray, AW (2000). Whither genomics? *Genome Biology*, **1**, comment003.1–003.6.

Muse, SV and Gaut, BS (1994). A likelihood approach for comparing synonymous and nonsynonymous nucleotide substitution rates, with application to chloroplast genome. *Molecular Biology and Evolution*, **11**, 715–724.

Nachman, MW (2001). Single nucleotide polymorphisms and recombination rate in humans. *Trends in Genetics*, **17**, 481–485.

Nachman, MW, Boyer, SN, and Aquadro, CF (1994). Nonneutral evolution at the mitochondrial NADH dehydrogenase subunit 3 gene in mice. *Proceedings of the National Academy of Sciences USA*, **91**, 6364–6368.

Nadeau, JH and Sankoff, D (1998). Counting on comparative maps. *Trends in Genetics*, **14**, 495–501.

Nager, RG, Keller, LF, and Van Noordwijk, AJ (2000). Understanding natural selection on traits that are influenced by environmental conditions. In TA Mousseau, B Sinervo, and J Endler, eds. *Adaptive genetic variation in the wild*, pp. 95–115. Oxford University Press, New York.

Naveira, H and Fontdevila, A (1985). The evolutionary history of *Drosophila buzzatii*. IX. High frequencies of new chromosome rearrangements induced by introgressive hybridization. *Chromosome*, **91**, 87–94.

Nei, M (1973). Analysis of gene diversity in subdivided populations. *Proceedings of the National Academy of Sciences USA*, **70**, 3321–3323.

Nei, M (1987). *Molecular evolutionary genetics*. Columbia University Press, New York.

Nei, M, Gu, X, and Sitnikova, T (1997). Evolution by the birth-and-death process in multigene families of the vertebrate immune system. *Proceedings of the National Academy of Sciences USA*, **94**, 7799–7806.

Nei, M, Rogozin, IB, and Piontkivska, H (2000). Purifying selection and birth-and-death evolution in the ubiquitin gene family. *Proceedings of the National Academy of Sciences USA*, **97**, 10866–10871.

Nersting, LG and Arctander, P (2001). Phylogeography and conservation of impala and greater kudu. *Molecular Ecology*, **10**, 711–719.

Nichols, RA and Hewitt, GM (1994). The genetic consequences of long distance dispersal during colonization. *Heredity*, **72**, 312–317.

Nogrady, T, Wallace, RL, and Snell, TW (1993). *Guides to the identification of the microinvertebrates of the continental waters of the world (Rotifera, Vol. 1)*. SPB Academic Publishing, Dordrecht.

Normark, BB (1999). Evolution in a putatively ancient asexual aphid lineage: recombination and rapid karyotype change. *Evolution*, **53**, 1458–1469.

Normark, BB and Moran, NA (2000). Testing for the accumulation of deleterious mutations in asexual genomes using molecular sequences. *Journal of Natural History*, **34**, 1719–1729.

Nornes, S, Clarkson, M, Mikkola, I, Pedersen, M, Bardsley, A, Martinez, JP, Krauss, S, and Johansen, T (1998). Zebrafish contains two *Pax6* genes involved in eye development. *Mechanisms of Development*, **77**, 185–196.

Nowak, MA and May, RM (1994). Superinfection and the evolution of virulence. *Proceedings of the Royal Society of London B*, **255**, 81–89.

Nowak, MA, Boerlijst, MC, Cooke, J, and Maynard Smith, J (1997). Evolution of genetic redundancy. *Nature*, **388**, 167–170.

Nurminsky, D, De Aguiar, D, Bustamante, C, and Hartl, DL (2001). Chromosomal effects of rapid gene evolution in *Drosophila melanogaster*. *Science*, **291**, 128–130.

Ochman, H and Moran, NA (2001). Genes lost and genes found: evolution of bacterial pathogenesis and symbiosis. *Science*, **292**, 1096–1099.

Ochman, H, Lawrence, JG, and Groisman, EA (2000). Lateral gene transfer and the nature of bacterial innovation. *Nature*, **405**, 299–304.

Ohnishi, O (1977a). Spontaneous and ethyl methanesulfonate induced mutations controlling viability in *Drosophila melanogaster*. I. Recessive lethal mutations. *Genetics*, **87**, 519–527.

Ohnishi, O (1977b). Spontaneous and ethyl methanesulfonate-induced mutations controlling viability in *Drosophila melanogaster*. II. Homozygous effects of polygenic mutations. *Genetics*, **87**, 529–545.

Ohnishi, O (1977c). Spontaneous and ethyl methane sulfonate-induced mutations controlling viability in *Drosophila melanogaster*. III. Heterozygous effect of polygenic mutations. *Genetics*, **87**, 547–556.

Ohno, S (1970). *Evolution by gene duplication*. Springer-Verlag, Berlin.

Ohta, T (1972). Population size and rate of evolution. *Journal of Molecular Evolution*, **1**, 305–314.

Ohta, T (1973). Slightly deleterious mutant substitutions in evolution. *Nature*, **246**, 96–98.

Ohta, T (1975). Statistical analyses of *Drosophila* and human protein polymorphisms. *Proceedings of the National Academy of Sciences USA*, **72**, 3194–3196.

Ohta, T (1992). The nearly neutral theory of molecular evolution. *Annual Review of Ecology and Systematics*, **23**, 263–286.

Ohta, T (1995). Synonymous and nonsynonymous substitutions in mammalian genes and the nearly neutral theory. *Journal of Molecular Evolution*, **40**, 56–63.

Ohta, T (1997). Role of random drift in the evolution of interactive systems. *Journal of Molecular Evolution*, **44**, S9–S14.

Ohta, T and Kimura, M (1971). On the constancy of the evolutionary rate of cistrons. *Journal of Molecular Evolution*, **1**, 18–25.

Ohta, T and Tachida, H (1990). Theoretical study of near neutrality. I. Heterozygosity and rate of mutant substitution. *Genetics*, **126**, 219–229.

O'Neill, CM and Bancroft, I (2000). Comparative physical mapping of segments of the genome of *Brassica oleracea* var. *alboglabra* that are homoeologous to sequenced regions of chromosomes 4 and 5 of *Arabidopsis thaliana*. *Plant Journal*, **23**, 233–243.

O'Neill, RJW, O'Neill MJ, and Graves JAM (1998). Undermethylation associated with retroelement activation and chromosome remodelling in an interspecific mammalian hybrid. *Nature*, **393**, 68–72.

O'Neill, SL, Hoffmann, AA, and Werren, JH (eds). (1997). *Influential passengers: inherited microorganisms and arthropod reproduction*. Oxford University Press, Oxford.

O'Riain, JM, Jarvis, JUM, and Faulkes, CG (1996). A dispersive morph in the naked mole-rat. *Nature*, **380**, 619–621.

Orr, HA (1996). Dobzhansky, Bateson, and the genetics of speciation. *Genetics*, **144**, 1331–1335.

Orr, HA (1997). Haldane's rule. *Annual Review of Ecology and Systematics*, **28**, 195–218.

Orr, MR and Smith, TB (1998). Ecology and speciation. *Trends in Ecology and Evolution*, **13**, 502–506.

Ortega, JC and Bekoff, M (1987). Avian play: comparative evolutionary and developmental trends. *Auk*, **104**, 338–341.

Panksepp, J (1998). *Affective neuroscience*. Oxford University Press, New York.

Pantazidis, A, Labrador, M, and Fontdevila, A (1999). The retrotransposon *Osvaldo* from *Drosophila buzzatii* displays all structural features of a functional retrovirus. *Molecular Biology and Evolution*, **16**, 909–921.

Paracer, S and Ahmadjian, V (2000). *Symbiosis: an introduction to biological associations*. Oxford University Press, Oxford.

Parciak, W (2002). Environmental variation in seed number, size, and dispersal of a fleshy-fruited plant. *Ecology*, **83**, 780–793.

Parker, HR, Philipp, DP, and Whitt, GS (1985). Relative developmental success of interspecific *Lepomis* hybrids as an estimate of gene regulatory divergence between species. *Journal of Experimental Zoology*, **233**, 451–466.

Parker, ST and McKinney, ML (1999). *Origins of intelligence: the evolution of cognitive development in monkeys, apes, and humans*. Johns Hopkins University Press, Baltimore.

Patthy, L (1999). *Protein evolution*. Blackwell Science, Oxford, UK.

Peeters, M, Vincent, R, Perret, JL, Lasky, M, Patrel, D, Liegeois, F, Courgnaud, V, Seng, R, Matton, T, Molinier, S, *et al.* (1999). Evidence for differences in MT2 cell tropism according to genetic subtypes of HIV-1: syncytium-inducing variants seem rare among subtype C HIV-1 viruses. *Journal of Acquired Immune Deficiency Syndromes and Human Retrovirology*, **20**, 115–121.

Pellis, SM (1993). Sex and the evolution of play fighting: a review and model based on the behavior of muroid rodents. *Play Theory and Research*, **1**, 55–75.

Pellis, SM and Iwaniuk, AN (1999a). The problem of adult play fighting: a comparative analysis of play and courtship in primates. *Ethology*, **105**, 783–806.

Pellis, SM and Iwaniuk, AN (1999b). The roles of phylogeny and sociality in the evolution of social play in muroid rodents. *Animal Behaviour*, **58**, 361–373.

Pellis, SM and Iwaniuk, AN (2000). Adult-adult play in primates: comparative analyses of its origin, distribution, and evolution. *Ethology*, **106**, 1083–1104.

Pellis, SM and Pellis, VC (1998). Play fighting of rats in comparative perspective: a schema for neurobehavioral analysis. *Neuroscience and Biobehavioral Reviews*, **23**, 87–101.

Penton-Voak, IS and Perrett, DI (2001). Male facial attractiveness: perceived personality and shifting female preferences for male traits across the menstrual cycle. *Advances in the Study of Behavior*, **30**, 219–259.

Perelson, AS, Neumann, AU, Markowitz, M, Leonard, JM, and Ho, DD (1996). HIV-1 dynamics in vivo: virion clearance rate, infected cell life-span, and viral generation time. *Science*, **271**, 1582–1586.

Pérez-Lezaun, A, Calafell, F, Mateu, E, Comas, D, Ruiz-Pacheco, R, and Bertranpetit, J (1997). Microsatellite variation and the differentiation of modern humans. *Human Genetics*, **99**, 1–7.

Perrett, DI, May, KA, and Yoshikawa, S (1994). Facial shape and judgements of female attractiveness. *Nature*, **368**, 239–242.

Perrett, DI, Lee, KJ, Penton-Voak, I, Rowland, D, Yoshikawa, S, Burt, DM, Henzi, SP, Castles, DL, and Akamatsu, S (1998). Effects of sexual dimorphism on facial attractiveness. *Nature*, **394**, 884–887.

Perrin, N and Goudet, J (2001). Inbreeding, kinship, and the evolution of dispersal. In J Clobert, E Danchin, AA Dhondt, and JD Nichols, eds. *Dispersal*, pp. 123–142. Oxford University Press, New York.

Peterson, MA and Denno, RF (1997). The influence of intraspecific variation in dispersal strategies on the genetic structure of planthopper populations. *Evolution*, **5**, 1189–1206.

Peterson, MA and Denno, RF (1998). The influence of dispersal and diet breadth on patterns of genetic isolation by distance in phytophagous insects. *American Naturalist*, **152**, 428–446.

Pfrender, ME and Lynch, M (2000). Quantitative genetic variation in *Daphnia*: temporal changes in genetic architecture. *Evolution*, **54**, 1502–1509.

Pianka, ER (2000). *Evolutionary ecology*, 6th Edition. Benjamin-Cummings, Addison-Wesley-Longman, San Francisco.

Piatigorsky, J and Wistow, G (1991). The recruitment of crystallins: new functions precede gene duplication. *Science*, **252**, 1078–1079.

Pinker, S (1997). *How the mind works*. Norton, New York.

Piot, P, Bartos, M, Ghys, PD, Walker, N, and Schwartländer, B (2001). The global impact of HIV/AIDS. *Nature*, **410**, 968–973.

Platz, JE and Conlon, JM (1997). and turn back. *Nature*, **389**, 246.

Pollock, DD, Eisen, JA, Doggett, NA, and Cummings, MP (2000). A case for evolutionary genomics and the comprehensive examination of sequence biodiversity. *Molecular Biology and Evolution*, **17**, 1776–1788.

Poon, A and Otto, SP (2000). Compensating for our load of mutations: freezing the meltdown of small populations. *Evolution*, **54**, 1467–1479.

Pope, KO, D'Hondt, SL, and Marshall, CR (1998). Meteorite impact and the mass extinction of species at the Cretaceous/Tertiary boundary. *Proceedings of the National Academy of Sciences USA*, **95**, 11028–11029.

Postlethwait, JH, Woods, IG, Ngo-Hazelett, P, Yan, YL, Kelly, PD, Chu, F, Huang, H, Hill-Force, A, and Talbot, WS (2000). Zebrafish comparative genomics and the origins of vertebrate chromosomes. *Genome Research*, **10**, 1890–1902.

Poundstone, W (1985). *The recursive universe: cosmic complexity and the limits of scientific knowledge*. Morrow, New York.

Pourriot, R and Snell, TW (1983). Resting eggs of rotifers. *Hydrobiologia*, **104**, 213–224.

Powell, JR (1983). Interspecific cytoplasmic gene flow in the absence of nuclear gene flow: evidence from *Drosophila*. *Proceedings of the National Academy of Sciences USA*, **80**, 492–495.

Power, TG (2000). *Play and exploration in children and animals*. Lawrence Erlbaum Associates, Mahwah, NJ.

Preston, BL, Snell, TW, and Dusenbery, DB (1999). The effects of sublethal pentachlorophenol exposure on predation risk in freshwater rotifer species. *Aquatic Toxicology*, **47**, 93–105.

Preston, BL, Snell, TW, Robinson, TL, and Dingmann, BJ (2000). Use of the freshwater rotifer *Brachionus calyciflorus* in a screening assay for potential endocrine disruptors. *Environmental Toxicology and Chemistry*, **19**, 2923–2928.

Price, EO (1984). Behavioral aspects of domestication. *Quarterly Review of Biology*, **59**, 1–32.

Przeworski, M, Hudson, RR, and DiRienzo, A (2000). Adjusting the focus on human variation. *Trends in Genetics*, **16**, 296–302.

Przeworski, M, Wall, J, and Andolfatto, P (2001). Recombination and the frequency spectrum in *Drosophila melanogaster* and *Drosophila simulans*. *Molecular Biology and Evolution*, **18**, 291–298.

Pybus, OG, Rambaut, A, and Harvey, PH (2000). An integrated framework for the inference of viral population history from reconstructed genealogies. *Genetics*, **155**, 1429–1437.

Queller, DC (1994). Genetic relatedness in viscous populations. *Evolutionary Ecology*, **8**, 70–73.

Queller, DC (1997). Cooperators since life began. *Quarterly Review of Biology*, **72**, 184–188.

Quint, E, Zerucha, T, and Ekker, M (2000). Differential expression of orthologous *Dlx* genes in zebrafish and mice: implications for the evolution of the *Dlx* homeobox gene family. *Journal of Experimental Zoology*, **288**, 235–241.

Quiñones-Mateu, ME, Ball, SC, Marozsan, AJ, Torre, VS, Albright, JL, Vanham, G, van der Groen, G, Colebunders, RL, and Arts, EJ (2000). A dual infection/competition assay shows a correlation between *ex vivo* human immunodeficiency type 1 fitness and disease progression. *Journal of Virology*, **74**, 9222–9233.

Quiros, CF, Grellet, F, Sadowski, J, Suzuki, T, Li, G, and Wroblewski, T (2001). *Arabidopsis* and *Brassica* comparative genomics: sequence, structure and gene content in the *ABI-Rps2-Ck1* chromosomal segment and related regions. *Genetics*, **157**, 1321–1330.

Rambaut, A, Robertson, DL, Pybus, OG, Peeters, M, and Holmes, EC (2001). Phylogeny and the origin of HIV-1. *Nature*, **410**, 1047–1048.

Ramos-Onsins, S and Aguadé, M (1998). Molecular evolution of the *Cecropin* multigene family in *Drosophila*: functional genes *vs* pseudogenes. *Genetics*, **150**, 157–171.

Rand, DM and Kann, LM (1996). Excess amino acid polymorphism in mitochondrial DNA: contrasts among genes from *Drosophila*, mice, and humans. *Molecular Biology and Evolution*, **13**, 735–748.

Rasmussen, AS, Janke, A, and Arnason, A (1998). The mitochondrial DNA molecule of the hagfish (*Myxine glutinosa*) and vertebrate phylogeny. *Journal of Molecular Evolution*, **46**, 382–388.

Raup, DM (1991). *Extinction: bad genes or bad luck*? Norton, New York.

Razin, A (1998). CpG methylation, chromatin structure and gene silencing-a three-way connection. *EMBO Journal*, **17**, 4905–4908.

Rhodes, G and Tremewan, T (1996). Averageness, exaggeration, and facial attractiveness. *Psychological Science*, **7**, 105–110.

Rhodes, G, Proffitt, F, Grady, JM, and Sumich, A (1998). Facial symmetry and the perception of beauty. *Psychonomic Bulletin & Review*, **5**, 659–669.

Rhodes, G, Sumich, A, and Byatt, G (1999). Are average facial configurations only attractive because of their symmetry? *Psychological Science*, **10**, 52–58.

Rhodes, G, Zebrowitz, LA, Clark, A, Kalick, SM, Hightower, A, and McKay, R (2001). Do facial averageness and symmetry signal health? *Evolution and Human Behavior*, **22**, 31–46.

Rhomberg, LR, Joseph, S, and Singh, RS (1985). Seasonal variation and clonal selection in cyclically parthenogenetic rose aphids (*Macrosiphum rosae*). *Canadian Journal of Genetics and Cytology*, **27**, 224–232.

Ricchetti, M, Fairhead, C, and Dujon, B (1999). Mitochondrial DNA repairs double-strand breaks in yeast chromosomes. *Nature*, **402**, 96–100.

Rice, WR and Hostert, EE (1993). Laboratory experiments on speciation: what have we learned in 40 years? *Evolution*, **47**, 1637–1653.

Rich, SM and Ayala, FJ (1998). The recent origin of allelic variation in antigenic determinants of *Plasmodium falciparum*. *Genetics*, **150**, 515–517.

Rich, SM and Ayala, FJ (1999). Reply to Saul. *Parasitology Today*, **15**, 39–40.

Rich, SM and Ayala, FJ (2000). Population structure and recent evolution of *Plasmodium falciparum*. *Proceedings of the National Academy of Sciences USA*, **97**, 6994–7001.

Rich, SM, Hudson, RR, and Ayala, FJ (1997). *Plasmodium falciparum* antigenic diversity: evidence of clonal population structure. *Proceedings of the National Academy of Sciences USA*, **94**, 13040–13045.

Rich, SM, Licht, MC, Hudson, RR, and Ayala, FJ (1998). Malaria's eve: evidence of a recent population bottleneck throughout the world populations of *Plasmodium falciparum*. *Proceedings of the National Academy of Sciences USA*, **95**, 4425–4430.

Rich, SM, Ferreira, MU, and Ayala, FJ (2000). The origin of antigenic diversity in *Plasmodium falciparum*. *Parasitology Today*, **16**, 390–396.

Ricklefs, RE and Fallon, SM (2002). Diversification and host switching in avian malaria parasites. *Proceedings of the Royal Society of London B*, **269**, 885–892.

Ridley, M (1996). *Evolution*, 2nd Edition. Blackwell Science, Cambridge.

Rieseberg, LH (2001). Chromosomal rearrangements and speciation. *Trends in Ecology and Evolution*, **16**, 351–358.

Rieseberg, LH and Noyes, RD (1998). Genetic map-based studies of reticulate evolution in plants. *Trends in Plant Science*, **3**, 254–259.

Rieseberg, LH, Sinervo, B, Linder, CR, Ungerer, MC, and Arias, DM (1996). Role of gene interactions in hybrid speciation: evidence from ancient and experimental hybrids. *Science*, **272**, 741–745.

Rincón-Limas, DE, Lu, CH, Canal, I, Calleja, M, Rodríguez-Esteban, C, Izpisúa-Belmonte, JC, and Botas, J (1999). Conservation of the expression and function of apterous orthologs in *Drosophila* and mammals. *Proceedings of the National Academy of Sciences USA*, **96**, 2165–2170.

Rispe, C, Pierre, JS, Simon, J-C, and Gouyon, PH (1998). Models of sexual and asexual coexistence in aphids based on constraints. *Journal of Evolutionary Biology*, **11**, 685–701.

Rispe, C, Bonhomme, J, and Simon, J-C (1999). Extreme life-cycle and sex ratio variation among sexually produced clones of the aphid *Rhopalosiphum padi* L. (Homoptera: Aphididae). *Oikos*, **86**, 254–264.

Roberts, RG, Flannery, TF, Ayliffe, LK, Yoshida, H, Olley, JM, Prideaux, GJ, Laslett, GM, Baynes, A, Smith, MA, Jones, R *et al.* (2001). New ages for the last Australian megafauna: continent-wide extinction about 46,000 years ago. *Science*, **292**, 1888–1892.

Roff, DA (1990). The evolution of flightlessness in insects. *Ecological Monographs*, **60**, 389–421.

Rogers, AR and Harpending H (1992). Population growth makes waves in the distribution of pairwise genetic differences. *Molecular Biology and Evolution*, **9**, 552–569.

Ronce, O, Olivieri, I, Clobert, J, and Danchin, E (2001). Perspectives on the study of dispersal evolution. In J Clobert, E Danchin, AA Dhondt, and JD Nichols, eds. *Dispersal*, pp. 341–357. Oxford University Press, New York.

Rosenberg, A (1990). Is there an evolutionary biology of play? In M Bekoff and D Jamieson, eds. *Interpretation and explanation in the study of animal behavior* (*Vol. 1: Interpretation, intentionality, and communication*), pp. 180–197. Westview Press, Boulder.

Rousset, F (2001). Genetic approaches to the estimation of dispersal rates. In J Clobert, E Danchin, AA Dhondt, and JD Nichols, eds. *Dispersal*, pp. 18–28. Oxford University Press, New York.

Rowland-Jones, SL (1998). Survival with HIV infection: good luck or good breeding? *Trends in Genetics*, **14**, 343–345.

Rowland-Jones, SL, Sutton, J, Ariyoshi, K, Dong, T, Gotch, F, McAdam, S, Whitby, D, Sabally, S, Gallimore, A, Corrah, T, *et al.* (1995). HIV-specific cytotoxic T-cells in HIV-exposed but uninfected Gambian women. *Nature Medicine*, **1**, 59–64.

Rozas, J and Rozas, R (1999). DnaSP version 3: an integrated program for molecular population genetics and molecular evolution analysis. *Bioinformatics*, **15**, 174–175.

Rozas, J, Gullaud, M, Blandin, G, and Aguadé, M (2001). DNA variation at the *rp49* gene region of *Drosophila simulans*: evolutionary inferences from an unusual haplotype structure. *Genetics*, **158**, 1147–1155.

Roze, D and Michod, RE (2001). Mutation load, multilevel selection and the evolution of propagule size during the origin of multicellularity. *American Naturalist*, **158**, 638–654.

Rozycka, M, Collins, N, Stratton, MR, and Wooster, R (2000). Rapid detection of DNA sequence variants by conformation-sensitive capillary electrophoresis *Genomics*, **70**, 34–40.

Rudel, T, Schmid, A, Benz, R, Kolb, HA, Lang, F, and Meyer, TF (1996). Modulation of *Neisseria* porin (PorB) by cytosolic ATP/GTP of target cells: parallels between pathogen accommodation and mitochondrial endosymbiosis. *Cell*, **85**, 391–402.

Rundle, HD, Mooers, AØ, and Whitlock, MC (1998). Single founder-flush events and the evolution of reproductive isolation. *Evolution*, **52**, 1850–1855.

Rutherford, SL and Lindquist, S (1998). Hsp 90 as a capacitor for morphological evolution. *Nature*, **396**, 336–342.

Ruvolo, M (1997). Molecular phylogeny of the hominoids: inferences from multiple independent DNA sequence data sets. *Molecular Biology and Evolution*, **14**, 248–265.

Sagan, L (1967). On the origin of mitosing cells. *Journal of Theoretical Biology*, **14**, 225–275.

SanMiguel, P, Gaut, BS, Tikhonov, A, Nakajima Y, and Bennetzen, JL (1998). The paleontology of intergene retrotransposons of maize. *Nature Genetics*, **20**, 43–45.

Santer, B and Lampert, W (1995). Summer diapause in cyclopoid copepods: adaptive response to a food bottleneck? *Journal of Animal Ecology*, **64**, 600–613.

Sapir, T, Horesh, D, Caspi, M, Atlas, R, Burgess, HA, Wolf, SG, Francis, F, Chelly, J, Elbaum, M, Pietrokovski, S, *et al.* (2000). Doublecortin mutations cluster in evolutionarily conserved functional domains. *Human Molecular Genetics*, **9**, 703–712.

Sarich, VM and Wilson, AC (1967). Immunological time scale for hominid evolution. *Science*, **158**, 1200–1203.

Saul, A (1999). Circumsporozoite polymorphisms, silent Mutations and the evolution of *Plasmodium falciparum*. *Parasitology Today*, **15**, 38–39.

Saunders, WB, Work, DM, and Nikolaeva, SV (1999). Evolution of complexity in Paleozoic ammonoid sutures. *Science*, **286**, 760–763.

Sawyer, SA, Dykhuizen, DE, and Hartl, HL (1987). A confidence interval for the number of selectively neutral amino acid polymorphisms. *Proceedings of the National Academy of Sciences USA*, **84**, 6225–6228.

Schilthuizen, M (2004). Land-snail conservation in Borneo: limestone outcrops act as arks. *Journal of Conchology*, Special Publication no 3, 149–154.

Schilthuizen, M, Vermeulen, JJ, Davison, GWH, and Gittenberger, E (1999). Population structure in a snail species from isolated Malaysian limestone hills, inferred from ribosomal DNA sequences. *Malacologia*, **41**, 283–296.

Schlichting, CD and Pigliucci, M (1998). *Phenotypic evolution. A reaction norm perspective*. Sinauer, Sunderland, MA.

Schliekelman, P, Garner, C, and Slatkin, S (2001). Natural selection and resistance to HIV. *Nature*, **411**, 545–546.

Schluter, D and Nagel, LM (1996). Parallel speciation by natural selection. *American Naturalist*, **146**, 292–301.

Schmidt, ER (1984). Clustered and interspersed repetitive DNA sequence family in *Chironomus*. *Journal of Molecular Biology*, **178**, 1–15.

Schneider, C, Cunningham, M, and Moritz, C (1998). Comparative phylogeography and the history of endemic vertebrates in the Wet Tropics rainforests of Australia. *Molecular Ecology*, **7**, 487–498.

Schneider, TD and Stephens, RM (1990). Sequence logos: a new way to display consensus sequences. *Nucleic Acids Research*, **18**, 6097–6100.

Schröeder, D, Deppisch, H, Obermayer, M, Krohne, G, Stakebrandt, E, Holldobler, B, Goebel, W, and Gross, R (1996). Intracellular endosymbiotic bacteria of *Camponotus* species (carpenter ants): systematics, evolution and ultrastructural characterization. *Molecular Microbiology*, **21**, 479–489.

Schug, J and Overton, GC (1997). Modeling transcription factor binding sites with Gibbs sampling and minimum description length encoding. *ISMB Proceedings*, **5**, 268–271.

Schultz, ST, Lynch, M, and Willis, JH (1999). Spontaneous deleterious mutation in *Arabidopsis thaliana*. *Proceedings of the National Academy of Sciences USA*, **96**, 11393–11398.

Schwenk, K and Spaak, P (1995). Evolutionary and ecological consequences of interspecific hybridization in cladocerans. *Experientia*, **51**, 465–481.

Schwenk, K, Posada, D, and Hebert, PDN (2000). Molecular systematics of European *Hyalodaphnia*: the role of contemporary hybridization in ancient species. *Proceedings of the Royal Society of London B*, **267**, 1833–1842.

Scott, B (1997). Diversity in central Australian land snails. *Memoirs of the Museum of Victoria*, **56**, 435–439.

Seddon, JM, Reeve, NJ, Santucci, F, and Hewitt, GM (2001). DNA footprints of European hedgehogs, *Erinaceus europaeus* and *E. concolor*: Pleistocene refugia, postglacial expansion and colonization routes. *Molecular Ecology*, **10**, 2187–2198.

Sepkoski, JJ (1984). A kinetic model of Phanerozoic taxonomic diversity. III. Post-Paleozoic families and mass extinction. *Paleobiology*, **10**, 246–267.

Sepkoski, JJ (1996). Competition in macroevolution: the double wedge revisited. In D Jablonski, DH Erwin, and J Lipps, eds. *Evolutionary paleobiology*, pp. 211–255. University of Chicago Press, Chicago.

Serra, M and Carmona, MJ (1993). Mixis strategies and resting egg production of rotifers living in temporally-varying habitats. *Hydrobiologia*, **255/256**, 117–126.

Serra, M and King, CE (1999). Optimal rates of bisexual reproduction in cyclical parthenogens with density-dependent growth. *Journal of Evolutionary Biology*, **12**, 263–271.

Servedio, MR and Kirkpatrick, M (1997). The effects of gene flow on reinforcement. *Evolution*, **51**, 1764–1772.

Shabalina, SA and Kondrashov, A (1999). Pattern of selective constraint in *C. elegans* and *C. briggsae* genomes. *Genetical Research*, **74**, 23–30.

Shabalina, SA, Ogurtsov, AY, Kondrashov, VA, and Kondrashov, AS (2001). Selective constraints in intergenic regions of human and mouse genomes. *Trends in Genetics*, **17**, 373–376.

Shapira, SK and Finnerty, VG (1986). The use of genetic complementation in the study of eukaryotic macromolecular evolution: rate of spontaneous gene duplication at two loci of *Drosophila melanogaster*. *Molecular Biology and Evolution*, **23**, 159–167.

Sharp, PM, Bailes, E, Chaudhuri, RR, Rodenburg, CM, Santiago, MO, and Hahn, BH (2001). The origins of acquired immune deficiency syndrome viruses: where and when? *Philosophical Transactions of the Royal Society of London B*, **356**, 867–876.

Shaw, DD (1994). Centromeres: moving chromosomes through space and time. *Trends in Ecology and Evolution*, **9**, 170–175.

Shaw, DD, Wilkinson, P, and Coates, DJ (1983). Increased chromosomal mutation rates after hybridization between two subspecies of grasshoppers. *Science*, **220**, 1165–1167.

Shaw, DD, Marchant, AD, Contreras, N, Arnold, ML, Groeters, F, and Kohlman, AC (1993). Genomic and environmental determinants of a narrow hybrid zone: cause or coincidence? In R Harrison, ed. *Hybrid zones and the evolutionary process*, pp. 165–195. Oxford University Press, New York.

Shaw, FH, Geyer, CJ, and Shaw, RG (2002). A comprehensive model of mutations affecting fitness and inferences for *Arabidopsis thaliana*. *Evolution*, **56**, 453–463.

Shaw, PJ, Wratten, NS, McGregor, AP, and Dover, GA (2002). Co-evolution in bicoid-dependent promoters and the inception of regulatory incompatibilities among species of higher Diptera. *Evolution & Development*, **4**, 265–277.

Shaw, RG, Byers, DL, and Darmo, E (2000). Spontaneous mutational effects on reproductive traits of *Arabidopsis thaliana*. *Genetics*, **155**, 369–378.

Sherman, IW (1998). A brief history of malaria and the discovery of the parasite's life cycle. In IW Sherman, ed. *Malaria: parasite biology, pathogenesis, and protection*, pp. 3–10. American Society of Microbiology Press, Washington, DC.

Shi, Y, Simpson, PC, Scherer, JR, Wexler, D, Skibola, C, Smith, MT, and Mathies, RA (1999). Radial capillary array electrophoresis microplate and scanner for high-performance nucleic acid analysis. *Analytical Chemistry*, **71**, 5354–5361.

Shigenobu, S, Watanabe, H, Hattori, M, Sakaki, Y, and Ishikawa, H (2000). Genome sequence of the endocellular bacterial symbiot of aphids *Buchnera* sp. APS. *Nature*, **407**, 81–86.

Shiino, T, Kato, K, Kodata, N, Miyakuni, T, Takebe, Y, and Sato, H (2000). A group of V3 sequences from human immunodeficiency virus type 1 subtype E non-syncytium-inducing, CCR5-using variants are resistant to positive selection pressure. *Journal of Virology*, **74**, 1069–1078.

Shikano, S, Luckinbill, LS, and Kurihara, Y (1990). Changes of traits in a bacterial population associated with protozoal predation. *Microbial Ecology*, **20**, 75–84.

Shu, DG, Luo, HL, Conway Morris, S, Zhang, XL, Hu, SX, Chen, L, Han, J, Zhu, M, Li, Y, and Chen, LZ (1999). Lower Cambrian vertebrates from south China. *Nature*, **402**, 42–46.

Sibley, CG and Ahlquist, JE (1990). *Phylogeny and classification of birds: a study in molecular evolution*. Yale University Press, New Haven, CT.

Sicheritz-Pontén, T and Andersson, SGE (2001). A phylogenomic approach to microbial evolution. *Nucleic Acids Research*, **29**, 545–552.

Signor, PW and Lipps, JH (1982). Sampling bias, gradual extinction patterns and catastrophes in the fossil record. In LT Silver and PH Schultz, eds. *Geological implications of impact hypothesis of large asteroids and comets on the earth*, pp. 283–290. Special Paper, Geological Society of America, Boulder, CO.

Silva, FJ, Latorre, A, and Moya, A (2001). Genome size reduction through multiple events of gene disintegration in *Buchnera* APS. *Trends in Genetics*, **17**, 615–618.

Simmons, MJ and Crow, JF (1977). Mutations affecting fitness in *Drosophila* populations. *Annual Review of Genetics*, **11**, 49–78.

Simon, J-C and Hebert, PDN (1995). Patterns of genetic variation among Canadian populations of the bird cherry-oat aphid, *Rhopalosiphum padi* L. (Homoptera. Aphididae). *Heredity*, **74**, 346–353.

Simon, J-C, Leterme, N, and Latorre, A (1999a). Molecular markers linked to breeding system differences in segregating and natural populations of the cereal aphid *Rhopalosiphum padi*. *Molecular Ecology*, **8**, 965–973.

Simon, J-C, Baumann, S, Sunnucks, P, Hebert, PDN, Pierre, JS, Le Gallic, JF, and Dedryver, CA (1999b). Reproductive mode and population genetic structure of the cereal aphid *Sitobion avenae* studied using phenotypic and microsatellite markers. *Molecular Ecology*, **8**, 531–545.

Simon, J-C, Rispe, C, and Sunnucks, P (2002). Ecology and evolution of sex in aphids. *Trends in Ecology and Evolution*, **17**, 34–39.

Siviy, SM (1998). Neurobiological substrates of play behavior: glimpses into the structure and function of mammalian playfulness. In M Bekoff and JA Byers, eds. *Animal play: evolutionary, comparative, and ecological perspectives*, pp. 221–242. Cambridge University Press, Cambridge.

Slatkin, M (1985). Gene flow in natural populations. *Annual Review of Ecology and Systematics*, **16**, 393–430.

Small, S, Kraut, R, Hoey, T, Warrior, R, and Levine, M (1991). Transcriptional regulation of a pair-rule stripe in *Drosophila*. *Genes & Development*, **5**, 827–839.

Smith, AB, Gale, AS, and Monks, NEA (2001). Sea-level change and rock-record bias in the Cretaceous: a problem for extinction and biodiversity studies. *Paleobiology*, **27**, 241–253.

Snell, TW (1987). Sex, population dynamics and resting egg production in rotifers. *Hydrobiologia*, **144**, 105–111.

Snell, TW (1998). Chemical ecology of rotifers. *Hydrobiologia*, **387/388**, 267–276.

Snell, TW and Boyer, E (1988). Thresholds for mictic reproduction in the rotifer *Brachionus plicatilis*. *Journal of Experimental Marine Biology and Ecology*, **124**, 73–85.

Snell, TW and Childress, MJ (1987). Aging and loss of fertility in male and female *Brachionus plicatilis* (Rotifera). *International Journal of Invertebrate Reproduction and Development*, **12**, 103–110.

Snell, TW and Garman, BL (1986). Encounter probabilities between male and female rotifers. *Journal of Experimental Marine Biology and Ecology*, **97**, 221–230.

Snell, TW and Hoff, FH (1987). Fertilization and male fertility in the rotifer *Brachionus plicatilis*. *Hydrobiologia*, **147**, 329–334.

Snell, TW and Serra, M (2000). Using probability of extinction to evaluate the ecological significance of toxicant effects. *Environmental Toxicology and Chemistry*, **19**, 2357–2363.

Snell, TW, Burke, BE, and Messur, SD (1983). Size and distribution of resting eggs in a natural population of the rotifer *Brachionus plicatilis*. *Gulf Research Reports*, **7**, 285–287.

Snell, TW, Serra, M, and Carmona, MJ (1999). Toxicity and sexual reproduction in rotifers: reduced resting egg production and heterozygosity loss. In VE Forbes, ed. *Genetics and ecotoxicology*, pp. 169–185. Taylor and Francis, Philadelphia.

Solé, RV, Montoya, JM, and Erwin, DH (2002). Recovery after mass extinction: evolutionary assembly in large-scale biosphere dynamics. *Philosophical Transactions of the Royal Society of London B*, **357**, 697–707.

Solem, A (1988). Maximum in the minimum: biogeography of land snails from the Ningbing Ranges and Jeremiah Hills, northeast Kimberley. *Journal of the Malacological Society of Australia*, **9**, 59–113.

Solem, A (1993). Camaenid land snails from Western and central Australia (Mollusca: Pulmonata: Camaenidae); VI. Taxa from the Red Centre. *Records of the Western Australian Museum, Supplement*, **43**, 983–1459.

Solignac, M and Monnerot, M (1986). Race formation, speciation, and introgression within *Drosophila simulans*, *D. mauritiana*, and *D. sechellia* inferred from mitochondrial DNA analysis. *Evolution*, **40**, 531–539.

Southwood, TRE (1962). Migration of terrestrial arthropods in relation to habitat. *Biological Reviews*, **27**, 171–214.

Spencer, M, Colegrave, N, and Schwartz, SS (2001). Hatching fraction and timing of resting stage production in seasonal environments: effects of density dependence and uncertain season length. *Journal of Evolutionary Biology*, **14**, 357–367.

Spielman, A, Kitron, U, and Pollack, RJ (1993). Time limitation and the role of research in the worldwide attempt to eradicate malaria. *Journal of Medical Entomology*, **30**, 6–19.

Spinka, M, Newberry, RC, and Bekoff, M (2001). Mammalian play: training for the unexpected. *Quarterly Review of Biology*, **76**, 141–168.

Spofford, JB (1969). Heterosis and the evolution of duplications. *American Naturalist*, **103**, 407–432.

Springer, MS, Westerman, M, Kavanagh, JR, Burk, A, Woodburne, MO, Kao, DJ, and Krajewski, C (1998). The origin of the Australian marsupial fauna and the phylogenetic affinities of the enigmatic monito del monte and marsupial mole. *Proceedings of the Royal Society of London B*, **265**, 2381–2386.

Stanley, SM (1973). An ecological theory for the sudden origin of multicellular life in the Late Precambrian. *Proceedings of the National Academy of Sciences USA*, **70**, 1486–1489.

Stanojevic, D, Hoey, T, and Levine, M (1989). Sequence-specific DNA-binding activities of the gap proteins encoded by hunchback and Kruppel in *Drosophila*. *Nature*, **341**, 331–335.

Stanojevic, D, Small, S, and Levine, M (1991). Regulation of a segmentation stripe by overlapping activators and repressors in the *Drosophila* embryo. *Science*, **254**, 1385–1387.

Stauffer, C, Latakos, F, and Hewitt, GM (1999). Phylogeography and postglacial colonization routes of *Ips typographicus* L. (Coleoptera, Scolytidae). *Molecular Ecology*, **8**, 763–773.

Stearns, SC (1989). The evolutionary significance of reaction norms. *BioScience*, **39**, 436–446.

Stearns, SC (1992). *The evolution of life histories*. Oxford University Press, New York.

Stephan, W (1994). Effects of genetic recombination and population subdivision on nucleotide sequence variation in *Drosophila ananassae*. In B Golding, ed. *Non-neutral evolution: theories and molecular data*, pp. 57–66. Chapman and Hall, New York.

Stern, DL (1998). A role of Ultrabithorax in morphological differences between *Drosophila* species. *Nature*, **396**, 463–466.

Stern, DL and Foster, WA (1998). The evolution of soldiers in aphids. *Biological Reviews*, **71**, 27–79.

Stibor, H and Lüning, J (1994). Predator-induced phenotypic variation in the pattern of growth and reproduction in *Daphnia hyalina* (Crustacea; Cladocera). *Functional Ecology*, **8**, 97–101.

Stone JR and Wray G (2001). Rapid evolution of *cis*-regulatory sequences via local point mutations. *Molecular Biology and Evolution*, **18**, 1764–1770.

Storfer, A (1999). Gene flow and local adaptation in a sunfish–salamander system. *Behavioral Ecology and Sociobiology*, **46**, 273–279.

Stormo, GD (2000). DNA binding sites: representation and discovery. *Bioinformatics*, **16**, 16–23.

Strassmann, JE and Bernasconi, G (1999). Cooperation among unrelated individuals: the ant foundress case. *Trends in Ecology and Evolution*, **14**, 477–482.

Sturtevant, AH (1939). High mutation frequency induced by hybridization. *Proceedings of the National Academy of Sciences USA*, **25**, 308–310.

Sucena, E and Stern, DL (2000). Divergence of larval morphology between *Drosophila sechellia* and its sibling species caused by *cis*-regulatory evolution of ovo/shaven baby. *Proceedings of the National Academy of Sciences USA*, **97**, 4530–4534.

Sullender, BW and Crease, TJ (2001). The behavior of a *Daphnia pulex* transposable element in cyclically and obligately parthenogenetic populations. *Journal of Molecular Evolution*, **53**, 63–69.

Sunnucks, P, De Barro, PJ, Lushai, G, Maclean, N, and Hales, DF (1997). Genetic structure of an aphid studied using microsatellite: cyclic parthenogenesis, differentiated lineages, and host specialization. *Molecular Ecology*, **6**, 1059–1073.

Sutton-Smith, B (1999). Evolving a consilience of play definitions: playfully. In S Reifel, ed. *Play and culture studies (Vol. 2)*, pp. 239–256. Ablex, Stamford, CT.

Sutton-Smith, B (2003). Play as a parody of emotional invulnerability. In DE Lytle, ed. *Play and culture studies (Vol. 5*: Play and educational theory and practice), pp. 3–17. Praeger, Westport, CT.

Symons, D (1995). Beauty is in the adaptations of the beholder: the evolutionary psychology of human female sexual attractiveness. In PR Abramson and SD Pinkerton, eds. *Sexual nature/sexual culture*, pp. 80–118. University of Chicago Press, Chicago.

Tachida, H (1991). A study on a nearly neutral mutation model in finite populations. *Genetics*, **128**, 183–192.

Tagle, DA, Koop, BF, Goodman, M, Slightom, JL, Hess, DL, and Jones, RT (1988). Embryonic epsilon and gamma globin genes of a prosimian primate (*Galago crassicaudatus*): nucleotide and amino acid sequences, developmental regulation and phylogenetic footprints. *Journal of Molecular Biology*, **203**, 439–455.

Tajima, F (1989). Statistical method for testing the neutral mutation hypothesis by DNA polymorphism. *Genetics*, **123**, 585–595.

Tamas, I, Klasson, LM, Sandström, JP, and Anderson, SGE (2001). Mutualists and parasites: how to paint yourself into a (metabolic) corner. *FEBS Letters*, **498**, 135–139.

Tamas, I, Klasson, LM, Canback, B, Naslund, AK, Eriksson, AS, Wernegreen, JJ, Sandstrom, JP, Moran, NA, and Andersson, SG (2002). 50 million years of genomic stasis in endosymbiotic bacteria. *Science*, **296**, 2376–2379.

Tanabe, K, Mackay, M, Goman, M, and Scaife, JG (1987). Allelic dimorphism in a surface antigen gene of the malaria parasite *Plasmodium falciparum*. *Journal of Molecular Biology*, **195**, 273–287.

Tautz, D (1992). Redundancies, development and the flow of information. *BioEssays*, **14**, 263–266.

Tautz, D and Nigro, L (1998). Microevolutionary divergence pattern of the segmentation gene hunchback in *Drosophila*. *Molecular Biology and Evolution*, **15**, 1403.

Taylor, BE and Gabriel, W (1992). To grow or not to grow: optimal resource allocation for *Daphnia*. *American Naturalist*, **139**, 248–266.

Taylor, DJ, Finston, TL, and Hebert, PDN (1998). Biogeography of a widespread freshwater crustacean: pseudocongruence and cryptic endemism in the North American *Daphnia laevis* complex. *Evolution*, **52**, 1648–1670.

Taylor, DJ, Crease, TJ, and Brown, WM (1999). Phylogenetic evidence for a single long-lived clade of crustacean cyclic parthenogens and its implications for the evolution of sex. *Proceedings of the Royal Society of London B*, **266**, 791–797.

Taylor, JS, van de Peer, Y, and Meyer, A (2001). Genome duplication, divergent resolution and speciation. *Trends in Genetics*, **17**, 299–301.

Taylor, PD (1992). Altruism in viscous populations—an inclusive fitness model. *Evolutionary Ecology*, **6**, 352–356.

Templeton, AR (1998). Species and speciation. Geography, population structure, ecology, and gene trees. In DJ Howard and SJ Berlocher, eds. *Endless forms. Species and speciation*, pp. 32–43. Oxford University Press, Oxford.

Templeton, AR (1999). Uses of evolutionary theory in the Human Genome Project. *Annual Review of Ecology and Systematics*, **30**, 23–49.

The International SNP Map Working Group (2001). A map of human genome sequence variation containing 1.42 million single nucleotide polymorphisms. *Nature*, **409**, 928–933.

Thompson, JN (1998). Rapid evolution as an ecological process. *Trends in Ecology and Evolution*, **13**, 329–332.

Thompson, KV (1998). Self assessment in juvenile play. In M Bekoff and JA Byers, eds. *Animal play: evolutionary, comparative, and ecological perspectives*, pp. 183–204. Cambridge University Press, Cambridge.

Thompson, RB (1991). *A guide to the geology and landforms of central Australia*. Northern Territory Geological Survey, Alice Springs.

Thornhill, R (1997). The concept of an evolved adaptation. In GR Bock and G Cardew, eds. *Characterizing human psychological adaptations*, pp. 4–22. John Wiley & Sons, New York.

Thornhill, R and Gangestad, SW (1993). Human facial beauty: averageness, symmetry and parasite resistance. *Human Nature*, **4**, 237–269.

Thornhill, R and Gangestad, SW (1999a). Facial attractiveness. *Trends in Cognitive Sciences*, **3**, 452–460.

Thornhill, R and Gangestad, SW (1999b). The scent of symmetry: a human sex pheromone that signals fitness? *Evolution and Human Behavior*, **20**, 175–201.

Thornhill, R and Gangestad, SW (2003). Do women have evolved adaptation for extra-pair copulation?

In E Voland and K Grammer, eds. *Evolutionary aesthetics* pp. 341–368. Springer-Verlag, Heidelberg, Germany.

Thornhill, R and Grammer, K (1999). The body and face of woman: one ornament that signals quality? *Evolution and Human Behavior*, **20**, 105–120.

Thornhill, R and Møller, AP (1997). Developmental stability, disease and medicine. *Biological Reviews*, **72**, 497–528.

Thornton, JW and DeSalle, R (2000). Gene family evolution and homology: genomics meets phylogenetics. *Annual Review of Genomics and Human Genetics*, **1**, 41–73.

Tinbergen, N (1951). *The study of instinct*. Clarendon Press, Oxford.

Tinbergen, N (1963). On aims and methods of ethology. *Zeitschrift für Tierpsychologie*, **20**, 410–433.

Tishkoff, SA, Varkonyi, R, Cahinhinan, N, Abbes, S, Argyropoulos, G, Destro-Bisol, G, Drousiotou, A, Dangerfield, B, Lefranc, G, Loiselet, J, *et al.* (2001). Haplotype diversity and linkage disequilibrium at human G6PD: recent origin of alleles that confer malarial resistance. *Science*, **293**, 455–462.

Tohyama, Y, Ichimiya, T, Kasama-Yoshida, H, Cao, Y, Hasegawa, M, Kojima, H, and Kurihara, T (2000). Phylogenetic relation of lungfish indicated by the amino acid sequence of myelin DM20. *Molecular Brain Research*, **80**, 256–259.

Tollrian, R and Dodson, SI (1999). Inducible defenses in Cladocera: constraints, costs, and multiple predator environments. In R Tollrian and CD Harvell, eds. *The ecology and evolution of inducible defenses*, pp. 177–202. Princeton University Press, Princeton, NJ.

Tollrian, R and Harvell, CD (eds). (1999). *The ecology and evolution of inducible defences*. Princeton University Press, Princeton.

Trigg, P and Kondrachine, A (1998). The current global malaria situation. In IW Sherman, ed. *Malaria: parasite biology, pathogenesis, and protection*, pp. 11–22. American Society of Microbiology Press, Washington, DC.

Trivers, RL (1971). The evolution of reciprocal altruism. *Quarterly Review of Biology*, **46**, 35–57.

Trivers, RL (1985). *Social evolution*. Benjamin/Cummings, Menlo Park, CA.

Trueb, L and Cloutier, R (1991). A phylogenetic investigation of the inter- and intrarelationships of the Lissamphibia (Amphibia: Temnospondyli). In HP Schultze and L Trueb, eds. *Origins of the major groups of tetrapods: Controversies and consensus*, pp. 223–313. Cornell University Press, Ithaca.

Underhill, PA, Jin, L, Lin, AA, Mehdi, SQ, Jenkins, T, Vollrath, D, Davis, RW, Cavalli-Sforza, LL, and Oefner, PJ (1997). Detection of numerous Y chromosome biallelic polymorphisms by denaturing high-performance liquid chromatography. *Genome Research*, **7**, 996–1005.

Underhill, PA, Shen, P, Lin, AA, Jin, L, Passarino, G, Yang, WH, Kauffman, E, Bonné-Tamir, B, Bertranpetit, J, Francalacci, P, *et al.* (2000). Y chromosome sequence variation and the history of human populations. *Nature Genetics*, **26**, 358–361.

Uyenoyama, MK and Feldman, MW (1984). Theories of kin and group selection: a population genetics perspective. *Theoretical Population Biology*, **38**, 87–102.

van Alphen, JJM and Seehausen, O (2001). Sexual selection, reproductive isolation, and the genic view of speciation. *Journal of Evolutionary Biology*, **14**, 874–875.

van Baalen, M and Rand, DA (1998). The unit of selection in viscous populations and the evolution of altruism. *Journal of Theoretical Biology*, **193**, 631–648.

Van de Peer, Y, Taylor, JS, Braasch, I, and Meyer, A (2001). The ghost of selection past: rates of evolution and functional divergence of anciently duplicated genes. *Journal of Molecular Evolution*, **53**, 436–446.

van Ham, RCHJ, González-Candelas, F, Silva, FJ, Sabater-Muñoz, B, Moya, A, and Latorre, A (2000). Postsymbiotic plasmid acquisition and evolution of the repA1-replicon in *Buchnera aphidicola*. *Proceedings of the National Academy of Sciences USA*, **97**, 10855–10860.

van Ham, RCHJ, Kamerbeek, J, Palacios, C, Rausell, C, Abascal, F, Bastolla, U, Fernandez, JM, Jimenez, JM, Postigo, L, Silva, FJ, *et al.* (2003). Reductive genome evolution in *Buchnera aphidicola*. *Proceedings of the National Academy of Science USA*, **100**, 581–586.

Vanoverbeke, J and De Meester, L (1997). Among-populational genetic differentiation in a cyclical parthenogen and its relation to geographic distance and clonal diversity. *Hydrobiologia*, **360**, 135–142.

Vassilieva, LL and Lynch, M (1999). The rate of spontaneous mutation for life-history traits in *Caenorhabditis elegans*. *Genetics*, **151**, 119–129.

Vassilieva, LL, Hook, AM, and Lynch, M (2000). The fitness effects of spontaneous mutations in *Caenorhabditis elegans*. *Evolution*, **54**, 1234–1246.

Venkatesh, B, Erdmann, MV, and Brenner, S (2001). Molecular synapomorphies resolve evolutionary relationships of extant jawed vertebrates. *Proceedings of the National Academy of Sciences USA*, **98**, 11382–11387.

Venter, JC, Adams, MD, Myers, EW, Li, PW, Mural, RJ, Sutton, GG, Smith, HO, Yandell, M, Evans, CA, Holt, RA, *et al.* (2001). The sequence of the human genome. *Science*, **291**, 1304–1351.

Via, S (1999). Reproductive isolation between sympatric races of pea aphids. I. Gene flow restriction and habitat choice. *Evolution*, **53**, 1446–1457.

Vigilant, L, Stoneking, M, Harpending, H, Hawkes, K, and Wilson, AC (1991). African populations and the evolution of human mitochondrial DNA. *Science*, **253**, 1503–1507.

Vilà, M, Weber, E, and D'Antonio, C (2000). Conservation implications of invasion by plant hybridization. *Biological Invasions*, **2**, 207–217.

Vivier, E and Desportes, I (1988). Apicomplexa. In L Margulis, O Corliss, M Melkonia, and DJ Chapman, eds. *Handbook of protoctista*. Jones and Bartlett, Boston, pp. 549–573.

Volkman, SK, Barry, AE, Lyons, EJ, Nielsen, KM, Thomas, SM, Choi, M, Thakore, SS, Day, KP, Wirth, DF, and Hartl, DL (2001). Recent origin of *Plasmodium falciparum* from a single progenitor. *Science*, **293**, 482–484.

von Dassow, G, Meir, E, Munro, EM, and Odell, GM (2000). The segment polarity network is a robust developmental module. *Nature*, **406**, 188–192.

Von Dohlen, CD and Moran, NA (2000). Molecular data support a rapid radiation of aphids in the Cretaceous and multiple origins of host alternation. *Biological Journal of the Linnean Society*, **71**, 689–717.

Wadell, PJ, Cao, Y, Hauf, J, and Hasegawa, M (1999). Using novel phylogenetic methods to evaluate mammalian mtDNA, including amino acid-invariant sites-LogDet plus site stripping, to detect internal conflicts in the data, with special reference to the positions of the hedgehog, armadillo, and elephant. *Systematic Biology*, **48**, 31–53.

Wagner, A (1998). The fate of duplicated genes: loss or new function? *BioEssays*, **20**, 785–788.

Wagner, A (2000a). The role of population size, pleiotropy and fitness effects of mutations in the evolution of overlapping gene functions. *Genetics*, **154**, 1389–1401.

Wagner, A (2000b). Inferring lifestyle from gene expression patterns. *Molecular Biology and Evolution*, **17**, 1985–1987.

Wagner, DL and Liebherr, JK (1992). Flightlessness in insects. *Trends in Ecology and Evolution*, **7**, 216–220.

Wagner, WL and Funk, VA (eds). (1995). *Hawaiian biogeography: evolution on a hot spot archipelago*. Smithsonian Institution Press, Washington and London.

Wallace, AR (1852). On the monkeys of the Amazon. *Proceedings of the Zoological Society of London*, **20**, 107–110.

Wallace, AR (1901). *Darwinism*. 3rd Edition. MacMillan, London.

Walsh, B (2001). Quantitative genetics in the age of genomics. *Theoretical Population Biology*, **59**, 175–184.

Walsh, JB (1995). How often do duplicated genes evolve new functions? *Genetics*, **110**, 345–364.

Wang, RL, Stec, DS, Hey, J, Lukka, M, and Doebley, J (1999). The limits of selection during maize domestication. *Nature*, **398**, 236–239.

Watson, DM (1998). Kangaroos at play: play behaviour in the Macropodoidea. In M Bekoff and JA Byers, eds. *Animal play: evolutionary, comparative, and ecological perspectives*, pp. 61–95. Cambridge University Press, Cambridge.

Watterson, GA (1983). On the time for gene silencing at duplicate loci. *Genetics*, **105**, 745–766.

Weaver, TA (2000). High-throughput SNP discovery and typing for genome-wide genetic analysis. *New Technologies for Life Sciences: A Trends Guide*, 36–42.

Weider, LJ, Lampert, W, Wessels, M, Colbourne, JK, and Limburg, P (1997). Long-term genetic shifts in a microcrustacean egg bank associated with anthropogenic changes in the Lake Constance ecosystem. *Proceedings of the Royal Society of London B*, **264**, 1613–1618.

Weider, LJ, Hobœk, A, Colbourne, JK, Crease, TJ, Dufresne, F, and Hebert, PDN (1999a). Holarctic phylogeography of an asexual species complex I. Mitochondrial DNA variation in arctic *Daphnia*. *Evolution*, **53**, 777–792.

Weider, LJ, Hobœk, A, Hebert, PDN, and Crease, TJ (1999b). Holarctic phylogeography of an asexual species complex II. Allozyme variation and clonal structure in Arctic *Daphnia*. *Molecular Ecology*, **8**, 1–13.

Weinreich, DM and Rand, DM (2000). Contrasting patterns of nonneutral evolution in proteins encoded in nuclear and mitochondrial genomes. *Genetics*, **156**, 385–399.

Weiss, RA (2001). The Leeuwenshoek lecture 2001. Animal origins of human infectious disease. *Philosophical Transactions of the Royal Society of London B*, **356**, 957–977.

Weisser, WW, Braendle, CG, and Minoretti, N (1999). Predator-induced morphology shift in the pea aphid. *Proceedings of the Royal Society of London B*, **266**, 1175–1181.

Welburn, SC, Barcinski, MA, and Williams, GT (1999). Programmed cell death in Trypanosomatids. *Parasitology Today*, **13**, 22–26.

Werth, CR and Windham, MD (1991). A model for divergent, allopatric speciation of polyploid pteridophytes resulting from silencing of duplicate-gene expression. *American Naturalist*, **137**, 515–526.

West, SA, Lively, CM, and Read, AF (1999). A pluralistic approach to sex and recombination. *Journal of Evolutionary Biology*, **12**, 1003–1012.

Westin, J and Lardelli, M (1997). Three novel *notch* genes in zebrafish: implications for vertebrate *Notch* gene evolution and function. *Development Genes and Evolution*, **207**, 51–63.

Whitlock, MC and McCauley, DE (1999). Indirect measures of gene flow and migration: $F_{ST} \neq 1/(4\mathrm{Nm} + 1)$. *Heredity*, **82**, 117–125.

Williams, D, Dunkerley, D, DeDecker, P, Kershaw, P, and Chappell, M (1998). *Quaternary environments*. Arnold, London.

Wilson, AC, Maxson, LR, and Sarich, VM (1974). Two types of molecular evolution: evidence from studies of interspecific hybridization. *Proceedings of the National Academy of Sciences USA*, **71**, 2843–2847.

Wilson, AC, Carlson, SS, and White, TJ (1977). Biochemical evolution. *Annual Review of Biochemistry*, **46**, 573–639.

Wilson, DS, Pollock, GB, and Dugatkin, LA (1992). Can altruism evolve in purely viscous populations? *Evolutionary Ecology*, **6**, 331–341.

Wloch, DM, Szafraniec, K, Borts, RH, and Korona, R (2001). Direct estimate of the mutation rate and the distribution of fitness effects in the yeast *Saccharomyces cerevisiae*. *Genetics*, **159**, 441–452.

Woese, CR (1998). The universal ancestor. *Proceedings of the National Academy of Sciences USA*, **95**, 6854–6859.

Wolfe, KH (2001). Yesterday's polyploids and the mystery of diploidization. *Nature Reviews Genetics*, **2**, 333–341.

Wood, R (1999). *Reef evolution*. Oxford University Press, Oxford.

Wright, S (1938). Size of population and breeding structure in relation to evolution. *Science*, **87**, 430–431.

Wright, S (1951). The genetical structure of populations. *Annals of Eugenics*, **15**, 323–354.

Wright, S (1969). *Evolution and the genetics of populations* (*Vol. 2: The theory of gene frequencies*). University of Chicago Press, Chicago.

Wu, C-I (2001). The genic view of the process of speciation. *Journal of Evolutionary Biology*, **14**, 851–865.

Wu, LY, Thompson, DK, Li, GS, Hurt, RA, Tiedje, JM, and Zhou, JZ (2001). Development and evaluation of functional gene arrays for detection of selected genes in the environment. *Applied and Environmental Microbiology*, **67**, 5780–5790.

Xu, PX, Zhang, X, Heaney, S, Yoon, A, Michelson, AM, and Maas, RL (1999). Regulation of Pax6 expression is conserved between mice and flies. *Development*, **126**, 383–395.

Yang, HP, Tanikawa, AY, Van Voorhies, WA, Silva, JC, and Kondrashov, AS (2001a). Whole-genome effects of ethyl methanesulfonate-induced mutation on nine quantitative traits in outbred *Drosophila melanogaster*. *Genetics*, **157**, 1257–1265.

Yang, HP, Tanikawa, AY, and Kondrashov, AS (2001b). Molecular nature of 11 spontaneous *de novo* mutations in *Drosophila melanogaster*. *Genetics*, **157**, 1285–1292.

Yang, YW, Lai, KN, Tai, PY, and Li, WH (1999). Rates of nucleotide substitution in angiosperm mitochondrial DNA sequences and dates of divergence between *Brassica* and other angiosperm lineages. *Journal of Molecular Evolution*, **48**, 597–604.

Yang, Z and Nielsen, R (1998). Synonymous and nonsynonymous rate variation in nuclear genes of mammals. *Journal of Molecular Evolution*, **46**, 409–418.

Zardoya, R and Meyer, A (1996). Evolutionary relationships of the coelacanth, lungfishes, and tetrapods based on the 28S ribosomal RNA gene. *Proceedings of the National Academy of Sciences USA*, **93**, 5449–5454.

Zardoya, R and Meyer, A (1998). Complete mitochondrial genome suggests diapsid affinities of turtles. *Proceedings of the National Academy of Sciences USA*, **95**, 14226–14231.

Zardoya, R and Meyer, A (2001a). On the origin of and phylogenetic relationships among living amphibians. *Proceedings of the National Academy of Sciences USA*, **98**, 7380–7383.

Zardoya, R and Meyer, A (2001b). Vertebrate phylogeny: limits of inference of mitochondrial genome and nuclear rDNA sequence data due to an adverse phylogenetic signal/noise ratio. In PE Ahlberg, ed. *Major events in early vertebrate evolution*, pp. 106–118. Taylor and Francis, London.

Zardoya, R, Cao, Y, Hasegawa, M, and Meyer, A (1998). Searching for the closest living relative(s) of tetrapods through evolutionary analyses of mitochondrial and nuclear data. *Molecular Biology and Evolution*, **15**, 506–517.

Zeng, L-W, Comeron, JM, Chen, B, and Kreitman, M (1998). The molecular clock revisited: the rate of synonymous vs. replacement change in *Drosophila*. *Genetica*, **102/103**, 369–382.

Zeyl, C and De Visser, JAGM (2001). Estimates of the rate and distribution of fitness effects of spontaneous mutation in *Saccharomyces cerevisiae*. *Genetics*, **157**, 53–61.

Zhang, P, Chopra, S, and Peterson, T (2000). A segmental gene duplication generated differentially expressed *myb*-homologous genes in maize. *Plant Cell*, **12**, 2311–2322.

Zhao, XP, Si, Y, Hanson, RE, Crane, CF, Price, HJ, Stelly, DM, Wendel, JF, and Paterson, AH (1998). Dispersed repetitive DNA has spread to new genomes since polyploid formation in cotton. *Genome Research*, **8**, 479–492.

Zhao, Z, Jin, L, Fu, Y-X, Ramsay, M, Jenkins, T, Leskinen, E, Pamilo, P, Trexler, M, Patthy, L, Jorde, LB, *et al.* (2000). Worldwide DNA sequence variation in a 10-kilobase noncoding region of human chromosome 22. *Proceedings of the National Academy of Sciences USA*, **97**, 11354–11358.

Zuk, M and Kolluru, GM (1998). Exploitation of sexual signals by predators and parasitoids. *Quarterly Review of Biology*, **73**, 415–438.

# Index

Numbers in italics refer to glossary entry

Printed in the United States of America/BNB